Silicon Heterostructure Devices

Silicon Heterostructure Devices

Edited by
John D. Cressler

CRC Press
Taylor & Francis Group
Boca Raton London New York

CRC Press is an imprint of the
Taylor & Francis Group, an **Informa** business

The material was previously published in *Silicon Heterostructure Handbook: Materials, Fabrication, Devices, Circuits and Applications of SiGe and Si Strained-Layer Epitaxy* © Taylor and Francis 2005.

CRC Press
Taylor & Francis Group
6000 Broken Sound Parkway NW, Suite 300
Boca Raton, FL 33487-2742

© 2008 by Taylor & Francis Group, LLC
CRC Press is an imprint of Taylor & Francis Group, an Informa business

No claim to original U.S. Government works

10 9 8 7 6 5 4 3 2 1

International Standard Book Number-13: 978-1-4200-6690-6 (Hardcover)

Library of Congress Cataloging-in-Publication Data

Silicon heterostructure devices / editor, John D. Cressler.
    p. cm.
  Includes bibliographical references and index.
  ISBN 978-1-4200-6690-6 (alk. paper)
  1. Bipolar transistors. 2. Heterostructures. 3. Bipolar integrated circuits--Design and construction.
I. Cressler, John D.

TK7871.96.B55S54 2008
621.3815'28--dc22   2007030748

Visit the Taylor & Francis Web site at
http://www.taylorandfrancis.com

and the CRC Press Web site at
http://www.crcpress.com

For the tireless efforts
Of the many dedicated scientists and engineers
Who helped create this field and make it a success.
I tip my hat, and offer sincere thanks from all of us
Who have benefitted from your keen insights and imaginings.

And ...

For Maria:
My beautiful wife, best friend, and soul mate for these 25 years.
For Matthew John, Christina Elizabeth, and Joanna Marie:
God's awesome creations, and our precious gifts.
May your journey of discovery never end.

He Whose Heart Has Been Set
On The Love Of Learning And True Wisdom
And Has Exercised This Part of Himself,
That Man Must Without Fail Have Thoughts
That Are Immortal And Divine,
If He Lay Hold On Truth.

Plato

*Εκείνος που έχει δώσει την
ψυχή του στην Αγάπη για Μάθηση
και Αληθινή Σοφία,
και έχει Ασκηθεί για τούτο,
Ένας τέτοιος μόνο Άνδρας μπορεί
το δίχως άλλο να κάνει σκέψεις
Αθάνατες και Θείες,
Εάν στηριχθεί στην Αλήθεια.*

Πλάτωνας

Foreword

Progress in a given field of technology is both desired and expected to follow a stable and predictable long-term trajectory. Semilog plots of technology trends spanning decades in time and orders of magnitude in value abound. Perhaps the most famous exemplar of such a technology trajectory is the trend line associated with Moore's law, where technology density has doubled every 12 to 18 months for several decades. One must not, however, be lulled into extrapolating such predictability to other aspects of semiconductor technology, such as device performance, or even to the long-term prospects for the continuance of device density scaling itself. New physical phenomena assert themselves as one approaches the limits of a physical system, as when device layers approach atomic dimensions, and thus, no extrapolation goes on indefinitely.

Technology density and performance trends, though individually constant over many years, are the result of an enormously complex interaction between a series of decisions made as to the layout of a given device, the physics behind its operation, manufacturability considerations, and its extensibility into the future. This complexity poses a fundamental challenge to the device physics and engineering community, which must delve as far forward into the future as possible to understand when physical law precludes further progress down a given technology path. The early identification of such impending technological discontinuities, thus providing time to ameliorate their consequences, is in fact vital to the health of the semiconductor industry. Recently disrupted trends in CMOS microprocessor performance, where the "value" of processor-operating frequency was suddenly subordinated to that of integration, demonstrate the challenges remaining in accurately assessing the behavior of future technologies. However, current challenges faced in scaling deep submicron CMOS technology are far from unique in the history of semiconductors.

Bipolar junction transistor (BJT) technology, dominant in high-end computing applications during the mid-1980s, was being aggressively scaled to provide the requisite performance for future systems. By the virtue of bipolar transistors being vertical devices rather than lateral (as CMOS is), the length scale of bipolar transistors is set by the ability to control layer thicknesses rather than lateral dimensions. This allowed the definition of critical device dimensions, such as base width, to values far below the limits of optical lithography of the day. Although great strides in device performance had been made by 1985, with unity gain cutoff frequencies (f_T) in the range 20–30 GHz seemingly feasible, device scaling was approaching limits at which new physical phenomena became significant. Highly scaled silicon BJTs, having base widths below 1000 Å, demonstrated inordinately high reverse junction leakage. This was due to the onset of band-to-band tunneling between heavily doped emitter and base regions, rendering such devices unreliable. This and other observations presaged one of the seminal technology discontinuities of the past decade, silicon–germanium (SiGe) heterojunction bipolar transistor (HBT) technology being the direct consequence.

Begun as a program to develop bipolar technology with performance capabilities well beyond those possible via the continued scaling of conventional Si BJTs, SiGe HBT technology has found a wealth of applications beyond the realm of computing. A revolution in bipolar fabrication methodology, moving from device definition by implantation to device deposition and definition by epitaxy, accompanied by the exploitation of bandgap tailoring, took silicon-based bipolar transistor performance to levels never anticipated. It is now common to find SiGe HBTs with performance figures in excess of 300 GHz for both f_T and f_{max}, and circuits operable at frequencies in excess of 100 GHz.

A key observation is that none of this progress occurred in a vacuum, other than perhaps in the field of materials deposition. The creation of a generation of transistor technology having tenfold improved performance would of itself have produced far less ultimate value in the absence of an adequate eco-system to enable its effective creation and utilization. This text is meant to describe the eco-system that developed around SiGe technology as context for the extraordinary achievement its commercial rollout represented.

Early SiGe materials, of excellent quality in the context of fundamental physical studies, proved near useless in later device endeavors, forcing dramatic improvements in layer control and quality to then enable further development. Rapid device progress that followed drove silicon-based technology (recall that SiGe technology is still a silicon-based derivative) to unanticipated performance levels, demanding the development of new characterization and device modeling techniques. As materials work was further proven SiGe applications expanded to leverage newly available structural and chemical control.

Devices employing ever more sophisticated extensions of SiGe HBT bandgap tailoring have emerged, utilizing band offsets and the tailoring thereof to create SiGe-based HEMTs, tunneling devices, mobility-enhanced CMOS, optical detectors, and more to come. Progress in these diverse areas of device design is timely, as I have already noted the now asymptotic nature of performance gains to be had from continued classical device scaling, leading to a new industry focus on innovation rather than pure scaling. Devices now emerging in SiGe are not only to be valued for their performance, but rather their variety of functionality, where, for example, optically active components open up the prospect of the seamless integration of broadband communication functionality at the chip level.

Access to high-performance SiGe technology has spurred a rich diversity of exploratory and commercial circuit applications, many elaborated in this text. Communications applications have been most significantly impacted from a commercial perspective, leveraging the ability of SiGe technologies to produce extremely high-performance circuits while using back level, and thus far less costly, fabricators than alternative materials such as InP, GaAs, or in some instances advanced CMOS.

These achievements did not occur without tremendous effort on the part of many workers in the field, and the chapters in this volume represent examples of such contributions. In its transition from scientific curiosity to pervasive technology, SiGe-based device work has matured greatly, and I hope you find this text illuminating as to the path that maturation followed.

Bernard S. Meyerson
IBM Systems and Technology Group

Preface

While the idea of cleverly using silicon–germanium (SiGe) and silicon (Si) strained-layer epitaxy to practice bandgap engineering of semiconductor devices in the highly manufacturable Si material system is an old one, only in the past decade has this concept become a practical reality. The final success of creating novel Si heterostructure transistors with performance far superior to their Si-only homojunction cousins, while maintaining strict compatibility with the massive economy-of-scale of conventional Si integrated circuit manufacturing, proved challenging and represents the sustained efforts of literally thousands of physicists, electrical engineers, material scientists, chemists, and technicians across the world.

In the electronics domain, the fruit of that global effort is SiGe heterojunction bipolar transistor (SiGe HBT) BiCMOS technology, and strained Si/SiGe CMOS technology, both of which are at present in commercial manufacturing worldwide and are rapidly finding a number of important circuit and system applications. As with any new integrated circuit technology, the industry is still actively exploring device performance and scaling limits (at present well above 300 GHz in frequency response, and rising), new circuit applications and potential new markets, as well as a host of novel device and structural innovations. This commercial success in the electronics arena is also spawning successful forays into the optoelectronics and even nanoelectronics fields. The Si heterostructure field is both exciting and dynamic in its scope.

The implications of the Si heterostructure success story contained in this book are far-ranging and will be both lasting and influential in determining the future course of the electronics and optoelectronics infrastructure, fueling the miraculous communications explosion of the twenty-first century. While several excellent books on specific aspects of the Si heterostructures field currently exist (for example, on SiGe HBTs), this is the first reference book of its kind that "brings-it-all-together," effectively presenting a comprehensive perspective by providing very broad topical coverage ranging from materials, to fabrication, to devices (HBT, FET, optoelectronic, and nanostructure), to CAD, to circuits, to applications. Each chapter is written by a leading international expert, ensuring adequate depth of coverage, up-to-date research results, and a comprehensive list of seminal references. A novel aspect of this book is that it also contains "snap-shot" views of the industrial "state-of-the-art," for both devices and circuits, and is designed to provide the reader with a useful basis of comparison for the current status and future course of the global Si heterostructure industry.

This book is intended for a number of different audiences and venues. It should prove to be a useful resource as:

1. A hands-on reference for practicing engineers and scientists working on various aspects of Si heterostructure integrated circuit technology (both HBT, FET, and optoelectronic), including materials, fabrication, device physics, transistor optimization, measurement, compact modeling and device simulation, circuit design, and applications
2. A hands-on research resource for graduate students in electrical and computer engineering, physics, or materials science who require information on cutting-edge integrated circuit technologies

3. A textbook for use in graduate-level instruction in this field
4. A reference for technical managers and even technical support/technical sales personnel in the semiconductor industry

It is assumed that the reader has some modest background in semiconductor physics and semiconductor devices (at the advanced undergraduate level), but each chapter is self-contained in its treatment.

In this age of extreme activity, in which we are all seriously pressed for time and overworked, my success in getting such a large collection of rather famous people to commit their precious time to my vision for this project was immensely satisfying. I am happy to say that my authors made the process quite painless, and I am extremely grateful for their help. The list of contributors to this book actually reads like a global "who's who" of the silicon heterostructure field, and is impressive by any standard. I would like to formally thank each of my colleagues for their hard work and dedication to executing my vision of producing a lasting Si heterostructure "bible." In order of appearance, the "gurus" of our field include:

Guofu Niu, Auburn University, USA
David R. Greenberg, IBM Thomas J. Watson Research Center, USA
Jae-Sung Rieh, Korea University, South Korea
Greg Freeman, IBM Microelectronics, USA
Andreas Stricker, IBM Microelectronics, USA
Kern (Ken) Rim, IBM Thomas J. Watson Research Center, USA
Scott E. Thompson, University of Florida, USA
Sanjay Banerjee, University of Texas at Austin, USA
Soichiro Tsujino, Paul Scherrer Institute, Switzerland
Detlev Grützmacher, Paul Scherrer Institute, Switzerland
Ulf Gennser, CNRS-LPN, France
Erich Kasper, University of Stuttgart, Germany
Michael Oehme, University of Stuttgart, Germany
Eugene A. Fitzgerald, Massachusetts Institute of Technology, USA
Robert Hull, University of Virginia, USA
Kang L. Wang, University of California at Los Angeles, USA
S. Tong, University of California at Los Angeles, USA
H.J. Kim, University of California at Los Angeles, USA
Lorenzo Colace, University "Roma Tre," Italy
Gianlorenzo Masini, University "Roma Tre," Italy
Gaetano Assanto, University "Roma Tre," Italy
Wei-Xin Ni, Linköping University, Sweden
Anders Elfving, Linköping University, Sweden
Douglas J. Paul, University of Cambridge, United Kingdom
Michael Schröter, University of California at San Diego, USA
Ramana M. Malladi, IBM Microelectronics, USA

I would also like to thank my graduate students and post-docs, past and present, for their dedication and tireless work in this fascinating field. I rest on their shoulders. They include: David Richey, Alvin Joseph, Bill Ansley, Juan Roldán, Stacey Salmon, Lakshmi Vempati, Jeff Babcock, Suraj Mathew, Kartik Jayanaraynan, Greg Bradford, Usha Gogineni, Gaurab Banerjee, Shiming Zhang, Krish Shivaram, Dave Sheridan, Gang Zhang, Ying Li, Zhenrong Jin, Qingqing Liang, Ram Krithivasan, Yun Luo, Tianbing Chen, Enhai Zhao, Yuan Lu, Chendong Zhu, Jon Comeau, Jarle Johansen, Joel Andrews, Lance Kuo, Xiangtao Li, Bhaskar Banerjee, Curtis Grens, Akil Sutton, Adnan Ahmed, Becca Haugerud, Mustayeen Nayeem, Mustansir Pratapgarhwala, Guofu Niu, Emery Chen, Jongsoo Lee, and Gnana Prakash.

Finally, I am grateful to Tai Soda at Taylor & Francis for talking me into this project, and supporting me along the way. I would also like to thank the production team at Taylor & Francis for their able assistance (and patience!), especially Jessica Vakili.

The many nuances of the Si heterostructure field make for some fascinating subject matter, but this is no mere academic pursuit. In the grand scheme of things, the Si heterostructure industry is already reshaping the global communications infrastructure, which is in turn dramatically reshaping the way life on planet Earth will transpire in the twenty-first century and beyond. The world would do well to pay attention. It has been immensely satisfying to see both the dream of Si/SiGe bandgap engineering, and this book, come to fruition. I hope our efforts please you. Enjoy!

John D. Cressler
Editor

Editor

John D. Cressler received a B.S. in physics from the Georgia Institute of Technology (Georgia Tech), Atlanta, Georgia, in 1984, and an M.S. and Ph.D. in applied physics from Columbia University, New York, in 1987 and 1990. From 1984 to 1992 he was on the research staff at the IBM Thomas J. Watson Research Center in Yorktown Heights, New York, working on high-speed Si and SiGe bipolar devices and technology. In 1992 he left IBM Research to join the faculty at Auburn University, Auburn, Alabama, where he served until 2002. When he left Auburn University, he was Philpott–Westpoint Stevens Distinguished Professor of Electrical and Computer Engineering and director of the Alabama Microelectronics Science and Technology Center.

In 2002, Dr. Cressler joined the faculty at Georgia Tech, where he is currently Ken Byers Professor of Electrical and Computer Engineering. His research interests include SiGe devices and technology; Si-based RF/microwave/millimeter-wave mixed-signal devices and circuits; radiation effects; device-circuit interactions; noise and linearity; reliability physics; extreme environment electronics, 2-D/3-D device-level simulation; and compact circuit modeling. He has published more than 350 technical papers related to his research, and is author of the books *Silicon-Germanium Heterojunction Bipolar Transistors*, Artech House, 2003 (with Guofu Niu), and *Reinventing Teenagers: The Gentle Art of Instilling Character in Our Young People*, Xlibris, 2004 (a slightly different genre!).

Dr. Cressler was Associate Editor of the *IEEE Journal of Solid-State Circuits* (1998–2001), Guest Editor of the *IEEE Transactions on Nuclear Science* (2003–2006), and Associate Editor of the *IEEE Transactions on Electron Devices* (2005–present). He served on the technical program committees of the IEEE International Solid-State Circuits Conference (1992–1998, 1999–2001), the IEEE Bipolar/BiCMOS Circuits and Technology Meeting (1995–1999, 2005–present), the IEEE International Electron Devices Meeting (1996–1997), and the IEEE Nuclear and Space Radiation Effects Conference (1999–2000, 2002–2007). He currently serves on the executive steering committee for the IEEE Topical Meeting on Silicon Monolithic Integrated Circuits in RF Systems, as international program advisor for the IEEE European Workshop on Low-Temperature Electronics, on the technical program committee for the IEEE International SiGe Technology and Device Meeting, and as subcommittee chair of the 2004 Electrochemical Society Symposium of SiGe: Materials, Processing, and Devices. He was the Technical Program Chair of the 1998 IEEE International Solid-State Circuits Conference, the Conference Co-Chair of the 2004 IEEE Topical Meeting on Silicon Monolithic Integrated Circuits in RF Systems, and the Technical Program Chair of the 2007 IEEE Nuclear and Space Radiation Effects Conference.

Dr. Cressler was appointed an IEEE Electron Device Society Distinguished Lecturer in 1994, an IEEE Nuclear and Plasma Sciences Distinguished Lecturer in 2006, and was awarded the 1994 Office of Naval Research Young Investigator Award for his SiGe research program. He received the 1996 C. Holmes

MacDonald National Outstanding Teacher Award by Eta Kappa Nu, the 1996 Auburn University Alumni Engineering Council Research Award, the 1998 Auburn University Birdsong Merit Teaching Award, the 1999 Auburn University Alumni Undergraduate Teaching Excellence Award, an IEEE Third Millennium Medal in 2000, and the 2007 Georgia Tech Outstanding Faculty Leadership in the Development of Graduate Students Award. He is an IEEE Fellow.

On a more personal note, John's hobbies include hiking, gardening, bonsai, all things Italian, collecting (and drinking!) fine wines, cooking, history, and carving walking sticks, not necessarily in that order. He considers teaching to be his vocation. John has been married to Maria, his best friend and soul-mate, for 25 years, and is the proud father of three budding scholars: Matt, Christina, and Jo-Jo.

Dr. Cressler can be reached at School of Electrical and Computer Engineering, 777 Atlantic Drive, N.W., Georgia Institute of Technology, Atlanta, GA 30332-0250 U.S.A. or cressler@ece.gatech.edu http://users.ece.gatech.edu/~cressler/

Contents

1

The Big Picture

John D. Cressler

Georgia Institute of Technology

1.1 The Communications Revolution

We are at a unique juncture in the history of humankind, a juncture that amazingly we engineers and scientists have dreamed up and essentially created on our own. This pivotal event can be aptly termed the "Communications Revolution," and the twenty-first century, our century, will be the era of human history in which this revolution plays itself out.

This communications revolution can be functionally defined and characterized by the pervasive acquisition, manipulation, storage, transformation, and transmission of "information" on a global scale. This information, or more generally, knowledge, in its infinitely varied forms and levels of complexity, is gathered from our analog sensory world, transformed in very clever ways into logical "1"s and "0"s for ease of manipulation, storage, and transmission, and subsequently regenerated into analog sensory output for our use and appreciation. In 2005, this planetary communication of information is occurring at a truly mind-numbing rate, estimates of which are on the order of 80 Tera-bits/sec (10^{12}) of data transfer across the globe in 2005 solely in wired and wireless voice and data transmission, 24 hours a day, 7 days a week, and growing exponentially. The world is quite literally abuzz with information flow—communication.* It is for the birth of the Communications Revolution that we humans likely will be remembered for 1000 years hence. Given that this revolution is happening during the working careers of most of us, I find it a wonderful time to be alive, a fact of which I remind my students often.

Here is my point. No matter how one slices it, at the most fundamental level, it is semiconductor devices that are powering this communications revolution. Skeptical? Imagine for a moment that one could flip a switch and instantly remove all of the integrated circuits (ICs) from planet Earth. A moment's reflection will convince you that there is not a single field of human endeavor that would not come to a grinding halt, be it commerce, or agriculture, or education, or medicine, or entertainment. Life as we in the first world know it in 2005 would simply cease to exist. And yet, remarkably, the same result would not have been true 50 years ago; even 20 years ago. Given the fact that we humans have been on planet Earth in our present form for at least 1 million years, and within communities

---

*I have often joked with my students that it would be truly entertaining if the human retina was sensitive to longer wavelengths of electromagnetic radiation, such that we could "see" all the wireless communications signals constantly bathing the planet (say, in greens and blues!). It might change our feelings regarding our ubiquitous cell phones!

having entrenched cultural traditions for at least 15,000 years, this is truly a remarkable fact of history. A unique juncture indeed.

Okay, hold on tight. It is an easy case to make that the semiconductor silicon (Si) has single-handedly enabled this communications revolution.* I have previously extolled at length the remarkable virtues of this rather unglamorous looking silver-grey element [1], and I will not repeat that discussion here, but suffice it to say that Si represents an extremely unique material system that has, almost on its own, enabled the conception and evolving execution of this communications revolution. The most compelling attribute, by far, of Si lies in the economy-of-scale it facilitates, culminating in the modern IC fabrication facility, effectively enabling the production of gazillions of low-cost, very highly integrated, remarkably powerful ICs, each containing millions of transistors; ICs that can then be affordably placed into widgets of remarkably varied form and function.†

So what does this have to do with the book you hold in your hands? To feed the emerging infrastructure required to support this communications revolution, IC designers must work tirelessly to support increasingly higher data rates, at increasingly higher carrier frequencies, all in the design space of decreasing form factor, exponentially increasing functionality, and at ever-decreasing cost. And by the way, the world is going portable and wireless, using the same old wimpy batteries. Clearly, satisfying the near-insatiable appetite of the requisite communications infrastructure is no small task. Think of it as job security!

For long-term success, this quest for more powerful ICs must be conducted within the confines of conventional Si IC fabrication, so that the massive economy-of-scale of the global Si IC industry can be brought to bear. Therein lies the fundamental motivation for the field of Si heterostructures, and thus this book. Can one use clever nanoscale engineering techniques to custom-tailor the energy bandgap of fairly conventional Si-based transistors to: (a) improve their performance dramatically and thereby ease the circuit and system design constraints facing IC designers, while (b) performing this feat without throwing away all the compelling economy-of-scale virtues of Si manufacturing? The answer to this important question is a resounding "YES!" That said, getting there took time, vision, as well as dedication and hard work of literally thousands of scientists and engineers across the globe.

In the electronics domain, the fruit of that global effort is silicon–germanium heterojunction bipolar transistor (SiGe HBT) bipolar complementary metal oxide semiconductor (BiCMOS) technology, and is in commercial manufacturing worldwide and is rapidly finding a number of important circuit and system applications. In 2004, the SiGe ICs, by themselves, are expected to generate US\$1 billion in revenue globally, with perhaps US\$30 billion in downstream products. This US\$1 billion figure is projected to rise to US\$2.09 billion by 2006 [2], representing a growth rate of roughly 42% per year, a remarkable figure by any economic standard. The biggest single market driver remains the cellular industry, but applications in optical networking, hard disk drives for storage, and automotive collision-avoidance radar systems are expected to represent future high growth areas for SiGe. And yet, in the beginning of 1987, only 18 years ago, there was no such thing as a SiGe HBT. It had not been demonstrated as a viable concept. An amazing fact.

In parallel with the highly successful development of SiGe HBT technology, a wide class of "transport enhanced" field effect transistor topologies (e.g., strained Si CMOS) have been developed as a means to boost the performance of the CMOS side of Si IC coin, and such technologies have also recently begun

---

*The lone exception to this bold claim lies in the generation and detection of coherent light, which requires direct bandgap III–V semiconductor devices (e.g., GaAs of InP), and without which long-haul fiber communications systems would not be viable, at least for the moment.

†Consider: it has been estimated that in 2005 there are roughly 20,000,000,000,000,000,000 (2×10^{19}) transistors on planet Earth. While this sounds like a large number, let us compare it to some other large numbers: (1) the universe is roughly 4.2×10^{17} sec old (13.7 billion years), (2) there are about 1×10^{21} stars in the universe, and (3) the universe is about 4×10^{23} miles across (15 billion light-years)! Given the fact that all 2×10^{20} of these transistors have been produced since December 23, 1947 (following the invention of the point-contact transistor by Bardeen, Brattain, and Shockley), this is a truly remarkable feat of human ingenuity.

to enter the marketplace as enhancements to conventional core CMOS technologies. The commercial success enjoyed in the electronics arena has very naturally also spawned successful forays into the optoelectronics and even nanoelectronics fields, with potential for a host of important downstream applications.

The Si heterostructure field is both exciting and dynamic in its scope. The implications of the Si heterostructure success story contained in this book are far-ranging and will be both lasting and influential in determining the future course of the electronics and optoelectronics infrastructure, fueling the miraculous communications explosion of our twenty-first century. The many nuances of the Si heterostructure field make for some fascinating subject matter, but this is no mere academic pursuit. As I have argued, in the grand scheme of things, the Si heterostructure industry is already reshaping the global communications infrastructure, which is in turn dramatically reshaping the way life of planet Earth will transpire in the twenty-first century and beyond. The world would do well to pay close attention.

1.2 Bandgap Engineering in the Silicon Material System

As wonderful as Si is from a fabrication viewpoint, from a device or circuit designer's perspective, it is hardly the ideal semiconductor. The carrier mobility for both electrons and holes in Si is comparatively small compared to their III–V cousins, and the maximum velocity that these carriers can attain under high electric fields is limited to about 1×10^7 cm/sec under normal conditions, relatively "slow." Since the speed of a transistor ultimately depends on how fast the carriers can be transported through the device under sustainable operating voltages, Si can thus be regarded as a somewhat "meager" semiconductor. In addition, because Si is an indirect gap semiconductor, light emission is fairly inefficient, making active optical devices such as diode lasers impractical (at least for the present). Many of the III–V compound semiconductors (e.g., GaAs or InP), on the other hand, enjoy far higher mobilities and saturation velocities, and because of their direct gap nature, generally make efficient optical generation and detection devices. In addition, III–V devices, by virtue of the way they are grown, can be compositionally altered for a specific need or application (e.g., to tune the light output of a diode laser to a specific wavelength). This atomic-level custom tailoring of a semiconductor is called bandgap engineering, and yields a large performance advantage for III–V technologies over Si [3]. Unfortunately, these benefits commonly associated with III–V semiconductors pale in comparison to the practical deficiencies associated with making highly integrated, low-cost ICs from these materials. There is no robust thermally grown oxide for GaAs or InP, for instance, and wafers are smaller with much higher defect densities, are more prone to breakage, and are poorer heat conductors (the list could go on). These deficiencies translate into generally lower levels of integration, more difficult fabrication, lower yield, and ultimately higher cost. In truth, of course, III–V materials such as GaAs and InP fill important niche markets today (e.g., GaAs metal semiconductor field effect transistor (MESFETs) and HBTs for cell phone power amplifiers, AlGaAs- or InP-based lasers, efficient long wavelength photodetectors, etc.), and will for the foreseeable future, but III–V semiconductor technologies will never become mainstream in the infrastructure of the communications revolution if Si-based technologies can do the job.

While Si ICs are well suited to high-transistor-count, high-volume microprocessors and memory applications, RF, microwave, and even millimeter-wave (mm-wave) electronic circuit applications, which by definition operate at significantly higher frequencies, generally place much more restrictive performance demands on the transistor building blocks. In this regime, the poorer intrinsic speed of Si devices becomes problematic. That is, even if Si ICs are cheap, they must deliver the required device and circuit performance to produce a competitive system at a given frequency. If not, the higher-priced but faster III–V technologies will dominate (as they indeed have until very recently in the RF and microwave markets).

The fundamental question then becomes simple and eminently practical: is it possible to improve the performance of Si transistors enough to be competitive with III–V devices for high-performance applications, while preserving the enormous yield, cost, and manufacturing advantages associated with conventional Si fabrication? The answer is clearly "yes," and this book addresses the many nuances associated with using SiGe and Si-strained layer epitaxy to practice bandgap engineering in the Si material system, a process culminating in, among other things, the SiGe HBT and strained Si CMOS, as well as a variety of other interesting electronic and optoelectronic devices built from these materials. This totality can be termed the "Si heterostructures" field.

1.3 Terminology and Definitions

A few notes on modern usage and pronunciation in this field are in order (really!). It is technically correct to refer to silicon–germanium alloys according to their chemical composition, $Si_{1-x}Ge_x$, where x is the Ge mole fraction. Following standard usage, such alloys are generally referred to as "SiGe" alloys. Note, however, that it is common in the material science community to also refer to such materials as "Ge:Si" alloys.

A SiGe film that is carbon doped (e.g., less than 0.20% C) in an attempt to suppress subsequent boron out-diffusion (e.g., in HBTs) is properly referred to as a SiGe:C alloy, or simply SiGeC (pronounced "silicon germanium carbon," *not* "silicon germanium carbide"). This class of SiGe alloys should be viewed as optimized SiGe alloys, and are distinct from SiGe films with a much higher C content (e.g., 2% to 3% C) that might be used, for instance, to lattice-match SiGeC alloys to Si.

Believe it or not, this field also has its own set of slang pronunciations. The colloquial usage of the pronunciation \'sig-ee\ to refer to "silicon–germanium" (begun at IBM in the late 1990s) has come into vogue (heck, it may make it to the dictionary soon!), and has even entered the mainstream IC engineers's slang; pervasively.*

In the electronics domain, it is important to be able to distinguish between the various SiGe technologies as they evolve, both for CMOS (strained Si) and bipolar (SiGe HBT). Relevant questions in this context include: Is company X's SiGe technology more advanced than company Y's SiGe technology? For physical as well as historical reasons, one almost universally defines CMOS technology (Si, strained Si, or SiGe), a lateral transport device, by the drawn lithographic gate length (the CMOS technology "node"), regardless of the resultant intrinsic device performance. Thus, a "90-nm" CMOS node has a drawn gate length of roughly 90 nm. For bipolar devices (i.e., the SiGe HBT), however, this is not so straightforward, since it is a vertical transport device whose speed is not nearly as closely linked to lithographic dimensions.

In the case of the SiGe HBT it is useful to distinguish between different technology generations according to their resultant *ac* performance (e.g., peak common-emitter, unity gain cutoff frequency (f_T), which is (a) easily measured and unambiguously compared technology to technology, and yet is (b) a very strong function of the transistor vertical doping and Ge profile and hence nicely reflects the degree of sophistication in device structural design, overall thermal cycle, epi growth, etc.) [1]. The peak f_T generally nicely reflects the "aggressiveness," if you will, of the transistor scaling which has been applied to a given SiGe technology. A higher level of comparative sophistication can be attained by also invoking the maximum oscillation frequency (f_{max}), a parameter which is well correlated to both intrinsic profile and device parasitics, and hence a bit higher on the ladder of device performance metrics, and thus more representative of actual large-scale circuit performance. The difficulty in this case is that f_{max} is far more ambiguous than f_T, in the sense that it can be inferred from various gain definitions (e.g., U vs. MAG), and in practice power gain data are often far less ideal in its behavior over frequency, more sensitive to accurate deembedding, and ripe with extraction "issues."

We thus term a SiGe technology having a SiGe HBT with a peak f_T in the range of 50 GHz as "first generation;" that with a peak f_T in the range of 100 GHz as "second generation;" that with a peak f_T in the range of 200 GHz as "third generation;" and that with a peak f_T in the range of 300 GHz as "fourth generation." These are loose definitions to be sure, but nonetheless useful for comparison purposes.

---

*I remain a stalwart holdout against this snowballing trend and stubbornly cling to the longer but far more satisfying "silicon–germanium."

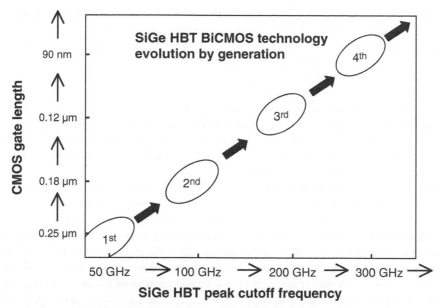

FIGURE 1.1 Evolution of SiGe HBT BiCMOS technology generations, as measured by the peak cutoff frequency of the SiGe HBT, and the CMOS gate length.

A complicating factor in SiGe technology terminology results from the fact that most, if not all, commercial SiGe HBT technologies today also contain standard Si CMOS devices (i.e., SiGe HBT BiCMOS technology) to realize high levels of integration and functionality on a single die (e.g., single-chip radios complete with RF front-end, data converters, and DSP). One can then speak of a given generation of SiGe HBT BiCMOS technology as the most appropriate intersection of both the SiGe HBT peak f_T and the CMOS technology node (Figure 1.1). For example, for several commercially important SiGe HBT technologies available via foundry services, we have:

- IBM SiGe 5HP—50 GHz peak f_T SiGe HBT + 0.35 μm Si CMOS (first generation)
- IBM SiGe 7HP—120 GHz peak f_T SiGe HBT + 0.18 μm Si CMOS (second generation)
- IBM SiGe 8HP—200 GHz peak f_T SiGe HBT + 0.13 μm Si CMOS (third generation)
- Jazz SiGe 60—60 GHz peak f_T SiGe HBT + 0.35 μm Si CMOS (first generation)
- Jazz SiGe 120—150 GHz peak f_T SiGe HBT + 0.18 μm Si CMOS (second generation)
- IHP SiGe SGC25B—120 GHz peak f_T SiGe HBT + 0.25 μm Si CMOS (second generation)

All SiGe HBT BiCMOS technologies can thus be roughly classified in this manner. It should also be understood that multiple transistor design points typically exist in such BiCMOS technologies (multiple breakdown voltages for the SiGe HBT and multiple threshold or breakdown voltages for the CMOS), and hence the reference to a given technology generation implicitly refers to the most aggressively scaled device within that specific technology platform.

1.4 The Application Space

It goes without saying in our field of semiconductor IC technology that no matter how clever or cool a new idea appears at first glance, its long-term impact will ultimately be judged by its marketplace "legs" (sad, but true). That is, was the idea good for a few journal papers and an award or two, or did someone actually build something and sell some useful derivative products from it? The sad reality is that the semiconductor field (and we are by no means exceptional) is rife with examples of cool new devices that never made it past the pages of the IEDM digest! The ultimate test, then, is one of stamina. And sweat. Did the idea make it out of the research laboratory and into the hands of the manufacturing lines? Did it pass the qualification-checkered flag, have design kits built around it, and get delivered to real circuit designers who built ICs, fabricated them, and tested them? Ultimately, were the derivative ICs inserted into real systems—widgets—to garner leverage in this or that system metric, and hence make the products more appealing in the marketplace?

Given the extremely wide scope of the semiconductor infrastructure fueling the communications revolution, and the sheer volume of widget possibilities, electronic to photonic to optoelectronic, it is useful here to briefly explore the intended application space of Si heterostructure technologies as we peer out into the future. Clearly I possess no crystal ball, but nevertheless some interesting and likely lasting themes are beginning to emerge from the fog.

SiGe HBT BiCMOS is the obvious ground-breaker of the Si heterostructures application space in terms of moving the ideas of our field into viable products for the marketplace. The field is young, but the signs are very encouraging. As can be seen in Figure 1.2, there are at present count 25 + SiGe HBT industrial fabrication facilities on line in 2005 around the world, and growing steadily. This trend points to an obvious recognition that SiGe technology will play an important role in the emerging electronics infrastructure of the twenty-first century. Indeed, as I often point out, the fact that virtually every major player in the communications electronics field either: (a) has SiGe up and running in-house, or (b) is using someone else's SiGe fab as foundry for their designers, is a remarkable fact, and very encouraging in the grand scheme of things. As indicated above, projections put SiGe ICs at a US$2.0 billion level by 2006, small by percentage perhaps compared to the near trillion dollar global electronics market, but growing rapidly.

The intended application target? That obviously depends on the company, but the simple answer is, gulp, a little bit of everything! As depicted in Figure 1.3 and Figure 1.4, the global communications landscape is exceptionally diverse, ranging from low-frequency wireless (2.4 GHz cellular) to the fastest high-speed wireline systems (10 and 40 Gbit/sec synchronous optical network (SONET)). Core CMOS technologies are increasingly being pushed into the lower frequency wireless space, but the compelling drive to higher carrier frequencies over time will increasingly favor SiGe technologies.

At present, SiGe ICs are making inroads into: the cellular industry for handsets [global system for mobile communications—GSM, code division multiple access (CDMA), wideband CDMA (W-CDMA), etc.], even for power amplifiers; various wireless local area networks (WLAN) building blocks, from components to fully integrated systems ranging from 2.4 to 60 GHz and up; ultrawide band (UWB) components; global positioning systems (GPS); wireless base stations; a variety of wireline networking products, from 2.5 to 40 Gbit/sec (and higher); data converters (D/A and A/D); high-speed memories; a variety of instrumentation electronics; read-channel memory storage products; core analog functions (op amps, etc.); high-speed digital circuits of various flavors; radiation detector

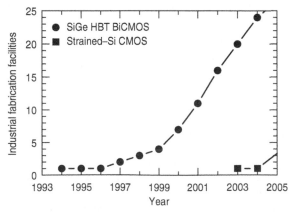

FIGURE 1.2   Number of industrial SiGe and strained Si fabrication facilities.

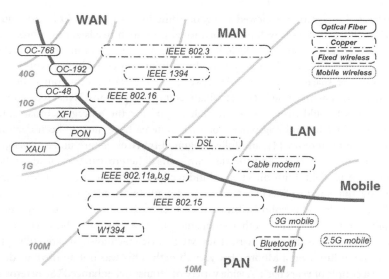

FIGURE 1.3 The global communications landscape, broken down by the various communications standards, and spanning the range of: wireless to wireline; fixed to mobile; copper to fiber; low data rate to broadband; and local area to wide area networks. WAN is *wide area network*, MAN is *metropolitan area network*, the so-called "last mile" access network, LAN is *local area network*, and PAN is *personal area network*, the emerging in-home network. (Used with the permission of Kyutae Lim.)

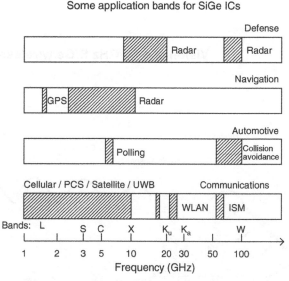

FIGURE 1.4 Some application frequency bands for SiGe integrated circuits.

electronics; radar systems (from 3 to 77 GHz and up); a variety space-based electronics components; and various niche extreme environment components (e.g., cryogenic (77 K) hybrid superconductor–semiconductor systems). The list is long and exceptionally varied—this is encouraging. Clearly, however, some of these components of "everything" are more important than others, and this will take time to shake out.

The strength of the BiCMOS twist to SiGe ICs cannot be overemphasized. Having both the high-speed SiGe HBT together on-chip with aggressively scaled CMOS allows one great flexibility in system design, the depths of which is just beginning to be plumbed. While debates still rage with respect to the most cost-effective partitioning at the chip and package level (system-on-a-chip versus system-in-a-package, etc.), clearly increased integration is viewed as a good thing in most camps (it is just a question of how much), and SiGe HBT BiCMOS is well positioned to address such needs across a broad market sector.

The envisioned high-growth areas for SiGe ICs over the new few years include: the cellular industry, optical networking, disk drives, and radar systems. In addition, potential high-payoff market areas span the emerging mm-wave space (e.g., the 60 GHz ISM band WLAN) for short range, but very high data rate (Gbit/sec) wireless systems. A SiGe 60 GHz single-chip/package transceiver (see Figure 1.5 for IBM's vision of such a beast) could prove to be the "killer app" for the emerging broadband multimedia market. Laughable? No. The building blocks for such systems have already been demonstrated using third-generation SiGe technology [4], and fully integrated transceivers are under development.

The rest of the potential market opportunities within the Si heterostructures field can be leveraged by successes in the SiGe IC field, both directly and indirectly. On the strained Si CMOS front, there are existent proofs now that strained Si is likely to become a mainstream component of conventional CMOS scaling at the 90-nm node and beyond (witness the early success of Intel's 90-nm logic technology built around uniaxially strained Si CMOS; other companies are close behind). Strained Si would seem to represent yet another clever technology twist that CMOS device technologists are pulling from their bag of tricks to keep the industry on a Moore's law growth path. This was not an obvious development (to me anyway) even a couple of years back. A wide variety of "transport enhanced" Si-heterostructure-based FETs have been demonstrated (SiGe-channel FETs, Si-based high electron mobility transistors (HEMTs), as well as both uniaxially and biaxially strained FETs, etc). Most of these devices, however, require complex substrate engineering that would have seemed to preclude giga-scale integration level needs for microprocessor-level integration. Apparently not so. The notion of using Si heterostructures (either

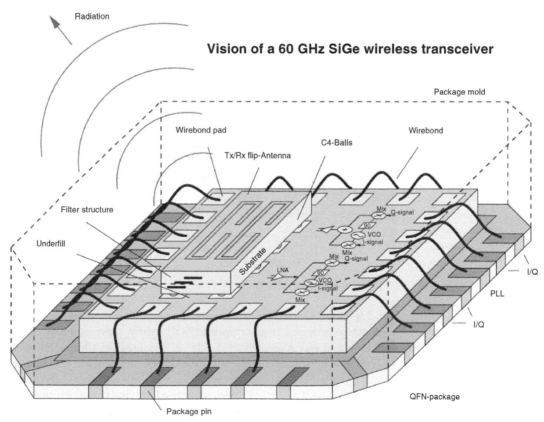

FIGURE 1.5 Vision for a single-chip SiGe mm-wave transceiver system. (Used with the permission of Ullrich Pfeiffer.)

uniaxial or biaxial strain or both) to boost conventional CMOS performance appears to be an appealing path for the future, a natural merging point I suspect for SiGe strained layers found in SiGe HBT BiCMOS (which to date contains only conventional Si CMOS) and strained Si CMOS.

From the optoelectronics camp, things are clearly far less evolved, but no less interesting. A number of functional optoelectronic devices have been demonstrated in research laboratories. Near-term successes in the short wavelength detector arena and light emitting diodes (LEDs) are beginning to be realized. The achievement of successful coherent light emission in the Si heterostructure system (e.g., via quantum cascade techniques perhaps) would appear to be the "killer app" in this arena, and research in this area is in progress. More work is needed.

1.5 Performance Limits and Future Directions

We begin with device performance limits. Just how fast will SiGe HBTs be 5 years from now? Transistor-level performance in SiGe HBTs continues to rise at a truly dizzying pace, and each major conference seems to bear witness to a new performance record (Figure 1.6). Both first- and second-generation SiGe HBT BiCMOS technology is widely available in 2005 (who would have thought even 3 years ago that fully integrated 100+ GHz Si-based devices would be "routine" on 200 mm wafers?), and even at the 200 GHz (third-generation) performance level, six companies (at last count) have achieved initial technology demonstrations, including IBM (Chapter 7), Jazz (Chapter 8), IHP (Chapter 11), ST Microelectronics (Chapter 12), Hitachi (Chapter 9), and Infineon (Chapter 10). (See *Fabrication of SiGe HBT BiCMOS Technology* for these chapters.) Several are now either available in manufacturing, or are very close (e.g., [5]). At press time, the most impressive new stake-in-the-ground is the report (June 2004) of the newly optimized "SiGe 9T" technology, which simultaneously achieves 302 GHz peak f_T and 306 GHz peak f_{max}, a clear record for any Si-based transistor, from IBM (Figure 1.7) [6]. This level of ac performance was achieved at a BV_{CEO} of 1.6 V, a BV_{CBO} of 5.5 V, and a current gain of 660. Noise measurements on these devices yielded NF_{min}/G_{assoc} of 0.45 dB/14 dB and 1.4 dB/8 dB at 10 and 25 GHz, respectively. Measurements of earlier (unoptimized) fourth-generation IBM SiGe HBTs have yielded record values of 375 GHz peak f_T [7] at 300 K and above 500 GHz peak f_T at 85 K. Simulations suggest that THz-level (1000 GHz) intrinsic transistor performance is not a laughable proposition in SiGe HBTs (Chapter 16). This fact still amazes even me, the eternal optimist of SiGe performance! I,

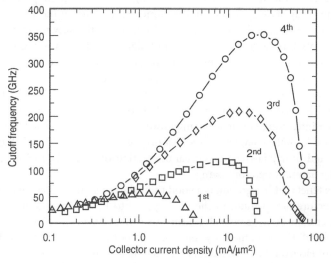

FIGURE 1.6 Measured cutoff frequency as a function of bias current density for four different SiGe HBT technology generations.

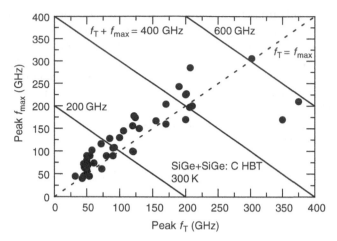

FIGURE 1.7 Measured maximum oscillation frequency versus cutoff frequency for a variety of generations of SiGe HBT BiCMOS technology shown in Figure 1.1.

for one, firmly believe that we will see SiGe HBTs above-500 GHz peak f_T and f_{max} fully integrated with nanometer-scale (90 nm and below) Si CMOS (possibly strained Si CMOS) within the next 3 to 5 years.

One might logically ask, particularly within the confines of the above discussion on ultimate market relevance, why one would even attempt to build 500 GHz SiGe HBTs, other than to win a best-paper award, or to trumpet that "because-it's-there" Mount Everest mentality we engineers and scientists love so dearly. This said, if the future "killer app" turns out to be single-chip mm-wave transceiver systems with on-board DSP for broadband multimedia, radar, etc., then the ability of highly scaled, highly integrated, very high performance SiGe HBTs to dramatically enlarge the circuit/system design space of the requisite mm-wave building blocks may well prove to be a fruitful (and marketable) path.

Other interesting themes are emerging in the SiGe HBT BiCMOS technology space. One is the very recent emergence of complementary SiGe (C-SiGe) HBT processes (npn + pnp SiGe HBTs). While very early pnp SiGe HBT prototypes were demonstrated in the early 1990s, only in the last 2 years or so have fully complementary SiGe processes been developed, the most mature of which to date is the IHP SGC25C process, which has 200 GHz npn SiGe HBTs and 80 GHz pnp SiGe HBTs (Chapter 11, see *Fabrication of SiGe HBT BiCMOS Technology*). Having very high-speed pnp SiGe HBTs on-board presents a fascinating array of design opportunities aimed particularly at the analog/mixed-signal circuit space. In fact, an additional emerging trend in the SiGe field, particularly for companies with historical pure analog circuit roots, is to target lower peak f_T, but higher breakdown voltages, while simultaneously optimizing the device for core analog applications (e.g., op amps, line drivers, data converters, etc.), designs which might, for instance, target better noise performance, and higher current gain-Early voltage product than mainstream SiGe technologies. One might even choose to park that SiGe HBT platform on top of thick film SOI for better isolation properties (Chapter 13, see *Fabrication of SiGe HBT BiCMOS Technology*). Another interesting option is the migration of high-speed vertical SiGe HBTs with very thin film CMOS-compatible SOI (Chapter 5, see *Fabrication of SiGe HBT BiCMOS Technology*). This technology path would clearly favor the eventual integration of SiGe HBTs with strained Si CMOS, all on SOI, a seemingly natural migratory path.

If one accepts the tenet that integration is a good thing from a system-level perspective, the Holy Grail in the Si heterostructure field would, in the end, appear to be the integration of SiGe HBTs for RF through mm-wave circuitry (e.g., single-chip mm-wave transceivers complete with on-chip antennae), strained Si CMOS for all DSP and memory functionality, both perhaps on SOI, Si-based light emitters, SiGe HBT modulator electronics, and detectors for such light sources, together with on-chip waveguides to steer the light, realized all on one Si wafer to produce a "Si-based optoelectronic superchip" [8], that could do-it-all. These diverse blocks would be optional plug-in modules around a core SiGe

HBT + strained Si CMOS IC technology platform, perhaps with flip-chip (or other) packaging techniques to join different sub-die to the main superchip (e.g., for a Si-based detector or laser).

I know, I know. It is not obvious that even if each of these blocks could be realized, that it would make economic sense to do so for real systems. I have no quarrel with that. I think such a Si-based superchip is a useful paradigm, however, to bind together all of the clever objects we wish to ultimately build with Si heterostructures, from electronic to photonic, and maintain the vision of the one overarching constraint that guides us as we look forward—keep whatever you do compatible with high-volume manufacturing in Si fabrication facilities if you want to shape the path of the ensuing communications revolution. This Si-based superchip clearly remains a dream at present. A realizable dream? And if realizable, commercially viable? Who knows? Only time will tell. But it is fun to think about.

As you peruse this book you hold in your hands, which spans the whole Si heterostructure research and development space, from materials, to devices, to circuit and system applications, I think you will be amazed at both the vision, cleverness, and smashing successes of the many scientists and engineers who make up our field. Do not count us out! We are the new architects of an oh-so-very-interesting future.

References

1. JD Cressler and G Niu. *Silicon–Germanium Heterojunction Bipolar Transistors*. Boston, MA: Artech House, 2003.
2. "SiGe devices market to hit $2 billion in 2006," article featured on CompoundSemicoductor.net, http://compoundsemiconductor.net/articles/news/8/3/22/1.
3. F Capasso. Band-gap engineering: from physics and materials to new semiconductor devices. *Science*, 235:172–176, 1987.
4. S Reynolds, B Floyd, U Pfeiffer, and T. Zwick. 60 GHz transciever circuits in SiGe bipolar technology. Technical Digest of the IEEE International Solid-State Circuits Conference, San Francisco, 2004, pp 442–443.
5. AJ Joseph, D Coolbaugh, D Harame, G Freeman, S Subbanna, M Doherty, J Dunn, C Dickey, D Greenberg, R Groves, M Meghelli, A Rylyakov, M Sorna, O Schreiber, D Herman, and T Tanji. 0.13 μm 210 GHz f_T SiGe HBTs—expanding the horizons of SiGe BiCMOS. Technical Digest of the IEEE International Solid-State Circuits Conference, San Francisco, 2002, pp 180–182.
6. J-S Rieh, D Greenberg, M Khater, KT Schonenberg, J-J Jeng, F Pagette, T Adam, A Chinthakindi, J Florkey, B Jagannathan, J Johnson, R Krishnasamy, D Sanderson, C Schnabel, P Smith, A Stricker, S Sweeney, K Vaed, T Yanagisawa, D Ahlgren, K Stein, and G Freeman. SiGe HBTs for millimeter-wave applications with simultaneously optimized f_T and f_{max}. Proceedings of the IEEE Radio Frequency Integrated Circuits (RFIC) Symposium, Fort Worth, 2004, pp 395–398.
7. JS Rieh, B Jagannathan, H Chen, KT Schonenberg, D Angell, A Chinthakindi, J Florkey, F Golan, D Greenberg, S-J Jeng, M Khater, F Pagette, C Schnabel, P Smith, A Stricker, K Vaed, R Volant, D Ahlgren, G Freeman, K Stein, and S Subbanna. SiGe HBTs with cutoff frequency of 350 GHz. Technical Digest of the IEEE International Electron Devices Meeting, San Francisco, 2002, pp 771–774.
8. R Soref. Silicon-based photonic devices. Technical Digest of the IEEE International Solid-State Circuits Conference, 1995, pp 66–67.

2

A Brief History of the Field

John D. Cressler
Georgia Institute of Technology

In the historical record of any field of human endeavor, being "first" is everything. It is often said that "hindsight is 20–20," and it is tempting in many cases to ascribe this or that pivotal event as "obvious" or "easy" once the answer is known. Anyone intimately involved in a creative enterprise knows, however, that it is never easy being first, and often requires more than a little luck and maneuvering. Thus the triumphs of human creativity, the "firsts," should be appropriately celebrated. Still, later chroniclers often gloss over, and then eventually ignore, important (and sometimes very interesting) twists and turns, starts and stops, of the winners as well as the second and third place finishers, who in the end may in fact have influenced the paths of the winners, sometimes dramatically. The history of our field, for instance, is replete with interesting competitive battles, unusual personalities and egos, no small amount of luck, and various other fascinating historical nuances.

There is no concise history of our field available, and while the present chapter is not intended to be either exhaustive or definitive, it represents my firm conviction that the history of any field is both instructive and important for those who follow in the footsteps of the pioneers. Hopefully this brief history does not contain too many oversights or errors, and is offered as a step in the right direction for a history of pivotal events that helped shape the Si heterostructures field.

2.1 Si–SiGe Strained Layer Epitaxy

The field of Si-based heterostructures solidly rests on the shoulders of materials scientists and crystal growers, those purveyors of the semiconductor "black arts" associated with the deposition of pristine films of nanoscale dimensionality onto enormous Si wafers with near infinite precision. What may seem routine today was not always so. The Si heterostructure story necessarily begins with materials, and circuit designers would do well to remember that much of what they take for granted in transistor performance owes a great debt to the smelters of the crystalline world. Table 2.1 summarizes the key steps in the development of SiGe–Si strained layer epitaxy.

Given that Ge was the earliest and predominant semiconductor pursued by the Bell Laboratories transistor team, with a focus on the more difficult to purify Si to come slightly later, it is perhaps not surprising that the first study of SiGe alloys, albeit unstrained bulk alloys, occurred as early as 1958 [1]. It was recognized around 1960 [2] that semiconductor epitaxy* would enable more robust and controllable transistor fabrication. Once the move to Si-based processing occurred, the field of Si epitaxy was

---

*The word "epitaxy" (or just "epi") is derived from the Greek word *epi*, meaning "upon" or "over."

TABLE 2.1   Milestones in the Development of SiGe–Si Strained Layer Epitaxy

Historical Event	Year	Ref.
First investigation of the bandgap of unstrained SiGe alloys	1958	[1]
First epitaxially grown layer to be used in a transistor	1960	[2]
First investigation of high-temperature Si epitaxy	1963	[3]
Concept of critical thickness for epitaxial strained layers	1963	[4]
Energy minimization approach for critical thickness	1963	[5]
Force-balance approach for critical thickness	1974	[6]
First growth of SiGe strained layers	1975	[7]
First growth of SiGe epitaxy by MBE	1984	[8]
First stability calculations of SiGe strained layers	1985	[9]
First measurements of energy bandgap in SiGe strained layers	1985	[10,11]
First growth of Si epitaxy by LRP-CVD	1985	[12]
First 2D electron gas in the SiGe system	1985	[13]
First growth of Si epitaxy by UHV/CVD	1986	[14]
First measurements of band alignments in SiGe–Si	1986	[15]
First growth of SiGe epitaxy by UHV/CVD	1988	[16]
First step-graded relaxed SiGe substrate	1988	[16]
First growth of SiGe epitaxy by LRP-CVD	1989	[17]
First growth of Si epitaxy by AP-CVD	1989	[18]
First 2D hole gas in the SiGe system	1989	[19]
First growth of SiGe epitaxy by AP-CVD	1991	[20]
First majority hole mobility measurements in SiGe	1991	[21]
First minority electron mobility measurements in SiGe	1992	[22]
First growth of lattice-matched SiGeC alloys	1992	[23]
First growth of SiGe layers with carbon doping	1994	[24]
First stability calculations to include a Si cap layer	2000	[25]

launched, the first serious investigation of which was reported in 1963 [3]. Early Si epitaxy was exclusively conducted under high-temperature processing conditions, in the range of 1100°C, a temperature required to obtain a chemically pure and pristine growth interface on the Si host substrate for the soon-to-be-grown crystalline Si epi. High-temperature Si epi has been routinely used in basically this same form for over 40 years now, and represents a mature fabrication technique that is still widely practiced for many types of Si devices (e.g., high-speed bipolar transistors and various power devices).

Device engineers have long recognized the benefits of marrying the many virtues of Si as a host material for manufacturing electronic devices, with the bandgap engineering principles routinely practiced in the III–V system. Ultimately this requires a means by which one can perform epitaxial deposition of thin Si layers on large Si substrates, for both p- and n-type doping of arbitrary abruptness, with very high precision, across large wafers, and doping control at high dynamic range. Only a moment's reflection is required to appreciate that this means the deposition of the Si epi must occur at very low growth temperatures, say 500°C to 600°C (not "low" per se, but low compared to the requisite temperatures needed for solid-state diffusion of dopants in Si). Such a low-temperature Si epi would then facilitate the effective marriage of Si and Ge, two chemically compatible elements with differing bandgaps, and enable the doping of such layers with high precision, just what is needed for device realizations. Clearly the key to Si-based bandgap engineering, Si-heterostructures, our field, is the realization of device quality, low-temperature Si epi (and hence SiGe epi), grown pseudomorphically* on large Si host substrates. Conquering this task proved to be remarkably elusive and time consuming.

In the III–V semiconductor world, where very low processing temperatures are much easier to attain, and hence more common than for Si, the deposition of multiple semiconductors on top of one another proved quite feasible (e.g., GaAs on InP), as needed to practice bandgap engineering, for instance,

---

*The word "pseudo" is derived from the Greek word *pseudēs*, meaning "false," and the word "morphic" is derived from the Greek word *morphē*, meaning "form." Hence, pseudomorphic literally means *false-form*.

resulting in complex material composites having differing lattice constants in intimate physical contact. To accommodate the differing lattice constants while maintaining the crystallinity of the underlying films, strain is necessarily induced in the composite film, and the notion of a film "critical thickness," beyond which strain relaxation occurs via fundamental thermodynamic driving forces, was defined as early as 1963 [4], as were the energy minimization techniques needed for calculating such critical thicknesses [5]. Alternative "force-balance" techniques for addressing the so-called stability issues in strained layer epitaxy came from the III–V world in 1974, and were applied to SiGe strained layer epitaxy in 1985 [9]. Interestingly, however, research continues today on stability in complicated (e.g., compositionally graded) SiGe films, and only very recently have reasonably complete theories been offered which seem to match well with experiment [25].

The first reported growth of SiGe strained layers was in 1975 in Germany [7], but the field did not begin to seriously heat up until the early 1980s, when several teams pioneered the application of molecular beam epitaxy (MBE) to facilitate materials studies of device-quality strained SiGe on Si in 1984 [8]. Optical studies on these films resulted in encouraging findings concerning the beneficial effects of strain on the band-edge properties of SiGe [10,11], paving the way for serious contemplation of devices built from such materials. Parallel paths toward other low-temperature Si epi growth techniques centered on the ubiquitous chemical vapor deposition (CVD) approach were simultaneously pursued, culminating in the so-called limited-reaction-processing CVD (LRP-CVD) technique (Si epi in 1985 [12], and SiGe epi in 1989 [17]), the ultrahigh-vacuum CVD (UHV/CVD) technique (Si epi in 1986 [14] and SiGe epi in 1988 [16]), and various atmospheric pressure CVD (AP-CVD) techniques (e.g., Si epi in 1989 [18], and SiGe epi in 1991 [20]). These latter two techniques, in particular, survive to this day, and are widely used in the SiGe heterojunction bipolar transistor (HBT) industry.

Device-quality SiGe–Si films enabled a host of important discoveries to occur, which have important bearing on device derivatives, including the demonstration of both two-dimensional electron and hole gases [13,19], and the fortuitous observation that step-graded SiGe buffer layers could be used to produce device-quality strained Si on SiGe, with its consequent conduction band offsets [16]. This latter discovery proved important in the development of SiGe–Si heterostructure-based FETs. Both majority and minority carrier mobility measurements occurred in the early 1990s [21,22], although reliable data, particularly involving minority carriers, remain sparse in the literature. Also in the early 1990s, experiments using high C content as a means to relieve strain in SiGe and potentially broaden the bandgap engineering space by lattice-matching SiGe:C materials to Si substrates (a path that has to date not borne much fruit, unfortunately), while others began studying efficacy of C-doping of SiGe, a result that ultimately culminated in the wide use today of C-doping for dopant diffusion suppression in SiGe:C HBTs [23,24].

The Si–SiGe materials field continues to evolve. Commercial single wafer (AP-CVD) and batch wafer (UHV/CVD) Si–SiGe epi growth tools compatible with 200 mm (and soon 300 mm) Si wafers exist in literally dozens of industrial fabrication facilities around the world, and SiGe growth can almost be considered routine today in the ease in which it can be integrated into CMOS-compatible fabrication processes. It was clearly of paramount importance in the ultimate success of our field that some of the "black magic" associated with robust SiGe film growth be removed, and this, thankfully, is the case in 2005.

2.2　SiGe HBTs

Transistor action was first demonstrated by Bardeen and Brattain in late December of 1947 using a point contact device [26]. Given all that has transpired since, culminating in the Communications Revolution, which defines our modern world (refer to the discussion in Chapter 1), this pivotal event surely ranks as one of the most significant in the course of human history—bold words, but nevertheless true. This demonstration of a solid-state device exhibiting the key property of amplification (power gain) is also unique in the historical record for the precision with which we can locate it in time—December 23,

1947, at about 5 p.m. Not to be outdone, Shockley rapidly developed a theoretical basis for explaining how this clever object worked, and went on to demonstrate the first true bipolar junction transistor (BJT) in 1951 [27]. The first BJT was made, ironically in the present context, from Ge. The first silicon BJT was made by Teal in 1954 using grown junction techniques. The first diffused silicon BJT was demonstrated in 1956 [28], and the first epitaxially grown silicon BJT was reported in 1960, see Ref. [2].

The concept of the HBT is surprisingly an old one, dating in fact to the fundamental BJT patents filed by Shockley in 1948 [29]. Given that the first bipolar transistor was built from Ge, and III–V semiconductors were not yet on the scene, it seems clear that Shockley envisioned the combination of Si (wide bandgap emitter) and Ge (narrow bandgap base) to form a SiGe HBT. The basic formulation and operational theory of the HBT, for both the traditional wide bandgap emitter plus narrow bandgap base approach found in most III–V HBTs, as well as the drift-base (graded) approach used in SiGe HBTs today, was pioneered by Kroemer, and was largely in place by 1957 [30–32]. It is ironic that Kroemer in fact worked hard early on to realize a SiGe HBT, without success, ultimately pushing him toward the III–V material systems for his heterostructure studies, a path that proved in the end to be quite fruitful for him, since he shared the Nobel Prize in physics in 2000 for his work in (III–V) bandgap engineering for electronic and photonic applications [33]. While III–V HBT (e.g., AlGaAs–GaAs) demonstrations began appearing in the 1970s, driven largely by the needs for active microwave components in the defense industry, reducing the SiGe HBT to practical reality took 30 years after the basic theory was in place due to material growth limitations. As pointed out [34] the semiconductor device field is quite unique in the scope of human history because "science" (theoretical understanding) preceded the "art" (engineering and subsequent technological advancement). Once device-quality SiGe films were finally achieved in the mid-1980s, however, progress was quite rapid. Table 2.2 summarizes the key steps in the evolution of SiGe HBTs.

The first functional SiGe HBT was demonstrated by an IBM team in December 1987 at the IEDM [35]. The pioneering result showed a SiGe HBT with functional, albeit leaky, dc characteristics; but it was a SiGe HBT, it worked (barely), and it was the first.* It is an often overlooked historical point, however, that at least four independent groups were simultaneously racing to demonstrate the first functional SiGe HBT, all using the MBE growth technique: the IBM team [35], a Japanese team [62], a Bell Laboratories team [63], and a Linköping University team [64]. The IBM team is fairly credited with the victory, since it presented (and published) its results in early December of 1987 at the IEDM (it would have been submitted to the conference for review in the summer 1987) [35]. Even for the published journal articles, the IBM team was the first to submit its paper for review (on November 17, 1987) [65]. All four papers appeared in print in the spring of 1988. Other groups soon followed with more SiGe HBT demonstrations.

The first SiGe HBT demonstrated using (the ultimately more manufacturable) CVD growth technique followed shortly thereafter, in 1989, first using LRP-CVD [17], and then with UHV/CVD [36]. Worldwide attention became squarely focused on SiGe technology, however, in June 1990 at the IEEE VLSI Technology Symposium with the demonstration of a non-self-aligned UHV/CVD SiGe HBT with a peak cutoff frequency of 75 GHz [37,38]. At that time, this SiGe HBT result was roughly twice the performance of state-of-the-art Si BJTs, and clearly demonstrated the future performance potential of the technology (doubling of transistor performance is a rare enough event that it does not escape significant attention!). Eyebrows were raised, and work to develop SiGe HBTs for practical circuit applications began in earnest in a large number of industrial and university laboratories around the world.[†]

The feasibility of implementing pnp SiGe HBTs was also demonstrated in June 1990 [40]. In December 1990, the simplest digital circuit, an emitter-coupled-logic (ECL) ring oscillator, using

---

*An interesting historical perspective of early SiGe HBT development at IBM is contained in Ref. [61].

[†]A variety of zero-Dt, mesa-isolated, III–V-like high-speed SiGe HBTs were reported in the early 1990s (e.g., Ref. [66]), but we focus here on fully integrated, CMOS-compatible SiGe HBT technologies, because they are inherently more manufacturable, and hence they are the only ones left standing today, for obvious reasons.

TABLE 2.2 Milestones in the Development of SiGe HBTs

Historical Event	Year	Ref.
First demonstration of transistor action	1947	[26]
Basic HBT concept	1948	[29]
First demonstration of a bipolar junction transistor	1951	[27]
First demonstration of a silicon bipolar transistor	1956	[28]
Drift-base HBT concept	1954	[30]
Fundamental HBT theory	1957	[31,32]
First epitaxial silicon transistors	1960	[2]
First SiGe HBT	1987	[35]
First ideal SiGe HBT grown by CVD	1989	[17]
First SiGe HBT grown by UHV/CVD	1989	[36]
First high-performance SiGe HBT	1990	[37,38]
First self-aligned SiGe HBT	1990	[39]
First SiGe HBT ECL ring oscillator	1990	[39]
First pnp SiGe HBT	1990	[40]
First operation of SiGe HBTs at cryogenic temperatures	1990	[41]
First SiGe HBT BiCMOS technology	1992	[42]
First LSI SiGe HBT integrated circuit	1993	[43]
First SiGe HBT with peak f_T above 100 GHz	1993	[44,45]
First SiGe HBT technology in 200-mm manufacturing	1994	[46]
First SiGe HBT technology optimized for 77 K	1994	[47]
First radiation tolerance investigation of SiGe HBTs	1995	[48]
First report of low-frequency noise in SiGe HBTs	1995	[49]
First SiGe:C HBT	1996	[50]
First high-power SiGe HBTs	1996	[51,52]
First sub-10 psec SiGe HBT ECL circuits	1997	[53]
First high-performance SiGe:C HBT technology	1999	[54]
First SiGe HBT with peak f_T above 200 GHz	2001	[55]
First SiGe HBT with peak f_T above 300 GHz	2002	[56]
First complementary (npn + pnp) SiGe HBT technology	2003	[57]
First C-SiGe technology with npn and pnp f_T above 100 GHz	2003	[58]
First vertical SiGe HBT on thin film (CMOS compatible) SOI	2003	[59]
First SiGe HBT with both f_T and f_{max} above 300 GHz	2004	[60]

self-aligned, fully integrated SiGe HBTs was produced [39]. The first SiGe BiCMOS technology (SiGe HBT + Si CMOS) was reported in December 1992 [42]. Theoretical predictions of the inherent ability of SiGe HBTs to operate successfully at cryogenic temperatures (in contrast to Si BJTs) were first confirmed in 1990 [41], and SiGe HBT profiles optimized for the liquid nitrogen temperature environment (77 K) were reported in 1994 [48]. The first LSI SiGe HBT circuit (a 1.2 Gsample/sec 12-bit digital-to-analog converter—DAC) was demonstrated in December 1993 [43]. The first SiGe HBTs with frequency response greater than 100 GHz were described in December 1993 by two independent teams [44,45], and the first SiGe HBT technology entered commercial production on 200-mm wafers in December 1994 [46].

The first report of the effects of ionizing radiation on advanced SiGe HBTs was made in 1995 [48]. Due to the natural tolerance of epitaxial-base bipolar structures to conventional radiation-induced damage mechanisms without any additional radiation-hardening process changes, SiGe HBTs are potentially very important for space-based and planetary communication systems applications, spawning an important new sub-discipline for SiGe technology. The first demonstration that epitaxial SiGe strained layers do not degrade the superior low-frequency noise performance of bipolar transistors occurred in 1995, opening the way for very low-phase noise frequency sources [49].

Carbon-doping of epitaxial SiGe layers as a means to effectively suppress boron out-diffusion during fabrication has rapidly become the preferred approach for commercial SiGe technologies, particularly those above first-generation performance levels. Carbon-doping of SiGe HBTs has its own interesting history, dating back to the serendipitous discovery [50] in 1996 that incorporating small amounts of C into a SiGe epi layer strongly retards (by an order of magnitude) the diffusion of the boron (B) base layer during subsequent thermal cycles. Given that maintaining a thin base profile during fabrication is perhaps the most challenging aspect of building a manufacturable SiGe technology, it is somewhat surprising that it took so long for the general adoption of C-doping as a key technology element. I think it is fair to say that most SiGe practitioners at that time viewed C-doping with more than a small amount of skepticism, given that C can act as a deep trap in Si, and C contamination is generally avoided at all costs in Si epi processes, particularly for minority carrier devices such as the HBT. At the time of the discovery of C-doping of SiGe in 1996, most companies were focused on simply bringing up a SiGe process and qualifying it, relegating the potential use of C to the back burner. In fairness, most felt that C-doping was not necessary to achieve first-generation SiGe HBT performance levels. The lone visionary group to solidly embrace C-doping of SiGe HBTs at the onset was the IHP team in Germany, whose pioneering work eventually paid off and began to convince the skeptics of the merits of C-doping. The minimum required C concentration for effective out-diffusion suppression of B was empirically established to be in the vicinity of 0.1% to 0.2% C (i.e., around $1 \times 10^{20}\,\mathrm{cm}^{-3}$). Early on, much debate ensued on the physical mechanism of *how* C impedes the B diffusion process, but general agreement for the most part now exists and is discussed in Chapter 11 (see *SiGe and Si Strained-Layer Epitaxy for Silicon Heterostructure Devices*). The first high-performance, fully integrated SiGe:C HBT technology was reported in 1999 [54].

The first "high-power" SiGe HBTs (S band, with multiwatt output power) were reported in 1996 using thick collector doping profiles [51,52]. The 10-psec ECL circuit performance barrier was broken in 1997 [53]. The 200-GHz peak f_T performance barrier was broken in November 2001 for a non-self-aligned device [55], and for a self-aligned device in February 2002 [67]. By 2004, a total of six industrial laboratories had achieved 200 GHz performance levels. A SiGe HBT technology with a peak f_T of 350 GHz (375 GHz values were reported in the IEDM presentation) was presented in December 2002 [56], and this 375 GHz f_T value remains a record for room temperature operation (it is above 500 GHz at cryogenic temperatures), and an optimized version with both f_T and f_{max} above 300 GHz was achieved in June 2004 [60]. This combined level of 300+ GHz for both f_T and f_{max} remains a solid record for any Si-based semiconductor device.

Other recent and interesting developments in the SiGe HBT field include the first report of a complementary (npn + pnp) SiGe HBT (C-SiGe) technology in 2003 [57], rapidly followed by a C–SiGe technology with f_T for both the npn and pnp SiGe HBTs above 100 GHz [58]. In addition, a novel vertical npn SiGe HBT has been implemented in thin-film (120 nm) CMOS-compatible SOI [59]. Besides further transistor performance enhancements, other logical developments to anticipate in this field include the integration of SiGe HBTs with strained-Si CMOS for a true all-Si-heterostructure technology.

Not surprisingly, research and development activity involving SiGe HBTs, circuits built from these devices, and various SiGe HBT technologies, in both industry and at universities worldwide, has grown very rapidly since the first demonstration of a functional SiGe HBT in 1987, only 18 years in the past.

2.3 SiGe–Strained Si FETs and Other SiGe Devices

The basic idea of using an electric field to modify the surface properties of materials, and hence construct a "field-effect" device, is remarkably old (1926 and 1935), predating even the quest for a solid-state amplifier [68]. Given the sweeping dominance of CMOS technology in the grand scheme of the electronics industry today, it is ironic that the practical demonstration of the BJT preceded that of the MOSFET by 9 years. This time lag from idea to realization was largely a matter of dealing with the many perils associated with obtaining decent dielectric materials in the Si system—doubly ironic given that Si has such a huge natural advantage over all other semiconductors in this regard. Bread-and-butter notions of ionic contamination, de-ionized water, fixed oxide charge, surface state passivation, and clean-room techniques in semiconductor fabrication had to be learned the hard way. Once device-quality SiO_2 was obtained in the late 1950s, and a robust gate dielectric could thus be fabricated, it was not long until the first functional MOSFET was demonstrated in 1960 [69]. The seemingly trivial (remember, however, that hindsight is 20–20!) connection of n-channel and p-channel MOSFETs to form low-power CMOS in 1963 [70] paved the way (eventually) to the high-volume, low-cost, highly integrated microprocessor, and the enormous variety of computational engines that exist today as a result.

Like their cousin, the SiGe HBT, SiGe–strained Si FETs did not get off the ground until the means for accomplishing the low-temperature growth of Si epitaxy could be realized. Once that occurred in the mid-1980s the field literally exploded. Table 2.3 summarizes the milestones in the evolution of SiGe–strained Si FETs, as well as a veritable menagerie of other electronic and optoelectronic components built from SiGe–strained Si epitaxy.

It was discovered as early as 1971 that direct oxidation of SiGe was a bad idea for building gate dielectrics [71]. Given that gate oxide quality, low-temperature deposited oxides, did not exist in the mid-1980s, the earliest FET demonstrations were modulation-doped, Schottky-gated, FETs, and both n-channel and p-channel SiGe MODFETs were pioneered as early as 1986 using MBE-grown material [72,73]. Before the SiGe MOSFET field got into high gear in the 1990s, a variety of other novel device demonstrations occurred, including: the first SiGe superlattice photodetector [74], the first SiGe Schottky barrier diodes (SBD) in 1988 [75], the first SiGe hole-transport resonant tunneling diode (RTD) in 1988 [76], and the first SiGe bipolar inversion channel FET (BiCFET) in 1989, a now-extinct dinosaur [77]. Meanwhile, early studies using SiGe in conventional CMOS gate stacks to minimize dopant depletion effects and tailor work functions, a fairly common practice in CMOS today, occurred in 1990 [78], and the first SiGe waveguides on Si substrates were produced in 1990 [79].

The first functional SiGe channel pMOSFET was published in 1991, and shortly thereafter, a wide variety of other approaches aimed at obtaining the best SiGe pMOSFETs (see, for instance, Refs. [93–95]). The first electron-transport RTD was demonstrated in 1991 [81], and the first LED in SiGe

TABLE 2.3 Milestones in the Development of SiGe–Strained Si FETs and Other Devices

Historical Event	Year	Ref.
Field effect device concept	1926	[68]
First Si MOSFET	1960	[69]
First Si CMOS	1963	[70]
First oxidation study of SiGe	1971	[71]
First SiGe nMODFET	1986	[72]
First SiGe pMODFET	1986	[73]
First SiGe photodetector	1986	[74]
First SiGe SBD	1988	[75]
First SiGe hole RTD	1988	[76]
First SiGe BiCFET	1989	[77]
First SiGe gate CMOS technology	1990	[78]
First SiGe waveguide	1990	[79]
First SiGe pMOSFET	1991	[80]
First SiGe electron RTD	1991	[81]
First SiGe LED	1991	[82]
First SiGe solar cell	1992	[83]
First a-SiGe phototransistor	1993	[84]
First SiGe pMOSFET on SOI	1993	[85]
First strained Si pMOSFET	1993	[86]
First strained Si nMOSFET	1994	[87]
First SiGe:C pMOSFET	1996	[88]
First SiGe pFET on SOS	1997	[89]
First submicron strained Si MOSFET	1998	[90]
First vertical SiGe pFET	1998	[91]
First strained Si CMOS technology	2002	[92]

also in 1991 (a busy year for our field). In 1992, the first a-SiGe solar cell was discussed [83], and in 1993, the first high-gain a-SiGe phototransistor [84]. The first SiGe pMOSFETs using alternate substrate materials were demonstrated, first in SOI in 1993 [85], and then on sapphire in 1997 [88], the first SiGe:C channel pMOSFET was demonstrated in 1996 [89], and the first vertical SiGe FET was published in 1998 [92].

Because of the desire to use Si-based bandgap engineering to improve not only the p-channel MOSFET, but also the n-channel MOSFET, research in the early- to mid-1990s in the FET field began to focus on strained Si MOSFETs on relaxed SiGe layers, with its consequent improvement in both electron and hole transport properties. This work culminated in the first strained Si pMOSFET in 1993 [87], and the first stained Si nMOSFET in 1994 [88], and remains an intensely active research field today. Key to the eventual success of strained Si CMOS approaches was that significant mobility enhancement could be achieved in both nFETs and pFETs down to very short (sub-micron) gate lengths, and this was first demonstrated in 1998 [90]. Strained Si CMOS at the 90-nm node and below is rapidly becoming mainstream for most serious CMOS companies, and the first commercial 90 nm strained Si CMOS technology platform was demonstrated by Intel in 2002 [91]. At last count, there were upwards of a half-dozen companies (e.g., Texas Instruments and IBM) also rapidly pushing toward 90 nm (and below) strained Si CMOS technologies, utilizing a variety of straining techniques, and thus it would appear that strained Si CMOS will be a mainstream IC technology in the near future, joining SiGe HBT BiCMOS technology. This is clearly outstanding news for our field. The merger of SiGe HBTs with strained Si CMOS would be a near-term logical extension.

References

1. R Braunstein, AR Moore, and F Herman. Intrinsic optical absorption in germanium–silicon alloys. *Physical Review B* 32:1405–1408, 1958.
2. HC Theuerer, JJ Kleimack, HH Loar, and H Christensen. Epitaxial diffused transistors. *Proceedings of the IRE* 48:1642–1643, 1960.
3. BA Joyce and RR Bradley. Epitaxial growth of silicon from the pyrolysis of monosilane on silicon substrates. *Journal of the Electrochemical Society* 110:1235–1240, 1963.
4. JH van der Merwe. Crystal interfaces. Part I. Semi-infinite crystals. *Journal of Applied Physics* 34:117–125, 1963.
5. JH van der Merwe. Crystal interfaces. Part II. Finite overgrowths. *Journal of Applied Physics* 34:123–127, 1963.
6. JW Matthews and AE Blakeslee. Defects in epitaxial multilayers: I. Misfit dislocations in layers. *Journal of Crystal Growth* 27:118–125, 1974.
7. E Kasper, HJ Herzog, and H Kibbel. A one-dimensional SiGe superlattice grown by UHV epitaxy. *Journal of Applied Physics* 8:1541–1548, 1975.
8. JC Bean, TT Sheng, LC Feldman, AT Fiory, and RT Lynch. Pseudomorphic growth of Ge_xSi_{1-x} on silicon by molecular beam epitaxy. *Applied Physics Letters* 44:102–104, 1984.
9. R People and JC Bean. Calculation of critical layer thickness versus lattice mismatch for Ge_xSi_{1-x}/Si strained layer heterostructures. *Applied Physics Letters* 47:322–324, 1985.
10. R People. Indirect bandgap of coherently strained $Si_{1-x}Ge_x$ bulk alloys on $\langle 0\,0\,1 \rangle$ silicon substrates. *Physical Review B* 32:1405–1408, 1985.
11. DV Lang, R People, JC Bean, and AM Sergent. Measurement of the bandgap of Ge_xSi_{1-x}/Si strained-layer heterostructures. *Applied Physics Letters* 47:1333–1335, 1985.
12. JF Gibbons, CM Gronet, and KE Williams. Limited reaction processing: silicon epitaxy. *Applied Physics Letters* 47:721–723, 1985.
13. G Abstreiter, H Brugger, T Wolf, H Joke, and HJ Kerzog. Strain-induced two-dimensional electron gas in selectively doped Si/Si_xGe_{1-x} superlattices. *Physical Review* 54:2441–2444, 1985.
14. BS Meyerson. Low-temperature silicon epitaxy by ultrahigh vacuum/chemical vapor deposition. *Applied Physics Letters* 48:797–799, 1986.

15. R People and JC Bean. Band alignments of coherently strained Ge_xSi_{1-x}/Si heterostructures on $\langle 0\,0\,1 \rangle$ Ge_ySi_{1-y} substrates. *Applied Physics Letters* 48:538–540, 1986.

16. BS Meyerson, KJ Uram, and FK LeGoues. Cooperative phenomena is silicon/germanium low temperature epitaxy. *Applied Physics Letters* 53:2555–2557, 1988.

17. CA King, JL Hoyt, CM Gronet, JF Gibbons, MP Scott, and J Turner. $Si/Si_{1-x}/Ge_x$ heterojunction bipolar transistors produced by limited reaction processing. *IEEE Electron Device Letters* 10:52–54, 1989.

18. TO Sedgwick, M Berkenbilt, and TS Kuan. Low-temperature selective epitaxial growth of silicon at atmospheric pressure. *Applied Physics Letters* 54:2689–2691, 1989.

19. PJ Wang, FF Fang, BS Meyerson, J Mocera, and B Parker. Two-dimensional hole gas in $Si/Si_{0.85}Ge_{0.15}$ modulation doped heterostructures. *Applied Physics Letters* 54:2701–2703, 1989.

20. P Agnello, TO Sedgwick, MS Goorsky, J Ott, TS Kuan, and G Scilla. Selective growth of silicon–germanium alloys by atmospheric-pressure chemical vapor deposition at low temperatures. *Applied Physics Letters* 59:1479–1481, 1991.

21. T Manku and A Nathan. Lattice mobility of holes in strained and unstrained $Si_{1-x}Ge_x$ alloys. *IEEE Electron Device Letters* 12:704–706, 1991.

22. T Manku and A Nathan. Electron drift mobility model for devices based on unstrained and coherently strained $Si_{1-x}Ge_x$ grown on $\langle 0\,0\,1 \rangle$ silicon subtrate. *IEEE Transactions on Electron Devices* 39:2082–2089, 1992.

23. K Erbel, SS Iyer, S Zollner, JC Tsang, and FK LeGoues. Growth and strain compensation effects in the ternary $Si_{1-x-y}Ge_xC_y$ alloy system. *Applied Physics Letters* 60:3033–3035, 1992.

24. HJ Osten, E Bugiel, and P Zaumseil. Growth of inverse tetragonal distorted SiGe layer on Si(0 0 1) by adding small amounts of carbon. *Applied Physics Letters* 64:3440–3442, 1994.

25. A Fischer, H-J Osten, and H Richter. An equilibrium model for buried SiGe strained layers. *Solid-State Electronics* 44:869–873, 2000.

26. J Bardeen and WH Brattain. The transistor, a semi-conductor triode. *Physical Review* 71:230–231, 1947.

27. W Shockley, M Sparks, and GK Teal. p–n junction transistors. *Physical Review* 83:151–162, 1951.

28. M Tanenbaum and DE Thomas. Diffused emitter and base silicon transistors. *Bell System Technical Journal* 35:23–34, 1956.

29. See, for instance, W Shockley. U.S. Patents 2,502,488, 2,524,035, and 2,569,347.

30. H Kroemer. Zur theorie des diffusions und des drifttransistors. Part III. *Archiv der Elektrischen Ubertragungstechnik* 8:499–504, 1954.

31. H Kroemer. Quasielectric and quasimagnetic fields in nonuniform semiconductors. *RCA Review* 18:332–342, 1957.

32. H Kroemer. Theory of a wide-gap emitter for transistors. *Proceedings of the IRE* 45:1535–1537, 1957.

33. B Brar, GJ Sullivan, and PM Asbeck. Herb's bipolar transistors. *IEEE Transactions on Electron Devices* 48:2473–2476, 2001.

34. RM Warner. Microelectronics: Its unusual origin and personality. *IEEE Transactions on Electron Devices* 48:2457–2467, 2001.

35. SS Iyer, GL Patton, SL Delage, S Tiwari, and J.M.C. Stork. Silicon–germanium base heterojunction bipolar transistors by molecular beam epitaxy. Technical Digest of the IEEE International Electron Devices Meeting, San Francisco, 1987, pp. 874–876.

36. GL Patton, DL Harame, JMC Stork, BS Meyerson, GJ Scilla, and E Ganin. Graded-SiGe-base, poly-emitter heterojunction bipolar transistors. *IEEE Electron Device Letters* 10:534–536, 1989.

37. GL Patton, JH Comfort, BS Meyerson, EF Crabbé, E de Frésart, JMC Stork, JY-C Sun, DL Harame, and J Burghartz. 63-75 GHz f_T SiGe-base heterojunction-bipolar technology. Technical Digest IEEE Symposium on VLSI Technology, Honolulu, 1990, pp. 49–50.

38. GL Patton, JH Comfort, BS Meyerson, EF Crabbé, GJ Scilla, E de Frésart, JMC Stork, JY-C Sun, DL Harame, and J Burghartz. 75 GHz f_T SiGe base heterojunction bipolar transistors. *IEEE Electron Device Letters* 11:171–173, 1990.

39. JH Comfort, GL Patton, JD Cressler, W Lee, EF Crabbé, BS Meyerson, JY-C Sun, JMC Stork, P-F Lu, JN Burghartz, J Warnock, K Kenkins, K-Y Toh, M D'Agostino, and G Scilla. Profile leverage in a self-aligned epitaxial Si or SiGe-base bipolar technology. Technical Digest IEEE International Electron Devices Meeting, Washington, 1990, pp. 21–24.

40. DL Harame, JMC Stork, BS Meyerson, EF Crabbé, GL Patton, GJ Scilla, E de Frésart, AA Bright, C Stanis, AC Megdanis, MP Manny, EJ Petrillo, M Dimeo, RC McIntosh, and KK Chan. SiGe-base PNP transistors fabrication with n-type UHV/CVD LTE in a "NO DT" process. Technical Digest IEEE Symposium on VLSI Technology, Honolulu, 1990, pp. 47–48.

41. EF Crabbeé, GL Patton, JMC Stork, BS Meyerson, and JY-C Sun. Low temperature operation of Si and SiGe bipolar transistors. Technical Digest IEEE International Electron Devices Meeting, Washington, 1990, pp. 17–20.

42. DL Harame, EF Crabbé, JD Cressler, JH Comfort, JY-C Sun, SR Stiffler, E Kobeda, JN Burghartz, MM Gilbert, J Malinowski, and AJ Dally. A high-performance epitaxial SiGe-base ECL BiCMOS technology. Technical Digest IEEE International Electron Devices Meeting, Washington, 1992, pp. 19–22.

43. DL Harame, JMC Stork, BS Meyerson, KY-J Hsu, J Cotte, KA Jenkins, JD Cressler, P Restle, EF Crabbé, S Subbanna, TE Tice, BW Scharf, and JA Yasaitis. Optimization of SiGe HBT technology for high speed analog and mixed-signal applications. Technical Digest IEEE International Electron Devices Meeting, San Francisco, 1993, pp. 71–74.

44. E Kasper, A Gruhle, and H Kibbel. High speed SiGe-HBT with very low base sheet resistivity. Techncial Digest IEEE International Electron Devices Meeting, San Francisco, 1993, pp. 79–81.

45. EF Crabbé, BS Meyerson, JMC Stork, and DL Harame. Vertical profile optimization of very high frequency epitaxial Si- and SiGe-base bipolar transistors. Technical Digest IEEE International Electron Devices Meeting, Washington, 1993, pp. 83–86.

46. DL Harame, K Schonenberg, M Gilbert, D Nguyen-Ngoc, J Malinowski, S-J Jeng, BS Meyerson, JD Cressler, R Groves, G Berg, K Tallman, K Stein, G Hueckel, C Kermarrec, T Tice, G Fitzgibbons, K Walter, D Colavito, T Houghton, N Greco, T Kebede, B Cunningham, S Subbanna, JH Comfort, and EF Crabbé. A 200 mm SiGe-HBT technology for wireless and mixed-signal applications. Technical Digest IEEE International Electron Devices Meeting, Washington, 1994, pp. 437–440.

47. JD Cressler, EF Crabbé, JH Comfort, JY-C Sun, and JMC Stork. An epitaxial emitter cap SiGe-base bipolar technology for liquid nitrogen temperature operation. *IEEE Electron Device Letters* 15:472–474, 1994.

48. JA Babcock, JD Cressler, LS Vempati, SD Clark, RC Jaeger, and DL Harame. Ionizing radiation tolerance of high performance SiGe HBTs grown by UHV/CVD. *IEEE Transactions on Nuclear Science* 42:1558–1566, 1995.

49. LS Vempati, JD Cressler, RC Jaeger, and DL Harame. Low-frequency noise in UHV/CVD Si- and SiGe-base bipolar transistors. Proceedings of the IEEE Bipolar/BiCMOS Circuits and Technology Meeting, Minnneapolis, 1995, pp. 129–132.

50. L Lanzerotti, A St Amour, CW Liu, JC Sturm, JK Watanabe, and ND Theodore. Si/Si$_{1-x-y}$Ge$_x$C$_y$/Si heterojunction bipolar transistors. *IEEE Electron Device Letters* 17:334–337, 1996.

51. A Schüppen, S Gerlach, H Dietrich, D Wandrei, U Seiler, and U König. 1-W SiGe power HBTs for mobile communications. *IEEE Microwave and Guided Wave Letters* 6:341–343, 1996.

52. PA Potyraj, KJ Petrosky, KD Hobart, FJ Kub, and PE Thompson. A 230-Watt S-band SiGe heterojunction junction bipolar transistor. *IEEE Transactions on Microwave Theory and Techniques* 44:2392–2397, 1996.

53. K Washio, E Ohue, K Oda, M Tanabe, H Shimamoto, and T Onai. A selective-epitaxial SiGe HBT with SMI electrodes featuring 9.3-ps ECL-Gate Delay. Technical Digest IEEE International Electron Devices Meeting, San Francisco, 1997, pp. 795–798.

54. HJ Osten, D Knoll, B Heinemann, H Rücker, and B Tillack. Carbon doped SiGe heterojunction bipolar transistors for high frequency applications. Proceedings of the IEEE Bipolar/BiCMOS Circuits and Technology Meeting, Minneapolis, 1999, pp. 109–116.

55. SJ Jeng, B Jagannathan, J-S Rieh, J Johnson, KT Schonenberg, D Greenberg, A Stricker, H Chen, M Khater, D Ahlgren, G Freeman, K Stein, and S Subbanna. A 210-GHz f_T SiGe HBT with non-self-aligned structure. *IEEE Electron Device Letters* 22:542–544, 2001.

56. JS Rieh, B Jagannathan, H Chen, KT Schonenberg, D Angell, A Chinthakindi, J Florkey, F Golan, D Greenberg, S-J Jeng, M Khater, F Pagette, C Schnabel, P Smith, A Stricker, K Vaed, R Volant, D Ahlgren, G Freeman, K Stein, and S Subbanna. SiGe HBTs with cut-off frequency of 350 GHz. Technical Digest of the IEEE International Electron Devices Meeting, San Francisco, 2002, pp. 771–774.

57. B El-Kareh, S Balster, W Leitz, P Steinmann, H Yasuda, M Corsi, K Dawoodi, C Dirnecker, P Foglietti, A Haeusler, P Menz, M Ramin, T Scharnagl, M Schiekofer, M Schober, U Schulz, L Swanson, D Tatman, M. Waitschull, JW Weijtmans, and C Willis. A 5V complementary SiGe BiCMOS technology for high-speed precision analog circuits. Proceedings of the IEEE Bipolar/BiCMOS Circuits and Technology Meeting, Toulouse, 2003, pp. 211–214.

58. B Heinemann, R Barth, D Bolze, J Drews, P Formanek, O Fursenko, M Glante, K Glowatzki, A Gregor, U Haak, W Höppner, D Knoll, R Kurps, S Marschmeyer, S Orlowski, H Rücker, P Schley, D Schmidt, R Scholz, W Winkler, and Y Yamamoto. A complementary BiCMOS technology with high speed npn and pnp SiGe:C HBTs. Technical Digest of the IEEE International Electron Devices Meeting, Washington, 2003, pp. 117–120.

59. J Cai, M Kumar, M Steigerwalt, H Ko, K Schonenberg, K Stein, H Chen, K Jenkins, Q Ouyang, P Oldiges, and T Ning. Vertical SiGe-base bipolar transistors on CMOS-compatible SOI substrate. Proceedings of the IEEE Bipolar/BiCMOS Circuits and Technology Meeting, Toulouse, 2003, pp. 215–218.

60. J-S Rieh, D Greenberg, M Khater, KT Schonenberg, J-J Jeng, F Pagette, T Adam, A Chinthakindi, J Florkey, B Jagannathan, J Johnson, R Krishnasamy, D Sanderson, C Schnabel, P Smith, A Stricker, S Sweeney, K Vaed, T Yanagisawa, D Ahlgren, K Stein, and G Freeman. SiGe HBTs for millimeter-wave applications with simultaneously optimized f_T and f_{max}. Proceedings of the IEEE Radio Frequency Integrated Circuits (RFIC) Symposium, Fort Worth, 2004, pp. 395–398.

61. DL Harame and BS Meyerson. The early history of IBM's SiGe mixed signal technology. *IEEE Transactions on Electron Devices* 48:2555–2567, 2001.

62. T Tatsumi, H Hirayama, and N Aizaki. $Si/Ge_{0.3}Si_{0.7}$ heterojunction bipolar transistor made with Si molecular beam epitaxy. *Applied Physics Letters* 52:895–897, 1988.

63. H Temkin, JC Bean, A Antreasyan, and R Leibenguth. Ge_xSi_{1-x} strained-layer heterostructure bipolar transistors. *Applied Physics Letters* 52:1089–1091, 1988.

64. D-X Xu, G-D Shen, M Willander, W-X Ni, and GV Hansson. $n-Si/p-Si_{1-x}/n-Si$ double-heterojunction bipolar transistors. *Applied Physics Letters* 52:2239–2241, 1988.

65. GL Patton, SS Iyer, SL Delage, S Tiwari, and JMC Stork. Silicon–germanium-base heterojunction bipolar transistors by molecular beam epitaxy. *IEEE Electron Device Letters* 9:165–167, 1988.

66. A Gruhle, H Kibbel, U König, U Erben, and E Kasper. MBE-Grown Si/SiGe HBTs with high β, f_T, and f_{max}. *IEEE Electron Device Letters* 13:206–208, 1992.

67. AJ Joseph, D Coolbaugh, D Harame, G Freeman, S Subbanna, M Doherty, J Dunn, C Dickey, D Greenberg, R Groves, M Meghelli, A Rylyakov, M Sorna, O Schreiber, D Herman, and T Tanji. 0.13 μm 210 GHz f_T SiGe HBTs—expanding the horizons of SiGe BiCMOS. Technical Digest IEEE International Solid-State Circuits Conference, San Francisco, 2002, pp. 180–182.

68. H. Lilienfeld Patent, 1926; O. Heil, British patent number 439,457, 1935.

69. D Khang and MM Atalla. Silicon–silicon dioxide field induced surface devices. Solid State Research Conference, Pittsburgh, 1960.

70. FM Wanlass and CT Sah. Nanowatt logic using field-effect metal-oxide-semiconductor triodes (MOSTs). IEEE International Solid-State Circuits Conference, Philadelphia, 1963, pp. 32–33.

71. P Balk. Surface properties of oxidized germanium-doped silicon. *Journal of the Electrochemical Society* 118:494–495, 1971.

72. H Daembkes, H-J Herzog, H Jorke, H. Kibbel, and E Kasper. The n-channel SiGe/Si modulation doped field-effect transistor. *IEEE Transactions on Electron Devices* 33:633–638, 1986.

73. TP Pearsall and JC Bean. Enhancement and depletion-mode p-channel Ge_xSi_{1-x} modulation-doped field effect transistor. *IEEE Electron Device Letters* 7:308–310, 1986.

74. H Temkin, TP Pearsall, JC Bean, RA Logan, and S. Luryi. Ge_xSi_{1-x} strained-layer superlattice waveguide photodetectors operating near 1.3 μm. *Applied Physics Letters* 48:963–965, 1986.

75. RD Thompson, KN Tu, J Angilello, S Delage, and SS Iyer. Interfacial reaction between Ni and MBE grown SiGe alloys. *Journal of the Electrochemical Society* 135:3161–3163, 1988.

76. HC Liu, D Landheer, M Buchmann, and DC Houghton. Resonant tunneling diode in the $Si_{1-x}Ge_x$ system. *Applied Physics Letters* 52:1809–1811, 1988.

77. RC Taft, JD Plummer, and SS Iyer. Demonstration of a p-channel BiCFET in the Ge_xSi_{1-x}/Si system. *IEEE Electron Device Letters* 10:14–16, 1989.

78. TJ King, JR Pfriester, JD Scott, JP McVittie, and KC Saraswat. A polycrystalline SiGe gate CMOS technology. Technical Digest of the IEEE International Electron Devices Meeting, Washington, 1990, pp. 253–256.

79. RA Soref, F Namavar, and JP Lorenzo. Optical waveguiding in a single-crystal layer of germanium–silicon grown on silicon. *Optics Letters* 15:270–272, 1990.

80. DK Nayak, JCS Woo, JS Park, KL Wang, and KP MacWilliams. Enhancement-mode quantum-well Ge_xSi_{1-x} PMOS. *IEEE Electron Device Letters* 12:154–156, 1991.

81. K Ismail, BS Meyerson, and PJ Wang. Electron resonant tunneling in Si/SiGe double barrier diodes. *Applied Physics Letters* 59:973–975, 1991.

82. DC Houghton, JP Noel, and NL Rowell. Electroluminescence and photoluminesence from SiGe alloys grown on (1 0 0) silicon by MBE. *Materials Science and Engineering B* 9:237–244, 1991.

83. DS Chen, JP Conde, V Chu, S Aljishi, JZ Liu, and S Wagner. Amorphous silicon–germanium thin-film photodetector array. *IEEE Electron Device Letters* 13:5–7, 1992.

84. S-B Hwang, YK Fang, K-H Chen, C-R Liu, J-D Hwang, and M-H Chou. An a-Si:H/a-Si, Ge:H bulk barrier phototransistor with a-SiC:H barrier enhancement layer for high-gain IR optical detector. *IEEE Transactions on Electron Devices* 40:721–726, 1993.

85. DK Nayak, JCS Woo, GK Yabiku, KP MacWilliams, JS Park, and KL Wang. High mobility GeSi PMOS on SIMOX. *IEEE Electron Device Letters* 14:520–522, 1993.

86. DK Nayak, JCS Woo, JS Park, KL Wang, and KP MacWilliams. High-mobility p-channel metal-oxide semiconductor field-effect transistor on strained Si. *Applied Physics Letters* 62:2853–2855, 1993.

87. J Welser, JL Hoyt, and JF Gibbons. Electron mobility enhancement in strained-Si n-type metal-oxide semiconductor field-effect transistors. *IEEE Electron Device Letters* 15:100–102, 1994.

88. SK Ray, S John, S Oswal, and SK Banerjee. Novel SiGeC channel heterojunction pMOSFET. Technical Digest of the IEEE International Electron Devices Meeting, Washington, 1996, pp. 261–264.

89. SJ Mathew, WE Ansley, WB Dubbelday, JD Cressler, JA Ott, JO Chu, PM Mooney, KL Vavanagh, BS Meyerson, and I Lagnado. Effect of Ge profile on the frequency response of a SiGe pFET on sapphire technology. Technical Digest of the IEEE Device Research Conference, Boulder, 1997, pp. 130–131.

90. K Rim, JL Hoyt, and JF Gibbons. Transconductance enhancement in deep submicron strained-Si n-MOSFETs. Technical Digest of the IEEE International Electron Devices Meeting, Washington, 1998, pp. 707–710.

91. KC Liu, SK Ray, SK Oswal, and SK Banerjee. $Si_{1-x}Ge_x$/Si vertical pMOSFET fabricated by Ge ion implantation. *IEEE Electron Device Letters* 19:13–15, 1998.

92. S Thompson, N. Anand, M Armstrong, C Auth, B Arcot, M Alavi, P Bai, J Bielefeld, R Bigwood, J Brandenburg, M Buehler, S Cea, V Chikarmane, C Choi, R Frankovic, T Ghani, G Glass, W Han, T Hoffmann, M Hussein, P Jacob, A Jain, C Jan, S Joshi, C Kenyon, J Klaus, S Klopcic, J Luce, Z Ma, B McIntyre, K Mistry, A Murthy, P Nguyen, H Pearson, T Sandford, R Schweinfurth, R Shaheed, S Sivakumar, M Taylor, B Tufts, C Wallace, P Wang, C Weber, and M Bohr. A 90 nm logic technology featuring 50 nm strained silicon channel transistors, 7 layers of Cu interconnects, low k ILD, and 1 μm² SRAM Cell. Technical Digest of the IEEE International Electron Devices Meeting, Washington, 2002, pp. 61–64.

93. VP Kesan, S Subbanna, PJ Restle, MJ Tejwani, JM Aitken, SS Iyer, and JA Ott. High performance 0.25 μm p-MOSFETs with silicon–germanium channels for 300 K and 77 K operation. Technical Digest of the IEEE International Electron Devices Meeting, San Francisco, 1991, pp. 25–28, 1991.

94. S Verdonckt-Vanderbroek, E Crabbé, BS Meyerson, DL Harame, PJ Restle, JMC Stork, AC Megdanis, CL Stanis, AA Bright, GMW Kroesen, and AC Warren. High-mobility modulation-doped, graded SiGe-channel p-MOSFETs. *IEEE Electron Device Letters* 12:447–449, 1991.

95. S Verdonckt-Vanderbroek, E Crabbé, BS Meyerson, DL Harame, PJ Restle, JMC Stork, and JB Johnson. SiGe-channel heterojunction p-MOSFETs. *IEEE Transactions on Electron Devices* 41:90–102, 1994.

3

Overview: SiGe HBTs

John D. Cressler
Georgia Institute of Technology

SiGe HBTs are far and away the most mature Si heterostructure devices and not surprisingly the most completely researched and discussed in the technical literature. That is not to say that we completely understand the SiGe HBT, and new effects and nuances of operation are still being uncovered year-by-year as transistor scaling advances and application targets march steadily upward in frequency and sophistication. There is still much to learn. Nevertheless, a large body of literature on SiGe HBT operation does exist, across an amazingly diverse set of topics, ranging from basic transistor physics, to noise, to radiation effects, to simulation. This section's comprehensive treatment of SiGe HBTs begins with Chapter 4, "Device Physics," by J.D. Cressler of Georgia Tech., and addresses perturbations to that first-order theory in Chapter 5, "Second-Order Effects," by J.D. Cressler of Georgia Tech. Chapters 6 to 9 address mixed-signal noise and linearity in SiGe HBTs, including: Chapter 6, "Low-Frequency Noise," by G. Niu of Auburn University; Chapter 7, "Broadband Noise," by D. Greenberg of IBM Microelectronics; Chapter 8, "Microscopic Noise Simulation," by G. Niu of Auburn University; and Chapter 9, "Linearity," by G. Niu of Auburn University.

The very recent development of complementary (npn + pnp) SiGe technologies for high-speed analog circuits makes the discussion in Chapter 10, "pnp SiGe HBTs," by J.D. Cressler of Georgia Tech particularly relevant. Chapter 11, "Temperature Effects," by J.D. Cressler of Georgia Tech addresses the impact of bandgap engineering on device behavior across temperature, as well as the inherent advantages enjoyed by SiGe HBTs for cryogenic electronics. The important and very recently emerging application associated with space-borne electronics operating in a hostile radiation-rich environment are addressed in Chapter 12, "Radiation Effects," by J.D. Cressler of Georgia Tech.

Reliability issues, of key importance to the deployment of SiGe HBT circuits and systems, are covered in Chapter 13, "Reliability Issues," by J.D. Cressler of Georgia Tech, and the related and important topic of thermal phenomena are treated in Chapter 14, "Self-Heating and Thermal Effects," by J.-S. Rieh of Korea University. Finally, subtleties associated with device-level (one-dimensional through three-dimensional) simulation of SiGe HBTs is presented in Chapter 15, "Device-Level Simulation," by G. Niu of Auburn University, and this section concludes with a look at the ultimate limits of SiGe HBTs in Chapter 16, "Performance Limits," by G. Freeman of IBM Microelectronics. In addition to this substantial collection of material, and the numerous references contained in each chapter, a number of review articles and books detailing the operation and modeling of SiGe HBTs exist, including Refs. [1–13].

References

1. SS Iyer, GL Patton, JMC Stork, BS Meyerson, and DL Harame. Heterojunction bipolar transistors using Si–Ge alloys. *IEEE Transactions on Electron Devices* 36:2043–2064, 1989.
2. GL Patton, JMC Stork, JH Comfort, EF Crabbé, BS Meyerson, DL Harame, and JY-C Sun. SiGe-base heterojunction bipolar transistors: physics and design issues. Technical Digest of the IEEE International Electron Devices Meeting, Washington, 1990, pp. 13–16.
3. C Kermarrec, T Tewksbury, G Dawe, R Baines, B Meyerson, D Harame, and M Gilbert. SiGe HBTs reach the microwave and millimeter-wave frontier. Proceedings of the IEEE Bipolar/BiCMOS Circuits and Technology Meeting, Minneapolis, 1994, pp. 155–162.
4. JD Cressler. Re-engineering silicon: Si–Ge heterojunction bipolar technology. *IEEE Spectrum*, 1995, pp. 49–55.
5. DL Harame, JH Comfort, JD Cressler, EF Crabbé, JY-C Sun, BS Meyerson, and T Tice. Si/SiGe epitaxial-base transistors: Part I—Materials, physics, and circuits. *IEEE Transactions on Electron Devices* 40:455–468, 1995.
6. DL Harame, JH Comfort, JD Cressler, EF Crabbé, JY-C Sun, BS Meyerson, and T Tice. Si/SiGe epitaxial-base transistors: Part II—Process integration and analog applications. *IEEE Transactions on Electron Devices* 40:469–482, 1995.
7. JD Cressler. SiGe HBT technology: a new contender for Si-based RF and microwave circuit applications. *IEEE Transactions of Microwave Theory and Techniques* 46:572–589, 1998.
8. A Gruhle. Prospects for 200 GHz on silicon with SiGe heterojunction bipolar transistors. Proceedings of the IEEE Bipolar/BiCMOS Circuits and Technology Meeting, Minneapolis, 2001, pp. 19–25.
9. DL Harame, DC Ahlgren, DD Coolbaugh, JS Dunn, G Freeman, JD Gillis, RA Groves, GN Henderson, RA Johnson, AJ Joseph, S Subbanna, AM Victor, KM Watson, CS Webster, and PJ Zampardi. Current status and future trends of SiGe BiCMOS technology. *IEEE Transactions on Electron Devices* 48:2575–2594, 2001.
10. JS Yuan. *SiGe, GaAs, and InP Heterojunction Bipolar Transistors*. New York, NY: John Wiley & Sons, 1999.
11. JD Cressler and G Niu. *Silicon–Germanium Heterojunction Bipolar Transistors*. Boston, MA: Artech House, 2003.
12. R Singh, DL Harame, and MM Oprysko. *Silicon Germanium: Technology, Modeling, and Design*. Piscataway, NJ: IEEE Press, 2004.
13. P Ashburn. *SiGe Heterojunction Bipolar Transistors*. New York, NY: John Wiley & Sons, 2004.

4

Device Physics

John D. Cressler

Georgia Institute of Technology

4.1 Introduction

The essential differences between the SiGe HBT and the Si BJT are best illustrated by considering a schematic energy band diagram. For simplicity, we consider an ideal, graded-base SiGe HBT with constant doping in the emitter, base, and collector regions. In such a device construction, the Ge content is linearly graded from 0% near the metallurgical emitter–base (EB) junction to some maximum value of Ge content near the metallurgical collector–base (CB) junction, and then rapidly ramped back down to 0% Ge. The resultant overlaid energy band diagrams for both the SiGe HBT and the Si BJT, biased identically in forward-active mode, are shown in Figure 4.1. Observe in Figure 4.1 that a Ge-induced reduction in base bandgap occurs at the EB edge of the quasi-neutral base ($\Delta E_{g,Ge}\,(x = 0)$), and at the CB edge of the quasi-neutral base ($\Delta E_{g,Ge}\,(x = W_b)$). This grading of the Ge across the neutral base induces a built-in quasi-drift field (($\Delta E_{g,Ge}\,(x = W_b) - \Delta E_{g,Ge}(x = 0))/W_b$) in the neutral base. In this chapter, we examine the impact of Ge on the dc and ac properties of the transistor—the essential devices physics of the SiGe HBT.

4.2 An Intuitive Picture

To intuitively understand how these band-edge changes affect the dc operation of the SiGe HBT, first consider the operation of the Si BJT. When V_{BE} is applied to forward bias the EB junction, electrons are injected from the electron-rich emitter into the base across the EB potential barrier (refer to Figure 4.1). The injected electrons diffuse across the base, and are swept into the electric field of the CB junction, yielding a useful collector current. At the same time, the applied forward bias on the EB junction produces a back-injection of holes from the base into the emitter. If the emitter region is doped heavily with respect to the base, however, the density of back-injected holes will be small compared to the forward-injected electron density, and hence a finite current gain $\beta \propto n/p$ results.

As can be seen in Figure 4.1, the introduction of Ge into the base region has two tangible dc consequences: (1) the potential barrier to injection of electrons from emitter into the base is decreased. Intuitively, this will yield exponentially more electron injection for the same applied V_{BE}, translating into

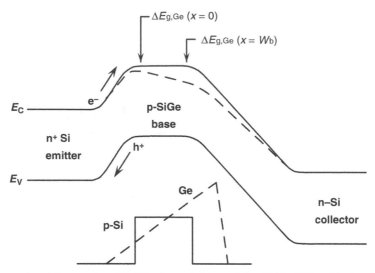

FIGURE 4.1   Energy band diagram for a Si BJT and a graded-base SiGe HBT, both biased in forward active mode at low injection.

higher collector current, and hence higher current gain, provided the base current remains unchanged. Given that band-edge effects generally couple strongly to transistor properties, we naively expect a strong dependence of J_C on Ge content. Of practical consequence, the introduction of Ge effectively decouples the base doping from the current gain, thereby providing device designers with much greater flexibility than in Si BJT design. If, for instance, the intended circuit application does not require high current gain (as a rule of thumb, $\beta = 100$ is usually sufficient for most circuits), we can effectively trade the higher gain induced by the Ge band offset for a higher base-doping level, leading to lower net base resistance, and hence better dynamic switching and noise characteristics. (2) The presence of a finite Ge content in the CB junction will positively influence the output conductance of the transistor, yielding higher Early voltage. While it is more difficult to physically visualize why this is the case, in essence, the smaller base bandgap near the CB junction effectively weights the base profile (through the integral of intrinsic carrier density across the base), such that the backside depletion of the neutral base with increasing applied V_{CB} (Early effect) is suppressed compared to a comparably doped Si BJT. This translates into a higher Early voltage compared to a Si BJT.

To intuitively understand how these band-edge changes affect the ac operation of the SiGe HBT, first consider the dynamic operation of the Si BJT. Electrons injected from the emitter into the base region must diffuse across the base (for constant doping), and are then swept into the electric field of the CB junction, yielding a useful (time-dependent) collector current. The time it takes for the electrons to traverse the base (base transit time) is significant, and typically is the limiting transit time that determines the overall transistor ac performance (e.g., peak f_T). At the same time, the applied forward bias on the EB junction dynamically produces a back-injection of holes from the base into the emitter. For fixed collector bias current, this dynamic storage of holes in the emitter (emitter charge storage delay time) is reciprocally related to the ac current gain of the transistor (β_{ac}).

As can be seen in Figure 4.1, the introduction of Ge into the base region has an important ac consequence, since the Ge-gradient-induced drift field across the neutral base is aligned in a direction (from collector to emitter) such that it will accelerate the injected minority electrons across the base. We are thus able to add a large drift field component to the electron transport, effectively speeding up the diffusive transport of the minority carriers and thereby decreasing the base transit time. Even though the band offsets in SiGe HBTs are typically small by III–V technology standards, the Ge grading over the short distance of the neutral base can translate into large electric fields. For instance, a linearly graded Ge profile with a modest peak Ge content of 10%, graded over a 50-nm neutral base width, yields $75\,\text{mV/50}\,\text{nm} = 15\,\text{kV/cm}$ electric field, sufficient to accelerate the electrons to near saturation velocity ($\nu_s \simeq 1 \times 10^7\,\text{cm/sec}$). Because the base transit time typically limits the frequency response of a Si BJT, we would expect that the frequency response should be significantly improved by introducing this Ge-induced drift field. In addition, we know that the Ge-induced band offset at the EB junction will exponentially enhance the collector current density (and thus β) of a SiGe HBT compared to a comparably constructed Si BJT. Since the emitter charge storage delay time is reciprocally related to β, we would also expect the frequency response to a SiGe HBT to benefit from this added emitter charge storage delay time advantage.

4.3 Current Gain

To understand the inner workings of the SiGe HBT, we must first formally relate the changes in the collector current density and hence current gain to the physical variables of this problem. It is also instructive to carefully compare the *differences* between a comparably constructed SiGe HBT and a Si BJT. In the present analysis, the SiGe HBT and Si BJT are taken to be of identical geometry, and it is assumed that the emitter, base, and collector-doping profiles of the two devices are identical, apart from the Ge in the base of the SiGe HBT. For simplicity, a Ge profile that is linearly graded from the EB to the CB junction is assumed (as depicted in Figure 4.2). The resultant expressions can be applied to a wide variety of practical SiGe profile designs, ranging from constant (box) Ge profiles, to triangular (linearly graded) Ge profiles, and including the intermediate case of the Ge trapezoid (a combination of box and linearly graded profiles) [2]. Unless otherwise stated, this analysis assumes standard low-injection conditions, negligible bulk and surface recombination, Boltzmann statistics, and holds for npn SiGe HBTs.

The theoretical consequences of the Ge-induced bandgap changes to J_C can be derived in closed form for a constant base-doping profile ($p_b(x) = N_{ab}^-(x) = N_{ab}^- = \text{constant}$) by considering the generalized Moll–Ross collector current density relation (refer to Appendices A.2 and A.3), which holds for low injection in the presence of both nonuniform base doping and nonuniform base bandgap at fixed V_{BE} and temperature (T) [3]

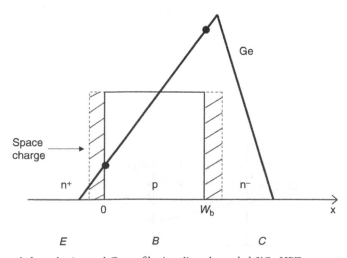

FIGURE 4.2 Schematic base doping and Ge profiles in a linearly graded SiGe HBT.

$$J_C = \frac{q(e^{qV_{BE}/kT} - 1)}{\displaystyle\int_0^{W_b} \frac{p_b(x)\,dx}{D_{nb}(x)\,n_{ib}^2(x)}}, \tag{4.1}$$

where $x = 0$ and $x = W_b$ are the neutral base boundary values on the EB and CB sides of the base, respectively. In this case, the base doping is constant, but both n_{ib} and D_{nb} are position-dependent; the former through the Ge-induced band offset, and the latter due to the influence of the (position-dependent) Ge profile on the electron mobility ($D_{nb} = kT/q\mu_{nb} = f(Ge)$). Note that J_C depends only on the Ge-induced changes in the base bandgap. In general, the intrinsic carrier density in the SiGe HBT can be written as

$$n_{ib}^2(x) = (N_C N_V)_{SiGe}(x)e^{-E_{gb}(x)/kT}, \tag{4.2}$$

where $(N_C N_V)_{SiGe}$ accounts for the (position-dependent) Ge-induced changes associated with both the conduction and valence band effective density-of-states. In Equation 4.2, the SiGe base bandgap can be broken into its various contributions (as depicted in Figure 4.3).

In Figure 4.3, E_{gbo} is the Si bandgap under low-doping (1.12 eV at 300 K), ΔE_{gb}^{app} is the heavy-doping-induced apparent bandgap narrowing in the base region, $\Delta E_{g,Ge}(0)$ is the Ge-induced band offset at $x = 0$, and $\Delta E_{g,Ge}(W_b)$ is the Ge-induced band offset at $x = W_b$. We can thus write $E_{gb}(x)$ as

$$E_{gb}(x) = E_{gbo} - \Delta E_{gb}^{app} + [\Delta E_{g,Ge}(0) - \Delta E_{g,Ge}(W_b)]\frac{x}{W_b} - \Delta E_{g,Ge}(0). \tag{4.3}$$

Substitution of Equation 4.3 into Equation 4.2 gives

$$n_{ib}^2(x) = \gamma n_{io}^2 e^{\Delta E_{gb}^{app}/kT} e^{[\Delta E_{g,Ge}(W_b) - \Delta E_{g,Ge}(0)]x/(W_b kT)} e^{\Delta E_{g,Ge}(0)/kT}, \tag{4.4}$$

where we have made use of the fact that for Si, we can define a low-doping intrinsic carrier density for Si as

$$n_{io}^2 = N_C N_V e^{-E_{go}/kT}, \tag{4.5}$$

and we have defined an "effective density-of-states ratio" between SiGe and Si according to [4]

$$\gamma = \frac{(N_C N_V)_{SiGe}}{(N_C N_V)_{Si}} < 1. \tag{4.6}$$

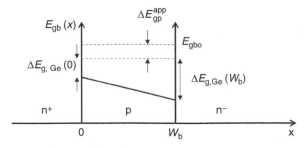

FIGURE 4.3 Schematic base bandgap in a linearly graded SiGe HBT.

Equation 4.4 can be inserted into the generalized Moll–Ross relation (4.1) to obtain

$$J_C = \frac{q\tilde{D}_{nb}}{N_{ab}^-} \frac{(e^{qV_{BE}/kT} - 1)\tilde{\gamma} n_{io}^2 e^{\Delta E_{gb}^{app}/kT} e^{\Delta E_{g,Ge}(0)/kT}}{\int_0^{W_b} e^{-[\Delta E_{g,Ge}(W_b) - \Delta E_{g,Ge}(0)](x/W_b kT)} dx}, \quad (4.7)$$

where we have defined $\tilde{D}_{nb}$ and $\tilde{\gamma}$ to be position-averaged quantities across the base profile, according to

$$\tilde{D}_{nb} = \frac{\displaystyle\int_0^{W_b} \frac{dx}{n_{ib}^2(x)}}{\displaystyle\int_0^{W_b} \frac{dx}{D_{nb}(x) n_{ib}^2(x)}}. \quad (4.8)$$

Using standard integration techniques, and defining

$$\Delta E_{g,Ge}(\text{grade}) = \Delta E_{g,Ge}(W_b) - \Delta E_{g,Ge}(0), \quad (4.9)$$

we get

$$J_{C,SiGe} = \frac{q\tilde{D}_{nb}}{N_{ab}^-} \frac{(e^{qV_{BE}/kT} - 1)\tilde{\gamma} n_{io}^2 e^{\Delta E_{gb}^{app}/kT} e^{\Delta E_{g,Ge}(0)/kT}}{\dfrac{W_b kT}{\Delta E_{g,Ge}(\text{grade})} \left\{ 1 - e^{-\Delta E_{g,Ge}(\text{grade})/kT} \right\}}. \quad (4.10)$$

Finally, by defining a minority electron diffusivity ratio between SiGe and Si as

$$\tilde{\eta} = \frac{(\tilde{D}_{nb})_{SiGe}}{(D_{nb})_{Si}}, \quad (4.11)$$

we obtain the final expression for $J_{C,SiGe}$ [2,5]

$$J_{C,SiGe} = \frac{qD_{nb}}{N_{ab}^- W_b} (e^{qV_{BE}/kT} - 1) n_{io}^2 e^{\Delta E_{gb}^{app}/kT} \left\{ \frac{\tilde{\gamma}\tilde{\eta} \Delta E_{g,Ge}(\text{grade})/kT e^{\Delta E_{g,Ge}(0)/kT}}{1 - e^{-\Delta E_{g,Ge}(\text{grade})/kT}} \right\}. \quad (4.12)$$

Within the confines of our assumptions stated above, this can be considered an exact result. As expected from our intuitive discussion of the band diagram, observe that J_C in a SiGe HBT depends exponentially on the EB boundary value of the Ge-induced band offset, and is linearly proportional to the Ge-induced bandgap grading factor. Given the nature of an exponential dependence, it is obvious that strong enhancement in J_C for fixed V_{BE} can be obtained for small amounts of introduced Ge, and that the ability to engineer the device characteristics to obtain a desired current gain is easily accomplished. Note as well that the thermal energy (kT) resides in the denominator of the Ge-induced band offsets. This is again expected from a simple consideration of how band-edge effects generally couple to the device transport equations. The inherent temperature dependence in SiGe HBTs will be revisited in detail in Chapter 11 [6].

If we consider a comparably constructed SiGe HBT and Si BJT with identical emitter contact technology, and further assume that the Ge profile on the EB side of the neutral base does not extend into the emitter enough to change the base current density, our experimental expectations are that for a comparably constructed SiGe HBT and Si BJT, the J_B should be comparable between the two devices, while J_C at fixed V_{BE} should be enhanced for the SiGe HBT. Figure 4.4 confirms this expectation experimentally. In this case, we note that the ratio of the current gain between an identically constructed SiGe HBT and a Si BJT can be written as

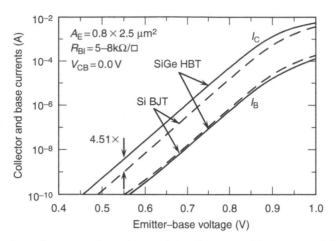

FIGURE 4.4 Comparison of current–voltage characteristics of a comparably constructed SiGe HBT and Si BJT. (From JD Cressler and G Niu. *Silicon–Germanium Heterojunction Bipolar Transistors*. Boston, MA: Artech House, 2003. With permission.)

$$\frac{\beta_{\text{SiGe}}}{\beta_{\text{Si}}} \cong \frac{J_{\text{C,SiGe}}}{J_{\text{C,Si}}}, \tag{4.13}$$

and thus we can define a SiGe current gain enhancement factor as

$$\left.\frac{\beta_{\text{SiGe}}}{\beta_{\text{Si}}}\right|_{V_{\text{BE}}} \equiv \Xi = \left\{\frac{\tilde{\gamma}\tilde{\eta}\Delta E_{\text{g,Ge}}(\text{grade})/kT e^{\Delta E_{\text{g,Ge}}(0)/kT}}{1 - e^{-\Delta E_{\text{g,Ge}}(\text{grade})/kT}}\right\}. \tag{4.14}$$

Typical experimental results for Ξ are shown for a comparably constructed SiGe HBT and Si BJT in Figure 4.5.

Based on the analysis above, we can make several observations regarding the effects of Ge on the collector current and hence current gain of a SiGe HBT:

- The presence of any Ge, in whatever shape, in the base of a bipolar transistor will enhance J_C at fixed V_{BE} (hence β) over a comparably constructed Si BJT.
- The J_C enhancement depends exponentially on the EB boundary value of Ge-induced band offset, and linearly on the Ge grading across the base. This observed dependence will play a role in understanding the best approach to profile optimization.
- In light of that, for two Ge profiles of constant stability, a box Ge profile is better for current gain enhancement than a triangular Ge profile, everything else being equal.
- The Ge-induced J_C enhancement is thermally activated (exponentially dependent on reciprocal temperature), and thus cooling will produce a strong magnification of the enhancement.

Relevant approximations and solutions for other types of Ge profiles are discussed at length in Ref. [1].

4.4 Output Conductance

The dynamic output conductance ($\partial I_C/\partial V_{\text{CE}}$ at fixed V_{BE}) of a transistor is a critical design parameter for many analog circuits. Intuitively, from the transistor output characteristics, we would like the output current to be independent of the output voltage, and thus ideally have zero output conductance (infinite output resistance). In practice, of course, this is never the case. As we increase V_{CB}, we deplete the neutral base from the backside, thus moving the neutral base boundary value ($x = W_b$) inward. Since

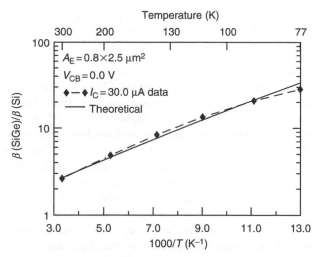

FIGURE 4.5 Measured and calculated current gain ratio as a function of reciprocal temperature for a comparably constructed SiGe HBT and Si BJT. (From JD Cressler and G Niu. *Silicon–Germanium Heterojunction Bipolar Transistors.* Boston, MA: Artech House, 2003. With permission.)

W_b determines the minority carrier density on the CB side of the neutral base, the slope of the minority electron profile, and hence the collector current, necessarily rises [7]. Thus, for finite base doping, I_C must increase as V_{CB} increases, giving a finite output conductance. This mechanism is known as the "Early effect," and for experimental convenience, we define the Early voltage (V_A) as

$$V_A = J_C(0) \left\{ \frac{\partial J_C}{\partial V_{CB}} \bigg|_{V_{BE}} \right\}^{-1} - V_{BE} \simeq J_C(0) \left\{ \frac{\partial J_C}{\partial W_b} \bigg|_{V_{BE}} \frac{\partial W_b}{\partial V_{CB}} \right\}^{-1}, \tag{4.15}$$

where $J_C(0) = J_C(V_{CB} = 0\ \text{V})$. The Early voltage is a simple and convenient measure of the change in output conductance with changing V_{CB}.

Simultaneously maintaining high current gain, high frequency response, and high V_A is particularly challenging in a Si BJT. For a Si BJT, we can use Equation 4.1 together with Equation 4.15 to obtain

$$V_{A,Si} = \frac{\int_0^{W_b} p_b(x)\,dx}{p_b(W_b) \left\{ \frac{\partial W_b}{\partial V_{CB}} \right\}} = \frac{Q_b(0)}{C_{cb}}, \tag{4.16}$$

where $Q_b(0)$ is the total base charge at $V_{CB} = 0\ \text{V}$, C_{cb} is the collector–base depletion capacitance, and we have assumed that V_{BE} is negligible compared to V_{CB}. Note that C_{cb} is dependent on both the ionized collector doping (N_{dc}^+) and the ionized base doping (N_{ab}^-). To estimate the sensitivity of V_A on N_{dc}^+ and N_{ab}^-, we can consider a Si BJT with constant base and collector-doping profiles. In this case, we can write

$$V_{A,Si} = -W_b(0) \left\{ \frac{\partial W_b}{\partial V_{CB}} \bigg|_{V_{BE}} \right\}^{-1}, \tag{4.17}$$

where $W_b(0)$ is the neutral base width at $V_{CB} = 0\ \text{V}$. The dependence of W_b on voltage and doping can be obtained from [8]

$$W_{\rm b} \simeq W_{\rm m} - \sqrt{\left(\frac{2\varepsilon}{q}\right)(\phi_{\rm bi} + V_{\rm CB})\left\{\frac{N_{\rm dc}^+}{N_{\rm ab}^-(N_{\rm ab}^- + N_{\rm dc}^+)}\right\}}, \qquad (4.18)$$

where $W_{\rm m}$ is the metallurgical base width, and $\phi_{\rm bi}$ is the CB junction built-in voltage. Using Equation 4.17 and Equation 4.18 we can calculate $V_{\rm A}$ as a function of doping. Cleary, if we fix $N_{\rm ab}^-$, increasing $N_{\rm dc}^+$ degrades $V_{\rm A}$, physically because the amount of backside neutral base depletion per unit bias is enhanced for a higher collector doping. If we instead fix $N_{\rm dc}^+$, increasing $N_{\rm ab}^-$ rapidly increases $V_{\rm A}$, which makes intuitive sense given that the base is much more difficult to deplete as the base doping increases, everything else being equal. In real Si BJT designs, a given device generally has a specified collector-to-emitter breakdown voltage ($BV_{\rm CEO}$) determined by the circuit requirements. To first order, this $BV_{\rm CEO}$ sets the collector-doping level. While this may appear to favor achieving a high $V_{\rm A}$, we must recall that the current gain is reciprocally related to the integrated base charge (refer to Equation 4.1).

Hence, increasing $N_{\rm ab}^-$ to improve $V_{\rm A}$ results in a strong decrease in β. In addition, for a Si BJT, for a fixed base width, increasing $N_{\rm ab}^-$ will degrade the cutoff frequency of the transistor (due to the reduction in the minority electron mobility). We might imagine that we can then increase $N_{\rm dc}^+$ to buy back the ac performance lost, this in turn degrades $V_{\rm A}$. This "catch-22" represents a fundamental problem in Si BJT design: it is inherently difficult to simultaneously obtain high $V_{\rm A}$, high β, and high $f_{\rm T}$. In practice one must then find some compromise design for $V_{\rm A}$, β, and $f_{\rm T}$, and in the process the performance capabilities of a given analog circuit suffer. Intuitively, this Si BJT design constraint occurs because β and $V_{\rm A}$ are both coupled to the base-doping profile. The introduction of Ge into the base region of a Si BJT can favorably alter this constraint by effectively decoupling β and $V_{\rm A}$ from the base-doping profile.

To formally obtain $V_{\rm A}$ in a SiGe HBT, we begin by combining Equation 4.1 with Equation 4.15 to obtain [9]

$$V_{\rm A,SiGe} = \frac{-\displaystyle\int_0^{W_{\rm b}} \frac{p_{\rm b}(x)\,{\rm d}x}{D_{\rm nb}(x)n_{\rm ib}^2(x)}}{\dfrac{\partial}{\partial V_{\rm CB}}\left\{\displaystyle\int_0^{W_{\rm b}} \frac{p_{\rm b}(x)\,{\rm d}x}{D_{\rm nb}(x)n_{\rm ib}^2(x)}\right\}}, \qquad (4.19)$$

from which we can write

$$V_{\rm A,SiGe} = \left\{\frac{-D_{\rm nb}(W_{\rm b})n_{\rm ib}^2(W_{\rm b})}{p_{\rm b}(W_{\rm b})}\int_0^{W_{\rm b}} \frac{p_{\rm b}(x)\,{\rm d}x}{D_{\rm nb}(x)n_{\rm ib}^2(x)}\right\}\left[\frac{\partial W_{\rm b}}{\partial V_{\rm CB}}\right]^{-1}. \qquad (4.20)$$

Comparing Equation 4.16 and Equation 4.20 we can see that the fundamental difference between $V_{\rm A}$ in a SiGe HBT and a Si BJT arises from the variation of $n_{\rm ib}^2$ as a function of position (the variation of $W_{\rm b}$ with $V_{\rm CB}$ is, to first order, similar between SiGe and Si devices). Observe that if $n_{\rm ib}$ is position-independent (i.e., for a box Ge profile), then Equation 4.20 collapses to Equation 4.16 and there is no $V_{\rm A}$ enhancement due to Ge (albeit there will obviously still be a strong β enhancement). On the other hand, if $n_{\rm ib}$ is position-dependent (i.e., in a linearly graded Ge profile), $V_{\rm A}$ will depend exponentially on the difference in bandgap between $x = W_{\rm b}$ and that region in the base where $n_{\rm ib}$ is smallest. That is, the base profile is effectively "weighted" by the increasing Ge content on the collector side of the neutral base, making it harder to deplete the neutral base for a given applied $V_{\rm CB}$, all else being equal, effectively increasing the Early voltage of the transistor.

For a linearly graded Ge profile, we can use Equation 4.4 and Equation 4.20 to obtain the ratio of $V_{\rm A}$ between a comparably constructed SiGe HBT and Si BJT (Θ) to be [10]

$$\left.\frac{V_{\rm A,SiGe}}{V_{\rm A,Si}}\right|_{V_{\rm BE}} \equiv \Theta \simeq e^{\Delta E_{\rm g,Ge}({\rm grade})/kT}\left[\frac{1 - e^{-\Delta E_{\rm g,Ge}({\rm grade})/kT}}{\Delta E_{\rm g,Ge}({\rm grade})/kT}\right]. \qquad (4.21)$$

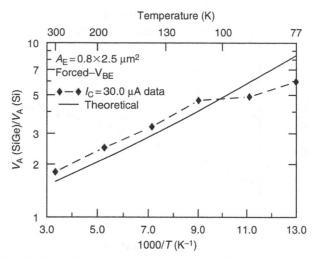

FIGURE 4.6 Measured and calculated Early voltage ratio as a function of reciprocal temperature for a comparably constructed SiGe HBT and Si BJT. (From JD Cressler and G Niu. *Silicon–Germanium Heterojunction Bipolar Transistors*. Boston, MA: Artech House, 2003. With permission.)

The important result is that the V_A ratio between a SiGe HBT and a Si BJT is an exponential function of Ge-induced bandgap grading across the neutral base. Typical experimental results for Θ are shown for a comparably constructed SiGe HBT and Si BJT in Figure 4.6.

4.5 Current Gain—Early Voltage Product

In light of the discussion above regarding the inherent difficulties in obtaining high V_A simultaneously with high β, one conventionally defines a figure-of-merit for analog circuit design: the so-called "βV_A" product. In a conventional Si BJT, a comparison of Equation 4.1 and Equation 4.16 shows that βV_A is to first-order independent of the base profile, and is thus not favorably impacted by conventional technology scaling, as for instance, the transistor frequency response would be. For a SiGe HBT, however, both β and V_A are decoupled from the base profile, and can be independently tuned by changing the Ge profile shape. By combining Equation 4.14 and Equation 4.21 we find that the ratio of βV_A between a comparably constructed SiGe HBT and Si BJT can be written as [9]

$$\frac{\beta V_{A,\text{SiGe}}}{\beta V_{A,\text{Si}}} = \tilde{\gamma}\tilde{\eta}e^{\Delta E_{g,\text{Ge}}(0)/kT}e^{\Delta E_{g,\text{Ge}}(\text{grade})/kT}. \tag{4.22}$$

Typical experimental results for the βV_A ratio for a comparably constructed SiGe HBT and Si BJT are shown in Figure 4.7.

Observe that βV_A is a thermally activated function of *both* the Ge-induced band offset at the EB junction and the Ge-induced grading across the neutral base. As can be seen in Figure 4.7, βV_A in a SiGe HBT is significantly improved over a comparably designed Si BJT, regardless of the Ge profile shape chosen, although the triangular Ge profile remains the profile shape of choice for both V_A and βV_A optimization. Due to their thermally activated nature, both V_A and βV_A are strongly enhanced with cooling, yielding enormous values ($\beta V_A > 10^4$) at 77 K for a 10% Ge triangular profile [10].

Based on the analysis above, we can make several observations regarding the effects of Ge on both the Early voltage and current gain—Early voltage product in SiGe HBTs:

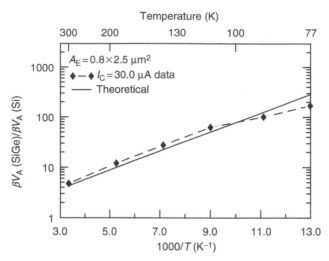

FIGURE 4.7 Measured and calculated ratio of the current gain–Early voltage product ratio as a function of reciprocal temperature for a comparably doped SiGe HBT and Si BJT. (From JD Cressler and G Niu. *Silicon–Germanium Heterojunction Bipolar Transistors.* Boston, MA: Artech House, 2003. With permission.)

- Unlike for J_C, only the presence of a larger Ge content at the CB side of the neutral base than at the EB side of the neutral base (i.e., finite Ge grading) will enhance V_A at fixed V_{BE} over a comparably constructed Si BJT.
- This V_A enhancement depends exponentially on the Ge grading across the base. This observed dependence will play a role in understanding the best approach to profile optimization, generally favoring strongly graded (triangular) profiles.
- In light of this, for two Ge profiles of constant stability, a triangular Ge profile is better for Early voltage enhancement than a box Ge profile is, everything else being equal.
- The Ge-induced V_A enhancement is thermally activated (exponentially dependent on reciprocal temperature), and thus cooling will produce a strong magnification of the enhancement.
- Given that β and V_A have the exact opposite dependence on Ge grading and EB Ge offset, the βV_A product in a SiGe HBT enjoys an ideal win–win scenario. Putting any Ge into the base region of a device will exponentially enhance this key analog figure-of-merit, a highly favorable scenario given the discussion above of inherent difficulties of achieving high βV_A in a Si BJT.
- A reasonable compromise Ge profile design that balances the dc optimization needs of β, V_A, and βV_A would be a Ge trapezoid, with a small (e.g., 3% to 4%) Ge content at the EB junction, and a larger (e.g., 10% to 15%) Ge content at the CB junction (i.e., finite Ge grading).

Relevant approximations and solutions for other types of Ge profiles are discussed at length in Ref. [1].

4.6 Charge Modulation Effects

At a deep level, transistor action, be it for a bipolar or field-effect transistor, is physically realized by voltage modulation of the charges inside the transistor, that in turn leads to voltage modulation of the output current. The voltage modulation of the charges results in a capacitive current, which increases with frequency. The bandwidth of the transistor is thus ultimately limited by various charge-storage effects in both the intrinsic and extrinsic device structure. Exact analysis of charge-storage effects requires the solution of semiconductor transport equations in the frequency domain. In practice, charge-storage effects are often taken into account by assuming that the charge distributions instantly follow the changes of terminal voltages under dynamic operation (i.e., a "quasi-static" assumption).

The first charge modulation effect in a SiGe HBT is the modulation of space charges associated with the EB and CB junctions. Voltage changes across the EB and CB junctions lead to changes of the space–charge (depletion) layer thicknesses and hence the total space charge. The capacitive behavior is similar to that of a parallel plate capacitor, because the changes in charge occur at the opposing faces of the space charge layer (which is depleted of carriers under reverse bias) to neutral region transition boundaries. The resulting capacitances are referred to as EB and CB "depletion" capacitances. Under high-injection conditions, the modulation of charges inside the space charge layer becomes significant. The resulting capacitance is referred to as the "transition" capacitance, and is important for the EB junction since it is forward biased. Under low-injection conditions, the CB capacitance is similar to that of a reverse biased pn junction, and is a function of the CB biasing voltage. At high injection, however, even in forward-active mode, the CB capacitance is also a function of the collector current, because of charge compensation by mobile carriers as well as base push-out at very high injection levels.

The second charge modulation effect is due to injected minority carriers in the neutral base and emitter regions. To maintain charge neutrality, an equal amount of excess majority carriers are induced by the injected minority carriers. Both minority and majority carriers respond to EB voltage changes, effectively producing an EB capacitance. This capacitance is historically referred to as "diffusion" capacitance, because it is associated with minority carrier diffusion in an ideal bipolar transistor with uniform base doping.

What is essential in order to achieve transistor action is modulation of the output current by an input voltage. The modulation of charge is just a means of modulating the current, and must be minimized in order to maintain ideal transistor action at high frequencies. For instance, a large EB diffusion capacitance causes a large input current, which increases with frequency, thus decreasing current gain at higher frequencies. At a fundamental level, for a given output current modulation, a decreased amount of charge modulation is desired in order to achieve higher operating frequency. A natural figure-of-merit for the efficiency of transistor action is the ratio of total charge modulation to the output current modulation

$$\tau_{ec} = \frac{\partial Q_n}{\partial I_C}, \tag{4.23}$$

which has dimensions of time and is thus called "transit time." Here, Q_n refers to the integral electron charge across the whole device, and can be broken down into various components for regional analysis. The partial derivative in Equation 4.23 indicates that there is modulation of both charge and current, and is thus necessary. A popular but *incorrect* definition of transit time leaves out the derivatives in Equation 4.23, and instead uses the simple ratio of charge to collector current [1]. The problem with this common formulation can be immediately deduced if we consider the resultant τ_{ec} of an npn bipolar transistor, where Q_n is dominated by the total number of emitter dopants. The use of $\tau_{ec} = Q_n/I_C$ thus leads to an incorrect transit time definition, since it produces a transit time that is independent of the base profile design, and is clearly non-physical. Equation 4.23 can be rewritten using the input voltage as an intermediate variable

$$\tau_{ec} = \frac{\partial Q_n/\partial V_{BE}}{\partial I_C/\partial V_{BE}} = \frac{g_m}{C_i}, \tag{4.24}$$

where C_i is the total input capacitance, and g_m is the transconductance. C_i can be divided into two components $C_{be} = \partial Q_n/\partial V_{BE}$ and $C_{bc} = \partial Q_n/\partial V_{BC}$. The transit times related to the neutral base and neutral emitter charge modulation are the base transit time and the emitter transit time, respectively. The base charge modulation required to produce a given amount of output current modulation can be decreased by introducing a drift field via Ge grading, thereby reducing the base transit time and extending transistor functionality to much higher frequencies. This Ge-grading-induced reduction in charge modulation is the fundamental reason why SiGe HBTs have better frequency response than Si BJTs. Ge grading is simply a convenient means by which we reduce the charge modulation.

4.7 AC Figures-of-Merit

For low injection, a key SiGe HBT ac figure-of-merit, the unity-gain cutoff frequency (f_T), can be written generally as

$$f_T = \frac{1}{2\pi\tau_{ec}} = \frac{1}{2\pi}\left[\frac{kT}{qI_C}(C_{te} + C_{tc}) + \tau_b + \tau_e + \frac{W_{CB}}{2v_{sat}} + r_c C_{tc}\right]^{-1}, \qquad (4.25)$$

where $g_m = qI_C/kT$ is the intrinsic transconductance at low injection ($g_m = \partial I_C/\partial V_{BE}$), C_{te} and C_{tc} are the EB and CB depletion capacitances, τ_b is the base transit time, τ_e is the emitter charge storage delay time, W_{cb} is the CB space–charge region width, v_{sat} is the saturation velocity, and r_c is the dynamic collector resistance. Physically, f_T is the common-emitter, unity-gain cutoff frequency ($h_{21} = 1$), and is conveniently measured using S-parameter techniques. A formal derivation is given in Ref. [1]. In Equation 4.25, τ_{ec} is the total emitter-to-collector delay time, and sets the ultimate limit of the switching speed of a bipolar transistor. Thus, we see that for fixed bias current, improvements in τ_b and τ_e due to the presence of SiGe will directly translate into an enhanced f_T and f_{max} of the transistor at fixed bias current.

In terms of transistor power gain (i.e., using the transistor to drive a "load"), one defines the "maximum oscillation frequency" figure-of-merit (f_{max}) by [11]

$$f_{max} = \sqrt{\frac{f_T}{8\pi C_{bc} r_b}}, \qquad (4.26)$$

where r_b is the small-signal base resistance, and C_{bc} is the total collector–base capacitance. A derivation of f_{max}, together with relevant assumptions and discussion, can be found in Ref. [1]. Physically, f_{max} is the common-emitter, unity power gain frequency, and can also be measured using S-parameter techniques. Clearly, f_{max} represents a "higher-order" (and therefore potentially more relevant to actual circuit applications) figure-of-merit than f_T, since the power gain depends not only on the intrinsic transistor performance (i.e., the device transit times), but also on the device parasitics associated with the process technology and its structural implementation. A larger f_T, a smaller r_b, and a smaller C_{bc} are clearly desired for increasing the maximum power gain and circuit operating frequency. Typical f_{max} data using the various definitions of power gain (i.e., U, MAG, MSG) for a second-generation SiGe HBT biased near peak f_T (120 GHz in this case) are shown in Figure 4.8.

4.8 Base and Emitter Transit Times

To understand the dynamic response of the SiGe HBT, and the role Ge plays in transistor frequency response, we must formally relate the changes in the base transit time and emitter transit time to the physical variables of this problem. It is also instructive to carefully compare the *differences* between a comparably constructed SiGe HBT and a Si BJT. In the present analysis, the SiGe HBT and the Si BJT are taken to be of identical geometry, and it is assumed that the emitter, base, and collector-doping profiles of the two devices are identical, apart from the Ge in the base of the SiGe HBT.

The theoretical consequences of the Ge-induced bandgap changes to the base transit time (τ_b) can be derived in closed-form for a constant base-doping profile ($p_b(x) = N_{ab}^-(x) = N_{ab}^- = $ constant) by considering the generalized Moll–Ross transit time relation, which holds for low injection in the presence of both non-uniform base doping and non-uniform base bandgap at fixed V_{BE} and T [3]

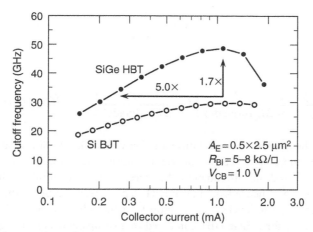

FIGURE 4.8 Measured comparison of unity gain cutoff frequency f_T as a function of bias current for a comparably constructed SiGe HBT and Si BJT. (From JD Cressler and G Niu. *Silicon–Germanium Heterojunction Bipolar Transistors*. Boston, MA: Artech House, 2003. With permission.)

$$\tau_b = \int_0^{W_b} \frac{n_{ib}^2(x)}{p_b(x)} \left[\int_0^{W_b} \frac{p_b(y)\mathrm{d}y}{D_{nb}(y)\, n_{ib}^2(y)} \right] \mathrm{d}x. \tag{4.27}$$

We can insert Equation 4.3 into Equation 4.2 to obtain Equation 4.4, and substitute Equation 4.4 into Equation 4.27 to obtain

$$\tau_{b,\mathrm{SiGe}} = \int_0^{W_b} \frac{n_{ib}^2(x)}{N_{ab}^-} \left\{ \int_z^{W_b} \frac{N_{ab}^-}{D_{nb}} \left[\frac{1}{\gamma n_{io}^2} e^{-\Delta E_{gb}^{app}/kT} e^{-\Delta E_{g,Ge}(0)/kT} \cdot e^{-\Delta E_{g,Ge}(\mathrm{grade})y/W_b kT} \mathrm{d}y \right] \right\} \mathrm{d}x. \tag{4.28}$$

Performing the first integration step yields

$$\tau_{b,\mathrm{SiGe}} = \int_0^{W_b} \frac{n_{ib}^2(x)}{N_{ab}^-} \left\{ \frac{-N_{ab}^- W_b}{\tilde{D}_{nb}\tilde{\gamma} n_{io}^2} \frac{kT}{\Delta E_{g,Ge}(\mathrm{grade})} e^{-\Delta E_{gb}^{app}/kT} e^{-\Delta E_{g,Ge}(0)/kT} \cdot \left[e^{-\Delta E_{g,Ge}(\mathrm{grade})/kT} - e^{-\Delta E_{g,Ge}(\mathrm{grade})x/W_b kT} \right] \right\} \mathrm{d}x \tag{4.29}$$

where we have accounted for the position dependence in both the mobility and the density-of-states product. Substitution of n_{ib}^2 from Equation 4.4 into Equation 4.29 and multiplying through gives

$$\tau_{b,\mathrm{SiGe}} = \left\{ \frac{W_b kT \tilde{\gamma} n_{io}^2}{\tilde{D}_{nb}\tilde{\gamma} n_{io}^2 \Delta E_{g,Ge}(\mathrm{grade})} \right\} \int_0^{W_b} \left[1 - e^{\Delta E_{g,Ge}(\mathrm{grade})x/W_b kT} e^{-\Delta E_{g,Ge}(\mathrm{grade})/kT} \right] \mathrm{d}x, \tag{4.30}$$

which can be integrated and evaluated to obtain, finally [2,5]

$$\tau_{b,\mathrm{SiGe}} = \frac{W_b^2}{\tilde{D}_{nb}} \frac{kT}{\Delta E_{g,Ge}(\mathrm{grade})} \left\{ 1 - \frac{kT}{\Delta E_{g,Ge}(\mathrm{grade})} \left[1 - e^{-\Delta E_{g,Ge}(\mathrm{grade})/kT} \right] \right\}. \tag{4.31}$$

As expected, we see that the base transit time in a SiGe HBT depends reciprocally on the amount of Ge-induced bandgap grading across the neutral base (i.e., for fixed base width, the band-edge-induced drift field). It is instructive to compare τ_b in a SiGe HBT with that of a comparably designed Si BJT. In the case of a Si BJT (trivially derived from Equation 4.27 for constant base doping and bandgap), we know that

$$\tau_{b,Si} = \frac{W_b^2}{2D_{nb}}, \tag{4.32}$$

and hence can write

$$\frac{\tau_{b,SiGe}}{\tau_{b,Si}} = \frac{2}{\tilde{\eta}} \frac{kT}{\Delta E_{g,Ge}(\text{grade})} \left\{ 1 - \frac{kT}{\Delta E_{g,Ge}(\text{grade})} \left[1 - e^{-\Delta E_{g,Ge}(\text{grade})/kT} \right] \right\}, \tag{4.33}$$

where we have used the ratio of electron diffusivities between SiGe and Si (Equation 4.11). Within the confines of our assumptions stated above, this can be considered an exact result. Figure 4.9 shows the theoretical calcuations based on this equation. As expected from our intuitive discussion of the band diagram, observe that τ_b and hence f_T in a SiGe HBT depend reciprocally on the Ge-induced bandgap grading factor, and hence for finite Ge grading across the neutral base, τ_b is less than unity, and thus we expect enhancement in f_T for a SiGe HBT compared to a comparably constructed Si BJT. Figure 4.10 confirms this expectation experimentally.

Based on the analysis above, we can make several observations regarding the effects of Ge on the frequency response of a SiGe HBT:

- For fixed bias current, the presence of Ge in the base region of a bipolar transistor affects its frequency response through the base and emitter transit times.

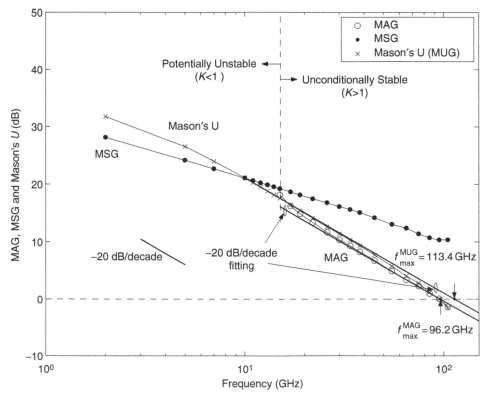

FIGURE 4.9 Measured MAG, MSG, and Mason's *U* versus frequency for a second-generation SiGe HBT biased near peak f_T. (From JD Cressler and G Niu. *Silicon–Germanium Heterojunction Bipolar Transistors*. Boston, MA: Artech House, 2003. With permission.)

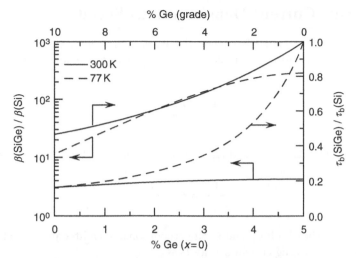

FIGURE 4.10 Theoretical calculations of base transit time ratio Γ as a function of Ge profile shape. (From JD Cressler and G Niu. *Silicon–Germanium Heterojunction Bipolar Transistors.* Boston, MA: Artech House, 2003. With permission.)

- The f_T enhancement for a SiGe HBT over a Si BJT depends reciprocally on the Ge grading across the base. This makes sense intuitively given the effects of the grading-induced drift field for the minority carrier transport. This observed dependence on Ge grading will play a role in understanding the best approach to profile optimization for a given application.
- For two Ge profiles of constant stability, a triangular Ge profile is better for cutoff frequency enhancement than a box Ge profile is, everything else being equal, *provided* τ_b is dominant over τ_e in determining f_T. While this is clearly the case in most first-generation SiGe HBTs, it is nonetheless conceivable that for a τ_e-dominated transistor, a more box-like Ge profile, which inherently favors β enhancement and hence τ_e improvement, might be a favored profile design for optimal frequency response. A compromise trapezoidal profile, which generally favors both τ_b and τ_e improvement, is a logical compromise profile design point. Such tradeoffs are obviously technology generation dependent.
- Given that f_T is improved across the entire useful range of I_C, the f_T versus power dissipation trade-off offers important opportunities for portable applications, where power minimization is often a premium constraint.
- The Ge-induced f_T enhancement depends strongly on temperature, and for τ_b and τ_e, is functionally positioned in a manner that will produce a magnification of f_T enhancement with cooling, in stark contrast to a Si BJT.

Relevant approximations and solutions for other types of Ge profiles are discussed at length in Ref. [1].

As can be seen in Figure 4.10, since f_T is increased across a large range of useful collector current, we can potentially gain dramatic savings in power dissipation for fixed frequency operation compared to a Si BJT. This power-for-performance tradeoff can in practice be even more important than the sheer increase in frequency response, particularly for portable applications. In this case, if we decided, for instance, to operate the transistor at a fixed frequency of 30 GHz, we could reduce the supply current by a factor of 5×. Note as well, that as for the collector current density expression (Equation 4.12), the thermal energy (kT) plays a key role in Equation 4.31, in this case residing in the numerator, and will thus have important favorable implications for SiGe HBT frequency response at cryogenic temperatures, as will be discussed in detail in Chapter 11.

4.9   Operating Current Density versus Speed

The fundamental nature of SiGe HBTs requires the use of high operating current density in order to achieve high speed. The operating current density dependence of f_T is best illustrated by examining the inverse of f_T [1]

$$\frac{1}{2\pi f_T} = \frac{C_{be} + C_{bc}}{g_m}. \tag{4.34}$$

Since $C_{be} = g_m \tau_f + C_{te}$, $C_{bc} = C_{tc}$, and $g_m = qI_C/kT$, Equation (4.34) can be rewritten as

$$\frac{1}{2\pi f_T} = \tau_f + \frac{kT}{qI_C} C_t, \tag{4.35}$$

where $C_t = C_{te} + C_{tc}$. Since both C_{te} and C_{tc} are proportional to emitter area, Equation (4.35) can be rewritten in terms of the biasing current density J_C as

$$\frac{1}{2\pi f_T} = \tau_f + \frac{kT}{qJ_C} C_t, \tag{4.36}$$

where $C_t = C_t/A_E$ is the total EB and CB depletion capacitances per unit emitter area, and $J_C = I_C/A_E$ is the collector operating current density. Thus, the cutoff frequency f_T is fundamentally determined by the biasing current density J_C, independent of the transistor emitter length. For very low J_C, the second term is very large, and f_T is very low regardless of the forward transit time τ_f. With increasing J_C, the second term decreases, and eventually becomes smaller than τ_f. At high J_C, however, base push-out (Kirk effect, refer to Chapter 5) occurs, and τ_f itself increases with J_C, leading to f_T roll-off. A typical f_T versus J_C characteristic is shown in Figure 4.11 for a first-generation SiGe HBT.

The values of τ_f and C_t can be easily extracted from a plot of $1/2\pi f_T$ versus $1/J_C$ (as shown in Figure 4.12). Near the peak f_T, the $1/2\pi f_T$ versus $1/J_C$ curve is nearly linear, indicating that C_t is close to constant for this biasing range at high f_T. Thus, C_t can be obtained from the slope, while τ_f can be determined from the y-axis intercept at infinite current ($1/J_C = 0$).

To improve f_T in a SiGe HBT, the transit time τ_f must be decreased by using a combination of vertical profile scaling as well as Ge grading across the base. At the same time, the operating current density J_C must be increased in proportion in order to make the second term in Equation (4.36) negligible compared to the first term (τ_f). That is, the high f_T potential of small τ_f transistors can only be realized by using sufficiently high operating current density. This is a fundamental criterion for high-speed SiGe HBT design. The higher the peak f_T, the higher the required operating J_C. For instance, the minimum required operating current density has increased from 1.0 mA/μm^2 for a first-generation SiGe HBT with 50-GHz peak f_T to 10 to 15 mA/μm^2 for >200-GHz peak f_T third-generation SiGe HBTs. Higher current density operation naturally leads to more severe self-heating effects, which must be appropriately dealt with in compact modeling and circuit design. Electromigration and other reliability constraints (refer to Chapter 13) associated with very high J_C operation have also produced an increasing need for copper metalization schemes.

In order to maintain proper transistor action under high J_C conditions, the collector doping must be increased in order to delay the onset of high injection effects. This requisite doping increase obviously reduces the breakdown voltage. At a fundamental level, trade-offs between breakdown voltage and speed are thus inevitable for all bipolar transistors (Si, SiGe, or III–V). Since the collector doping in SiGe HBT is typically realized by self-aligned collector implantation (as opposed to during epi growth in III–V), devices with multiple breakdown voltages (and hence multiple f_T) can be trivially obtained in the same fabrication sequence, giving circuit designers added flexibility.

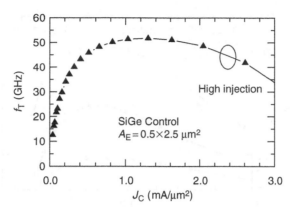

FIGURE 4.11 Typical measured $f_T - J_C$ characteristics of a first-generation SiGe HBT. (From JD Cressler and G Niu. *Silicon–Germanium Heterojunction Bipolar Transistors*. Boston, MA: Artech House, 2003. With permission.)

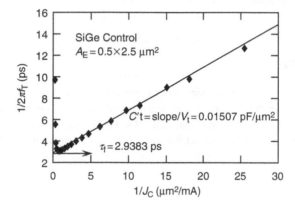

FIGURE 4.12 Illustration of C_t and τ_f extraction in a SiGe HBT.

Another closely related manifestation of Equation 4.36 is that the minimum required J_C to realize the full potential of a small τ_f transistor depends on C_t. Both C_{te} and C_{tc} thus must be minimized in the device and are usually addressed via a combination of structural design, ground-rule shrink, and doping profile tailoring via selective collector implantation. This reduction of C_{tc} is also important for increasing the power gain (i.e., f_{max}).

The record f_T in SiGe HBT technology stands at present at 350 GHz [12]. From today's vantage point, a combined 300+ GHz peak f_T/f_{max} appears to be a very realistic performance goal for fully integrated, commercial SiGe BiCMOS processes (see Chapter 16). Clearly, breakdown voltages must decrease as the transistor performance improves. For the case depicted in Figure 4.13 [13], the 50, 120, and 210 GHz SiGe HBTs have an associated BV_{CEO} of 3.3, 2.0, and 1.7 V, respectively. Achievable $f_T \times BV_{CEO}$ products in the 350 to 400 GHz V range are realistic goals. The sub-1.5 V breakdown voltages required to reach 300 GHz should not prove to be a serious limitation for many designs, given that BV_{CEO} does not present a hard boundary above which one cannot bias the transistor. Rather, one simply has to live with base current reversal and potential bias instabilities in this (above-BV_{CEO}) bias domain [14]. In addition, on-wafer breakdown voltage tuning will provide an additional level of flexibility for circuit designs needing larger operating headroom.

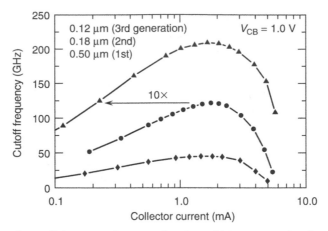

FIGURE 4.13 Measured cutoff frequency data as a function of bias current for three different SiGe HBT technology generations. (From G Freeman, B Jagannathan, S-J Jeng, J-S Rieh, A Stricker, D Ahlgren, and S Subbanna. *IEEE Trans. Electron Dev.* 50:645–655, 2003. With permission.)

4.10   Summary

In this chapter, I have detailed "first-order" device physics of SiGe HBTs, both from a dc and an ac perspective. This theoretical framework, while clearly simplistic in its assumptions, is nonetheless very useful in providing insight into the way that real SiGe HBTs operate, and how they are designed in practice in industry. Chapter 5 addresses additional subtle, but important, nuances in the operation of SiGe HBTs.

Acknowledgments

I am grateful to G. Niu, A. Joseph, D. Harame, G. Freeman, B. Meyerson, D. Herman, and the IBM SiGe team for their contributions. This work was supported by the Semiconductor Research Corporation, the GEDC at Georgia Tech, and IBM.

References

1. JD Cressler and G Niu. *Silicon–Germanium Heterojunction Bipolar Transistors*. Boston, MA: Artech House, 2003.
2. DL Harame, JH Comfort, JD Cressler, EF Crabbé, JY-C Sun, BS Meyerson, and T Tice. Si/SiGe epitaxial-base transistors—Part I: Materials, physics, and circuits. *IEEE Trans. Electron Dev.* 42:455–468, 1995.
3. H Kroemer. Two integral relations pertaining to electron transport through a bipolar transistor with a nonuniform energy gap in the base region. *Solid-State Electron.* 28:1101–1103, 1985.
4. EJ Prinz, PM Garone, PV Schwartz, X Xiao, and JC Sturm. The effect of emitter–base spacers and strain-dependent density-of-states in Si/SiGe/Si heterojunction bipolar transistors. Technical Digest of the IEEE International Electron Devices Meeting, Washington, 1989, pp. 639–642.
5. SS Iyer, GL Patton, JMC Stork, BS Meyerson, and DL Harame. Heterojunction bipolar transistors using Si–Ge alloys. *IEEE Trans. Electron Dev.* 36:2043–2064, 1989.
6. JD Cressler, JH Comfort, EF Crabbé, JMC Stork, and JY-C. Sun. On the profile design and optimization of epitaxial Si- and SiGe-base bipolar technology for 77 K applications—Part I: Transistor dc design considerations. *IEEE Trans. Electron Dev.* 40:525–541, 1993.
7. JM Early. Effects of space-charge layer widening in junction transistors. *Proc. IRE* 40:1401–1406, 1952.

8. DJ Roulston. *Bipolar Semiconductor Devices.* New York, NY: McGraw-Hill, 1990.

9. EJ Prinz and JC Sturm. Current gain–Early voltage products in heterojunction bipolar transistors with nonuniform base bandgaps. *IEEE Electron Dev. Lett.* 12:661–663, 1991.

10. AJ Joseph, JD Cressler, and DM Richey. Optimization of Early voltage for cooled SiGe HBT precision current sources. *J. Phys. IV* 6:125–129, 1995.

11. SJ Mason. Power gain in feedback amplifiers. *IRE Trans. Circuit Theory* CT-1:20–25, 1954.

12. J-S Rieh, B Jagannathan, H Chen, KT Schonenberg, D Angell, A Chinthakindi, J Florkey, F Golan, D Greenberg, S-J Jeng, M Khater, F Pagette, C Schnabel, P Smith, A Stricker, K Vaed, R Volant, D Ahlgren, G Freeman, K Stein, and S Subbanna. SiGe HBTs with cut-off frequency of 350 GHz. Technical Digest of the IEEE International Electron Devices Meeting, San Francisco, 2002, pp. 771–775.

13. G Freeman, B Jagannathan, S-J Jeng, J-S Rieh, A Stricker, D Ahlgren, and S Subbanna. Transistor design and application considerations for >200 GHz SiGe HBTs. *IEEE Trans. Electron Dev.* 50:645–655, 2003.

14. M Rickelt, HM Rein, and E Rose. Influence of impact-ionization-induced instabilities on the maximum usable output voltage of silicon bipolar transistors. *IEEE Trans. Electron Dev.* 48:774–783, 2001.

5

Second-Order Effects

John D. Cressler

Georgia Institute of Technology

5.1 Introduction

While second-order deviations from the first-order theory presented in Chapter 4 will always exist in SiGe HBTs, their specific impact on actual SiGe HBT devices and circuits is both profile-design and application dependent, and thus they must be carefully appreciated and kept at the back of the mind by designers.

We first analyze the so-called "Ge grading effect" associated with the position dependence of the Ge content across the neutral base found in SiGe designs. The influence of Ge grading effect on SiGe HBT properties is physically tied to the movement of emitter–base space–charge edge along the graded Ge profile with increasing base–emitter voltage. This Ge grading effect can present potential problems for circuit designs that require precise knowledge and control over the current dependence of both current gain and base–emitter voltage as a function of temperature. We then discuss the impact of neutral base recombination (NBR) on SiGe HBT operation. A finite trap density necessarily exists in the base region of all bipolar transistors, and while the impact is usually assumed to be negligible in Si BJTs, it can become important in SiGe HBTs, particularly when they are operated across a wide temperature range. NBR can strongly affect the output conductance (Early voltage) of SiGe HBTs, and is strongly dependent on the mode of base drive (i.e., whether the device is voltage- or current-driven), and hence the circuit application. Finally, we address high-injection heterojunction barrier effects (HBE) in SiGe HBTs. Barrier effects associated with the collector–base heterojunction under high current density operation are inherent to SiGe HBTs, and if not carefully controlled, can strongly degrade both dc and ac performance at the large current densities which SiGe HBTs are often operated. We conclude each section with a brief discussion of the implications and potential problems imposed by these design constraints on both device and circuit designers ("the bottom line").

5.2 Ge Grading Effect

To ensure the applicability of SiGe HBTs to precision analog circuits, parameter stability over both temperature and bias clearly must be ensured. Given the bandgap-engineered nature of the SiGe HBT, this can become an issue for concern, particularly for devices with non-constant (graded) Ge content across the neutral base. Even a cursory examination of the bias current dependence of the current gain in a graded-base SiGe HBT as a function of temperature, for instance, shows a profound functional difference from that of a Si BJT (Figure 5.1). In particular, for a graded-base SiGe HBT, the current gain peaks at low injection, and degrades significantly before the onset of high-injection effects. This

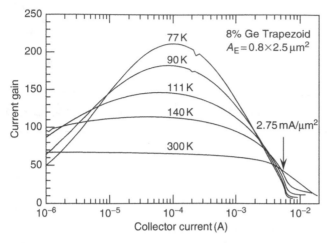

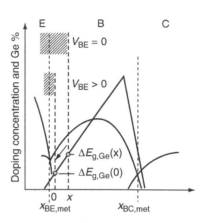

FIGURE 5.1 SiGe HBT current gain as a function of collector current at various temperatures, illustrating the Ge grading effect. (From JD Cressler and G Niu. *Silicon–Germanium Heterojunction Bipolar Transistors*. Boston, MA: Artech House, 2003. With permission.)

FIGURE 5.2 Schematic diagram of the base profile of a SiGe HBT, illustrating the physical origin of the Ge grading effect.

medium-injection "collapse" of β is clearly enhanced by cooling, and thus can be logically inferred to be the result of a band-edge phenomenon.

To understand the physical origin of this bias-dependent behavior in the current gain in SiGe HBTs, consider Figure 5.2, which shows a schematic doping and Ge profile in a graded-base SiGe HBT. As derived in Chapter 4, the collector current at any bias of a graded-base SiGe HBT is exponentially dependent upon the amount of Ge at the edge of the emitter–base (EB) space–charge region. Physically, as the collector current density increases, the base–emitter voltage must also increase, and hence from charge balance considerations the EB space–charge width necessarily contracts, thereby reducing the EB boundary value of the amount of Ge ($\Delta E_{g,Ge}(0)$), and producing a bias and temperature dependence different from that of a Si BJT [2]. Since this Ge grading effect is the physical result of the modulation of the base width with increasing base–emitter voltage ($W_b(V_{BE})$), it can be logically associated with the so-called inverse Early effect.

The dependence of the collector current density on the Ge profile shape in a SiGe HBT is given approximately by

$$J_{C,SiGe} \simeq \frac{qD_{nb}}{N_{ab}^- W_b}(e^{qV_{BE}/kT} - 1)n_{io}^2 e^{\Delta E_{gb}^{app}/kT}\left\{\tilde{\gamma}\tilde{\eta}\frac{\Delta E_{g,Ge}(grade)}{kT}e^{\Delta E_{g,Ge}(0)/kT}\right\}. \qquad (5.1)$$

The relationship between J_C and Ge profile shape in Equation 5.1 highlights the dependence of the collector current density on Ge profile design. Since $\Delta E_{g,Ge}(0)$ changes with increasing base–emitter bias, any changes in the amount of Ge seen by the device at that EB boundary will have a large impact due to the exponential relationship. Consequently, the more strongly graded the Ge profile, the more serious Ge grading effect can be expected to be.

Given that Ge grading effect in SiGe HBTs impacts the bias-current dependence of the current gain, a logical test-case circuit for examining the circuit-level influence of Ge grading effect is the ubiquitous bandgap reference (BGR) circuit, since its functionality relies heavily on the identical dependence of $V_{BE}(I_C)$ on temperature between transistors of differing size. Given two transistors with a (realistic) non-constant base doping, and biased at the same collector current, two SiGe HBTs with a sufficiently strongly graded Ge profile might be expected to "feel" the Ge ramp effect differently, since the voltage-induced space–charge width changes would differ slightly between the two. The conceivable result would

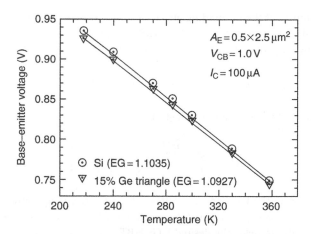

FIGURE 5.3 Base–emitter voltage as a function of temperature at fixed bias current for a Si BJT and a SiGe HBT. The lines represent a SPICE fit to the data using the EG fitting parameter. (From SL Salmon, JD Cressler, RC Jaeger, and DL Harame. *IEEE Trans. Electron Dev.* 47:292–298, 2000. With permission.)

be a slight mismatch in V_{BE} over temperature between the two transistors, thereby degrading the output voltage stability of the BGR circuit over temperature [3].

In the context of SiGe HBT-based BGR implementations, it is key that detailed knowledge of the impact of Ge profile shape on the temperature dependence of the base–emitter voltage exists. Clearly, V_{BE} in turn depends on the variation of I_C across the desired temperature range of interest (e.g., -55°C (218 K) to 85°C (358 K)). As can be observed in Figure 5.3, the differences in $V_{BE}(T)$ between a Si BJT and a SiGe HBT are small, but clearly observable, and must be more carefully examined.

The Ge grading effect in SiGe HBTs is primarily determined by the "steepness" of the Ge profile through the EB space–charge region, and the magnitude and shape of $N_{ab}^-(x)$ at the space–charge to quasineutral base boundary. We can roughly estimate the variation on the SiGe-to-Si current gain ratio ($\Xi = \beta_{SiGe}/\beta_{Si}$) with V_{BE} for varying amounts of Ge grading by considering a linearly graded SiGe HBT with uniform doping levels in the emitter and base regions. From Chapter 4, we have

$$\left. \frac{\beta_{SiGe}}{\beta_{Si}} \right|_{V_{BE}} \equiv \Xi = \left\{ \frac{\tilde{\gamma}\tilde{\eta}\Delta E_{g,Ge}(\text{grade})/kT e^{\Delta E_{g,Ge}(0)/kT}}{1 - e^{-\Delta E_{g,Ge}(\text{grade})/kT}} \right\}, \tag{5.2}$$

from which we can obtain

$$\frac{\partial \Xi}{\partial V_{BE}} = \left\{ \frac{-\Xi}{\phi_{bi,BE} - V_{BE}} \right\} \left\{ \frac{x_{pE}}{2W_{b0}} \right\} \left[\frac{\Delta E_{g,Ge}(\text{grade})/kT}{1 - e^{-\Delta E_{g,Ge}(\text{grade})/kT}} \right], \tag{5.3}$$

where $\phi_{bi,BE}$ is the built-in potential of the EB junction, x_{pE} is the EB space–charge width on the base side of the junction, and W_{b0} is the neutral base width at zero-bias. As $\Delta E_{g,Ge}(\text{grade})$ gets small (i.e., approaching a Ge box profile), the Ge grading effect becomes negligible, yielding a flat β versus I_C characteristic, as in a Si BJT. Equation 5.3 also predicts a weaker Ge grading effect in transistors with higher base doping, since x_{pE} becomes negligible with respect to W_{b0}. However, in practical SiGe HBT base profiles, which typically have a retrograded base doping level in the vicinity of the EB junction to reduce the EB electric field, the Ge grading effect is enhanced, since x_{pE} varies nonlinearly with V_{BE}. Finally, we note that due to the band-edge nature of Ge grading effect, its impact on device performance should be greatly magnified at reduced temperatures.

To determine the impact of the Ge grading effect on practical BGR circuits, we must recast the SiGe HBT collector current density into the familiar BGR design equation. First, the process-dependent parameters (B) and the Ge profile dependent terms (ξ) can be lumped together in I_C as

$$I_C(T) = \xi B T^m e^{-E_{g0}/kT} e^{E_{gb}^{app}/kT} e^{qV_{BE}/kT}, \tag{5.4}$$

and rewritten in terms of the base–emitter voltage as

$$V_{BE} = \frac{E_{g0}}{q} - \frac{E_{gb}^{app}}{q} + \frac{kT}{q} \ln\left\{\frac{I_C}{\xi B T^m}\right\}. \tag{5.5}$$

In practice, we can measure the base–emitter voltage at a reference temperature and collector current and solve for the lumped process parameters (B). Inserting the lumped parameters back into the original V_{BE} equation and simplifying yields the desired SiGe HBT result [3]

$$V_{BE,SiGe} = \frac{1}{q}\left\{E_{g0} - E_{gb}^{app} - \Delta E_{g,Ge}(0)\right\} - \frac{T}{qT_R}\left\{E_{g0} - E_{gb}^{app} - \Delta E_{g,Ge}(0)\right\} + \frac{T}{T_R}V_{BE,R}$$
$$+ \left\{\frac{kT}{q}\ln\frac{I_C}{I_{C,R}} - m\frac{kT}{q}\ln\frac{T}{T_R}\right\}. \tag{5.6}$$

The effects of Ge on the base–emitter voltage of the transistor can be gleaned directly from this more generalized result. Observe that the effective bandgap at the emitter–base junction is simply the Si result in the presence of doping-induced bandgap narrowing ($E_{g0} - E_{gb}^{app}$), minus the bandgap reduction due to the amount of Ge at the EB junction ($\Delta E_{g,Ge}(0)$). In addition, the shape of V_{BE} versus temperature in a SiGe HBT is changed from that of a Si BJT due to the addition of Ge, as is apparent in the last two terms of the equation. The ratio T/T_R enhances this difference between Si BJTs and SiGe HBTs. For temperatures near the reference temperature, the last two terms of Equation 5.6 have little effect on $V_{BE}(I_C,T)$, but as the temperature decreases, these effects can become more pronounced.

Note that the effective bandgap parameters ($E_{g0} - E_{gb}^{app} - \Delta E_{g,Ge}(0)$) and m correspond to the SPICE modeling parameters EG and XTI, respectively. The amount of curvature in V_{BE} versus temperature is affected by the addition of Ge, as is apparent in the last two terms of Equation 5.6. Assuming that the Ge grading ($\Delta E_{g,Ge}(\text{grade})$) does not change significantly with temperature, the deviation from linearity of V_{BE} versus temperature (i.e., V_{BE} curvature) using Equation 5.6 is actually reduced with increasing Ge grading across the base. In the curvature results presented, the deviation from linearity is calculated by drawing a line through the endpoints of V_{BE} across the relevant temperature range, and then subtracting the actual V_{BE} value from the value on the line at each temperature, according to

$$\Delta_{linearity}(T) = V_{BE}(T) - \left[V_{BE}(T_L) - \frac{V_{BE}(T_L) - V_{BE}(T_H)}{T_L - T_H}(T_L - T)\right], \tag{5.7}$$

where in this case $T_L = 218\,\text{K}$ (-55°C) and $T_H = 358\,\text{K}$ (85°C).

While this Ge-grading-induced V_{BE} curvature reduction might naively appear to be a good thing for BGR design, it in fact can worsen the performance of BGR circuits. Figure 5.4 shows the theoretical deviation from linearity that results from three different hypothetical Ge profiles: (1) no Ge grading; (2) 8.6% Ge grading; and (3) 18.6% Ge grading. Note that a box-shaped Ge profile (no Ge grading), in which the Ge concentration across the base is finite but constant, will have the same deviation from linearity as a Si BJT. Given sufficient Ge grading, it is clear that differences between Si BJTs and SiGe HBTs should be experimentally observable, and a combination of measurement and modeling results confirm this [1].

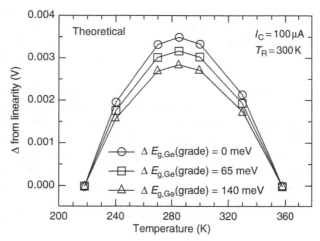

FIGURE 5.4 Theoretical dependence of the peak V_{BE} deviation from linearity as a function of temperature for various amounts of Ge grading. (From SL Salmon, JD Cressler, RC Jaeger, and DL Harame. *IEEE Trans. Electron Dev.* 47:292–298, 2000. With permission.)

When discussing any second-order effect in transistors, it is important to clearly understand both its physical origins and its potential implications for both device and circuit designers, so that it can be effectively "designed around." We can summarize these implications for Ge grading effect as follows:

- Ge grading effect is likely to be important only in precision analog circuits, not in digital or RF/ microwave circuits. While the BGR circuit is a natural candidate for observing Ge grading effect, any analog circuit that depends strongly on current gain across a wide bias range, or that requires the matching of V_{BE} between multiple devices across both bias and temperature, could be potentially affected.
- While the Ge grading effect exists only in compositionally graded Ge profiles, these graded profile designs typically achieve the best dc and ac performance, and thus represent the vast majority of commercially relevant SiGe technologies. As such, Ge grading effect should never be discounted.
- The impact of the Ge grading effect is expected to be highly dependent on the specifics of the Ge profile shape, and thus will vary from technology to technology. BGRs implemented in the first-generation SiGe HBT technology containing only modest amounts of Ge generally show little impact of Ge grading effect [4].
- Since the seriousness of the Ge grading effect depends on the Ge grading, it is a phenomenon that will generally worsen with technology scaling, since for constant strained layer stability, the peak Ge content in a SiGe HBT (and hence the grading across the neutral base) will naturally rise. This scaling-induced enhancement, however, will be at least partially offset by the natural increase in base doping with scaling.
- Conventional modeling methodologies employed in Gummel–Poon (SPICE) compact transistor models appear to adequately capture the Ge grading effect.
- Due to its thermally activated nature, cooling clearly exaggerates Ge grading effect, and thus is potentially important for precision analog circuits required to operate across a very wide temperature range.

5.3 Neutral Base Recombination

Physically, NBR in bipolar transistors involves the recombination of injected electrons transiting the neutral base with holes, via intermediate trap levels. Physically, NBR removes the desired injected electrons from the collector current via recombination (i.e., they do not exit the base), and increases the undesired hole density (required to support the recombination process), thereby degrading the base transport factor. Significant NBR thus leads to an increase in the base current and a simultaneous decrease in the collector current, thereby causing a substantial degradation in the current gain.

For fixed trap density, the impact of NBR on transistor characteristics, which is generally considered to be negligible in modern Si BJTs, can be exaggerated due to the presence of an increased total base minority carrier charge concentration (Q_{nb}) that participates with the trap recombination process. Because in a SiGe HBT the Ge-induced base bandgap reduction *exponentially* increases Q_{nb} compared to that in a comparably constructed Si BJT, one would naively expect that the NBR would be strongly enhanced in a SiGe HBT compared to a Si BJT, even at identical trap base density. This situation is also expected to become especially important as the temperature changes, due to the thermally activated nature of Q_{nb} in a SiGe HBT. It is essential, therefore, to understand the physical mechanism of NBR in SiGe HBTs, its impact on the transistor characteristics, and possible circuit implications.

For an npn bipolar transistor with negligible EB space–charge region recombination, I_B under arbitrary forward-active bias is the sum of the hole current back-injected into the emitter, the hole current due to impact ionization in the collector–base region, and the NBR current component under discussion. For small values of V_{CB}, the additional hole current due to impact-ionization is negligible and thus I_B is dominated by the other two components. As the electron diffusion length (L_{nb}) gets comparable to the neutral base width (W_b), the NBR component of I_B becomes increasingly important. With negligible NBR (the ideal case), I_B will be independent of V_{CB} for any given V_{BE}. However, under non-negligible NBR, any change in W_b with respect to L_{nb} will perturb the NBR component of I_B. Thus, an easy way to estimate the impact of NBR in a given transistor is to observe the rate of decrease in I_B with respect to varying V_{CB}, at a fixed V_{BE}. The base current in this case can be expressed as the sum of the drift-diffusion component and the NBR component as

$$J_B = J_{B,\text{diff}} + J_{nbr} = J_{b0}e^{qV_{BE}/kT} + J_{nbr,0}e^{qV_{BE}/kT}. \qquad (5.8)$$

Here J_{b0} is assumed to be independent of V_{CB}, while $J_{nbr,0}$ is a function of V_{CB}:

$$J_{nbr,0}(\text{Si}) = \frac{qD_{nb}n_{ib}^2}{N_{ab}L_{nb}}\left\{\frac{\cosh\chi_{Si}-1}{\sinh\chi_{Si}}\right\}, \qquad (5.9)$$

where $\chi = W_b/L_{nb}$. Since the diffusion component of I_B is independent of V_{CB}, the change in I_B with V_{CB} will only be due to the variation in J_{nbr} through the variations in W_b. Therefore, in general, the input conductance of the transistor (g_μ) can be written as

$$g_\mu = \left.\frac{\partial J_B}{\partial V_{CB}}\right|_{V_{BE}} = \left.\frac{\partial J_{nbr}}{\partial V_{CB}}\right|_{V_{BE}} = \left.\frac{\partial J_{nbr}}{\partial W_b}\right|_{V_{BE}}\left.\frac{\partial W_b}{\partial V_{CB}}\right|_{V_{BE}}. \qquad (5.10)$$

We can thus determine the g_μ in a Si BJT to be

$$g_\mu(\text{Si}) = \frac{qD_{nb}n_{ib}^2}{N_{ab}L_{nb}}e^{qV_{BE}/kT}\left\{\frac{\cosh\chi_{Si}-1}{\sinh^2\chi_{Si}}\right\}\frac{1}{L_{nb}}\left[\frac{\partial W_b}{\partial V_{CB}}\right]. \qquad (5.11)$$

A more convenient way to compare the variations in J_{nbr} between devices and across temperature is to normalize g_μ to the base current at $V_{CB} = 0$ V (g'_μ). Therefore, by rewriting Equation 5.11 using the expressions for both J_C and V_A derived in Chapter 4, we finally obtain g'_μ in a Si BJT

$$g'_\mu(\text{Si}) = \frac{g_\mu(\text{Si})}{I_{B,\text{Si}}(V_{BE}, V_{CB} = 0)} \approx \frac{-\beta_{\text{Si}}(V_{BE}, V_{CB} = 0)\chi^2_{\text{Si}}(\cosh \chi_{\text{Si}} - 1)}{V_{A,\text{Si}}(\text{forced-}V_{BE}) \sinh^2 \chi_{\text{Si}}} \tag{5.12}$$

where $\beta_{\text{Si}}(V_{BE}, V_{CB} = 0)$ represents the current gain measured at a given V_{BE} and $V_{CB} = 0\,\text{V}$ and $V_{A,\text{Si}}(\text{forced-}V_{BE})$ represents the Early voltage at a fixed V_{BE}. Observe that when χ_{Si} is zero (representing the ideal situation in the transistor with no NBR), Equation 5.12 predicts that g'_μ will be zero, as one would expect. On the other hand, when χ_{Si} becomes large (i.e., significant NBR is present), then Equation 5.12 yields an increased value of g'_μ.

The g'_μ in a SiGe HBT with a trapezoidal Ge profile is difficult to derive analytically. In order to qualitatively determine the device design parameters that strongly influence g'_μ, one can consider a simple box Ge profile. In this case, it is easily shown that g'_μ in a SiGe HBT is the same as Equation 5.12, except for the differences in β. This is expected, since both J_{nbr} and β in a SiGe HBT are determined primarily by the amount of Ge-induced bandgap reduction at the EB space–charge edge (i.e., $\Delta E_{g,\text{Ge}}(0)$). In general, however, the NBR current component and hence g'_μ will be a function of the amount of Ge introduced into the base region of a SiGe HBT (i.e., EB boundary value as well as Ge grading). In addition, since the SiGe-to-Si J_{nbr} ratio is effectively amplified by cooling, it is expected that the SiGe-to-Si g'_μ ratio will also exponentially increase with decreasing temperature.

Figure 5.5 compares the variation in the normalized-I_B as a function of V_{CB} for both the Si and SiGe transistors at 358 and 200 K, respectively. In this case, the transistors are biased in the low-injection region where their collector and base currents are ideal. One can clearly observe the decrease in I_B at low V_{CB} due to the modulation of the NBR current component for both transistors, at 358 and 200 K, respectively. The strong decrease in I_B at larger values of V_{CB} is due to an increase in the impact-ionization base current component. By observing the variation in I_B with V_{CB} in the low-V_{CB} range, one can easily conclude that the Si BJT shows a weak NBR component ($\approx 0.5\%$ decrease in I_B), while the SiGe HBT shows not only a larger NBR base current component, but also an increase in the NBR with cooling, as anticipated from theory. It is important to note here that, although the NBR component in the SiGe HBT is clearly larger than that in the Si BJT, the magnitude of the NBR component is nevertheless still quite small ($\approx 3\%$ of I_B at 200 K). The measured g'_μ in SiGe HBTs is not only expected to be larger than for a comparably constructed Si BJT, but also thermally activated due to the presence of Ge band-offsets in the base region. Figure 5.6 confirms this expectation for both triangular and trapezoidal profile SiGe HBTs.

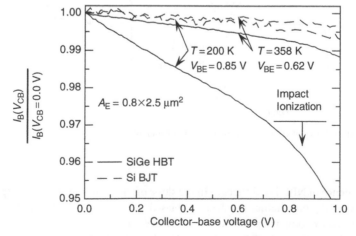

FIGURE 5.5 Measured normalized-I_B as a function of V_{CB} for a SiGe HBT and a Si BJT, at both 358 and 200 K. (From AJ Joseph, JD Cressler, RC Jaeger, DM Richey, and DL Harame. *IEEE Trans. Electron Dev.* 44:404–413, 1997. With permission.)

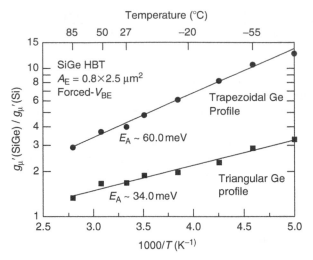

FIGURE 5.6 Measured ratio of g'_μ in a SiGe HBT and a Si BJT as a function of reciprocal temperature. (From AJ Joseph, JD Cressler, RC Jaeger, DM Richey, and DL Harame. *IEEE Trans. Electron Dev.* 44:404–413, 1997. With permission.)

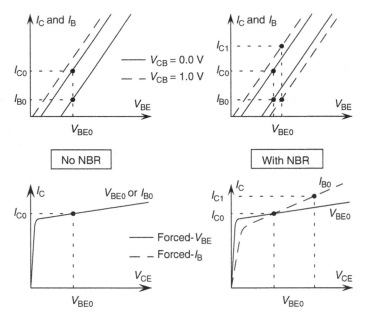

FIGURE 5.7 Illustration of the effects of NBR on the Gummel and output characteristics of a transistor.

A direct consequence of NBR is a difference in the slope of the common-emitter output characteristics of a transistor depending on whether the device is biased using forced-I_B or forced-V_{BE} conditions [1]. This can be explained by comparing the dc characteristics for a transistor under an ideal situation (no NBR) with that in the presence of NBR (see Figure 5.7). Without NBR, the increase in I_C with V_{CB} is the same whether the transistor is biased under forced-V_{BE} or forced-I_B input drive, yielding the same V_A for both conditions. In the presence of NBR, however, V_A measured using both techniques will differ because of the decrease in I_B with V_{CB}. In a forced-I_B situation, V_{BE} is allowed to change in such a way as to maintain constant I_B. Due to the fact that I_B decreases with increasing V_{CB} in the presence of NBR, V_{BE} is forced to increase so as to maintain constant I_B. This small increase in V_{BE} exponentially increases I_C, leading to a much smaller V_A. In a forced-V_{BE} situation, however, the I_C increase is due only to the decrease in W_b for an increase in V_{CB}, as one might expect in the ideal case. Thus, in the presence of NBR, V_A(forced-I_B) will be smaller than V_A(forced-V_{BE}), and the two quantities are related through g'_μ, according to Ref. [1]:

$$g'_\mu \approx \frac{1}{V_A(\text{forced-}V_{BE})} - \frac{1}{V_A(\text{forced-}I_B)}. \tag{5.13}$$

Since g'_μ is small in a well-made Si BJT, the difference between V_A(forced-V_{BE}) and V_A(forced-I_B) is expected to be small. Equation 5.13 predicts that in SiGe HBTs, however, the difference between V_A(forced-V_{BE}) and V_A(forced-I_B) will be greater because of the larger g'_μ compared to that in a Si BJT. Figure 5.8 shows V_A obtained for both Si and SiGe transistors using forced-I_B and forced-V_{BE} conditions as a function of reciprocal temperature. Observe that the V_A in a Si BJT, obtained using both techniques, yields similar results, thus confirming the presence of only a weak NBR component in the base current of these transistors. In the SiGe HBTs, however, we can clearly observe a quasi-exponential degradation of V_A(forced-I_B) compared to a quasi-exponential improvement in V_A(forced-V_{BE}) with cooling. While it is the bandgap grading in the SiGe HBT that increases the V_A(forced-V_{BE}), it is the amount of $\Delta E_{g,Ge}(x = 0)$ that causes the exponential degradation in V_A(forced-I_B) with cooling. From these experimental results, it is clear that such a strong temperature and input-bias dependent situation for SiGe HBTs could potentially have important consequences on the performance of SiGe HBT analog circuits that depend critically on the output conductance of the transistor. This anticipated impact on circuit performance is confirmed in Figure 5.9, which uses SPICE models designed to properly account for NBR in SiGe HBTs and carefully calibrated to data, to assess the impact of NBR on precision current sources (refer to the discussion in Ref. [1]).

The presumption in this section is that significant NBR exists in the SiGe HBTs under consideration.

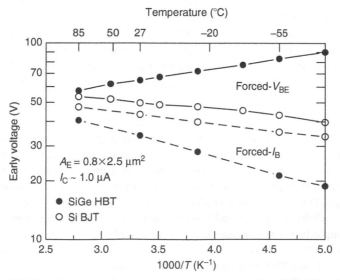

FIGURE 5.8 Measured Early voltage as a function of reciprocal temperature for a SiGe BJT and a Si BJT, using both forced-I_B and forced-V_{BE} techniques. (From AJ Joseph, JD Cressler, RC Jaeger, DM Richey, and DL Harame. *IEEE Trans. Electron Dev.* 44:404–413, 1997. With permission.)

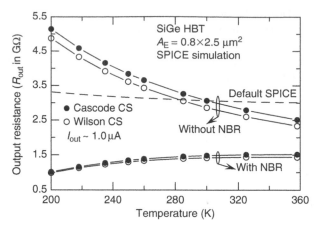

FIGURE 5.9 Comparison of calibrated SPICE modeling results both with NBR and without NBR for the output resistance of both Cascode and Wilson precision current sources as a function of temperature. (From AJ Joseph, JD Cressler, RC Jaeger, DM Richey, and DL Harame. *IEEE Trans. Electron Dev.* 44:404–413, 1997. With permission.)

In this situation we can say:

- An observable difference between $V_A(\text{forced-}V_{BE})$ and $V_A(\text{forced-}I_B)$ will exist in the SiGe HBT, and will be reflected in the output characteristics of the transistor.
- The measured $V_A(\text{forced-}V_{BE})$ value in the presence of NBR will be consistent with simple device theory (Chapter 4), but the $V_A(\text{forced-}I_B)$ will be degraded (lower) compared to simple theoretical expectations.
- This input-drive dependent V_A difference will be amplified in the SiGe HBT compared to a comparably constructed Si BJT. That is, Ge-induced bandgap engineering will always act to enhance the effects of NBR.
- This input-drive dependent V_A difference will get larger (worse) as the temperature decreases.
- Careful two-dimensional simulations can be used to identify the physical location of the traps responsible for the NBR component, and correlated with the fabrication process.
- Accurate compact modeling of SiGe HBTs for circuit design, which includes NBR can be accomplished using existing Si BJT models, but may require an additional parameter to account for the inherently different temperature dependence in V_A between a SiGe HBT and a Si BJT. Such NBR-compatible models can provide a detailed assessment of the role of NBR-induced V_A changes on particular circuits.

NBR, while clearly inherent to bipolar transistor operation because finite trap densities necessarily exist in semiconductor crystals, does not necessarily strongly perturb the characteristics of modern SiGe HBTs. The experimental results presented in this chapter show a significant NBR base current component, and thus are instructive for understanding and modeling NBR in SiGe HBTs, but we have also measured devices from the other SiGe technologies which do not show appreciable NBR-induced V_A changes. Thus, we do not consider NBR to be a "show-stopper" in SiGe HBTs, but rather something to be carefully monitored and assessed during technology development and qualification. In this case, a simple bench-top measurement of $I_B(V_{CB})/I_B$ as a function of V_{CB} at two different temperatures (e.g., 300 and 200 K) provides a simple and powerful tool for accurately assessing the presence of significant NBR in a given SiGe HBT technology generation. If present, appropriate steps can be taken to either try and correct the situation by process modification, or models can be developed which accurately account for the effect, thus ensuring that circuit designs are not negatively impacted.

5.4 Heterojunction Barrier Effects

In order to achieve maximum performance, SiGe HBTs must be biased at very high collector current densities (typically, above $1.0\,\text{mA}/\mu\text{m}^2$ for the first-generation SiGe HBTs). High-injection heterojunction barrier effects (HBE), which occur in all HBTs, can cause severe degradation in key transistor metrics such as, β, g_m, V_A, f_T, and f_{max}, especially at reduced temperatures. Careful transistor optimization is therefore required to delay the onset of the HBE to well above the current density levels required for normal circuit operation. Since the severity of the HBE is mainly determined by the amount of Ge-induced band offset at the SiGe–Si heterointerface and the collector doping level, one needs to carefully design the CB junction of the HBT. In order to delay the onset of Kirk effect and hence HBE, one can easily increase the collector doping level (N_{dc}). Increasing N_{dc}, however, decreases f_{max} and BV_{CEO} due to the increase in C_{CB} and the CB electric field, respectively, presenting serious design constraints.

The shape and position of the Ge profile in the CB region of a SiGe are critical in determining the characteristics of the onset of HBE and the rate of degradation in HBT characteristics with increasing J_C. While large Ge grading is desirable for increasing V_A, f_T, and f_{max} of a SiGe HBT, the increased Ge concentration at the CB junction increases the induced barrier associated with HBE. To reduce the impact of the barrier on device performance, one can either gradually decrease the Ge at the CB region or place the SiGe–Si heterointerface deeper inside the collector region, instead of having an abrupt SiGe–Si transition at the interface. Obviously, these methods lead to an increase in the total Ge content of the film, which imposes film stability (and hence manufacturing) constraints on the fabrication process. These device design trade-offs clearly indicate that there exists no specific design solution to completely eliminate HBE. One can, however, tailor the CB design to suit the application at hand, and offers testament to the versatility that can be achieved with bandgap engineering.

In Si BJTs operated under high injection in the collector, there are several phenomena that can cause the collector and base currents to deviate from their ideal low-injection behavior (i.e., I_C, $I_B \propto e^{qV_{BE}/KT}$), including Kirk effect [7], Webster–Rittner effect, the IR drop associated with the base and emitter resistances, and quasi-saturation due to collector resistance. Among these, Kirk effect (or "base push-out") is usually the most important in practical Si BJTs (and SiGe HBTs). The physical basis of Kirk effect lies in the fact that the increased minority carrier concentration in the CB region, at high injection, is sufficient to compensate for the doping-induced charge in the CB space–charge region, causing the space–charge region to first collapse, and then to be pushed deeper into the collector region as J_C (hence n_C) rises. The displacement of the CB space–charge region effectively increases the base width, which leads to a decrease in the collector current ($J_C \propto 1/W_b$), and an increase in the base transit time ($\tau_b \propto 1/W_b^2$), thus causing a premature degradation in both β and f_T.

The value of J_C at the onset of Kirk effect ($J_{C,\text{Kirk}}$) can be written generally as

$$J_{C,\text{Kirk}} \approx q v_s N_{dc}\left\{1 + \frac{2\varepsilon(V_{CB} + \phi_{bi})}{qN_{dc}W_{epi}^2}\right\}. \tag{5.14}$$

The direct relationship between the onset of Kirk effect and the collector doping level is obvious. To get a feel for the numbers, if we assume realistic values for the uniformly doped collector (e.g., $N_{dc} = 1\times10^{17}\,\text{cm}^{-3}$), and an epi-layer thickness of $0.5\,\mu\text{m}$, we thus expect from Equation 5.14 that the onset of Kirk effect will occur at approximately $1.6\,\text{mA}/\mu\text{m}^2$.

Since maximum device performance is achieved at large current densities, one usually needs to increase N_{dc} to provide additional immunity to Kirk effect, thereby increasing the CB electric field, and decreasing the CB breakdown voltage. Thus, a fundamental trade-off exists in Si BJTs between device performance (i.e., peak f_T) and maximum operating voltage (i.e., BV_{CEO}), as reflected in the so-called "Johnson-limit" [8].

In SiGe HBTs, the transition from a narrow bandgap SiGe base layer to the larger bandgap Si collector layer introduces a valence band offset at the SiGe–Si heterointerface. Since this band offset is masked by the band bending in the CB space–charge region during low-injection operation, it has negligible effect on the device characteristics. At high injection, however, the collapse of original CB electric field at the heterointerface exposes the offset, which opposes hole injection into the collector. The hole pile-up that occurs at the heterointerface induces a conduction band barrier that then opposes the electron flow from base to collector, causing an increase in the stored base charge which results in the sudden decrease in both f_T and f_{max}. The "pinning" of the collector current due to this induced conduction band barrier, and the simultaneous increase in the base current due to valence band offset, causes a rapid degradation in desirable characteristics of the SiGe HBT at a HBE onset current density, and can present serious device and circuit design issues. This effect was first reported in Ref. [9], and later addressed by other authors [10,11]. In addition, since the transport currents are thermally activated functions of the barrier height, it is expected that the HBE will have a much more pronounced impact at reduced temperatures, raising important questions about operation over a wide temperature range [11]. It is therefore essential that the collector profile and the Ge profile be designed properly to reduce the impact of HBE on circuit performance.

To experimentally investigate HBE, first-generation SiGe HBTs with three different Ge profiles were measured (a 15% Ge triangle, a 10% Ge trapezoid, and an 8% Ge trapezoid), along with a comparably designed Si BJT control. The collector profile was identical for all the transistors and was selectively implanted to simultaneously optimize f_T (at high J_C) while maintaining an acceptable BV_{CEO} of about 3.3 V. The Gummel characteristics of all the transistors are ideal across the measured temperature range of 200 to 358 K. While the SiGe HBTs and Si BJTs have differing current gains, as expected, a normalization of β as a function of J_C shows that there is a clear difference in high-injection behavior for the SiGe and Si devices, particularly at reduced temperatures. A sensitive test for clearly observing high-injection HBE in SiGe HBTs is to extract the transconductance (g_m) at high J_C from the Gummel characteristics, at high and low temperatures. As shown in Figure 5.10, a clear dip in the g_m at 200 K at J_C of about 2.0 mA/μm² can be clearly seen. By comparing g_m and β at $J_C = 2.0$ mA/μm² between the SiGe HBT and the Si BJT at 358 and 200 K, respectively, one can easily deduce that the differences are associated with the Ge profile, and hence are a signature of high-injection HBE. In addition, Figure 5.10 suggests that the trapezoidal Ge profiles show a weaker degradation in g_m at 200 K compared to the triangular Ge profile, because of the presence of a smaller Ge band offset in the CB junction (15% Ge versus 8% and 10% Ge, respectively), indicating that the specific design of the Ge profile plays a role, as expected.

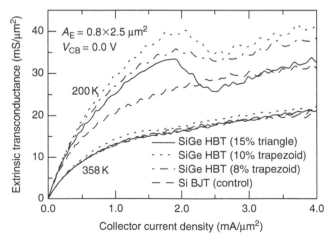

FIGURE 5.10 Extrinsic transconductance as a function of collector current density for a Si BJT and three SiGe HBT profiles, at 358 and 200 K. (From AJ Joseph, JD Cressler, DM Richey, DL Harame, and G Niu. *IEEE Trans. Electron Dev.* 46:1347–1356, 1999. With permission.)

To shed light on both the physics of HBE in SiGe HBTs, as well as to determine the optimum doping and Ge profiles for scaled SiGe HBTs, numerical device simulation is required. Figure 5.11 shows the electric field distribution in the base–collector region of both calibrated SiGe and Si transistors at low and high J_C. Observe that at low J_C, the CB built-in electric field entirely covers the SiGe–Si heterointerface. At high injection ($J_C = 4.0$ mA/μm^2, past peak f_T), however, the CB space–charge region is pushed deep into the collector region in both transistors due to Kirk effect, and in the SiGe HBT a barrier is formed at the original SiGe–Si heterointerface and can be clearly seen in the high-J_C field distribution.

Figure 5.12 shows the evolution of the induced conduction band barrier to electrons in the SiGe HBT as a function of J_C. Clearly, the electron barrier appears only at high injection and this can be correlated with the exposure of the SiGe–Si valence band offset (Figure 5.13). In addition, the magnitude of the induced conduction band barrier (ϕ_B) gets larger as the device is biased progressively into higher

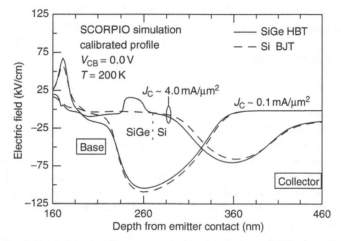

FIGURE 5.11 Simulated electric field distribution in a Si BJT and SiGe HBT at both low and high current density at 200 K. (From AJ Joseph, JD Cressler, DM Richey, DL Harame, and G Niu. *IEEE Trans. Electron Dev.* 46:1347–1356, 1999. With permission.)

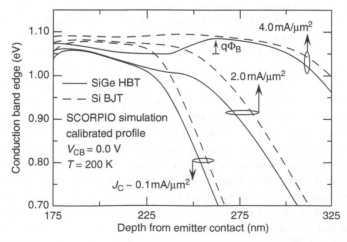

FIGURE 5.12 Simulated conduction band edge as a function of depth for a Si BJT and SiGe HBT, for various current densities at 200 K. (From AJ Joseph, JD Cressler, DM Richey, DL Harame, and G Niu. *IEEE Trans. Electron Dev.* 46:1347–1356, 1999. With permission.)

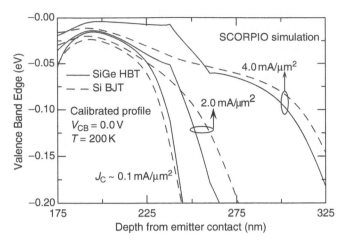

FIGURE 5.13 Simulated valence band edge as a function of depth for a Si BJT and SiGe HBT, for various current densities at 200 K. (From AJ Joseph, JD Cressler, DM Richey, DL Harame, and G Niu. *IEEE Trans. Electron Dev.* 46:1347–1356, 1999. With permission.)

injection, while at very large current densities ϕ_B eventually saturates. Although ϕ_B at a fixed J_C decreases with cooling due to the shift in operating point with temperature, its impact will be much greater at low temperatures due to its thermally activated nature as a band-edge phenomenon.

The sudden increase in J_B accompanying the barrier onset in a SiGe HBT is the result of the accumulation of holes in the base region due to HBE. At low injection, one clearly sees that the hole concentration in the base is unperturbed compared to a Si BJT. At high injection, however, not only is the hole profile pushed out into the collector region (Kirk effect) but also the presence of the barrier increases the hole concentration close to the CB junction.

A fundamental trade-off in collector profile design exists between maximizing both BV_{CEO} and f_{max} in SiGe HBTs. RF and microwave power amplifiers require large BV_{CEO}, and therefore the collector doping must be reduced. Obviously, such a reduction in N_{dc} will adversely affect the large-signal performance due to the premature onset of HBE. One can also, in principle, "tune" the barrier onset by properly adjusting the Ge retrograde profile shape. A higher Ge grading in the base region of a SiGe HBT provides better high-frequency performance throughout the temperature range. Increasing the Ge grading, however, necessarily increases the Ge content in the CB junction, which leads to a stronger barrier effect at high injection. In order to reduce the impact of barrier effect in such cases, one can either more gradually decrease the Ge or push the Ge deeper into the collector. In either case, however, one is limited by the amount of Ge that can be added because of the stability constraints of the SiGe films.

The successful insertion of SiGe HBTs into practical systems requires accurate compact circuit models for design. Because SiGe HBTs are typically modeled using Si BJT-based compact models (e.g., SPGP, VBIC, MEXTRAM, or HICUM), it is important to assess the accuracy of these models for capturing unique device phenomena such as high-injection HBE. In most compact models, the Kirk effect and HBE are lumped into a single function, assuming the Kirk effect and barrier effect occur simultaneously. This assumption, however, is no longer valid when the SiGe–Si heterojunction is located either in the neutral base region or deeper in the epitaxial collector. The latter, for instance, can be true in SiGe HBTs optimized for high breakdown voltage. Compact models can fail in this case to capture the functional form of the $f_T - J_C$ roll-off in SiGe HBTs (Figure 5.14). To accurately capture this phenomenon, a new transit time model that decouples the two effects is needed, and is discussed in detail in Ref. [12].

Due to the presence of SiGe–Si heterojunctions in SiGe HBTs, HBEs are inherent in SiGe HBT design and operation, and thus in some sense can be considered the most serious of the three second-order phenomena considered in this chapter. Given this situation, HBE must always be carefully "designed

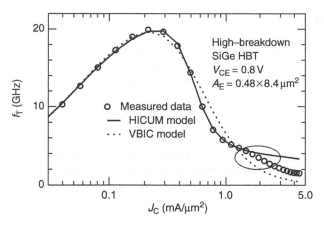

FIGURE 5.14   Comparison of measured f_T–J_C characteristics with the HICUM and VBIC compact models in a high breakdown voltage SiGe HBT. (From Q Liang, JD Cressler, G Niu, RR Murty, K Newton, and DL Harame. *IEEE Trans. Electron Dev.* 49:1807–1813, 2002. With permission.)

around." This is not overly difficult for low-BV_{CEO} transistors where the collector doping is relatively high, effectively retarding Kirk effect. For applications requiring higher breakdown voltage devices (e.g., power amplifiers), however, care must be taken to ensure that HBE do not adversely impact circuit designs, and that they are accurately modeled. For HBE in SiGe HBTs we can state the following:

- HBEs fall into two general categories: (1) induced barriers due to Ge misplacement; and (2) high-injection-induced barriers. The former can be corrected with proper growth and fabrication techniques, and are thus not inherent to a given SiGe technology. The latter, however, can be considered fundamental to the operation of SiGe HBTs, and must be carefully accounted for and accurately modeled by designers.
- High-J_C HBE causes a rapid degradation in β, g_m, and f_T once the barrier is induced. The critical onset current density for HBE ($J_{C,barrier}$) is thus a key device design parameter.
- HBEs are induced in the conduction band when the hole density in the pushed-out base under high-J_C is effectively blocked from moving into the collector by the SiGe–Si heterojunction. Both J_C and J_B are strongly affected.
- For low-breakdown voltage devices, HBE and Kirk effect generally occur at similar current densities. In higher breakdown voltage devices, or devices with deep SiGe–Si heterojunctions, however, the two effects can occur at very different current densities, producing unusual structure in the f_T–J_C characteristics, which first-order compact models do not accurately capture.
- Changes to the Ge retrograde can be used to effectively retard the onset of HBE, but at the expense of reduced film stability.
- Changes to the collector doping profile can be used to effectively retard the onset of HBE, but at the expense of increased CB capacitance and reduced breakdown voltage.
- The impact of HBE on device and circuit performance will rapidly worsen as the temperature decreases, because they are band-edge phenomena.

For any SiGe HBT technology generation, it is a prudent exercise to carefully characterize the transistors and assess the significance of HBE on the overall device response, and determine $J_{C,barrier}$. This is easily accomplished by plotting linear g_m on linear J_C at two temperatures (e.g., 300 and 200 K), and this knowledge can then be communicated to circuit designers. If $J_{C,barrier}$ is low enough for practical concerns, then Ge or collector profile modifications can be implemented to alleviate any problems. When moving to a new technology generation with different Ge and doping profiles, HBE should always be revisited.

5.5 Summary

Important second-order effects associated with: (1) Ge grading, (2) neutral base recombination, and (3) HBEs will always exist in SiGe HBTs, and their specific impact on actual SiGe HBT device and circuit operation is both profile-design- and application-dependent, and thus must be carefully considered by designers.

Acknowledgments

I am grateful to A. Joseph, S. Salmon, Q. Liang, G. Niu, D. Harame, G. Freeman, B. Meyerson, D. Herman, and the IBM SiGe team for their contributions. This work was supported by the Semiconductor Research Corporation, the GEDC at Georgia Tech, and IBM.

References

1. JD Cressler and G Niu. *Silicon–Germanium Heterojunction Bipolar Transistors.* Boston, MA: Artech House, 2003.
2. EF Crabbé, JD Cressler, GL Patton, JMC Stork, JH Comfort, and JY-C Sun. Current gain rolloff in graded-base SiGe heterojunction bipolar transistors. *IEEE Electron Dev. Lett.* 14:193–195, 1993.
3. SL Salmon, JD Cressler, RC Jaeger, and DL Harame. The impact of Ge profile shape on the operation of SiGe HBT precision voltage references. *IEEE Trans. Electron Dev.* 47:292–298, 2000.
4. HA Ainspan and CS Webster. Measured results on bandgap references in SiGe BiCMOS. *Electron. Lett.* 34:1441–1442, 1998.
5. AJ Joseph, JD Cressler, RC Jaeger, DM Richey, and DL Harame. Neutral base recombination and its impact on the temperature dependence of Early voltage and current gain—Early voltage product in UHV/CVD SiGe heterojunction bipolar transistors. *IEEE Trans. Electron Dev.* 44:404–413, 1997.
6. AJ Joseph, JD Cressler, and DM Richey. Optimization of Early voltage for cooled SiGe HBT precision current sources. *J. Phys. IV* 6:125–130, 1996.
7. CT Kirk. Theory of transistor cutoff frequency falloff at high current densities. *IRE Trans. Electron Dev.* 3:164–170, 1964.
8. EO Johnson. Physical limitations on frequency and power parameters of transistors. *RCA Rev.* 163–177, 1965.
9. S Tiwari. A new effect at high currents in heterostructure bipolar transistors. *IEEE Electron Dev. Lett.* 9:142–144, 1988.
10. PE Cottrell and Z Yu. Velocity saturation in the collector of Si/Ge_xSi_{1-x} HBTs. *IEEE Electron Dev. Lett.* 11:431–433, 1990.
11. AJ Joseph, JD Cressler, DM Richey, DL Harame, and G Niu. Optimization of SiGe HBTs for operation at high current densities. *IEEE Trans. Electron Dev.* 46:1347–1356, 1999.
12. Q Liang, JD Cressler, G Niu, RR Murty, K Newton, and DL Harame. A physics-based, high-injection transit-time model applied to barrier effects in SiGe HBTs. *IEEE Trans. Electron Dev.* 49:1807–1813, 2002.

6

Low-Frequency Noise

Guofu Niu
Auburn University

6.1 Background

Because of Si-based processing, the low-frequency noise in SiGe HBTs is comparable to that in Si BJTs, which is typically much better (lower) than in III–V HBTs [1–3]. Of particular importance is the $1/f$ noise or flicker noise, which dominates at low frequency, and can be upconverted to phase noise in RF oscillators through nonlinear I–V and C–V relationships inherent to the transistor. This results in noise sidebands on the carrier frequency, which fundamentally limits spectral purity. Low-frequency noise is also very important for wireless receivers utilizing zero or very low intermediate frequency (IF) architectures.

In this chapter, we will introduce the basics of $1/f$ noise measurement and modeling in SiGe HBTs, its dependence on technology scaling, implications of SiGe, and upconversion to phase noise. We will introduce a method to determine the maximum tolerable $1/f$ noise level for a given RF process, which can be used to aid process development, as the reduction of $1/f$ noise is quite challenging in manufacturing, particularly in scaled technologies with low thermal cycle.

6.2 Measurement Methods

The major $1/f$ noise source in SiGe HBTs is in the base current, as it is in typical polysilicon emitter Si BJTs. In an equivalent circuit representation, this is described using a noise *current* source placed between the internal base and emitter nodes. One can measure this noise current either indirectly through measuring the collector noise voltage or directly through measuring the base noise current.

The indirect method is relatively easier to implement in practice, and is illustrated in Figure 6.1. The potentiometers P_B and P_C set the dc bias at the base and collector, respectively. Batteries are the preferred power supplies as they have the least amount of spurious noise. The two large capacitors C_B and C_C provide dynamic ac grounding. They can be left out for simplicity, but care must be exercised in determining the effective source and load resistances seen by the transistor.

R_B is chosen to be much greater than transistor input impedance r_π such that the base noise current flows into the transistor for amplification. The voltage noise at R_C is further amplified by a low-noise preamp, and detected by a dynamic signal analyzer (DSA). The base current noise (S_{I_B}) is obtained from the measured collector voltage noise S_{V_C} as

$$S_{I_B} = \frac{S_{V_C}}{(R_C\beta)^2}. \tag{6.1}$$

Strictly speaking, the small-signal β should be used in Equation 6.1. If capacitor C_C is not used, the effective dynamic load resistance seen by the collector node is used in place of R_C. The base bias resistance R_B typically ranges from 50 kΩ to 10 MΩ, and the collector sampling resistance R_C is on the order of 2 kΩ. Because of the $R_B \gg r_\pi$ requirement, very large R_B is needed for measurement at low I_B values.

At low I_B, the base current noise can be directly measured using a current amplifier (as shown in Figure 6.2). A large bypass capacitance C_B short-circuits the noise from the base-biasing network, and creates a low impedance path for the base current $1/f$ noise. The key is to make sure that the input impedance of the current amplifier is much lower than the transistor input impedance (r_π), so that all of

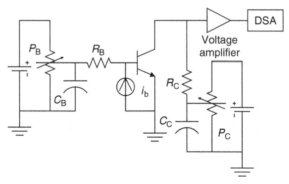

FIGURE 6.1 Indirect measurement of $1/f$ noise.

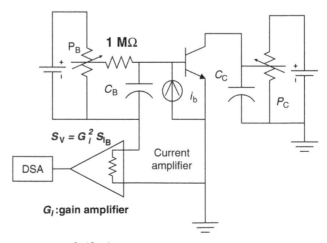

FIGURE 6.2 Direct measurement of $1/f$ noise.

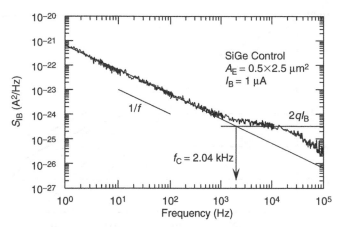

FIGURE 6.3 A typical low-frequency noise spectrum of a first generation SiGe HBT ($A_E = 0.5 \times 2.5 \, \mu m^2$, and $I_B = 1 \, \mu A$).

the base noise current flows into the current amplifier. The current amplifier output voltage is proportional to the base noise current, and the gain of the current amplifier has units of V/A.

Figure 6.3 shows a typical low-frequency base current noise spectrum (S_{I_B}) for a first-generation SiGe HBT. The noise spectrum shows a clear $1/f$ component as well as the $2qI_B$ shot noise level. The corner frequency f_C is determined from the intercept of the $1/f$ component and the $2qI_B$ shot noise level. At higher I_B values, the $2qI_B$ shot noise level cannot be directly observed for various reasons. The calculated $2qI_B$ value can be used to determine f_C in this case.

6.3 Physical Origins

The exact origins of $1/f$ noise are not well understood. A popular theory, proposed by McWhorter [4], describes $1/f$ noise as a superposition of individual generation–recombination (g–r) noise. Each interfacial trap generates g–r noise with a Lorentzian-shaped spectral density given by

$$S_{I,i} = \frac{A\tau_i}{1 + (2\pi f \pi_i^2)},$$ (6.2)

where A is the magnitude of the g–r noise, and τ_i is the time constant for the trapping–detrapping process. A large number of these traps and a particular statistical distribution of τ_i ($1/\tau$) give rise to $1/f$ spectral shape.

Figure 6.4 shows such an example of Lorentzian spectra (dashed lines), and the superpositioned spectrum, which is approximately $1/f$. This model assumes no interaction between trap levels at different energies. If the levels interact with each other, a Lorentzian spectrum instead of $1/f$ spectrum may be observed [5]. This model is expected to work well in devices with large emitter area, where a large number of traps exist. In small emitter area devices, however, one may expect to observe a Lorentzian shape of behavior.

Experimental data support the above trapping origin of $1/f$ noise in SiGe HBTs [6]. Figure 6.5a and b shows the low-frequency noise spectra measured on SiGe HBTs with small and large emitter areas, $0.24 \times 0.48 \, \mu m^2$ and $0.96 \times 1.6 \, \mu m^2$, respectively. Three samples are measured for each area. The low-frequency noise spectra show a strong deviation from $1/f$ behavior and a larger statistical scatter in the small devices. The average of all samples, shown as solid lines, shows a close to $1/f$ frequency dependence.

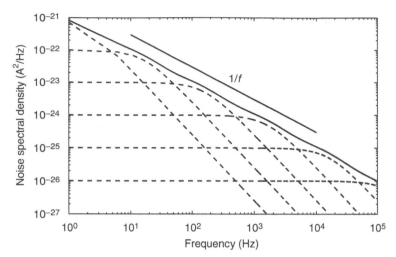

FIGURE 6.4 $1/f$ noise as a superposition of Lorentzians.

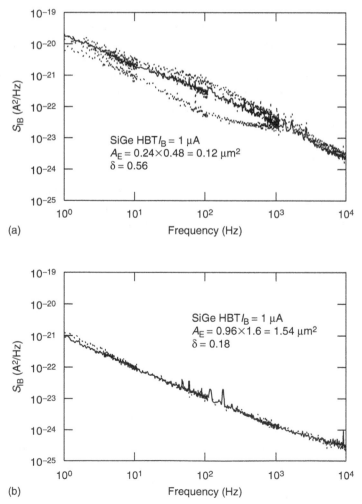

FIGURE 6.5   Low-frequency noise spectra in three samples with emitter area of (a) $A_E = 0.24 \times 0.48\,\mu m^2$ and (b) $A_E = 0.96 \times 1.6\,\mu m^2$.

6.4 SiGe Profile Impact and Modeling

1/f Noise K_F Factor

In general, S_{I_B} is related to I_B by

$$S_{I_B} = K_F \frac{I_B^\alpha}{f},\qquad(6.3)$$

where K_F and α correspond to the KF and AF model parameters used in SPICE. $\alpha = 1$ is often viewed as indication of carrier mobility fluctuations, and $\alpha = 2$ is often viewed as for carrier number fluctuations [7–13]. The α for typical SiGe HBTs is close to 2, and varies only slightly with SiGe profile and collector doping profile (2 + 0.2).

An often-asked question is whether the introduction of SiGe in transistor base affects the 1/f noise K_F factor. Assuming that the 1/f noise is solely a function of the number of minority carriers injected into the emitter, one may expect the same 1/f noise at a given V_{BE}, which means the same I_B for a SiGe HBT and its Si counterpart. Thus, the 1/f noise K factor is expected to be the same. This turns out to be true experimentally [14]. Figure 6.6 shows the S_{I_B} at 10 Hz versus I_B for three experimental SiGe HBTs and a comparably fabricated Si BJT. The SiGe HBTs include a 10% peak Ge profile control, a 14% peak Ge low-noise design (LN1), and an 18% peak Ge low-noise design (LN2). The Si BJT is an epi-base device. All the devices were fabricated in the same wafer lot.

A constant I_C comparison is more meaningful in the context of RFIC design, because many RF figures-of-merit fundamentally depend on I_C instead of I_B (e.g., f_T and f_{max}). In addition, NF_{min}, though dependent on I_B, is often compared at the same operating I_C as well. If we compare S_{I_B} at the same I_C, however, S_{I_B} is significantly *lower* (better) in SiGe HBTs than in Si BJTs, because of the lower I_B (higher β) found in SiGe HBTs, all else being equal. Since $S_{I_B} \propto I_C^2/\beta^2$, the S_{I_B} for the LN1 and LN2 SiGe HBTs should be naturally lower than for the SiGe control and Si BJT because of their higher β. This is confirmed by the measured data shown in Figure 6.7.

Geometry Dependence

The 1/f noise amplitude, as measured by the K_F factor, scales inversely with the total number of carriers in the noise-generating elements, according to Hooge's theory [15]. The 1/f noise generated by sources in the EB spacer oxide at the device periphery is inversely proportional to the emitter perimeter $P_E = W_E + L_E$, while the 1/f noise generated by sources located at the intrinsic EB interface (i.e., the emitter

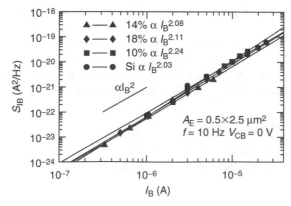

FIGURE 6.6 Measured S_{I_B} at 10 Hz as a function of I_B for the Si BJT, the SiGe control, and the two low-noise SiGe HBTs.

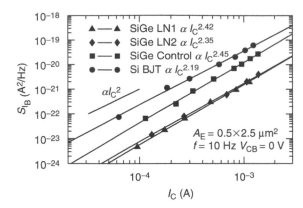

FIGURE 6.7 Measured S_{I_B} at 10 Hz as a function of I_C for the Si BJT, the SiGe control, and the two low-noise SiGe HBTs.

polysilicon–silicon interface) across the emitter window is inversely proportional to the emitter area $A_E = W_E L_E$. The K_F factor is often examined as a function of the emitter area, the emitter perimeter, or the perimeter-to-area ratio as a means of locating the contributing $1/f$ noise sources [9–13]. For instance, for fixed frequency, the combination of $1/A_E$ dependence with an I_B^2 bias dependence for S_{I_B} is consistent with a uniform area distribution of noise-generating traps across the emitter region. In practice, caution must be exercised in interpreting P_E or A_E scaling data, because test devices are often designed with the emitter width equal to the minimum feature size, and with an emitter length much larger than the emitter width. As a result, such data tend to scale with the emitter perimeter and area in a similar manner, making interpretation difficult. A wide distribution of device sizes and P_E/A_E ratios thus needs to be used when designing test structures for noise-scaling studies in order to make a clear distinction between P_E and A_E scaling in SiGe HBTs. For all the SiGe HBTs described in "$1/f$ Noise K_F Factor," the $1/f$ noise K_F factor is inversely proportional to A_E. Equation 6.3 can thus be rewritten as

$$S_{I_B} = \frac{K}{A_E} \frac{I_B^2}{f} = \frac{K}{\beta^2} \frac{1}{A_E} \frac{I_C^2}{f}, \tag{6.4}$$

where K is a factor independent of the emitter area and is defined as $K = K_F A_E$, where $\alpha = 2$ is assumed. Equation 6.4 is written as a function of I_C to facilitate technology comparisons for RFIC circuit design, for reasons discussed above. Because the K factor for low-frequency noise is approximately independent of base profile design, a higher β SiGe HBT has a lower S_{I_B}, and hence generates lower phase noise when used in RF amplifiers and oscillators. For a given operating current, a larger device can clearly be used to reduce S_{I_B}. This tactic, however, reduces f_T because of the lower J_C. The maximum device size one can use is usually limited by this f_T requirement. Optimum transistor sizing is thus important not only for reducing $\mathrm{NF}_{\min}$, but also for reducing phase noise [14].

$1/f$ Corner Frequency

Traditionally, $1/f$ noise performance is characterized by the corner frequency (f_C) figure-of-merit, defined to be the frequency at which the $1/f$ noise equals the shot noise level $2qI_B$. Equating 6.4 with $2qI_B$ leads to

$$f_C = \frac{KI_B}{2qA_E} = \frac{KJ_C}{2q\beta}, \tag{6.5}$$

where J_C is the collector current density, and β is the dc β. Equation 6.5 suggests that f_C is proportional to J_C and K, and inversely proportional to β. We note that this conclusion differs from that derived in Ref. [16]. The derivation in Ref. [16] showed that f_C is independent of bias current density, because $\alpha = 1$ was assumed (i.e., according to mobility fluctuation theory). This dependence of α, however, is not the case in typical SiGe HBTs, which show an α close to 2. Figure 6.8 shows the measured and modeled $f_C - J_C$ dependence for the devices used here. As expected, f_C is the lowest in the two low-noise SiGe HBTs, LN1 and LN2, and highest for the Si BJT. The modeling results calculated using Equation 6.5 fit the measured data well.

Impact of Collector–Base Junction Traps

A subtle effect in SiGe HBTs is the carrier traps near the SiGe–Si growth interface, which are also referred to as collector–base junction traps, because the SiGe–Si growth interface is right in the collector–base junction depletion layer. These traps contribute to a recombination current in the CB "depletion" layer, which is small in magnitude but strongly modulated by the CB voltage, and is responsible for the output conductance degradation under forced-I_B operation [17]. At high injection, when the electron concentration in the CB depletion layer becomes comparable to the depletion charge density, the electrical neutral base pushes out. The CB junction traps are now exposed to a large amount of electrons and holes, resulting in a large amount of recombination current. This effect is particularly severe in high-breakdown voltage (HBV) devices, in which collector doping is high and high injection occurs at lower current densities.

A natural question is whether these CB junction traps contribute to low-frequency noise. By comparing the base current and base current noise of standard and HBV devices, it was recently found that these CB junction traps indeed produce additional $1/f$ noise [18]. Conceptually, one can divide the total $1/f$ noise in a HBV device into two components, one due to EB junction traps and the other due to CB junction traps. The EB noise component is a function of the emitter injection component of the base current, and the CB noise component is a function of the CB recombination current. Figure 6.9 shows the two $1/f$ noise components as a function of their respective base current components [18]. The EB and CB $1/f$ noises clearly show different dependences on their respective base current components, indicating that the $1/f$ noise process at the CB junction traps is different from that in the EB junction. For circuit modeling, however, one may continue to simply model the total base current $1/f$ noise as a function of the total base current, despite that there are two different $1/f$ noise processes. An accurate model for the CB junction trap-induced recombination current is necessary and has yet to be developed.

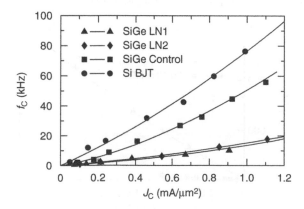

FIGURE 6.8 Measured and modeled f_C versus J_C for the standard breakdown voltage Si BJT, the SiGe control, and the two low-noise SiGe HBTs.

Technology Scaling

The $1/f$ noise K factor has recently been found to increase with technology scaling [19,20]. This is likely associated with the physical changes of the emitter–base junction composition during scaling. With scaling, a narrow and more heavily doped base profile is used, which requires a lower thermal cycle for the bipolar processing. The Ge grading often increases with scaling to create a higher accelerating electric field for minority carriers. The addition of carbon in scaled SiGe HBT technologies, though a small amount for suppressing boron outdiffusion, could inadvertently increase $1/f$ noise.

The increase of the $1/f$ noise K factor with scaling tends to increase $f_{C,1/f}$. However, depending on device design, β is often increased with scaling as well, which partially offsets the K factor increase. Figure 6.10 shows the measured f_C a function of J_C for HBTs with peak f_T of 50 and 120 GHz. The nature of bipolar transistor operation necessitates a higher operating J_C to realize the high-speed potential offered by scaling. A larger J_C range is thus used for the 120 GHz HBT. For a given J_C, an increase of $f_{C,1/f}$ is observed. At $J_C = 2.5$ mA/μm^2, the 120 GHz HBT shows a $f_{C,1/f}$ of 1.6 MHz, which is relatively high compared to a 50 GHz HBT at $J_C = 1$ mA/μm^2. Such an increase of $1/f$ corner frequency, however, does not necessarily cause an increase of the overall oscillator phase noise or the ultimate frequency synthesizer noise, due to different mechanisms of phase noise upconversion for the base current $1/f$ noise and base current shot noise [20].

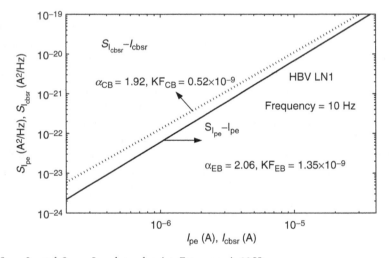

FIGURE 6.9 $S_{I_{pe}} - I_{pe}$ and $S_{I_{cbsr}} - I_{cbsr}$ dependencies. Frequency is 10 Hz.

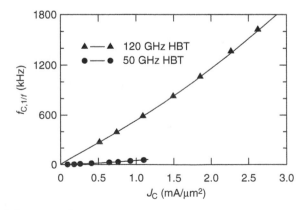

FIGURE 6.10   Measured $f_{C,1/f}$ versus J_C.

6.5   Oscillator and Synthesizer Phase Noise Issues

$1/f$ Noise in Oscillators

So far we have considered $1/f$ noise measured under a given dc biasing I_B. In oscillators, the situation is more complicated as [20]:

- The base terminal I_B is different from the base–emitter junction transport current I_{BE}, because of the capacitive charging and discharging of the junction capacitances. Only I_{BE} generates noise.
- The noise generating I_{BE} is "oscillating" by the very nature of oscillation.

Figure 6.11 compares the waveforms of the terminal I_B and internal noise generating I_{BE} in a 5.5 GHz SiGe HBT oscillator. Also shown is the V_{BE} waveform. Clearly, a large difference exists between I_{BE} and I_B.

Strictly speaking, one cannot simply determine the $1/f$ noise for an oscillating transistor from the small signal $1/f$ noise measurement. Measurement of $1/f$ noise with a periodic large signal biasing, however, is quite involved, particularly if high frequency is involved. In practice, the amount of $1/f$ noise for an oscillating transistor is assumed to be the same as the $1/f$ measured at a dc biasing current identical to the dc component of the oscillating noise generating current I_{BE}. This is implemented in CAD tools such as Agilent ADS.

Phase Noise Implications

In oscillators, $1/f$ noise is upconverted to a $1/f^3$ phase noise, while all the white noise (shot noise and thermal resistance noise) are upconverted to $1/f^2$ phase noises. Figure 6.12 shows simulated phase noise versus offset frequency for two HBTs with 50 and 120 GHz peak f_T. With device scaling from 50 to 120 GHz technology, the $1/f^3$ component increases by 11.4 dB, in part because of the increasing K factor. The $1/f^2$ phase noise resulting from upconversion of white noises, however, improves (decreases) by 7.2 dB with HBT scaling.

We now define the corner offset frequency, $f_{C,offset}$, using the intersect of the $1/f^3$ and $1/f^2$ phase noises. $f_{C,offset}$ is a direct measure of the importance of the phase noise upconverted from $1/f$ noise with respect to the phase noise upconverted from the white noise sources. $f_{C,offset}$ is 595.4 Hz and 40.8 kHz for the 50 and 120 GHz technologies (as can be seen from Figure 6.13). Note that $f_{C,offset}$ itself does not contain any information on either the $1/f^3$ or $1/f^2$ phase noise level.

A higher $f_{C,offset}$ does not necessarily mean higher phase noise. In this case, the $f_{C,offset}$ for the 120 GHz HBT is nearly 70× higher than for the 50 GHz HBT. The overall effect of scaling on oscillator phase

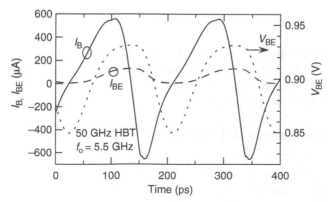

FIGURE 6.11   Comparison of terminal I_B and internal I_{BE} for a SiGe HBT in a 5.5 GHz oscillator. The internal V_{BE} is shown on the right y-axis.

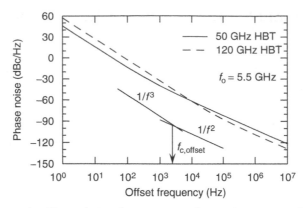

FIGURE 6.12 Phase noise of oscillators designed using HBTs with 50 and 120 GHz peak f_T.

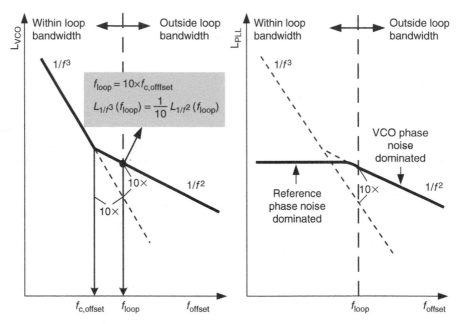

FIGURE 6.13 Illustration of the conversion process of VCO and reference phase noise to frequency synthesizer phase noise and the definition of K_{th}.

noise is a degradation at offsets below 10 kHz, but an improvement at higher offset frequencies. In a frequency synthesizer, if the loop bandwidth is much greater than 10 kHz, the overall synthesizer phase noise will improve with scaling, despite increased $1/f$ corner frequency, as the oscillator phase noise below 10 kHz is removed by loop feedback.

Synthesizer Phase Noise and Threshold *K*

In frequency synthesizers, the VCO phase noise within the loop bandwidth is suppressed by the loop feedback mechanism. The out-of-band phase noise of the VCO, however, directly translates into synthesizer out-of-band phase noise. From an application standpoint, if the loop bandwidth is sufficiently higher than the corner offset frequency, the $1/f^3$ phase noise can be completely suppressed by loop feedback. The out-of-band noise will then be the $1/f^2$ phase noise due to white noises. Using $10\times$

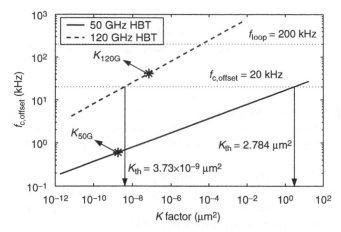

FIGURE 6.14 $f_{C,offset}$ versus K factor. K_{th} is determined for a loop bandwidth of 200 kHz. The actual K values are shown as "*".

as a criterion, the $1/f^3$ phase noise is only one tenth of the $1/f^2$ phase noise at the loop bandwidth offset frequency, for a $f_{C,offset}$ that is one tenth of the loop bandwidth (as shown in Figure 6.14).

For a given oscillator, $f_{C,offset}$ decreases linearly with K according to the analysis in Section 6.4. For a given process and loop bandwidth, a threshold K that makes $f_{C,offset}$ equal to one tenth of the loop bandwidth can be defined. Once $K < K_{th}$, the synthesizer phase noise no longer decreases with further decrease of K. One can also view this threshold K as the maximum tolerable K. This is very attractive from a semiconductor technology development standpoint, because the $1/f K$ factor is sensitive to defect level, and very challenging to minimize.

Figure 6.14 shows $f_{C,offset}$ versus $1/f$ noise K factor for two technologies of 50 and 120 GHz peak f_T. Assuming a loop bandwidth of 200 kHz, K_{th} is determined as K at which $f_{C,offset} = 20$ kHz, such that the $1/f^3$ phase noise is only 10% of the $1/f^2$ noise at 200 kHz. For the 50 GHz HBT, $K_{th} = 2.748 \, \mu m^2$, and the actual K ($2.0 \times 10^{-9} \, \mu m^2$) is well below K_{th}. Thus $1/f$ noise is not a concern for synthesizer phase noise, since practically all the $1/f^3$ phase noise is suppressed by loop feedback, and the in-band noise is limited by reference oscillator. For the 120 GHz HBT, $K_{th} = 3.73 \times 10^{-9} \, \mu m^2$, which is slightly smaller than the actual K of $8.6 \times 10^{-8} \, \mu m^2$. The combination of decreasing K_{th} and increasing K with scaling makes $1/f$ noise an increasingly important concern for phase noise of frequency synthesizers.

6.6   Summary

In this chapter, we have discussed the measurement, modeling, and system phase noise implications of $1/f$ noise. The $1/f$ noise K factor, which measures the amount of base current $1/f$ noise for a given I_B, is shown to be a better measure for phase noise than the traditional $1/f$ noise corner frequency. SiGe HBTs with high corner frequency can still show excellent phase noise performance when used in oscillators. For a given process, the $1/f K$ factor only needs to be below a certain threshold, and any further reduction of the K factor does not help in reducing system phase noise. The threshold (K_{th}) is shown to decrease with scaling. This, together with the increase of K factor with technology scaling, makes $1/f$ noise an increasingly important concern with further scaling.

Acknowledgments

The author would like to thank J. Tang for help in preparation of the manuscript. This work is supported by NSF under ECS-0112923 and ECS-0119623 and SRC under SRC-2001-NJ-937 and SRC-2003-NJ-1133.

References

1. L Vempati, JD Cressler, JA Babcock, RC Jaeger, and DL Harame. Low-frequency noise in UHV/CVD epitaxial Si and SiGe bipolar transistors. *IEEE J. Solid-State Circ.* 31:1458–1467, 1996.
2. B Van Haaren, M Regis, O Llopis, L Escotte, A Gruhle, C Mahner, R Plana, and J Graffeuil. Low-frequency noise properties of SiGe HBTs and application to ultra-low phase-noise oscillators. *IEEE Trans. Micro. Theory Tech.* 46:647–652, 1998.
3. DL Harame, DC Ahlgren, DD Coolbaugh, JS Dunn, GG Freeman, JD Gillis, RA Groves, GN Hendersen, RA Johnson, AJ Joseph, S Subbanna, AM Victor, KM Watson, CS Webster, and PJ Zampardi. Current status and future trends of SiGe BiCMOS Technology. *IEEE Trans. Electron Dev.* 48:2575–2594, 2001.
4. AL McWhorter. $1/f$ noise and germanium surface properties. *Semicond. Surf. Phys.* 207, 1980.
5. FN Hooge. $1/f$ noise sources. *IEEE Trans. Electron Dev.* 41:1926–1935, 1994.
6. Z Jin, JD Cressler, GF Niu, and AJ Joseph. Impact of geometrical scaling on low-frequency noise in SiGe HBTs. *IEEE Trans. Electron Dev.* 50:676–682, 2003.
7. LKJ Vandamme. Noise as a diagnostic tool for quality and reliability of electronic devices. *IEEE Trans. Electron Dev.* 41:2174–2187, 1994.
8. A Mounib, F Balestra, N Mathieu, J Brini, G Ghibaudo, A Chovet, A Chantre, and A Nouailhat. Low-frequency noise sources in polysilicon emitter BJTs: Influence of hot-electron-induced degradation and post-stress recovery. *IEEE Trans. Electron Dev.* 42:1647–1652, 1995.
9. MJ Deen, J Ilowski, and P Yang. Low frequency noise in polysilicon–emitter bipolar junction transistors. *J. Appl. Phys.* 77:6278–6285, 1995.
10. M Koolen and JCJ Aerts. The influence of non-ideal base current on $1/f$ noise behaviour of bipolar transistors. Proc. IEEE Bipolar/BiCMOS Circ. Tech. Meeting, September 1990, pp. 232–235.
11. HAW Markus and TGM Kleinpenning. Low-frequency noise in polysilicon emitter bipolar transistors. *IEEE Trans. Electron Dev.* 42:720–727, 1995.
12. P Llinares, D Celi, and O Roux-dit-Buisson. Dimensional scaling of $1/f$ noise in the base current of quasi self-aligned polysilicon emitter bipolar junction transistors. *J. Appl. Phys.* 82:2671–2675, 1997.
13. MJ Deen, SL Rumyantsev, and M Schroter. On the origin of $1/f$ noise in polysilicon emitter bipolar transistors. *J. Appl. Phys.* 85:1192–1195, 1999.
14. GF Niu, Z Jin, JD Cressler, R Rapeta, AJ Joseph, and DL Harame. Transistor noise in SiGe HBT RF technology. *IEEE J. Solid-State Circ.* 36:1424–1427, 2001.
15. A van der Ziel. *Noise in Solid State Devices and Circuits.* Wiley, New York, 1986.
16. LKJ Vandamme and G Trefan. Review of low-frequency noise in bipolar transistors over the last decade. Proc. IEEE Bipolar/BiCMOS Circ. Tech. Meeting, 2001, pp. 68–73.
17. GF Niu, JD Cressler, and AJ Joseph. Quantifying neutral base recombination and the effects of collector–base junction traps in UHV/CVD SiGe HBTs. *IEEE Trans. Electron Dev.* 45:2499–2503, 1998.
18. J Tang, GF Niu, AJ Joseph, and DL Harame. Impact of collector–base junction traps on low-frequency noise in high breakdown voltage SiGe HBTs. *IEEE Trans. Electron Dev.* 51:1475–1482, 2004.
19. JA Johansen, Z Jin, JD Cressler, and AJ Joseph. Geometry-dependent low-frequency noise variations in 120 GHz f_T/SiGe HBTs. Topical Meeting on Silicon Monolithic Integrated Circuits in RF Systems, 2003, pp. 57–59.
20. GF Niu, J Tang, Z Feng, AJ Joseph, and DL Harame. Scaling and technological limitations of $1/f$ noise and oscillator phase noise in SiGe HBTs. *IEEE Trans. Micro. Theory Tech.* 53:506–514, 2005.

7
Broadband Noise

David R. Greenberg
*IBM Thomas J. Watson
Research Center*

7.1 Introduction

The ultimate data rate of any communication channel is constrained by the bandwidth and the signal-to-noise ratio (SNR) of the channel [1]. As a result, the quantity of noise at any point in a signal chain places a direct limit on how many bits per second can pass through that point for a given bandwidth and power level. In a typical system such as a wireless receiver, noise can be introduced in a variety of ways, including by the active devices (both bipolar and MOS) used to implement the circuits. Noise may be added directly in band with the signal, such as in the case of a low-noise amplifier (LNA) at the front end of a receiver chain or in the baseband amplifier near the end of the chain. In addition, nonlinearities may allow noise to enter the channel band indirectly. For example, transistor noise in a voltage-controlled oscillator (VCO) circuit introduces phase noise in the output sinusoid. When serving as a local oscillator, such a signal passed to a mixer along with the data signal can cause out-of-band (e.g., adjacent channel) energy to fold in-band, contributing interference that degrades channel capacity in a manner analogous to noise. While the effects of noise can be countered by increasing the energy per bit, which translates into signal power for a given bit rate, many applications, particularly portable systems such as cellular phones, GPS receivers, and WLAN-capable PDAs, are constrained in transmitted or received signal power by battery life limits, transmitter–receiver distance, feasible antenna and package size or standards and regulatory requirements. Supply voltage and amplifier gain compression can impose an upper signal power limit as well. Thus, communication system designers seeking an attractive trade-off between throughput, signal power, and bit-error rate demand technologies capable of processing high-frequency signals while introducing as little noise as possible.

Both in discrete and integrated form, the SiGe HBT has emerged as an attractive low-noise solution, combining the performance of more expensive devices such as III–V FETs and HBTs with the process control, integration potential, and low cost associated with silicon [2,3]. The technology has been penetrating the market steadily, beginning with cellular telephony and wireless networking parts built in the 0.5-μm node for the 0.8 to 2.4 GHz range and progressing toward emerging applications such as 77 GHz automotive radar using the 0.13-μm node. At the same time, the SiGe HBT has proven amenable to accurate analysis and modeling, enabling the circuit designer to achieve the cost and time to market advantages of first-pass design success.

7.2   HBT Noise Mechanisms and Modeling

The SiGe HBT is a vertical device, with carrier transport confined largely to low-defect bulk material. As a result, high-frequency noise in the device under typical bias conditions is well described by a simple, one-dimensional physical model containing five fundamental noise sources (as illustrated in Figure 7.1). In forward bias, a voltage V_{BE} applied across the base–emitter junction creates a net flow of electrons over the junction energy barrier and into the base. For collector–base voltage V_{CB} well below the collector–base breakdown voltage BV_{CBO}, the collector current I_C consists primarily of this electron flow. Since electrons arrive at and cross this energy barrier according to a random distribution, I_C contains shot noise with noise power described by $\langle i_{njc} \rangle^2 = 2qI_CB$, where B is the bandwidth over which noise power is measured [4]. Under the same conditions, the base current I_B is composed primarily of holes injected from base to emitter and contains shot noise power similarly described by $\langle i_{njb} \rangle^2 = 2qI_BB$. Although involving a common junction, each carrier type moves and crosses the E–B junction independently of the other. Further, the space–charge and neutral–base recombination components that contribute to both I_C and I_B are quite low in modern, low-defect SiGe HBTs and may typically be neglected except at very low temperatures. As a result, noise components i_{njc} and i_{njb} are almost entirely statistically uncorrelated [5].

Resistors are another source of thermal noise and all physically realized HBTs contain a resistance R_B in series with the base, arising from both the nonzero lateral dimension of the pinched intrinsic base and from the extrinsic layers used to create electrical access and contact to this intrinsic region. The emitter polysilicon contributes a second parasitic series resistance R_E. These resistances generate noise components that can be described using noise voltage sources with powers $\langle |v_{nRB}| \rangle^2 = 4kTR_BB$ and $\langle |v_{nRE}| \rangle^2 = 4kTR_EB$ [4]. Although a noise voltage on the emitter terminal adds to both V_{BE} and V_{CE}, I_C in a HBT is quite insensitive to V_{CE} and thus the indirect contribution of R_E noise to I_C via the creation of V_{CE} noise can be neglected. As a result, R_B and R_E can be combined into a single noise voltage source $\langle |v_{nR}| \rangle^2 = 4kT(R_B + R_E)B \approx 4kTR_B$ for a modern SiGe HBT in which $R_E \ll R_B$. The overall noise voltage contribution from the collector resistance R_C is negligible since it is divided by the transistor voltage gain as seen from the vantage point of the other parasitic resistances.

At sufficiently high V_{CB}, electrons crossing the space–charge region from base to collector can acquire enough energy to excite a valence electron to the conduction band and create an electron–hole pair. Since the timing of an impact ionization event is statistical in nature, the resulting electrons contribute an additional component noise to I_C while the holes add a correlated component to I_B. This impact ionization noise is often neglected, but we will later demonstrate its effect on minimum noise figure (F_{min}) and describe a device design that reduces this component with few trade-offs [6].

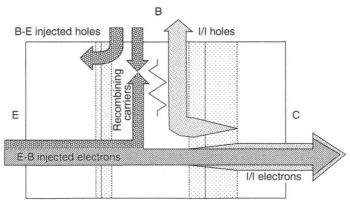

FIGURE 7.1   Schematic cross section of an idealized one-dimensional bipolar transistor illustrating the components of current flow which contribute to noise.

Relating device-modeling parameters such as R_B and device performance metrics such as f_T to circuit-level noise parameters F_{min}, R_n, and Γ_{opt} is essential for designing and optimizing both the structure and layout of a HBT. One method for doing so is to represent each physical noise source as a component in a small-signal equivalent circuit model and to then use this model to determine each noise parameter [7]. Figure 7.2 illustrates a common-emitter equivalent circuit noise model that is simple yet works remarkably well for capturing the essential bias and frequency behaviors of the HBT noise parameters. Collector and base shot noise are represented by noise current sources, while base resistance thermal noise is represented by a noise voltage source. For completeness at high V_{CB}, the impact ionization noise contribution to I_C is represented by current source i_{nii}.

A noisy two-port, such as a HBT, can be represented in an equivalent manner by two noise sources, generally correlated, connected to an ideal, noiseless two-port [8,9]. Figure 7.3 illustrates two possible options for doing so. Representation 1 features noise voltage source v_{nA} and noise current source i_{nA} connected to port 1, while Representation 2 drives ports 1 and 2 with noise current sources i_{n1} and i_{n2}, respectively. Either representation creates a path for calculating the noise source correlation matrix elements, such as $\langle|v_{nA}|\rangle^2$, $\langle|i_{nA}|\rangle^2$, $\langle|v_{nA}\,i_{nA}^*|\rangle^2$, and $\langle|i_{nA}\,v_{nA}^*|\rangle^2$ for Representation 1, in terms of the four noise parameters. Casting these same matrix elements in terms of the equivalent noise circuit model Y-parameters and noise sources, in turn, creates the bridge between the noise and device-modeling parameters [5,7,10].

Circuit designers turn first to the minimum noise figure F_{min} in evaluating the noise performance of an RF technology. Following the above methodology and employing the equivalent noise circuit model of Figure 7.2, F_{min} can be expressed in terms of the device-modeling parameters as [5,7,10–12]

$$F_{min} \cong 1 + \frac{n}{\beta} + \sqrt{\frac{2I_C}{V_T}(R_B + R_E)\left(\frac{f^2}{f_T^2} + \frac{1}{\beta}\right) + \frac{n^2}{\beta}} \tag{7.1}$$

which, after subtracting 1 from both sides and assuming that $(R_B + R_E) \approx R_B$, $1/\beta \ll 1$ and E–B junction ideality factor $n \approx 1$, simplifies to

$$F_{min} - 1 \cong \sqrt{\frac{2I_C}{V_T}R_B\left(\frac{f^2}{f_T^2} + \frac{1}{\beta}\right)} \tag{7.2}$$

This expression reveals exactly which device-modeling parameters are most critical to achieving low minimum noise figure: R_B, f_T, and β. The relative importance of each parameter is a function of frequency and of bias, which enters into the expression both explicitly as I_C and implicitly in the bias dependencies of f_T and, to a lesser extent, R_B.

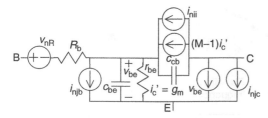

FIGURE 7.2 Small-signal equivalent circuit model for a bipolar transistor, including the dominant noise sources. Here, the total noise contributions of resistances R_B and R_E are lumped into R_B (see text for additional discussion).

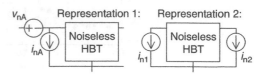

FIGURE 7.3 Two common representations of a noisy two-port as an ideal, noiseless two-port connected to two external noise sources.

Frequency sweeps are important for model assessment and evaluating a device against candidate applications as well as for providing a basis for interpolating or extrapolating to frequencies outside of the measured range. Equation 7.2 implies that the frequency dependence of F_{min} divides into two regimes, illustrated in the linear-scale (non-dB) graph of $(F_{min} - 1)$ versus frequency for a $(0.12 \times 4) \times 16\ \mu m^2$ emitter-area SiGe HBT (200 GHz peak f_T) plotted in Figure 7.4. For $f^2/f_T^2 \ll 1/\beta$ (equivalently $f^2 \ll f_T^2/\beta$), frequency and f_T drop out and F_{min} simplifies to

$$F_{min} - 1 \cong \sqrt{\frac{2I_C}{V_T\beta}R_B} \qquad (7.3)$$

In this low-frequency regime, F_{min} depends only on R_B, I_C, and β, with f_T not an explicit factor except in defining the boundaries of the regime. Low R_B and high β are key to achieving low F_{min} in this regime. The frequency dependence here is flat as long as I_C is held constant. However, analysis (see below) shows that obtaining strictly optimized F_{min} per frequency requires optimizing the bias for each frequency point, leading to F_{min} decreasing as the square root of frequency.

As frequency passes through the regime boundary, the behavior of F_{min} shifts to

$$F_{min} - 1 \cong f\sqrt{\frac{2}{V_T}R_B\left(\frac{I_C}{f_T^2}\right)} \qquad (7.4)$$

In this higher frequency regime ($f^2 \gg f_T^2/\beta$), F_{min} rises linearly with frequency, with a slope dependent on R_B and f_T and with β no longer an important concern. In contrast with the low-frequency regime, a value of I_C that minimizes F_{min} at one frequency also minimizes F_{min} over all frequencies, improving ease of test and increasing the utility of such a sweep in evaluating the device. Low R_B remains essential for achieving low F_{min} as at lower frequency but high f_T now becomes important as well.

When designing a low-noise circuit, a circuit designer will typically target an application-specific frequency and examine the bias dependences of the noise parameters in order to select an optimal bias. Although bias current I_C appears in Equation 7.2 directly, f_T contains an implicit I_C dependence as well, which may be expressed in reciprocal form as

$$\frac{1}{f_T} = 2\pi\left[\frac{V_T(C_{EB} + C_{CB})}{I_C} + \tau_F\right] \qquad (7.5)$$

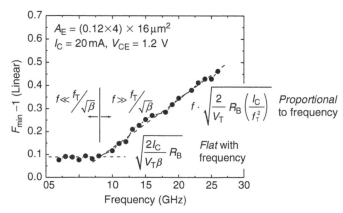

FIGURE 7.4 $F_{min} - 1$ (linear scale) versus frequency for a sample 0.13-μm node SiGe HBT, illustrating the characteristic behavior in the low-frequency ($f^2 \ll f_T^2/\beta$) and high-frequency ($f^2 \gg f_T^2/\beta$) regimes. (From D Greenberg. IEEE MTT-S International Microwave Symposium, Fort Worth, 2004. With permission.)

where input capacitance ($C_{EB} + C_{CB}$) sets the f_T versus I_C slope at low bias and τ_F determines the upper limit that would be approached by f_T at high I_C if there were no high-level injection or Kirk effect. This relationship may be substituted into Equation 7.2 to yield an expression containing fully explicit references to I_C, which may then be minimized with respect to I_C to yield a minimum value of F_{min} across all currents (an R_B versus I_C dependence may be inserted as well for those devices in which R_B varies significantly with bias over the current range of interest).

At low frequencies in which $f^2 \ll f_T^2/\beta$, Equation 7.3 suggests that F_{min} has no minimum and can be made arbitrarily low simply by reducing I_C. The equation contains a hidden dependency on f_T, however, in the form of the frequency regime boundary defining where the equation is valid. As a result, Equation 7.3 cannot be used directly to explore the F_{min} bias dependency at a given low frequency. Rather, this dependency must be determined from Equation 7.2 and Equation 7.5 minimization, with the final result simplified for the low-frequency regime.

In this manner, F_{min} is found to have a minimum at a bias of

$$\text{opt. } I_C = fV_T(C_{EB} + C_{CB})\sqrt{\beta} \tag{7.6}$$

This minimum depends on the capacitances looking into the base (which determine low-current f_T) and on β and is an increasing function of frequency. The actual minimum value of F_{min} at this optimal bias is given by

$$\text{opt. } F_{min} \cong \sqrt{f}\sqrt{\frac{8\pi^2 R_B(C_{EB} + C_{CB})}{\beta^{1/2}}} \tag{7.7}$$

Figure 7.5 illustrates this behavior using a graph of ($F_{min} - 1$) versus I_C for a $(0.2 \times 19.2) \times 2\,\mu m^2$ emitter-area device (120 GHz peak f_T) at 10 and 15 GHz. Good noise performance in this regime requires low R_B, a ubiquitous requirement, as well as high β and good low-current f_T optimization (low input capacitance).

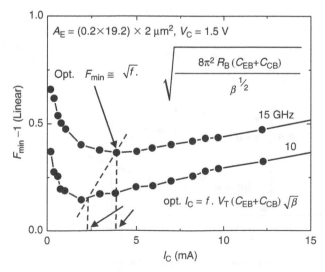

FIGURE 7.5 $F_{min} - 1$ (linear scale) versus I_C for a sample 0.18-μm node SiGe HBT illustrating the bias dependencies, lowest F_{min}, and noise-optimal I_C in the low frequency ($f^2 \ll f_T^2/\beta$) regime. (From D Greenberg. IEEE MTT-S International Microwave Symposium, Fort Worth, 2004. With permission.)

The bias dependence of F_{min} at higher frequencies differs significantly as noted earlier. While the optimal current for lowest F_{min} is frequency dependent at low frequencies, F_{min} is optimized by a single value of I_C for $f^2 \gg f_T^2/\beta$:

$$\text{opt. } I_C = \frac{V_T(C_{EB} + C_{CB})}{\tau_F} \tag{7.8}$$

This bias point is a function of both the input capacitances and the current-independent f_T transit time term. Indeed, this optimal I_C corresponds to a value of f_T equal to one half the limit approached by f_T asymptotically at high I_C neglecting high-level injection and the Kirk effect. Since peak f_T can approach 85% of this limit in an actual SiGe HBT optimized for f_T, the optimal noise bias can be well approximated by finding I_C corresponding to half of peak f_T. The F_{min} value at this I_C is an increasing function of frequency given by

$$\text{opt. } F_{min} \cong f \cdot 4\pi\sqrt{2R_B(C_{EB} + C_{CB})\tau_F} \tag{7.9}$$

Figure 7.6 illustrates this behavior using a graph of $(F_{min} - 1)$ versus I_C for a $(0.2 \times 19.2) \times 2\,\mu\text{m}^2$ emitter-area device (120 GHz peak f_T) at 15 and 20 GHz. Good noise performance in this regime requires low R_B and good optimization of both low- and high-current f_T (low input capacitance and short transit time τ_F). β does not affect noise in the higher frequency regime.

In each expression for F_{min}, a parameter that scales with emitter length L_E, such as I_C or $(C_{EB} + C_{CB})$, appears in product with R_B, a parameter that typically scales inversely with L_E in device layouts employing a base contact finger on each side of the emitter finger. As a result, F_{min} itself is generally insensitive to L_E.

The analysis used to connect F_{min} with device and modeling parameters also yields the remaining three noise parameters, which include the real and imaginary parts of the optimal source impedance as well as the noise resistance R_n. These parameters are key to designing an optimal source match.

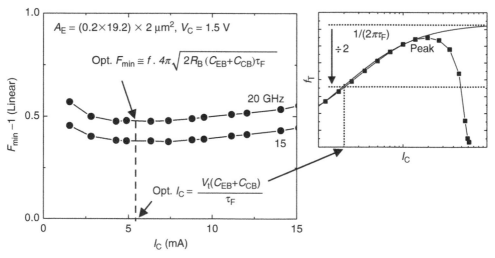

FIGURE 7.6 $F_{min} - 1$ (linear scale) versus I_C for a sample 0.18-μm node SiGe HBT illustrating the bias dependencies, lowest F_{min}, and noise-optimal I_C in the high-frequency ($f^2 \gg f_T^2/\beta$) regime. Inset connects noise-optimal I_C directly to the f_T versus I_C behavior (see text for details). (From D Greenberg. IEEE MTT-S International Microwave Symposium, Fort Worth, 2004. With permission.)

Under the same assumptions used to simplify the expressions for $F_{\min}$, R_n may be approximated as

$$R_n \cong V_T/2I_C + R_B \tag{7.10}$$

This is merely the sum of the small-signal impedance of the emitter–base diode (as seen from the emitter terminal and assuming $I_E \approx I_C$) and the base resistance. R_n is dominated by the former term at low currents and approaches R_B as bias increases. Low values of R_B, achieved through either process improvements or through layouts featuring long L_E, can result in quite low noise resistances and thus to devices that are very tolerant of mismatch and thus more readily able to achieve circuit noise figures close to $F_{\min}$ despite process variation.

Shifting the real or resistive portion of the optimal source impedance toward the desired system impedance (typically 50 Ω) is the most challenging aspect of achieving a good match using integrated passives with values of Q that are low compared to discrete off-chip equivalents. This impedance component may, under a wide range of conditions, be simplified to

$$R_{opt} \cong \frac{2R_n}{F_{\min} - 1} = \left(\frac{V_T}{2I_C} + R_B\right) \frac{2}{F_{\min} - 1} \tag{7.11}$$

While $F_{\min}$, and thus the second factor in this expression, does not change with emitter finger length L_E, both terms in the first factor scale inversely with this dimension, allowing the designer to target a value of R_{opt} as close to 50 Ω as is possible within limits set by additional sizing constraints such as available chip area and power budget (impacted by I_C).

7.3   Noise Performance Overview and Trends

Continuous progress in both vertical and lateral scaling as well as innovations in processing and structure have led to a steady increase in the peak f_T and f_{MAX} for the highest performance HBT at each SiGe BiCMOS lithography node [2,3,13–17,19–22]. These improvements in f_{MAX} have, in large part, stemmed from the steady reduction in normalized base resistance R_B illustrated in Figure 7.7 and

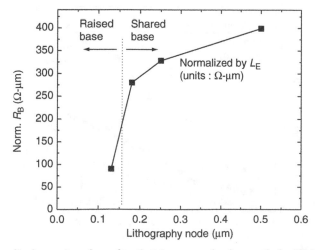

FIGURE 7.7 R_B (normalized to emitter finger length L_E) versus technology node for IBM SiGe HBTs, illustrating the improvement with generation from lateral scaling and structural innovation. (From D Greenberg, S Sweeney, B Jagannathan, G Freeman, and D Ahlgren. Proceedings of the Topical Meeting on Silicon Monolithic Integrated Circuits in RF Systems, Grainau, 2003. With permission.)

achieved through advancements such as the raised extrinsic base structure described elsewhere in this book [2,3]. The f_T and R_B trends have reaped parallel rewards in noise performance, driving F_{min} down and G_A up with each generation. The result is a selection of SiGe BiCMOS technologies available to the circuit designer at a variety of cost-performance points and enabling the feasible use of silicon technology to implement noise-sensitive applications at an increasingly broader range of frequencies.

The HBT performance at a given technology node is a function not only of the node, but of deliberate trade-offs designed to satisfy application requirements. A high-performance technology may not only a include a flagship HBT optimized for f_T and f_{MAX} but also additional variants trading some peak f_T for other properties such as higher breakdown voltage. Furthermore, an entire technology may be tuned for a market niche, such as high-volume consumer wireless applications for which low cost is paramount over maximum performance. Noise performance varies with choice of technology node and application optimization as well (as illustrated in Figure 7.8) [13]. The figure plots minimum F_{min} (optimized over I_C bias) versus frequency for a sampling of high-performance technologies at the 0.5-, 0.18-, and 0.13-μm nodes, together with a cost-reduced variant at the 0.18-μm node. A performance range selected from commercial data sheets for the current industry low-noise favorite, the GaAs PHEMT, is also indicated for reference.

With a peak f_T of 45 GHz and a normalized base resistance of 400 Ω μm, the 0.5-μm node maintains F_{min} values below 1 dB out to ~5 GHz and below 2 dB out to ~10 GHz. This level of performance is suitable for receiver front-ends for cost-sensitive, high-volume applications including GSM, CDMA, and 802.11b, as evidenced by the commercial availability of SiGe parts for these markets.

At the 0.18-μm node, vertical scaling increases the peak f_T of performance-optimized HBTs to 120 GHz while lateral scaling and process optimization reduce normalized R_B by 30%, to below 280 Ω μm. The scaling of both parameters has a direct impact on F_{min}, which decreases by more than 1 dB at 10 GHz and by more than 2 dB at 26 GHz [14]. This improvement means that F_{min} remains below 1 dB at ~12 GHz and below 2 dB at ~23 GHz. As a result, the 0.18-μm node can address higher application frequencies while providing for greater digital content integration due to improved digital logic performance and density.

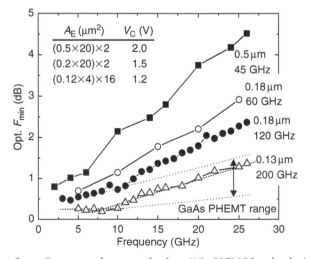

FIGURE 7.8 HBT noise figure F_{min} versus frequency for four SiGe BiCMOS technologies, including three high-performance (f_T emphasized) variants at the 0.5-, 0.18-, and 0.13-μm nodes as well as a cost-reduced (and slightly higher breakdown voltage) variant at the 0.18-μm node. (From D Greenberg. IEEE MTT-S International Microwave Symposium, Fort Worth, 2004. With permission.)

Although the 0.18-μm node HBT may be tuned for high f_T, cellular telephony, and wireless networking applications below 6 GHz may prioritize cost over high-frequency capability, such markets can be addressed through cost-reduced technology variants, which eliminate process steps and retune the collector doping for an application-optimized combination of peak f_T and breakdown voltage. Although these modifications reduce peak f_T by shifting the onset of the Kirk effect to lower I_C, values of f_T, and thus of F_{min}, at currents below the Kirk effect onset is impacted to a much lower extent. As a result, F_{min} for the cost-reduced process suffers only minor degradation (0.1 dB at 5 GHz and 0.5 dB at 26 GHz) compared with that of the high-performance process and still represents a significant advance over the 0.5-μm node.

The 0.13-μm node introduces a leap in noise performance, due not only to an increase in f_T to over 200 GHz but to a further 68% reduction in normalized R_B to 90 Ω μm as well [2,3]. This large R_B improvement results in part from lateral scaling but stems primarily from a shift to a new HBT design featuring a separate, raised polysilicon layer as the extrinsic base [2]. The use of a separate layer rather than a layer shared with the intrinsic base decouples R_B and C_{CB} and allows the introduction of much more extrinsic base doping without increasing C_{CB} and losing gain. As a result, F_{min} remains below 0.4 dB beyond 12 GHz and rises to only 1.3 dB at 26.5 GHz [16,17]. This level of performance falls within the range established using the data sheets for GaAs PHEMTs currently on the commercial market, placing silicon within one generation of this benchmark.

Guidelines drawn atop the 0.13 μm data in Figure 7.8 highlight the frequency dependencies expected from Equation 7.2 and illustrated in Figure 7.4. F_{min} is flat with frequency in the low-frequency regime and rises linearly with frequency once f^2 exceeds f_T^2/β. The flat region seems absent from the 0.5- and 0.18-μm nodes, but this is due simply to the lower f_T of these nodes, which positions the regime boundary below the measured frequency range.

The bias dependencies of both F_{min} and associated gain G_A are critical inputs to the design process. Figure 7.9 and Figure 7.10 explore the 0.18- and 0.13-μm nodes, plotting both F_{min} and G_A versus I_C in order to compare high-performance (hollow symbols) and cost-reduced (solid symbols) variants from this important perspective [14–17]. As described analytically in Equation 7.6 and Equation 7.8, F_{min} is indeed minimized at a particular bias for any given frequency, with this bias a function of frequency at low frequencies and independent of frequency at higher frequencies. Even at the highest plotted frequency of 25 GHz, this optimal collector bias for lowest F_{min} in the 0.18-μm node occurs at only 5 mA, which is a mere 8% of the peak f_T current for the illustrated device. This result emphasizes that noise performance is tied to low-current f_T and not to peak values.

Although F_{min} is minimized at a unique current, associated gain rises monotonically with current. This behavior requires that the designer selects an optimal balance between achieving lowest F_{min} and higher G_A based on the requirements of the application.

Reducing cost in the 0.18-μm node increases F_{min} somewhat, with this degradation an increasing function of frequency. Since low-current f_T is similar between the process variants, most of this difference is due to R_B, which benefits from a more complex, self-aligned structure in the high-performance variant. Despite the modestly increased F_{min} in the cost-reduced variant, however, G_A is actually improved as the result of the reduced collector–base capacitance C_{CB} that arises from reducing the collector doping to achieve higher breakdown voltage.

Measuring noise at frequencies beyond 26.5 GHz is challenging and the means to do so may not yet be available on many test benches. The excellent correlation in the HBT between data and analytical prediction for F_{min} versus frequency provides a means for accurate extrapolation of the data to frequencies beyond the testing boundaries, however, and thus a means for helping the designer evaluate the suitability of new technologies for emerging applications. As an example, Figure 7.11 plots $(F_{min} - 1)$ versus frequency on a linear scale for a 0.13-μm HBT, with the axis taken out to 80 GHz and with data plotted out to 26 GHz. F_{min} rises linearity with frequency above 8 GHz and is well-fit by a line using the least-squares method. The fit line may then be extended with reasonable accuracy for at least 1 to 2 octaves beyond the measured data and the resulting F_{min} values expressed in dB form for more familiar reference. This method predicts F_{min} values for the 0.13-μm HBT of 3.1 and 3.8 dB at 60 and 77 GHz,

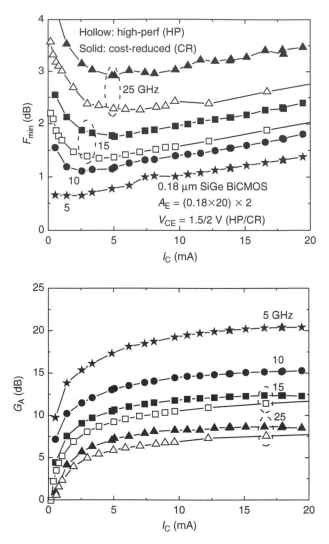

FIGURE 7.9 HBT noise figure F_{min} and associated gain G_A versus I_C at 5, 10, 15, and 26 GHz for high-performance and cost-reduced variants of 0.18-μm IBM SiGe BiCMOS technology ($A_E = 0.18 \times 20 \times 2 \, \mu m^2$). (From DR Greenberg, D Ahlgren, G Freeman, S Subbanna, V Radisic, DS Harvey, C Webster, and L Larson. Digest IEEE MTT-S International Microwave Symposium, Boston, 2000, pp. 9–12; D Greenberg, S Sweeney, C LaMothe, K Jenkins, D Friedman, B Martin Jr, G Freeman, D Ahlgren, S Subbanna, and A Joseph. Technical Digest IEEE International Electron Devices Meeting, Washington, DC, 2001, pp. 495–498. With permission.)

respectively [16,17]. Such levels of performance are suitable for applications such as 60 GHz wireless LAN and 77 GHz automotive radar, opening a high-frequency regime to silicon that was once the exclusive domain of III–V devices. Recent circuit results verifying these noise figure values appear later in the chapter [40].

It is significantly more difficult to achieve noise figures close to F_{min} after integrating a transistor into a circuit such as an LNA than when measuring the discrete device on the test bench. Part of this difficulty stems from the need for high Q passives to achieve the required source impedance match, especially when this match differs significantly from 50 Ω and resides near the outer edges of the Smith chart.

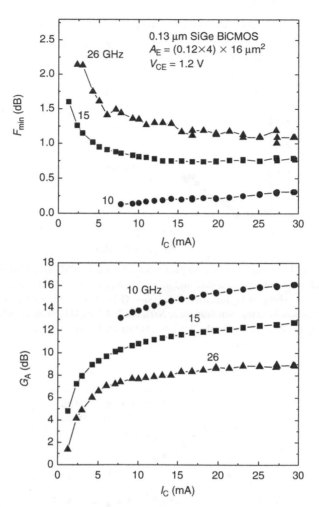

FIGURE 7.10 Noise figure F_{min} and associated gain G_A versus I_C at 10, 15, and 26 GHz for a 0.13-µm IBM SiGe HBT (noise-optimized layout with $A_E = 0.12 \times 4 \times 16\,\mu m^2$). (From DR Greenberg, B Jagannathan, S Sweeney, G Freeman, and D Ahlgren. Technical Digest IEEE International Electron Devices Meeting, San Francisco, 2002, pp. 787–790; D Greenberg, S Sweeney, B Jagannathan, G Freeman, and D Ahlgren. Proceedings of the Topical Meeting on Silicon Monolithic Integrated Circuits in RF Systems, Grainau, 2003. With permission.)

The challenge is compounded by process variation, which introduces a random degree of mismatch into each sample. The impact of this design challenge is observed most readily when designing with FETs, which suffer from relatively high noise resistance and from largely capacitive optimum source impedances with very high real components. The HBT, however, fares much better in this regard, leading to LNA circuits that approach the measured device F_{min} more closely. Figure 7.12 and Figure 7.13 illustrate the key matching parameters for a $(0.2 \times 4) \times 16\,\mu m^2$ HBT, plotting R_n versus I_C and showing Γ_{opt} in Smith chart form at 10, 15, 20, and 26 GHz [16,17]. The very low R_B at this technology node leads to correspondingly low R_n, an order of magnitude lower than observed in FETs at the same node. At the same time, Γ_{opt} remains comfortably away from the Smith chart edge, with the real portion of the optimum source impedance remaining within a factor of 5 of 50 Ω at the chosen device size. In practice, the L_E scaling required to design a value of R_{opt} close to 50 Ω at a target application frequency is typically well within feasible limits for the SiGe HBT.

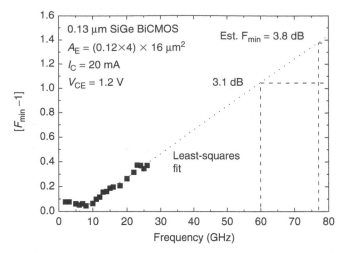

FIGURE 7.11 $F_{min} - 1$ (linear scale) versus frequency for a 0.13-μm IBM SiGe HBT illustrating a means for accurate extrapolation beyond the measured data and indicating F_{min} estimates of 3.1 and 3.8 dB at 60 and 77 GHz, respectively. (From DR Greenberg, B Jagannathan, S Sweeney, G Freeman, and D Ahlgren. Technical Digest IEEE International Electron Devices Meeting, San Francisco, 2002, pp. 787–790; D Greenberg, S Sweeney, B Jagannathan, G Freeman, and D Ahlgren. Proceedings of the Topical Meeting on Silicon Monolithic Integrated Circuits in RF Systems, Grainau, 2003. With permission.)

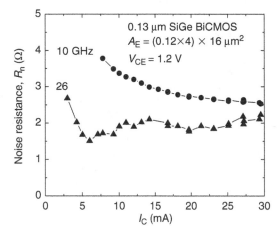

FIGURE 7.12 Noise resistance R_n versus I_C at 10 and 26 GHz for a 0.13-μm IBM SiGe HBT (noise-optimized layout with $A_E = 0.12 \times 4 \times 16 \, \mu m^2$). (From DR Greenberg, B Jagannathan, S Sweeney, G Freeman, and D Ahlgren. Technical Digest IEEE International Electron Devices Meeting, San Francisco, 2002, pp. 787–790; D Greenberg, S Sweeney, B Jagannathan, G Freeman, and D Ahlgren. Proceedings of the Topical Meeting on Silicon Monolithic Integrated Circuits in RF Systems, Grainau, 2003. With permission.)

7.4 Device Optimization for Low Noise

SiGe HBT technologies, designed to suit the widest possible range of applications, are already achieving outstanding levels of noise performance suitable for use to beyond 60 GHz. Still, the relationships between the noise and device parameters explored in Section 7.2 suggest the means for optimizing noise performance even further.

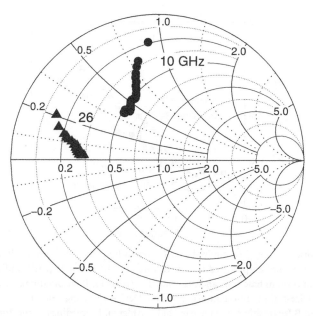

FIGURE 7.13 Γ_{opt} bias traces on a Smith chart at 10 and 26 GHz for a 0.13-μm IBM SiGe HBT (noise-optimized layout with $A_E = 0.12 \times 4 \times 16\,\mu m^2$). Position away from chart edge suggests relatively easy matching to 50 Ω. (From DR Greenberg, B Jagannathan, S Sweeney, G Freeman, and D Ahlgren. Technical Digest IEEE International Electron Devices Meeting, San Francisco, 2002, pp. 787–790; D Greenberg, S Sweeney, B Jagannathan, G Freeman, and D Ahlgren. Proceedings of the Topical Meeting on Silicon Monolithic Integrated Circuits in RF Systems, Grainau, 2003. With permission.)

The use conditions for a modern HBT typically fall within the domain of Equation 7.9, which relates bias-optimized F_{min} to the product of three parameters under the direct control of the device designer. At any given frequency, this F_{min} can be improved by reducing R_B, $(C_{EB} + C_{CB})$ or τ_F, with each playing an equal role.

The impact of R_B on F_{min} may be isolated empirically by varying the fabrication process so as to alter R_B while keeping all other design parameters fixed. Figure 7.14 plots F_{min} versus frequency for two variants of a 0.13-μm process employing a raised extrinsic base to decouple the trade-off between R_B and C_{CB}. Compared with baseline process no. 1, process no. 2 introduces significantly more extrinsic base doping and reduces R_B by 40% from 140 to 85 Ω μm [2,3]. This lone process improvement drops F_{min} by an average of 0.3 to 0.4 dB at frequencies above 8 GHz (below which F_{min} becomes too low for clean measurement) [16,17].

F_{min} may also be improved through the independent approach of reducing the several components of transit time, which limit f_T at the optimal I_C for lowest noise. Since F_{min} occurs well below peak f_T in a HBT, design choices aimed at peak f_T may be traded for improved low-current f_T instead. The SiGe composition profile in the base is one such design-tuning knob. In a graded-base HBT design optimized for operation at the higher currents required to reach peak f_T, significant SiGe mole fraction is maintained not only through the neutral base, but into the base–collector space–charge region as well, where the composition transitions back to silicon. This positions the barrier from the steep profile retrograde within the space–charge region where it causes minimal carrier pileup and thus minimal performance impact at high I_C. Although the total SiGe budget is limited by the maximum strain that can be tolerated in a base of given thickness, a noise-optimized HBT design, freed of the need to operate at high I_C, can partition this budget differently and shift SiGe mole fraction from the collector side of the base toward the emitter [18]. As illustrated in the Figure 7.15 (inset), this creates a narrower profile, positioning the retrograde barrier closer to the neutral base but allowing for either a larger emitter–base

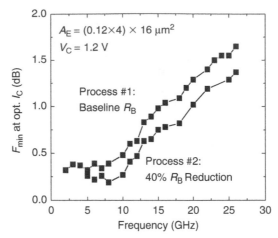

FIGURE 7.14 HBT noise figure F_{min} (optimized over I_C) versus frequency for both baseline and low-R_B variants of a 0.13-μm IBM SiGe device (noise-optimized layout with $A_E = 0.12 \times 4 \times 16 \, \mu m^2$), indicating the improvement possible from a 40% reduction in base resistance. (From DR Greenberg, B Jagannathan, S Sweeney, G Freeman, and D Ahlgren. Technical Digest IEEE International Electron Devices Meeting, San Francisco, 2002, pp. 787–790; D Greenberg, S Sweeney, B Jagannathan, G Freeman, and D Ahlgren. Proceedings of the Topical Meeting on Silicon Monolithic Integrated Circuits in RF Systems, Grainau, 2003. With permission.)

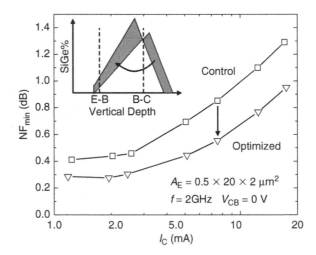

FIGURE 7.15 F_{min} versus I_C at 2 GHz comparing a graded-base SiGe HBT ($A_E = 0.5 \times 20 \times 2 \, \mu m^2$) with control SiGe base profile with a noise-optimized device in which the SiGe base profile has been modified as per the inset. (From G Niu, S Zhang, JD Cressler, AJ Joseph, JS Fairbanks, LE Larson, CS Webster, WE Ansley, and DL Harame. *IEEE Transactions on Electron Devices* 47:2037–2044, 2000. With permission.)

heterojunction or, in the case of a graded-profile design, for a steeper SiGe ramp. A larger emitter–base heterojunction can be leveraged to increase β (reducing F_{min} in the low-frequency regime) or to increase intrinsic base doping and thus to decrease R_B (reducing F_{min} at all frequencies). A steeper SiGe ramp creates a strong quasielectric field for electrons in the base, increasing their speed and decreasing τ_F. Low-noise profile optimization has indeed been demonstrated. Figure 7.15 compares F_{min} versus I_C at 2 GHz for a standard high-f_T (45 GHz) 0.5-μm device with a variant containing a noise-optimized graded SiGe base profile. The optimized profile reduces F_{min} by as much as 0.3 dB.

The data analysis presented to this point has ignored the effects of impact ionization, a source of noise included in the equivalent circuit model of Figure 7.2. This simplification is valid as long as the

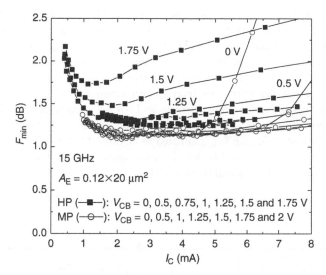

FIGURE 7.16 F_{min} versus I_C at 15 GHz for two 0.13-μm IBM SiGe HBT devices comparing a high-f_T baseline device and a modified-collector variant with reduced impact ionization noise. (From DR Greenberg, S Sweeney, G Freeman, and D Ahlgren. Digest IEEE MTT-S International Microwave Symposium, Philadelphia, 2003, pp. 113–116. With permission.)

collector–base and collector–emitter voltages remain well below the respective two-terminal breakdown voltages BV$_{CBO}$ and BV$_{CEO}$. As breakdown is approached, however, impact ionization can begin to contribute significantly to the noise figure. One goal in optimizing a HBT design for low noise is to suppress this phenomenon and its impact on noise. Reducing the collector doping is one approach to achieving this goal. As discussed in Section 7.3, lowering the collector doping decreases peak f_T by shifting the onset of the Kirk effect toward lower I_C yet has little impact on f_T or F_{min} prior to this onset. The f_T for a device optimized for the least impact ionization noise should therefore peak at a value of I_C at or just beyond the bias for best F_{min}. Such a design minimizes the collector doping and maximizes the breakdown voltage [6]. Figure 7.16 plots F_{min} versus I_C for both a high-f_T (200 GHz) HBT (with a BV$_{CEO}$ of 2 V) and a medium-performance (collector-doping-reduced) variant in the 0.13-μm node, comparing the noise performance as V_{CB} is stepped from 0 to 2 V (i.e., V_{CE} from $\sim$0.8 to $\sim$2.8 V).

A reduced collector doping does create a large V_{CB} dependence to the Kirk effect, requiring a V_{CB} greater than 0.5 V ($V_{CE} >1.3$ V) to position peak f_T current beyond the optimum F_{min} bias point. At larger V_{CB}, however, F_{min} for the medium-performance device maintains its value with further V_{CB} increases while F_{min} in the high-f_T device degrades significantly from impact ionization, particularly at higher I_C. At a V_{CB} value of 1 V, V_{CE} is approximately 1.8 V and thus below BV$_{CEO}$. Yet, noise in the high-f_T HBT has already degraded measurably. By the time V_{CB} reaches 1.75 V, impact ionization in the high-f_T HBT is severe and minimum F_{min} has increased by over 0.75 dB. The modified device shows no such excess noise contribution at these voltages, however, and actually enjoys improved associated gain from the reduction in C_{CB} resulting from the lower collector doping. Thus, the high-breakdown HBT variant offered in the device libraries of many SiGe technologies, while not providing the highest peak f_T, can actually be the preferred choice for low-noise operation at typical supply voltages.

7.5 Circuit Performance

Device performance on the test bench does not always translate into performance in an integrated circuit. For example, LNA performance can be limited by the performance of available inductors which, due to their finite Q, may not always be able to achieve a noise-optimal match and which contribute noise of their own. The matching characteristics of the SiGe HBT, however, help ease good LNA design compared with those of FETs. HBT LNA noise figures, while still higher than those of the discrete device, can nevertheless come quite close. Figure 7.17 surveys both noise figure and gain for published or commercially available LNAs built in a variety of technology nodes ranging from 0.5 to 0.13 μm [22–39]. The results at any given node vary greatly due to differences in design priorities and the learning curve associated with design in newly available technologies. However, the bounding limits approach discrete device test bench values and demonstrate the levels of performance possible in silicon over a wide range of frequencies. At the high end of the frequency scale, an LNA noise figure of 3.8 dB has been achieved at 60 GHz, with a corresponding gain of 14.5 dB (including the impact of the bondpads) [40]. This noise figure falls within 0.7 dB of the device F_{min} estimated in Figure 7.11. A more complete characterization of this result is shown in Figure 7.18, which plots the noise figure and gain as a function of frequency and presents a microphotograph of the actual measured chip.

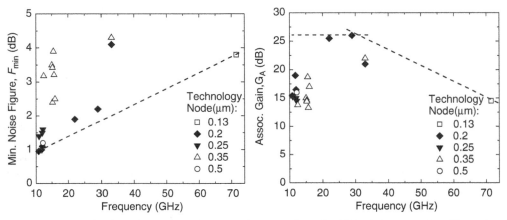

FIGURE 7.17 Noise figure and gain versus frequency for a sampling of published SiGe LNA data, illustrating the attained performance ranges for the 0.5- to 0.13-μm technology nodes. (From Refs. [22–39]. With permission.)

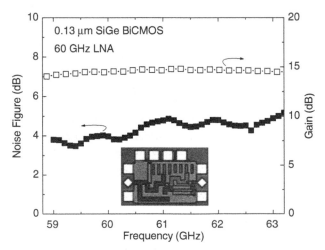

FIGURE 7.18 Noise figure and gain versus frequency for a fully integrated 60 GHz LNA (complete chip, including pad parasitics) built in IBM's 0.13-μm SiGe technology and demonstrating a sub-4 dB noise figure consistent with discrete device data extrapolation. (From S Reynolds, B Floyd, U Pfeiffer, and T Zwick. Proceedings of the IEEE International Solid-State Circuits Conference, San Francisco, 2004, pp. 442–443. With permission.)

7.6 Summary

Each generation of SiGe HBT technology has leveraged scaling, process, and layout innovations to improve low-noise performance. The key contributors to these advances have been the increase in f_T and decrease in R_B at low to moderate current levels, factors which dovetail naturally with the drive to increase f_{MAX}. As a result, the 0.5-μm generation has been able to penetrate the wireless telephony and networking markets in the 0.8 to 5.8 GHz bands while the latest 0.13-μm generation has demonstrated sub-4 dB performance in actual circuits in the 60 to 77 GHz regime targeted by emerging multimedia and automotive applications. Designers working on noise-sensitive systems have a choice of technology nodes with which to balance cost against performance and integration density. At the same time, the nature of the HBT as a vertical, bulk transistor has kept the device amenable to the detailed analysis and accurate modeling that has helped speed time to market through a strong record of first-pass design success. A variety of ideas have been proposed for optimizing noise still further, many of which will undoubtedly find their way into the technology as SiGe spreads out into the many untapped market niches.

Acknowledgments

This chapter would not have been possible without the outstanding and invaluable contributions toward both test and hardware development by S. Sweeney, S. Parker, G. Freeman, D. Ahlgren, A. Joseph, S. St. Onge, P. Cottrell, J. Dunn, D. Harame, and each of the world-class members of IBM's SiGe development and enablement teams.

References

1. CE Shannon. A mathematical theory of communication. *Bell System Technical Journal* 27:379–423, 623–656, 1948.
2. B Jagannathan, M Khater, F Pagette, J-S Rieh, D Angell, H Chen, J Florkey, F Golan, DR Greenberg, R Groves, S-J Jeng, J Johnson, E Mengistu, KT Schonenberg, CM Schnabel, P Smith, A Stricker, D Ahlgren, G Freeman, K Stein, and S Subbanna. Self-aligned SiGe NPN transistors with 285 GHz f_{MAX} and 207 GHz f_T in a manufacturable technology. *IEEE Electron Device Letters* 23:258–260, 2002.
3. A Joseph, D Coolbaugh, D Harame, G Freeman, S Subbanna, M Doherty, J Dunn, C Dickey, D Greenberg, R Groves, M Meghelli, A Rylyakov, M Sorna, O Schreiber, D Herman Jr, and T Tanji. 0.13 μm 210 GHz f_T SiGe HBTS—expanding the horizons of SiGe BiCMOS. Technical Digest IEEE International Solid-State Circuits Conference, San Francisco, 2002, pp. 180–181.
4. W Shockley, JA Copeland, and RP James. *Quantum Theory of Atoms, Molecules and the Solid State.* New York, NY: Academic Press, 1966, pp. 537–563.
5. G Niu, JD Cressler, S Zhang, WE Ansley, CS Webster, and DL Harame. A unified approach to RF and microwave noise parameter modeling in bipolar transistors. *IEEE Transactions on Electron Devices* 48:2568–2574, 2001.
6. DR Greenberg, S Sweeney, G Freeman, and D Ahlgren. Low-noise performance near BV_{CEO} in a 200 GHz SiGe technology at different collector design points. Digest IEEE MTT-S International Microwave Symposium, Philadelphia, 2003, pp. 113–116.
7. SP Voinigescu, MC Maliepaard, JL Showell, GE Babcock, D Marchesan, M Schroter, P Schvan, and DL Harame. A scalable high-frequency noise model for bipolar transistors with application to optimal transistor sizing for low-noise amplifier design. *IEEE Journal of Solid-State Circuits* 32:1430–1439, 1997.
8. H Rothe and W Dahlke. Theory of noisy fourpoles. Proceedings of the Institute of Radio Engineers 44:811–818, 1956.
9. HA Haus, WR Atkinson, WH Fonger, WW Mcleod, GM Branch, WA Harris, EK Stodola, WB Davenport Jr, SW Harrison, and TE Talpey. Representation of noise in linear twoports. *Proceedings of the Institute of Radio Engineers* 48:69–74, 1960.

10. O Shana'a, I Linscott, and L Tyler. Frequency-scalable SiGe bipolar RF front-end design. *IEEE Journal of Solid-State Circuits* 36:888–895, 2001.

11. G Niu, WE Ansley, S Zhang, JD Cressler, CS Webster, and RA Groves. Noise parameter optimization of UHV/CVD SiGe HBT's for RF and microwave applications. *IEEE Transactions on Electron Devices* 46:1589–1598, 1999.

12. S Zhang, G Niu, JD Cressler, AJ Joseph, G Freeman, and DL Harame. The effects of geometric scaling on the frequency response and noise performance of SiGe HBTs. *IEEE Transactions on Electron Devices* 49:429–435, 2002.

13. D Greenberg. SiGe HBT Noise Performance. Workshop: High frequency noise in advanced silicon-based devices. IEEE MTT-S International Microwave Symposium, Fort Worth, 2004.

14. DR Greenberg, D Ahlgren, G Freeman, S Subbanna, V Radisic, DS Harvey, C Webster, and L Larson. HBT low noise performance in a 0.18 μm SiGe BiCMOS technology. Digest IEEE MTT-S International Microwave Symposium, Boston, 2000, pp. 9–12.

15. D Greenberg, S Sweeney, C LaMothe, K Jenkins, D Friedman, B Martin Jr, G Freeman, D Ahlgren, S Subbanna, and A Joseph. Noise performance and considerations for integrated RF/analog/mixed-signal design in a high-performance SiGe BiCMOS technology. Technical Digest IEEE International Electron Devices Meeting, Washington, DC, 2001, pp. 495–498.

16. DR Greenberg, B Jagannathan, S Sweeney, G Freeman, and D Ahlgren. Noise performance of a low base resistance 200 GHz SiGe technology. Technical Digest IEEE International Electron Devices Meeting, San Francisco, 2002, pp. 787–790.

17. D Greenberg, S Sweeney, B Jagannathan, G Freeman, and D Ahlgren. Noise performance scaling in high-speed silicon RF technologies. Proceedings of the Topical Meeting on Silicon Monolithic Integrated Circuits in RF Systems, Grainau, 2003.

18. G Niu, S Zhang, JD Cressler, AJ Joseph, JS Fairbanks, LE Larson, CS Webster, WE Ansley, and DL Harame. Noise modeling and SiGe profile design trade-offs for RF applications. *IEEE Transactions on Electron Devices* 47:2037–2044, 2000.

19. B Mheen, D Suh, HS Kim, S-Y Lee, CW Park, SH Kim, K-H Shim, and J-Y Kang. RF performance trade-offs of SiGe HBT fabricated by reduced pressure chemical vapor deposition. Digest IEEE MTT-S International Microwave Symposium, Seattle, 2002, pp. 413–416.

20. FS Johnson, J Ai, S Dunn, B El Kareh, J Erdeljac, S John, K Benaissa, A Bellaour, B Benna, L Hodgson, G Hoffleisch, L Hutter, M Jaumann, R Jumpertz, M Mercer, M Nair, J Seitchik, C Shen, M Schiekofer, T Scharnagl, K Schimpf, U Schulz, B Staufer, L Stroth, D Tatman, M Thompson, B Williams, and K Violette. A highly manufacturable 0.25 mm RF technology utilizing a unique SiGe integration. Proceedings of the IEEE Bipolar/BiCMOS Circuits and Technology Meeting, Minneapolis, 2001, pp. 56–59.

21. JG Tartarin, R Plana, M Borgarino, H Lafontaine, M Regis, O Llopis, and S Kovacic. Noise properties in SiGe BiCMOS devices. IEEE International Symposium on High Performance Electron Devices for Microwave and Optoelectronic Applications, Piscataway, 1999, pp. 131–136.

22. J Bock, TF Meister, H Knapp, D Zoschg, H Schafer, K Aufinger, M Wurzer, S Boguth, M Franosch, R Stengl, R Schreiter, M Rest, and L Treitinger. SiGe bipolar technology for mixed digital and analogue RF applications. Technical Digest IEEE International Electron Devices Meeting, San Francisco, 2000, pp. 745–748.

23. G Schuppener, T Harada, and Y Li. A 23 GHz low-noise amplifier in SiGe heterojunction bipolar technology. Digest IEEE Radio Frequency Integrated Circuits Symposium, Phoenix, 2001, pp. 177–180.

24. T Nakatani, J Itoh, I Imanishi, and O Ishikawa. A wide dynamic range switched-LNA in SiGe BiCMOS. Digest IEEE MTT-S International Microwave Symposium, Phoenix, 2001, pp. 281–284.

25. M Kamat, P Ye, Y He, B Agarwal, P Good, S Lloyd, and A Loke. High performance low current CDMA receiver front end using 0.18 μm SiGe BiCMOS. Digest IEEE Radio Frequency Integrated Circuits Symposium, Philadelphia, 2003, pp. 23–26.

26. A Schmidt and S Catala. A universal dual band LNA implementation in SiGe technology for wireless applications. *IEEE Journal of Solid-State Circuits* 36:1127–1131, 2001.

27. J Lee, G Lee, G Niu, JD Cressler, JH Kim, JC Lee, B Lee, and NY Kim. The design of SiGe HBT LNA for IMT-2000 mobile application. Digest IEEE MTT-S International Microwave Symposium, Seattle, 2002, pp. 1261–1264.

28. M Racanelli, P Ma, and P Kempf. SiGe BiCMOS technology for highly integrated wireless transceivers. Digest IEEE GaAs Integrated Circuit Symposium, San Diego, 2003, pp. 183–186.

29. DYC Lie, J Kennedy, D Livezey, B Yang, T Robinson, N Sornin, T Beukema, LE Larson, A Senior, C Saint, J Blonski, N Swanberg, P Pawlowski, D Gonya, X Yuan, and H Zamat. A direct conversion W-CDMA front-end SiGe receiver chip. Digest IEEE Radio Frequency Integrated Circuits Symposium, Seattle, 2002, pp. 31–34.

30. G Zhang, A Dengi, and LR Carley. Automatic synthesis of a 2.1 GHz SiGe low noise amplifier. IEEE Radio Frequency Integrated Circuits Symposium, Seattle, 2002, pp. 125–128.

31. P-W Lee, H-W Chiu, T-L Hsieh, C-H Shen, G-W Huang, and S-S Lu. A SiGe low noise amplifier for 2.4/5.2/5.7GHz WLAN applications. Technical Digest IEEE International Solid-State Circuits Conference, San Francisco, 2003, pp. 349, 364–365.

32. P Chevalier, F De Pestel, H Ziad, M Fatkhoutdinov, J Ackaert, P Coppens, J Craninckx, S Guncer, A Pontioglu, P Tasci, F Yayil, BS Ergun, AI Kurhan, E Vestiel, E De Backer, J-L Loheac, and M Tack. An industrial 0.35 μm SiGe BiCMOS technology for 5 GHz WLAN featuring an improved selective epitaxial growth process. International Symposium on Electron Devices for Microwave and Optoelectronic Applications, Vienna, 2001, pp. 175–180.

33. B Foley, P Murphy, and A Murphy. A monolithic SiGe 5 GHz low noise amplifier and tuneable image-reject filter for wireless LAN applications. IEEE High Frequency Postgraduate Colloquium, Dublin, 2000, pp. 26–31.

34. E Imbs, I Telliez, S Detout, and Y Imbs. A low-cost-packaged 4.9–6 GHz LNA for WLAN applications. Digest IEEE MTT-S International Microwave Symposium, Philadelphia, 2003, pp. 1569–1572.

35. B Banerjee, B Matinpour, C-H Lee, S Venkataraman, S Chakraborty, and J Laskar. Development of IEEE802.11a WLAN LNA in silicon-based processes. Digest IEEE MTT-S International Microwave Symposium, Philadelphia, 2003, pp. 1573–1576.

36. J Sadowy, I Telliez, J Graffeuil, E Tournier, L Escotte, and R Plana. Low noise, high linearity, wide bandwidth amplifier using a 35 μm SiGe BiCMOS for WLAN applications. Digest IEEE Radio Frequency Integrated Circuits Symposium, Seattle, 2002, pp. 217–220.

37. S Chakraborty, SK Reynolds, H Ainspan, and J Laskar. Development of 5.8 GHz SiGe BiCMOS direct conversion receivers. Digest IEEE MTT-S International Microwave Symposium, Philadelphia, 2003, pp. 1551–1554.

38. J Bock, H Schafer, H Knapp, D Zoschg, K Aufinger, M Wurzer, S Boguth, M Rest, R Schreiter, R Stengl, and TF Meister. Sub 5 ps SiGe bipolar technology. Technical Digest IEEE International Electron Devices Meeting, San Francisco, 2002, pp. 763–766.

39. Y Li, M Bao, M Ferndahl, and A Cathelin. 23 GHz front-end circuits in SiGe BiCMOS technology. Digest IEEE Radio Frequency Integrated Circuits Symposium, Philadelphia, 2003, pp. 99–102.

40. S Reynolds, B Floyd, U Pfeiffer, and T Zwick. 60 GHz receiver circuits in SiGe bipolar technology. Proceedings of the IEEE International Solid-State Circuits Conference, San Francisco, 2004, pp. 442–443.

8

Microscopic Noise Simulation

Guofu Niu
Auburn University

8.1 Introduction

The major noise sources in a bipolar transistor are the base resistance *thermal* noise, or Johnson noise, the base current *shot* noise, and the collector current *shot* noise. The base resistance thermal noise is typically described by a noise voltage with a spectral density of $4kTR$, and the shot noise is described by a spectral density of $2qI$, with I as the DC base current or collector current. These descriptions are based on *macroscopic* views. The standard derivation of the magic $2qI$ shot noise assumes a Poisson stream of an elementary charge q. These charges need to overcome a potential barrier, and thus flow in a completely uncorrelated manner. In a bipolar transistor, the base current shot noise $2qI_B$ results from the flow of base *majority* holes across the EB junction potential barrier. The reason that I_B appears in the base shot noise is that the amount of hole current overcoming the EB barrier is determined by the *minority* hole current in the emitter, I_B. Similarly, the collector current shot noise results from the flow of emitter *majority* electrons over the EB junction potential barrier, and has a spectral density of $2qI_C$.

Surprisingly, however, both the $4kTR$ resistor noise and the $2qI$ shot noise can be attributed to the same physical origin at the *microscopic* level—the Brownian motion of electrons and holes, also referred to as *diffusion* noise as the same mechanism is responsible for diffusion. The thermal motion of carriers gives rise to fluctuations of carrier velocities, and hence fluctuations of current densities. The density of such current density fluctuation is $4q^2 n D_n$ according to microscopic treatment of carrier motion [1,2]. The current density fluctuation at each location propagates toward device terminals, giving rise to device terminal voltage or current noise. The problem of noise analysis is now equivalent to solving the transfer functions of noise propagation at each location and summing over the whole device space. These transfer functions can be solved analytically for ideal transistor operation with simplified boundary conditions, or numerically for arbitrary device structures, and the latter process is referred to as *microscopic noise simulation.*

Various mathematical methods have been developed, all based on the impedance field method developed by Schockley and his colleagues [1] and its various generalizations. A very satisfying early application is the successful derivation of the $4kTR$ Johnson noise, a macroscopic model result, from the microscopic $4q^2nD_n$ *noise source density*. The impedance field approach is equivalent to the Green's function based approaches [3], which can be rigorously derived from the general master equation. Efficient numerical algorithms have been developed, which enabled the recent implementation of noise analysis in commercial device simulators, e.g., DESSIS from ISE and Taurus from TMA.

8.2 Noise Source Density and Noise Propagation

The mathematical development of various microscopic noise simulation methods has been well treated in Refs. [4,5]. We will focus on the aspects of using noise simulations instead. The key to successful application of microscopic noise simulation is to understand the following three concepts:

1. Noise source density, which is a measure of the local current density noise due to velocity fluctuation. This is a scalar, like the electron density, and is a function of position.
2. Vector noise transfer function, or vector Green's function, which is the gradient of the scalar Green's function. This is a *vector* like the electric field.
3. Scalar Green's function, or scalar noise transfer function, is a ratio of the transistor terminal noise produced for a unity noise current injection into a location.

All the three quantities are functions of location, like electron density n, electric field E, and potential ψ. We now derive the collector shot noise using "microscopic" theory as an example to illustrate the above concepts [6]. Like any other analytical theories, simplifying assumptions are inevitable to arrive at manageable solutions in our derivation. Many such assumptions can be removed using numerical noise simulation. We will consider one-dimensional, and neglect carrier recombination, which will necessitate including frequency in the small signal diffusion equation, in order to minimize complexity.

Noise Source Density

Consider a one-dimensional semiconductor bar as shown in Figure 8.1. For purpose of analysis, we divide the bar into small sections. The current density is related to carrier concentration and velocity as

$$J_n = -qnv, \tag{8.1}$$

where n is electron concentration and v is the net velocity. Because of velocity fluctuation, there is random charge transport within an incremental section. This can be equivalently described by

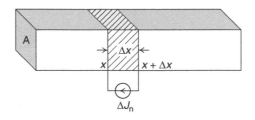

FIGURE 8.1 Equivalent circuit representation of the current density noise due to velocity fluctuations within an incremental section from x to $x + \Delta x$.

placing a noise current source across the incremental section. A fundamental result of velocity fluctuation is that the spectrum density of the ΔJ_n fluctuation for an incremental section between x and $x + \Delta x$ is given by

$$S_{\Delta J_n} = K_n \frac{1}{A\Delta x}. \tag{8.2}$$

The current density fluctuation increases with decreasing $A\Delta x$, the volume of the incremental section. Here A is the cross-sectional area of the one-dimensional bar, Δx is the length of the incremental section in question, and K_n is the electron *noise source density* [1,2]:

$$K_n = 4q^2 n D_n, \tag{8.3}$$

where n is electron concentration and D_n is electron diffusivity. Equation 8.2 and Equation 8.3 are physically plausible, and consistent with our physical intuitions that statistical fluctuation increases with decreasing size, increasing diffusivity as well as increasing electron concentration.

Now consider the neutral base of an ideal one-dimensional bipolar transistor. The purpose is to solve for the collector current noise resulting from current density fluctuations within all the incremental sections when the base and collector are AC shorted (as shown in Figure 8.2a).

Scalar and Vector Green's Functions

The first step is to consider the amount of J_C noise resulting from the current density noise within $(x, x + \Delta x)$, ΔJ_n, the spectral density of which is described by Equation 8.2. This is equivalent to injecting ΔJ_n at x and extracting ΔJ_n at $x + \Delta x$ (as shown in Figure 8.2b). If we define the current gain G as the ratio of the ΔJ_n observed at W_b to a current excitation at x *alone*, the net $\Delta J_n\ (W_b)$ can be written as

$$\Delta J_n(W_b) = \Delta J_n G(x) - \Delta J_n G(x + \Delta x) = -\Delta J_n G' \Delta x \tag{8.4}$$

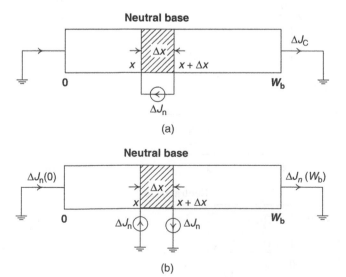

FIGURE 8.2 (a) Equivalent circuit for the the noise propagation from an incremental section $(x, x + \Delta x)$ toward the collector. (b) An alternative equivalent circuit for the noise propagation from an incremental section $(x, x + \Delta x)$ toward the collector.

where $G' \equiv dG(x)/dx$ is the *vector Green's function*. $G(x)$ is the *scalar Green's function*. The collector current density noise is essentially $\Delta J_n(W_b)$. Thus

$$\Delta S_{J_C} = S_{\Delta J_n}|G'|^2 \Delta x^2. \tag{8.5}$$

Similarly, one can solve for the open-circuit noise voltage with an open-circuit boundary condition. The $G(x)$ will then be the ratio of the voltage output at $x = W_b$ to a current excitation at x, an *impedance*. The vector defined by $G' = dG(x)/dx$ has the meaning of a field defined using the gradient of an impedance function, and is thus called *impedance field*. The open-circuit voltage and short-circuit current noises are equivalent to each other, similar to the equivalence between Norton and Thevenin equivalent circuits.

Green's Function Evaluation and Boundary Conditions

To evaluate $G(x)$, we evaluate the amount of ΔJ_n (W_b) produced by an excitation ΔJ_n at x *alone* (as shown in Figure 8.3). At the neutral base boundaries, we assume that the quasi-Fermi levels remain fixed at the levels set by the applied voltages, which implies that the minority carrier concentration changes are zero:

$$\Delta n(0) = 0, \tag{8.6}$$

$$\Delta n(W_b) = 0. \tag{8.7}$$

To avoid confusion, we will use x' to denote the distance in the neutral base, and use x to denote the injection point. Assuming constant diffusivity, zero base electric field, zero recombination, the minority electron current in a homogeneous p-type base under small signal excitations can be obtained from linearization of its large signal expression as

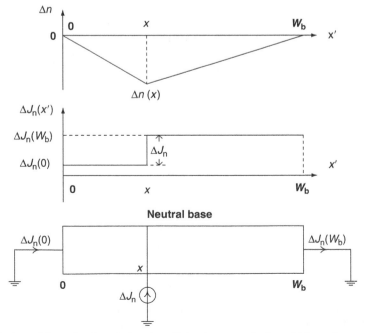

FIGURE 8.3 Solution of the scalar Green's function $G(x)$ using an excitation of ΔJ_n at $x' = x$.

$$\Delta J_n(x') = qD_n \frac{d\Delta n(x')}{dx'}, \tag{8.8}$$

$\Delta J_n(x')$ is constant from $x' = 0$ to x as there is no recombination, and must be equal to $\Delta J_n(0)$. Using Equation 8.6 and integrating Equation 8.8 from 0 to x,

$$\Delta J_n(0) \times (x - 0) = qD_n[\Delta n(x) - 0]. \tag{8.9}$$

$\Delta J_n(0)$ is now related to $\Delta n(x)$:

$$\Delta J_n(0) = \frac{qD_n \Delta n(x)}{x}. \tag{8.10}$$

Similarly, $\Delta J_n(x')$ is constant from x to W_b, and must be equal to $\Delta J_n(W_b)$. Using Equation 8.7 and integrating Equation 8.8 from $x' = x$ to W_b,

$$\Delta J_n(W_b) \times (W_b - x) = qD_n[0 - \Delta n(x)], \tag{8.11}$$

$\Delta J_n(W_b)$ is thus obtained as

$$\Delta J_n(W_b) = -\frac{qD_n \Delta n(x)}{W_b - x}. \tag{8.12}$$

A sudden change from $\Delta J_n(0)$ to $\Delta J_n(W_b)$ occurs at $x' = x$, because of the ΔJ_n injection:

$$\Delta J_n = \Delta J_n(W_b) - \Delta J_n(0) = qD_n(-1)\Delta n(x)\left(\frac{1}{W_b - x} + \frac{1}{x}\right). \tag{8.13}$$

Using Equation 8.12 and Equation 8.13, $G(x)$ is obtained as

$$G(x) = \frac{\Delta J_n(W_b)}{\Delta J_n} = \frac{\dfrac{1}{W_b - x}}{\dfrac{1}{W_b - x} + \dfrac{1}{x}} = \frac{x}{W_b}. \tag{8.14}$$

We note that $G(x)$ is dimensionless, as it represents a current gain. In this case, $G(x)$ is simply x/W_b. In general, however, $G(x)$ is a frequency-dependent complex number.

G' is readily evaluated as

$$G'(x) = \frac{dG(x)}{dx} = \frac{1}{W_b}. \tag{8.15}$$

$G'(x)$ is *position-independent* in this case, and therefore, the final noise for each section is determined by the noise source density.

Recall that the spectral density of ΔJ_n for the incremental section $(x, x+\Delta x)$ is given by Equation 8.2, which is rewritten as

$$S_{\Delta J_n} = 4q^2 D_n n(x) \frac{1}{A\Delta x}. \tag{8.16}$$

Substituting Equation 8.16 and Equation 8.15 into Equation 8.5,

$$\Delta S_{J_C} = 4q^2 D_n n(x) \frac{1}{A} \frac{1}{W_b^2} \Delta x. \tag{8.17}$$

Replacing Δx by dx, and integrating from 0 to W_b, one obtains the total noise S_{J_C} as

$$
\begin{aligned}
S_{J_C} &= \int_0^{W_d} dS_{J_C} \\
&= \int_0^{W_b} 4q^2 D_n n(x) \frac{1}{A} \frac{1}{W_b^2} dx \\
&= 4q^2 D_n \frac{1}{A} \frac{1}{W_b^2} \int_0^{W_b} n(x) dx = 4q^2 D_n \frac{1}{A} \frac{1}{W_b^2} \frac{n(0)}{2} W_b \\
&= 2q \frac{q D_n n(0)}{W_b} \frac{1}{A} = 2q J_C \frac{1}{A}
\end{aligned} \tag{8.18}
$$

where we used:

$$J_C = \frac{q D_n n(0)}{W_b}, \tag{8.19}$$

a direct result of linear minority carrier distribution in an ideal bipolar transistor. J_C is related to I_C by $I_C = A J_C$, thus:

$$S_{I_C} = A^2 S_{J_C} = 2q I_C. \tag{8.20}$$

The base current shot noise can be derived in the same manner:

$$S_{I_B} = 2q I_B. \tag{8.21}$$

Noise Concentration

In practice, one may only be interested in the terminal noise produced by a unit volume at a given location, that is, dS_{J_C}/dx in our one-dimensional bipolar example. Just like one can calculate the total number of electrons in the device by integrating electron concentration, one can calculate the total terminal noise current or voltage by integrating the noise concentration. A plot of the noise concentration immediately shows where most of the noise comes from within the physical structure of the device. Obviously, if one desires to understand the details of the noise concentration plot, one would have to examine the noise source density and the Green's functions.

8.3   Input Voltage and Current Noise Concentrations

The results of microscopic noise simulation are typically given for either the open-circuit noise voltages or the short-circuit noise currents. However, for noise optimization, the input noise current and voltage for a chain representation are the most convenient, and directly relate to NF_{min}, Y_{opt}, and R_n. The microscopic noise concentrations for the input noise current, voltage, and their correlation S_{i_a}, S_{v_a}, and $S_{i_a v_a^*}$ facilitate identification of major noise sources within the transistor physical structure, leading to device-level optimization (e.g., doping profile, Ge profile, and device layout) with respect to the noise parameters, and can be obtained as follows [7].

Consider the transistor as a noisy linear two port. The open-circuit noise voltage parameters are obtained by integrating the "noise concentration" over the device volume

$$S_n = \int_\Omega C_{S_n} d\Omega, \tag{8.22}$$

where n is v_1, v_2, or $v_1 v_2^*$. For instance $C_{S_{v_1}}$ is the "concentration" (volume density) of S_{v_1}, and has a unit of $V^2/Hz/cm^3$. $C_{S_{v_1}}$, $C_{S_{v_2}}$ and $C_{S_{v_1 v_2^*}}$ are solved in TCAD tools, including DESSIS and TAURUS [8,9]. In principle, the boundary conditions can be modified to directly solve for the "concentration" of the chain representation noise parameters S_{i_a}, S_{v_a}, and $S_{i_a v_a^*}$. This, however, has not been implemented in device simulators. The problem can be circumvented by postprocessing of $C_{S_{v_1}}$, $C_{S_{v_2}}$, and $C_{S_{v_1 v_2^*}}$ which requires no code development by TCAD vendors [7]:

$$\begin{bmatrix} C_{S_{v_a}} & C_{S_{v_a i_a^*}} \\ C_{S_{i_a v_a^*}} & C_{S_{i_a}} \end{bmatrix} = T \cdot \begin{bmatrix} C_{S_{v_1}} & C_{S_{v_1 v_2^*}} \\ C_{S_{v_2 v_1^*}} & C_{S_{v_2}} \end{bmatrix} T^\dagger, \tag{8.23}$$

where

$$T = \begin{bmatrix} 1 & -A_{11} \\ 0 & -A_{21} \end{bmatrix}, \tag{8.24}$$

where A_{11} and A_{21} are elements of the ABCD parameter matrix A. $T^\dagger$ is the transposed conjugate of T. Integration of the obtained chain representation noise concentrations $C_{S_{v_a}}$, $C_{S_{i_a}}$, and $C_{S_{i_a v_a^*}}$ over the whole device gives the transistor S_{v_a}, S_{i_a}, and $S_{i_a v_a^*}$, respectively. Each noise concentration consists of an electron contribution and a hole contribution, which account for electron and hole velocity fluctuations, respectively.

8.4 A SiGe HBT Example

One major benefit of simulation is that it offers insight into internal device operation, which can be used to develop better compact model. The same principle applies to noise simulation and noise modeling. We now examine the classical bipolar transistor noise model using the simulation results as a reference for a SiGe HBT.

Macroscopic Input Noise

Figure 8.4a and b shows the chain representation and an equivalent circuit containing the main macroscopic noise sources, respectively. Through noise-circuit analysis, S_{v_a}, S_{i_a}, and $S_{i_a v_a^*}$ are obtained as [10]

$$S_{v_a} = \frac{2qI_C}{|Y_{21}|^2} + 2qI_B r_b^2 + 4kTr_b, \tag{8.25}$$

$$S_{i_a} = 2qI_B + \frac{2qI_C}{|h_{21}|^2}, \tag{8.26}$$

$$S_{i_a v_a^*} = 2qI_C \frac{Y_{11}}{|Y_{21}|^2} + 2qI_B r_b, \tag{8.27}$$

where $h_{21} = Y_{21}/Y_{11}$. The Y parameters are for the whole transistor that includes both r_b and the intrinsic transistor.

Macroscopic and Microscopic Connections

The $4kTr_b$ terms in the model equations account for velocity fluctuations of holes in the base. One can therefore compare the $4kTr_b$ related terms with the integration of the base hole noise concentration. Similarly, the $2qI_B$ terms account for emitter minority hole velocity fluctuation, and the $2qI_C$ terms account for base minority electron velocity fluctuation [4]. Thus, connections between compact noise model and microscopic noise simulation can be established for S_{v_a}, S_{i_a}, and $S_{i_a v_a^*}$, as shown in Table 8.1. Here the superscripts e and h stand for electron and hole contributions, respectively.

Input Noise Voltage and Current

Figure 8.5a compares the modeled and simulated S_{v_a}, $S_{v_a}^e$, and $S_{v_a}^h$ for $J_C = 0.05\,\text{mA}/\mu\text{m}^2$. $S_{v_a}^e$ dominates over $S_{v_a}^h$. The model slightly underestimates $S_{v_a}^e$, and significantly underestimates $S_{v_a}^h$. The simulated $S_{v_a}^e$ and $S_{v_a}^h$ are both frequency dependent. Despite inaccurate modeling of $S_{v_a}^h$, the total S_{v_a} is well modeled, because of the dominance of $S_{v_a}^e$. At a higher J_C of $0.65\,\text{mA}/\mu\text{m}^2$, however, the hole contribution dominates over the electron contribution, as shown in Figure 8.5b. With increasing J_C, $S_{v_a}^e$ decreases, while $S_{v_a}^h$ stays about the same. The model underestimates $S_{v_a}^h$, and overestimates $S_{v_a}^e$. Observe that the simulated $S_{v_a}^h$ is frequency dependent, while the modeled $S_{v_a}^h$ ($4kTr_b$) is frequency independent.

Figure 8.6 and Figure 8.7 show the electron and hole noise concentration contours at 2 GHz ($C_{S_{v_a}}^e$ and $C_{S_{v_a}}^h$) at $J_C = 0.65\,\text{mA}/\mu\text{m}^2$. Both $C_{S_{v_a}}^e$ and $C_{S_{v_a}}^h$ are the highest in the SiGe base, indicating that transistor S_{v_a} mainly comes from the SiGe base. This provides guidelines to future noise model

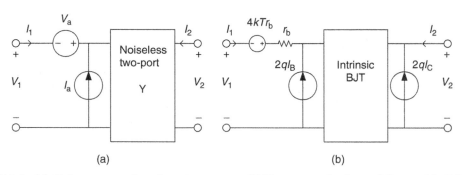

(a)   (b)

FIGURE 8.4 (a) Chain representation of a noisy two-port. (b) The essence of noise modeling used in CAD tools.

TABLE 8.1 Connections between Noise Modeling and Simulation

	Model	Simulation
$S_{v_a}^e$	$2qI_C/|Y_{21}|^2$	$\int_{\text{base}} C_{S_{v_a}}^e\, d\Omega$
$S_{v_a}^h$	$2qI_B r_b^2$	$\int_{\text{emitter}} C_{S_{v_a}}^h\, d\Omega$
	$4kTr_b$	$\int_{\text{base}} C_{S_{v_a}}^h\, d\Omega$
$S_{i_a}^e$	$2qI_C/|h_{21}|^2$	$\int_{\text{base}} C_{S_{i_a}}^e\, d\Omega$
$S_{i_a}^h$	$2qI_B$	$\int_{\text{emitter}} C_{S_{i_a}}^h\, d\Omega$
$S_{i_a v_a^*}^e$	$2qI_C Y_{11}/|Y_{21}|^2$	$\int_{\text{base}} C_{S_{i_a v_a^*}}^e\, d\Omega$
$S_{i_a v_a^*}^h$	$2qI_B r_b$	$\int_{\text{emitter}} C_{S_{i_a v_a^*}}^h\, d\Omega$

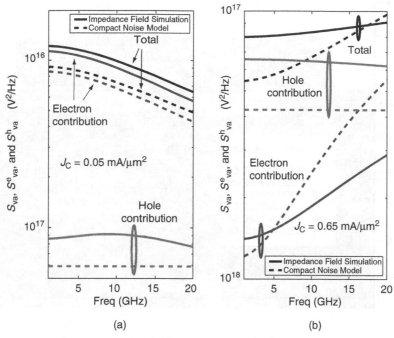

FIGURE 8.5 S_{v_a}, $S_{v_a}^e$, and $S_{v_a}^h$ versus frequency at (a) $J_C = 0.05 \, \mathrm{mA/\mu m^2}$. (b) $J_C = 0.65 \, \mathrm{mA/\mu m^2}$.

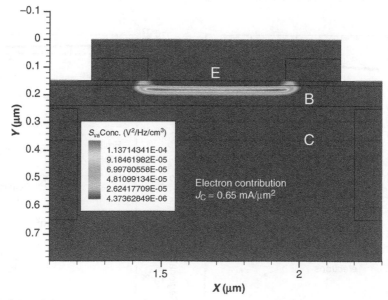

FIGURE 8.6 Two-dimensional distribution of $C_{S_{v_a}}^e$ (2 GHz, $J_C = 0.65 \, \mathrm{mA/\mu m^2}$).

development, that is, the transistor noise mainly originates from the EB junction. In the intrinsic base, and along the x-direction, $C_{S_{v_a}}^e$ is uniform, while $C_{S_{v_a}}^h$ is highly nonuniform, and shows a strong "base noise crowding" effect.

Figure 8.8a shows the integrals of $C_{S_{v_a}}^e$ in the base, emitter, collector, and p-substrate, together with the $2qI_C$ related term in the model at $J_C = 0.65 \, \mathrm{mA/\mu m^2}$. The model accounts for only the base contribution, which is reasonable, since the simulated base electron contribution overwhelmingly

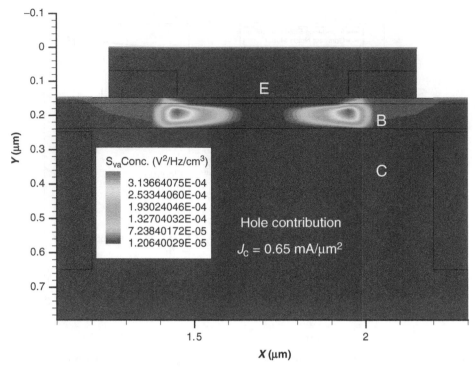

FIGURE 8.7   Two-dimensional distribution of $C_{S_{v_a}}^{h}$ (2 GHz, $J_C = 0.65\,\text{mA}/\mu\text{m}^2$).

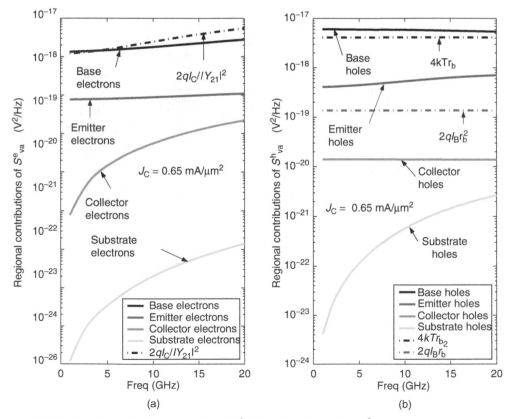

FIGURE 8.8   Regional contributions of $S_{v_a}^{e}$ (a) and $S_{v_a}^{h}$ (b) at $J_C = 0.65\,\text{mA}/\mu\text{m}^2$.

dominates over other electron contributions. The $2qI_C$ description, however, overestimates $S_{v_a}^e$, and thus a better description is required. Figure 8.8b shows the integrals of $C_{S_{v_a}}^h$ in the base, emitter, collector, and p-substrate. Also shown are the $2qI_B$ (emitter holes) and $4kT r_b$ (base holes) related terms accounted for in the model. The collector and substrate hole noises are indeed negligible. The noise from the base majority holes dominates over the noise from the emitter minority holes. The base majority hole noise contribution is less than predicted by $4kT r_b$, and frequency dependent as well. The noise from the emitter minority holes increases with frequency, and is underestimated by $2qI_B$ related term.

Figure 8.9a compares modeled and simulated S_{i_a}, $S_{i_a}^e$, and $S_{i_a}^h$ for $J_C = 0.05\,\text{mA}/\mu\text{m}^2$. $S_{i_a}^e$ decreases with frequency and is slightly underestimated (by the model). $S_{i_a}^h$ increases dramatically with frequency, and is significantly underestimated. At a higher J_C of $0.65\,\text{mA}/\mu\text{m}^2$, however, the $S_{i_a}^e$ discrepancy between model and simulation becomes much more pronounced (as shown in Figure 8.9b). Thus, for S_{i_a}, $2qI_C$ is not a good description for base minority electron noise. Like for $S_{v_a}^h$, the frequency dependence for $S_{i_a}^h$ is not accounted for in the model. $S_{i_a}^h$ dominates at lower frequencies, while $S_{i_a}^e$ becomes dominant at higher frequencies.

Figure 8.10a shows the integrals of $C_{S_{i_a}}^e$ in the base, emitter, collector, and p-substrate. $J_C = 0.65\,\text{mA}/\mu\text{m}^2$. The model only accounts for the base electron contribution, a $2qI_C/|h_{21}|^2$ term. Like for other noise parameters, the base minority electron contribution for $S_{i_a}^e$ is poorly modeled by the $2qI_C$ related term. Figure 8.10b shows the regional contributions of $S_{i_a}^h$. The model accounts for only the emitter hole contribution through the $2qI_B$ term. Even though the collector and substrate hole contributions are indeed negligible, the base hole contribution is not negligible at higher frequencies. This emitter contribution constitutes the main discrepancy for the total $S_{i_a}^h$ between modeling and simulation, and shows frequency dependence.

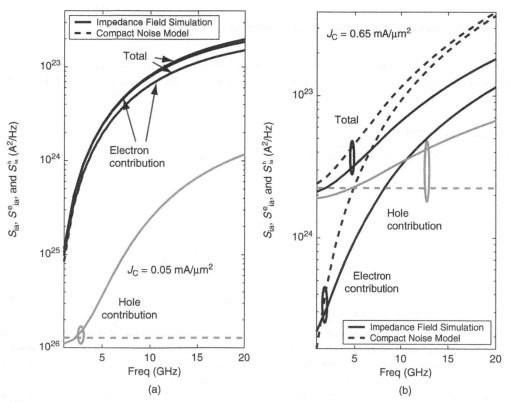

FIGURE 8.9 S_{i_a}, $S_{i_a}^e$, and $S_{i_a}^h$ versus frequency at (a) $J_C = 0.05\,\text{mA}/\mu\text{m}^2$. (b) $J_C = 0.65\,\text{mA}/\mu\text{m}^2$.

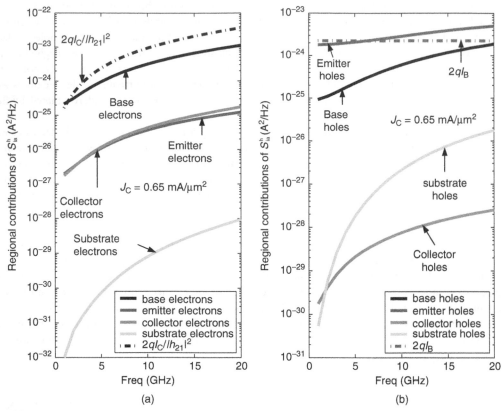

FIGURE 8.10 Regional contributions of $S_{i_a}^e$ (a) and $S_{i_a}^h$ (b) at $J_C = 0.65\,\text{mA}/\mu\text{m}^2$.

Similar analysis can be performed for $S_{i_a v_a^*}$. The results also show that the noise from the base minority electrons is poorly described by the model. Similar problems exist with $4kTr_b$ description of the base hole noise, and $2qI_C$ description of the base minority electron noise.

NF$_{\text{min}}$, Y_{opt}, and R_n

NF$_{\text{min}}$, Y_{opt}, and R_n are obtained from S_{v_a}, S_{i_a}, and $S_{i_a v_a^*}$ by [11]

$$\text{NF}_{\text{min}} = 10\log\left[1 + \frac{\sqrt{S_{v_a} S_{i_a} - [\Im(S_{i_a v_a^*})]^2} + \Re(S_{i_a v_a^*})}{2kT}\right], \tag{8.28}$$

$$Y_{\text{opt}} = \sqrt{\frac{S_{i_a}}{S_{v_a}} - \left[\frac{\Im(S_{i_a v_a^*})}{S_{v_a}}\right]^2} - j\frac{\Re(S_{i_a v_a^*})}{S_{v_a}}, \tag{8.29}$$

$$R_n = S_{v_a}/4kT. \tag{8.30}$$

To compare the impact of electron and hole noise on circuit-level noise parameters, we examine NF$_{\text{min}}^e$ and NF$_{\text{min}}^h$, defined as the NF$_{\text{min}}$ that the transistor would have when only electron velocity or only hole velocity fluctuates, respectively. NF$_{\text{min}}^e$ is obtained by substituting $S_{v_a}^e$, $S_{i_a}^e$, and $S_{i_a v_a^*}^e$ into Equation 8.7. NF$_{\text{min}}^h$ is obtained similarly. Since NF$_{\text{min}}$ is not a linear function of S_{v_a}, S_{i_a}, and $S_{i_a v_a^*}$, NF$_{\text{min}} \neq$ NF$_{\text{min}}^e$ + NF$_{\text{min}}^h$. Y_{opt}^e is similarly defined and obtained by substituting $S_{v_a}^e$, $S_{i_a}^e$, and $S_{i_a v_a^*}^e$ into Equation 8.8. Like NF$_{\text{min}}$, $Y_{\text{opt}} \neq Y_{\text{opt}}^e + Y_{\text{opt}}^h$.

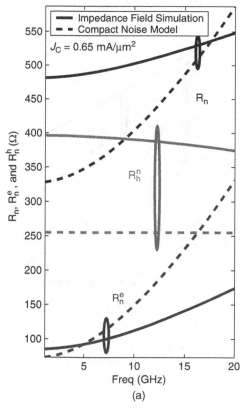
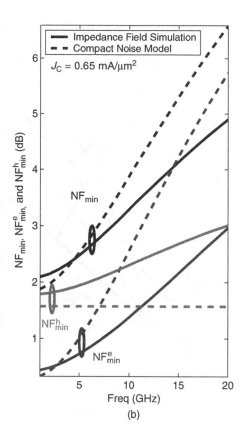

FIGURE 8.11 (a) R_n. (b) NF_{min}. $J_C = 0.65\,mA/\mu m^2$.

Figure 8.11a shows the simulated and modeled R_n, R_n^e, and R_n^h for $J_C = 0.65\,mA/\mu m^2$. Figure 8.11b shows the modeled and simulated NF_{min}, NF_{min}^e, and NF_{min}^h, also for $J_C = 0.65\,mA/\mu m^2$. Since $R_n = S_{v_a}/4kT$, which is a linear function, $R_n = R_n^e + R_n^h$. The problems with S_{v_a} modeling directly translate into R_n inaccuracy. Both NF_{min} and NF_{min}^e are overestimated by the model, and the discrepancies increase dramatically with frequency. NF_{min}^h is poorly modeled. Note that the frequency dependence of NF_{min}^h is not modeled.

Figure 8.12a and b shows the real and imaginary parts of Y_{opt} for $J_C = 0.65\,mA/\mu m^2$. Neither Y_{opt} nor Y_{opt}^e or Y_{opt}^h is well modeled. The discrepancies increase with frequency. The frequency dependence of Y_{opt}^h is not accounted for by the model. The discrepancies of R_n, NF_{min}, and Y_{opt} are all fundamentally caused by the inaccurate modeling of S_{v_a}, S_{i_a}, and $S_{i_a v_a^*}$. In particular, the description of base minority electron noise using $2qI_C$ is clearly responsible for the inaccuracy of the electron contributions, and the description of base majority hole noise using $4kTr_b$ is responsible for the inaccuracy of the hole contributions.

8.5 Summary

We have presented an overview of the concepts related to microscopic noise analysis. The key concepts involved are the noise source density, and the scalar and vector Green's functions that describe noise propagation. These concepts are illustrated using analytical treatment of the collector current shot noise. Two-dimensional microscopic noise simulation results on a SiGe HBT are presented, and used to examine the validity of the widely used macroscopic transistor noise model.

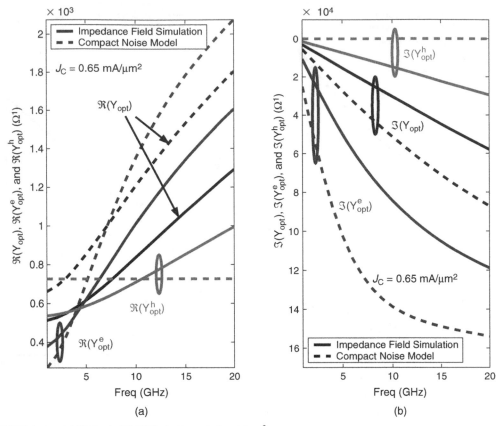

FIGURE 8.12   (a) $\Re(Y_{\mathrm{opt}})$. (b) $\Im(Y_{\mathrm{opt}})$. $J_C = 0.65\,\mathrm{mA/\mu m^2}$.

Acknowledgments

The author would like to thank Y. Cui for help in preparation of the manuscript. This work is supported by NSF under ECS-0112923 and ECS-0119623 and SRC under SRC-2001-NJ-937 and SRC-2003-NJ-1133.

References

1. W Shockley, J Copeland, and R James. The impedance field method of noise calculation in active semiconductor devices. *Quantum Theory of Atoms, Molecules, and the Solid-State.* Academic Press, New York, 1966, pp. 537–563.
2. A van der Ziel and KM van Vliet. H.f. thermal noise in space-charge limited solid state diodes—II. *Solid-State Electron.* 11:508–509, 1968.
3. KM van Vliet. General transistor theory of noise in PN junction-like devices—I. Three-dimensional Green's function formulation. *Solid-State Electron.* 15:1033–1053, 1972.
4. F Bonani, G Ghione, M Pinto, and R Smith. An efficient approach to noise analysis through multidimensional physics-based models. *IEEE Trans. Electron Dev.* 45:261–269, 1998.
5. F Bonani and G Ghione. *Noise in Semiconductor Devices: Modeling and Simulation.* Springer-Verlag, Heidelberg, 2001.
6. GF Niu. Bridging the gap between microscopic and macroscopic theories of noise in bipolar junction transistors. Tech. Digest of IEEE Topical Meeting on Silicon Monolithic Integrated Circuits in RF Systems. pp. 227–230, 2004.

7. Y Cui, GF Niu, Y Shi, and DL Harame. Spatial distribution of microscopic noise contributions in SiGe HBT. IEEE Topical Meeting on Silicon Monolithic Integrated Circuits in RF Systems, April 2003, pp. 170–173.
8. DESSIS device simulator. ISE Integrated Systems Engineering.
9. TAURUS-DEVICE simulator. Synopsys, Inc.
10. Y Cui, GF Niu, and DL Harame. An examination of bipolar transistor noise modeling and noise physics using microscopic noise simulation. Proceedings of the IEEE BCTM, September 2003, pp. 225–228.
11. HA Haus. Representation of noise in linear twoports. *Proc. IRE* 48:69–74, 1960.

9

Linearity

9.1   Introduction

"Linearity" is the counterpart of "distortion," or "nonlinearity," and refers to the ability of a device, circuit, or system to amplify input signals in a linear fashion. SiGe HBTs, like other semiconductor devices, are in general nonlinear elements. Most obviously, the collector current I_C depends on the base–emitter voltage V_{BE} exponentially, common to all bipolar transistors. This exponential I–V relation in fact represents the strongest nonlinearity found in nature, and underlies the "translinear principle," which enables a large variety of linear and nonlinear functions to be realized using bipolar transistors. Despite our intuition, the distortion in translinear circuits is *not caused* by the exponential I_C–V_{BE} relationship, but rather is due to the *departure* from it, by various means (i.e., series resistance, high-level injection, impact ionization, early effect, and inverse early effect).

SiGe HBTs can be used to build both "nonlinear" and "linear" circuits depending on the required application and the circuit topology used. In fact, transistor nonlinearity is both a blessing and a curse. Nonlinearity can be a blessing because we need nonlinearity to realize frequency translation; but can also be a curse, because it creates distortion in the various signals we are interested in preserving, amplifying, or transmitting. For instance, nonlinearity causes intermodulation (IM) of two adjacent strongly interfering signals at the input of a receiver, which can corrupt the nearby (desired) weak signal we are trying to receive. In the transmit path, nonlinearity in power amplifiers clips the large amplitude input, which causes power leaking (and thus interference) to adjacent channels in digitally modulated signals.

Perhaps surprisingly, SiGe HBTs exhibit excellent linearity in both small-signal (e.g., LNA) and large-signal (e.g., PA) RF circuits, despite their strong I–V and C–V nonlinearities. Clearly, the overall circuit linearity strongly depends on the interaction (and potential cancellation) between the various I–V and C–V nonlinearities, the linear elements in the device, as well as the source termination, the load termination, and any feedback present, intentional or parasitic. These issues can be best understood using Volterra series [1–3], a powerful formalism for analysis of nonlinear systems. In this chapter, we focus on the intermodulation linearity of SiGe HBTs, a major concern in RF circuits.

9.2 Basic Concepts

We first introduce basic nonlinearity concepts using simple power series, which strictly speaking only applies to a memory-less circuit. Under small-signal input, the output voltage $y(t)$ is related to its input voltage $x(t)$ by

$$y(t) = k_1 x(t) + k_2 x^2(t) + k_3 x^3(t). \tag{9.1}$$

Consider a two-tone input $x(t) = A\cos\omega_1 t + A\cos\omega_2 t$. The output has not only harmonics of ω_1 and ω_2, but also "intermodulation products" at $2\omega_1 - \omega_2$ and $2\omega_2 - \omega_1$. A full expansion of Equation 9.1 shows that the output contains signals at $\omega_1, \omega_2, 2\omega_1, 2\omega_2, 3\omega_1, 3\omega_2, \omega_1 + \omega_2, \omega_1 - \omega_2, 2\omega_2 - \omega_1, 2\omega_2 + \omega_1, 2\omega_1 - \omega_2$, and $2\omega_1 + \omega_2$.

When ω_1 and ω_2 are closely spaced, the third-order intermodulation products at $2\omega_2 - \omega_1$ and $2\omega_1 - \omega_2$ are the major concerns, because they are close to ω_1 and ω_2, and thus within the amplifier bandwidth. Consider a weak desired signal channel, and two nearby strong interferers at the input. One intermodulation product falls in band, and corrupts the desired component (as illustrated in Figure 9.1).

The fundamental signal and intermodulation products in the output are given by

$$y(t) = \left(k_1 A + \frac{3k_3 A^3}{4} + \frac{3k_3 A^3}{2}\right)\cos\omega_1 t + \cdots + \frac{3k_3 A^3}{4}\cos(2\omega_2 - \omega_1)t$$

$$+ \cdots \text{fundamental intermodulation.} \tag{9.2}$$

The ratio of the amplitude of the IM product to the amplitude of the fundamental output is defined as the "third-order intermodulation distortion" (IM3). Neglecting the higher order terms added to $k_1 A$, one has

$$\text{IM3} = \frac{3k_3 A^3}{4} \Big/ k_1 A = \frac{3}{4}\frac{k_3}{k_1}A^2. \tag{9.3}$$

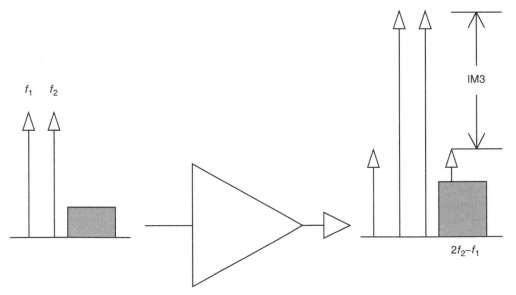

FIGURE 9.1 Illustration of the corruption of desired signals by the intermodulation product of two strong interferers.

For small A, the fundamental output at ω_1 grows *linearly* with A, while the IM product at $2\omega_2 - \omega_1$ grows as A^3. A 1-dB increase in the input results in a 1-dB increase of fundamental output, but a 3-dB increase of IM product. The extrapolation of the fundamental output and the IM3 versus the input intersect at a given input level, which is defined as the *input third-order intercept point* (IIP3). IIP3 is obtained from Equation 9.3 by letting IM3 $= 1$

$$\text{IIP3} = \sqrt{\frac{4}{3}\frac{k_1}{k_3}}. \tag{9.4}$$

IIP3 is a widely used figure-of-merit because it does not depend on the input signal level. Because IM3 grows with A^2 (Equation 9.3), IIP3 can be measured at a single input level A_0,

$$\text{IIP3}^2 = \frac{A_0^2}{\text{IM3}_0}, \tag{9.5}$$

where IM3_0 is the measured relative intermodulation distortion. Note that IIP3 and A_0 are voltages, and thus IIP3^2 and A_0^2 are measures of power. Taking 10 log on both sides, one has

$$20\log\text{IIP3} = 20\log A_0 - 10\log\text{IM3}_0. \tag{9.6}$$

Here, 20 log IIP3 is the power expressed in dB at the intercept point, and 20 log A_0 is now the input power level expressed in dB. The reference power level does not enter into the equation. Now Equation 9.6 can be rewritten in terms of power

$$P_{\text{IIP3}} = P_{\text{in}} + \frac{1}{2}\left(P_{\text{o,1st}} - P_{\text{o,3rd}}\right) \tag{9.7}$$

and

$$P_{\text{OIP3}} = P_{\text{o,1st}} + \frac{1}{2}\left(P_{\text{o,1st}} - P_{\text{o,3rd}}\right). \tag{9.8}$$

In practice, IIP3 and OIP3 are used to denote the power at the intercept point. The IP3 data in commercial load-pull systems and CAD tools (e.g., ADS) are defined for each input power level according to the equations above. The following is the sample output of a load-pull measurement on a SiGe HBT amplifier. The two tones are at 2.000 and 2.001 GHz (i.e., 1-MHz spacing).

P_{in} (dB m)	$P_{\text{out,1st}}$ (dB m)	Gain (dB)	$P_{\text{out,3rd}}$ (dB m)	P_{OIP3} (dB m)
−30.00	−11.72	18.28	−74.48	19.65
−29.00	−10.75	18.25	−72.68	20.20
−28.00	−9.74	18.26	−69.91	20.35
−27.00	−8.74	18.26	−67.24	20.51
−26.00	−7.72	18.28	−64.89	20.87
−25.00	−6.77	18.23	−62.28	20.98
−24.00	−5.74	18.26	−59.57	21.18
−23.00	−4.73	18.27	−57.15	21.47
−22.00	−3.75	18.25	−54.66	21.71

Frequency	2.00 GHz
Source state	1; #377
Source imp.	0.20–170.1
Load state	1; #550
Load imp.	0.70–14.9
V_c	3.001 V
V_b	0.816 V
Date	April 30, 1998
Time	16:22:43

Figure 9.2 shows the measured fundamental and third-order IM product power versus input power data for the above SiGe HBT amplifier, along with gain. The measured slope of the IM product curve deviates from 3:1, because of the "high" input power level used in the measurement. As a result, the IP3 numbers measured at different input powers are different, as can be seen from the data output. One would obtain an OIP3 of 35 dB m by simply extrapolating the linear portions of the measured $P_{out,1st}$ and $P_{out,3rd}$ data. The OIP3 based on a theoretical 3:1 slope at $P_{in} = -30$ dB m is only 20 dB m, however, and therefore, caution must be exercised in interpreting the IP3 numbers. The gain compression at very high input power level can also be clearly seen here.

Clearly IIP3 is an important figure-of-merit for front-end RF–microwave low-noise amplifiers, because they must contend with a variety of signals coming from the antenna. The interfering signals are often much stronger than the desired signal, thus generating strong intermodulation products that can corrupt the weak but desired signal. To some extent, IIP3 is a measure of the ability of a handset, for instance, not to "drop" a phone call in a crowded environment. For many LNA applications, IIP3 is just as important (if not more so) as the noise figure. The dc power consumption must also be kept very low because the LNA is likely to be continuously listening for transmitted signals of interest and hence continuously draining power. The power consumption aspect is taken into account by another figure-of-merit, the *linearity efficiency*, which is defined as IIP3/P_{dc}, where P_{dc} is the dc power dissipation. First-generation SiGe HBTs typically exhibit excellent linearity efficiencies above 10, which is competitive with III–V technologies. We note, however, that IIP3/P_{dc} is not adequate for describing the Class AB operating mode for transistors in the driver and output stage of power amplifiers [4].

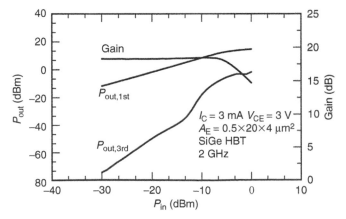

FIGURE 9.2 A typical P_{out} versus P_{in} curve for a first generation SiGe HBT ($I_C = 3$ mA and $V_{CE} = 3$ V). The input power at the 1-dB compression point is -3 dB m.

9.3 Physical Nonlinearities

The major physical nonlinearities in a SiGe HBT are depicted in Figure 9.3, including:

1. I_{CE}, the collector current transported from the emitter, $I_{CE} \propto \exp(V_{BE}/V_T)$.
2. I_{BE}, the hole injection current into the emitter. I_{BE} tracks I_{CE} through $I_{BE} = I_{CE}/\beta$.
3. I_{CB}, the avalanche multiplication current, and is a strong nonlinear function of both V_{BE} and V_{CB} [5]. The nonlinearity due to I_{CB} can be minimized using a low V_{CB}.
4. C_{BE}, the EB junction capacitance, including the diffusion capacitance and depletion capacitance. C_{BE} tracks I_{CE} at high J_C because diffusion charge is in proportion to the transport current I_{CE}.
5. C_{CB}, the CB junction capacitance. C_{CB} is much smaller than C_{BE}, because of reverse bias on the CB junction. The feedback position, however, makes C_{CB} important for circuit linearity. A medium value V_{CB} that makes C_{CB} more linear and keeps the avalanche (breakdown) current I_{CB} small is optimum for linearity [6]. Both the value and the bias dependence of C_{CB} matter for the overall transistor linearity [7].

Among all the nonlinearities, the $I_{CE}-V_{BE}$ nonlinearity often dominates in RF circuits. For relatively weak input signal, the nonlinear $I_{CE}-V_{BE}$ relation can be approximated by a Taylor expansion. A direct consequence of the $I_{CE}-V_{CE}$ nonlinearity is to make the effective transconductance a function of v_{be} (as opposed to a constant in a linear circuit)

$$g_{m,eff} = \frac{i_c}{v_{be}} = g_m \left(1 + \overbrace{\frac{1}{2}\frac{v_{be}}{V_t} + \frac{1}{6}\frac{v_{be}^2}{V_t^2}}^{\text{nonlinear contributions}} + \cdots \right). \tag{9.9}$$

Equation 9.9 indicates that the nonlinear contributions to $g_{m,eff}$ increase with the voltage drop across the EB junction v_{be}. In typical bipolar amplifiers, v_{be} decreases with increasing biasing current, making

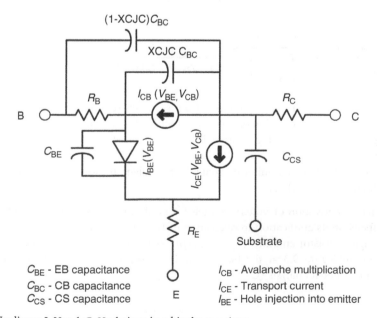

FIGURE 9.3 Nonlinear *I–V* and *C–V* relations in a bipolar transistor.

$g_{m,eff}$ closer to a constant like in a linear circuit. *Linearity of bipolar circuits can therefore be improved by increasing the biasing current.* The expense is, however, increased power consumption. Parasitic or intentionally used emitter resistance or inductance also helps improving linearity by decreasing v_{be} through negative feedback. All the capacitances, either internal or external C_{BE} and C_{CB}, help improving linearity through negative feedback at the expense of gain.

The nonlinear contributions from various I–V and C–V nonlinearities cancel out each other to certain extent depending on circuit design, which further improves circuit overall linearity. In particular, for bipolar transistors, the exponential I–V and exponential C–V nonlinearities can be engineered to cancel out through tuning of harmonic impedance termination, which can be understood using Volterra series, as described below.

9.4 Volterra Series Linearity Analysis

We now examine how the physical nonlinearities affect IP3 using Volterra series, a mathematical method of analyzing small-signal distortion. Compared to other distortion analysis methods, Volterra series allows us to easily identify the contribution of various individual nonlinearities, as well as identify the interaction between individual nonlinearities. Volterra series is an extension of the theory of linear systems to weakly nonlinear systems, its essence is summarized as follows:

- Volterra series approximate the output of a nonlinear system in a manner similar to the more familiar Taylor series approximation of analytical functions. Similarly, the analysis is applicable only to weak nonlinearities, or small inputs.
- The response of a nonlinear system to an input $x(t)$ is equal to the sum of the responses of a series of transfer functions of different orders ($H_1, H_2, \ldots, H_n$)

$$Y = H_1(x) + H_2(x) + H_3(x) + \cdots. \tag{9.10}$$

- In the time domain, H_n is described as an impulse response $h_i(\tau_1, \tau_2, \ldots, \tau_n)$. As in linear circuit analysis, frequency-domain representation is more convenient, and thus $H_n(s_1, \ldots, s_n)$, the nth-order transfer function or Volterra kernel in frequency domain, is obtained through a multi-dimensional Laplace transform of the time-domain impulse response

$$\overbrace{H_n(s_1, \cdots, s_n)}^{\text{frequency domain}} = \overbrace{\int_{-\infty}^{+\infty} \cdots \int_{-\infty}^{+\infty}}^{\text{Laplace transform}} \overbrace{h_n(\tau_1, \tau_2, \ldots, \tau_n)}^{\text{time domain}} e^{-(s_1\tau_1 + s_2\tau_2 + \cdots + s_n\tau_n)} d\tau_1 \cdots d\tau_n. \tag{9.11}$$

- Here, H_n takes n frequencies as the input, from $s_1 = j\omega_1$ to $s_n = j\omega_n$.
- The first-order transfer function or Volterra kernel $H_1(s)$ is essentially the transfer function of the small-signal linear circuit at dc bias. Higher order transfer functions represent higher order phenomena.
- Solving the output of a nonlinear circuit is equivalent to solving the Volterra series $H_1(s)$, $H_2(s_1,s_2)$, and $H_3(s_1,s_2,s_3), \ldots$

The mathematical derivation of Volterra kernels for nonlinear circuits has been well treated in Refs. [1,2]. We will focus on its application to calculation of IIP3 in SiGe HBTs.

Consider a single transistor amplifier shown in Figure 9.4. The first step is to linearize the large-signal equivalent circuit in Figure 9.3 at the bias point. The resulting linear circuit is then solved using compacted modified nodal analysis (CMNA) [8]:

$$Y(s) \cdot \vec{H}_1(s) = \vec{I}_1, \tag{9.12}$$

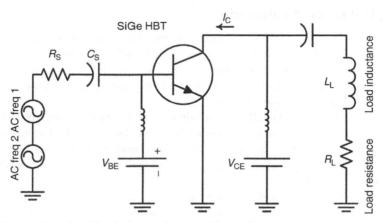

FIGURE 9.4 Circuit schematic of the single transistor amplifier used.

where $Y(s)$ is the CMNA [8] admittance matrix at frequency $s(j\omega)$, $\vec{H}_1(s)$ is the vector of first-order Volterra kernel transforms of the node voltages, and $\vec{I}_1$ is the vector of the node excitations. The admittance matrix Y and the excitation vector $\vec{I}_1$ are obtained by applying the Kirchoff's current law at every circuit node. The unknowns are the node voltages. The circuit output and the voltages that control nonlinearities can be expressed as a linear combination of the elements of $\vec{H}_1(s)$.

With $\vec{H}_1(s)$ solved, we now excite the same circuit using the second-order nonlinear current sources $\vec{I}_2$, which are functions of the first-order voltages that control individual nonlinearities, and the second-order derivatives of all the $I-V$ and $C-V$ nonlinearities. Every nonlinearity in the original circuit corresponds to a nonlinear current source in parallel with the corresponding linearized circuit element. The node voltages under such an excitation are the second-order Volterra kernels $\vec{H}_2(s_1, s_2)$:

$$Y(s_1 + s_2) \cdot \vec{H}_2(s_1, s_2) = \vec{I}_2 \tag{9.13}$$

where $Y(s_1 + s_2)$ is the same CMNA admittance matrix used in Equation 9.12, but evaluated at the frequency $s_1 + s_2$.

In a similar manner, the third-order Volterra kernels $\vec{H}_3$ can be solved as response to excitations specified in terms of the previously determined first- and second-order kernels:

$$Y(s_1 + s_2 + s_3) \cdot \vec{H}_2(s_1, s_2, s_3) = \vec{I}_3. \tag{9.14}$$

P_{out} versus P_{in}, the third-order input intercept (IIP3) at which the first- and third-order signals have equal power, and the (power) gain can then be obtained from $\vec{H}_3$ and $\vec{H}_1$. IIP3, the input power at which the fundamental output power equals the intermodulation output power, is obtained as [6]:

$$\text{IIP3} = \frac{1}{6R_S} \frac{|H_1(j\omega_1)|}{|H_3(j\omega_1, j\omega_1, -j\omega_2)|}, \tag{9.15}$$

where R_S is the source resistance. IIP3 is often expressed in dBm by $\text{IIP3}_{\text{dBm}} = 10 \log_{10}(10^3 \text{IIP3})$.

An inspection of the Volterra series procedure immediately shows that the Volterra kernels are strongly related to the properties of the circuit admittance matrix at various frequencies. For intermodulation ($2\omega_1 - \omega_2$), the circuit admittance matrix at the second harmonic frequency and at very low frequency ($\omega_1 - \omega_2$) are of particular importance. Harmonic tuning and low-frequency "traps" circuit techniques of linearity enhancement [9–11] are based on this concept.

Identifying Dominant Nonlinearity

An unique feature of the Volterra series approach is the ability to identify the dominant physical nonlinearity [2,6]. This is realized by turning on and off each individual nonlinearity-related nonlinear current sources in formulating the excitations for solving $\vec{H}_2$ and $\vec{H}_3$ [6]. An individual IIP3 is thus obtained for each nonlinearity. The individual nonlinearity that gives the lowest IIP3 (the worst linearity) is identified as the dominant nonlinearity.

We can then calculate the overall IIP3 by including all of the nonlinearities in the calculation of both $\vec{H}_2$ and $\vec{H}_3$. A comparison of the individual IIP3 and the overall IIP3 reveals the interaction between individual nonlinearities. As shown below, the overall IIP3 obtained by including all of the nonlinearities can be larger (better) than an individual IIP3, implying cancellation between individual nonlinearities [6,12].

We now consider the Volterra series linearity analysis of a SiGe HBT with 50 GHz peak f_T. The frequency is 2 GHz, and the tone spacing is 1 MHz. The SiGe HBT has four $A_E = 0.5 \times 20\ \mu m^2$ emitter fingers. $I_C = 3$ mA, $V_{CE} = 3$ V, $R_S = 50\ \Omega$, $C_S = 300$ pF, $R_L = 186\ \Omega$, and $L_L = 9$ nH.

Collector Current Dependence

Figure 9.5 shows the IIP3 and gain as a function of I_C up to 60 mA at which f_T and f_{max} peak. The collector biasing voltage is $V_{CE} = 3$ V. At low I_C (<5 mA), the exponential I_{CE}–V_{BE} nonlinearity ($\times$) yields the lowest individual IIP3, and hence is the dominant factor. For 5 mA < I_C < 25 mA, the I_{CB} nonlinearity due to avalanche multiplication ($\diamond$) dominates. For $I_C > 25$ mA, the C_{CB} nonlinearity due to the CB capacitance ($\triangledown$) dominates. Interestingly, the overall IIP3 obtained by including all of the nonlinearities is close to the lowest individual IIP3 for all the I_C in this case. The closeness indicates a weak interaction between individual nonlinearities.

The overall IIP3 increases with I_C for $I_C < 5$ mA when the exponential I_{CE} nonlinearity dominates. For $I_C > 5$ mA where the avalanche current (I_{CB}) nonlinearity dominates, the I_C dependence of the overall IIP3 is twofold:

1. The initial current for avalanche I_{CE} increases with I_C.
2. The avalanche multiplication factor $(M-1)$ decreases with I_C.

The increase of the avalanche IIP3 and hence the overall IIP3 for $I_C > 17$ mA is a result of the decrease of $M - 1$ with increasing J_C. For $I_C > 25$ mA, "the overall IIP3 becomes limited by the C_{CB} nonlinearity, and is approximately independent of I_C." The optimum biasing current is therefore $I_C = 25$ mA in this case ($V_{CE} = 3$ V). The use of a higher I_C only increases power consumption, and does not improve

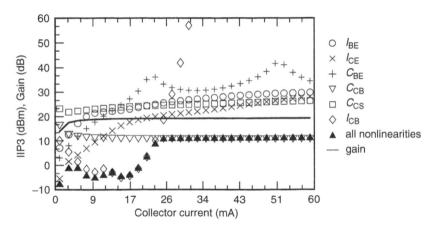

FIGURE 9.5 IIP3 and gain as a function of I_C.

linearity. The decrease of $M - 1$ with increasing J_C is therefore beneficial to the linearity of these SiGe HBTs.

The low-noise biasing J_C for this HBT is 0.1 to 0.2 mA/μm^2, which corresponds to a I_C of 4 to 8 mA in Figure 9.5. In this I_C range, IIP3 is limited by avalanche multiplication for the circuit configuration in Figure 9.4. To further improve IIP3, a lower collector doping is desired, provided that the noise performance is not inadvertently degraded. The noise figure, for instance, is relatively independent of the collector doping as long as Kirk effect does not occur at the J_C of interest [13]. Thus, there must exist an optimum collector-doping profile for producing low-noise transistors with the best linearity.

Collector Voltage Dependence

An alternative way of reducing avalanche is to decrease the collector biasing voltage, which, however, also reduces the output voltage swing and hence the dynamic range. Another disadvantage from a linearity standpoint is that the CB capacitance nonlinearity is increased. Therefore, one must carefully optimize the collector biasing voltage for optimum IIP3. Figure 9.6 shows the overall IIP3 as a function of V_{CE} up to 3.3 V, the BV_{CEO} of the transistor. A peak of IIP3 generally exists as V_{CE} increases. For $I_C = 10$ mA where noise figure is minimum, the optimum biasing V_{CE} is 2.4 V, yielding an IIP3 of 9 dB m. The IIP3 obtained (9 dB m) is 11 dB higher than the IIP3 at $V_{CE} = 3$ V (-2 dB m), illustrating the importance of biasing in determining linearity of these SiGe HBTs.

The biasing current and voltage has significant impact on transistor linearity. Figure 9.7 shows the IIP3 as a function of I_C for different V_{CE}. At sufficiently high I_C, IIP3 approaches a value that depends on V_{CE}. The threshold I_C where IIP3 reaches its maximum is higher for a higher V_{CE}. For a given V_{CE}, I_C must be above this threshold to achieve good IIP3. On the other hand, the use of an I_C well above the threshold does not further increase IIP3, and only increases power consumption. The optimum I_C is thus at the threshold value, which is 10 mA for $V_{CE} = 2$ V. Figure 9.8 shows the contours of IIP3 as a function of I_C and V_{CE}, which can be used for selection of biasing current and voltage.

Load Dependence and Nonlinearity Cancellation

Figure 9.9 shows the individual and overall IIP3 as a function of load resistance, together with gain. As expected, the gain varies with load, and peaks when the load is closest to conjugate matching. IIP3, however, is sensitive to load variation. The IIP3 with all nonlinearities (denoted by ▲) is noticeably

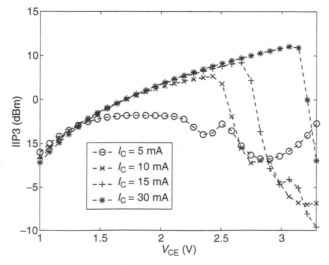

FIGURE 9.6 IIP3 as a function of V_{CE} for different I_C.

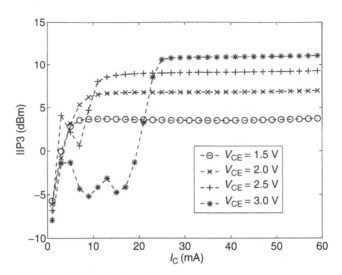

FIGURE 9.7 IIP3 as a function of I_C for different V_{CE}.

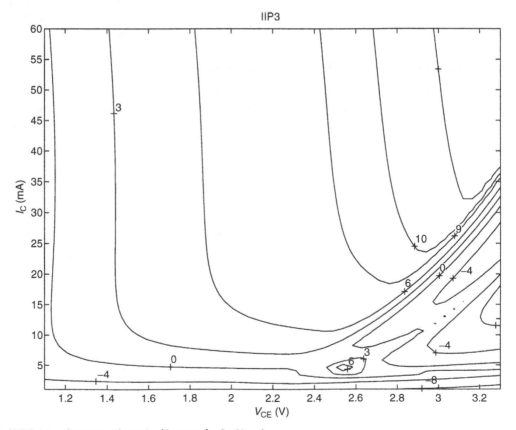

FIGURE 9.8 Contours of IIP3 in dB m on the I_C–V_{CE} plane.

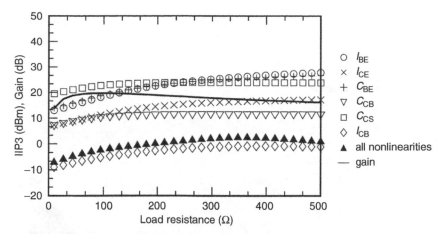

FIGURE 9.9 IIP3 and gain as a function of load resistance at $I_C = 13\,\text{mA}$ and $V_{CE} = 3\,\text{V}$.

higher than the IIP3 with the avalanche current (I_{CB}) nonlinearity alone (denoted by ◇). The interaction between individual nonlinearities has *improved* the overall linearity through cancellation. In this case, the two most dominating nonlinearities are the avalanche current I_{CB} nonlinearity and the C_{CB} nonlinearity. The cancellation between the I_{CB} and C_{CB} nonlinearities leads to an overall IIP3 value that is higher (better) than the IIP3 obtained using the I_{CB} nonlinearity alone. The degree of cancellation depends on the biasing, source and load conditions, as expected from the Volterra series theory.

Physically, the load dependence of linearity in these HBTs results from the CB feedback [6]. The first kind of such feedback is the CB capacitance C_{CB}, and the second kind is the avalanche multiplication current I_{CB} that flows from the collector to base. Both feedbacks are nonlinear, though the load dependence would still exist for linear CB feedback [7] (for instance, externally connected linear CB capacitance).

Linearity Limiting Factors

Figure 9.10 shows the dominant nonlinearity factor on the I_C–V_{CE} plane. The upper limit of I_C is where f_T reaches its peak value. Avalanche multiplication and C_{CB} nonlinearities are the dominant factors for most of the bias currents and voltages. Both avalanche multiplication and C_{CB} nonlinearities can be reduced by reducing the collector doping. This, however, conflicts with the need for high collector doping required to suppress Kirk effect and heterojunction barrier effects in SiGe HBTs. Therefore, multiple collector-doping profiles are needed to provide both high f_T devices and high IIP3 devices for different stages of the same circuit. Typical SiGe processes offer HBTs with different collector-doping profiles through selective collector implantations. Circuit designs could use the higher breakdown voltage devices when f_T is sufficient, which may provide better linearity.

9.5 Summary

Despite the strong *I–V* and *C–V* nonlinearities, SiGe HBTs can be used to design highly linear RF amplifiers. The avalanche multiplication nonlinearity can be minimized by proper choice of biasing current density and voltage. Linearity in general can be improved by increasing the biasing current, and parasitic or intentionally used feedbacks also help improving linearity. All the capacitances, either internal or external C_{BE} and C_{CB}, help improving linearity through negative feedback at the expense of gain. The nonlinear contributions from various *I–V* and *C–V* nonlinearities cancel out each other to certain extent, depending on the impedance of the linear circuit admittance matrix at both the

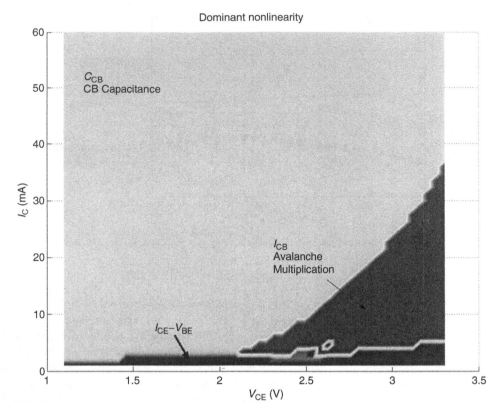

FIGURE 9.10 Dominant nonlinearity factor on the I_C–V_{CE} plane.

fundamental and harmonic frequencies, which can be utilized for further linearity improvement, e.g., through harmonic tuning. Both the C_{CB} and I_{CB} (avalanche) nonlinearities depend on the collector-doping profile, which can be optimized for linearity improvement. The higher breakdown voltage lower speed HBTs in commercial SiGe processes can be used in circuit design for linearity leverage.

Acknowledgments

The author would like to thank Q. Liang for help in preparation of the manuscript. This work is supported by NSF under ECS-0112923 and ECS-0119623 and SRC under SRC-2001-NJ-937 and SRC-2003-NJ-1133.

References

1. DD Weiner and JE Spina. *Sinusoidal Analysis and Modeling of Weakly Nonlinear Circuits*. Van Nostrand Reinhold, New York, 1980.
2. P Wambacq and W Sansen. *Distortion Analysis of Analog Integrated Circuits*. Kluwer Academic, Dordrecht, 1998.
3. S Narayanan. Transistor distortion analysis using Volterra series representation. *Bell Syst. Tech. J.* 46:991–1024, 1967.
4. H Jos. Technology developments driving evolution of cellular phone power amplifiers to integrated RF front-end modules. *IEEE J. Solid-State Circ.* 36:1382–1389, 2001.
5. GF Niu, JD Cressler, S Zhang, U Gogineni, and DC Ahlgren. Measurement of collector–base junction avalanche multiplication effect in advanced UHV/CVD SiGe HBTs. *IEEE Trans. Electron Dev.* 46:1007–1015, 1999.

6. GF Niu, QQ Liang, JD Cressler, C Webster, and DL Harame. RF linearity characteristics of SiGe HBTs. *IEEE Trans. Micro. Theory Tech.* 49:1558–1565, 2001.
7. GF Niu, JD Cressler, WE Ansley, CS Webster, R Anna, and N King. Intermodulation characteristics of UHV/CVD SiGe HBTs. Proc. IEEE Bipolar/BiCMOS Circ. Tech. Meeting, 1999, pp. 50–53.
8. G Gielen and W Sansen. *Symbolic Analysis for Automated Design of Analog Integrated Circuits.* Kluwer Academic, Dordrecht, 1991.
9. V Aparin and LE Larson. Analysis and reduction of cross-modulation distortion in CDMA receivers. *IEEE Trans. Micro. Theory Tech.* 35:1591–1602, 2003.
10. KL Fong. High-frequency analysis of linearity improvement technique of common-emitter transconductance stage using a low-frequency-trap network. *IEEE J. Solid-State Circ.* 35:1249–1252, 2000.
11. QQ Liang, JM Andrews, JD Cressler, and GF Niu. General Analysis of the impact of harmonic impedance on linearity, with applications to SiGe HBTs. Proc. of IEEE Bipolar/BiCMOS Circ. Tech. Meeting, pp. 48–51, 2004.
12. S Maas, B Nelson, and D Tait. Intermodulation in heterojunction bipolar transistors. *IEEE Trans. Micro. Theory Tech.* 40:442–448, 1992.
13. GF Niu, WE Ansley, S Zhang, JD Cressler, C Webster, and R Groves. Noise parameter optimization of UHV/CVD SiGe HBTs for RF and microwave applications. *IEEE Trans. Electron Dev.* 46:1347–1354, 1999.

10

pnp SiGe HBTs

John D. Cressler
Georgia Institute of Technology

10.1 Introduction

At present, SiGe technology development is almost exclusively centered on npn SiGe HBTs. For high-speed analog and mixed-signal circuit applications, however, it is well known that a complementary (npn + pnp) bipolar technology offers significant performance advantages over an npn-only technology [1]. Push–pull circuits, for instance, ideally require a high-speed vertical pnp transistor with comparable performance to the npn transistor [2]. The historical bias in favor of npn Si BJTs is due to the significantly larger minority electron mobility in the p-type base of an npn Si BJT, compared to the lower minority hole mobility in the n-type base of a pnp Si BJT. In addition, the valence band offset in SiGe strained layers is generally more conducive to npn SiGe HBT designs, because it translates into an induced conduction band offset and band grading that greatly enhance minority electron transport in the device, thereby significantly boosting transistor performance over a similarly constructed npn Si BJT. It has been shown that this band alignment is not as restrictive, however, as has been commonly assumed [3]. For a pnp SiGe HBT, on the other hand, the valence band offset directly results in a valence band barrier, *even at low injection*, which strongly degrades minority hole transport and thus limits the frequency response. Careful optimization to minimize these hole barriers in pnp SiGe HBTs is thus required, and has in fact yielded impressive device performance compared to Si pnp BJTs, as demonstrated in the pioneering work reported in Refs. [4–6].

Very recently, in fact, the first commercial complementary SiGe HBT BiCMOS technologies have been reported, in one case targeting high-speed and high-voltage analog circuit applications [7], and the other demonstrating impressive levels of pnp SiGe HBT performance, in this with a peak f_T/f_{max} of 80/120 GHz at a BV_{CEO} of 2.6 V [8]. Details of these impressive complementary SiGe technology demonstrations are given in Chapters 11 and 13 (see *Fabrication of SiGe HBT BiCMOS Technology*), respectively.

An analysis of the inherent profile design differences between npn and pnp SiGe HBTs is relevant in this context of complementary SiGe HBT technologies, as well as meaningful design guidelines for constructing pnp SiGe HBTs. Relevant questions include, for instance:

- How does SiGe npn and pnp profile design fundamentally differ?
- Can a single Ge profile design point be used for both npn and pnp transistors, for a given stability constraint?
- Is a graded-base Ge profile design preferable to a box-shaped Ge profile design for pnp HBTs?
- How much Ge retrograding in the collector–base junction is required to obtain acceptable SiGe pnp HBT performance?

These issues are addressed using calibrated device simulations to shed light on the fundamental SiGe profile design differences between npn SiGe HBT and pnp SiGe HBTs that might be encountered, for instance, in developing such a viable complementary SiGe HBT technology [9].

10.2 Simulation of pnp SiGe HBTs

One-dimensional MEDICI simulations [10] which were used to analyze the differences in intrinsic profile design between pnp and npn transistors are the central focus. To aid in interpretation of the results, simplistic hypothetical npn and pnp SiGe profiles with constant emitter, base, and collector doping, and a Ge content *not* subject to thermodynamic stability constraints, were initially adopted (Figure 10.1). These profile assumptions are clearly nonphysical for real SiGe technologies, but are very useful for comparing npn and pnp devices so that their differences can be more easily discriminated and not masked by doping-gradient-induced phenomena (stability issues will be addressed below). This artificial assumption on constant doping clearly yields ac performance numbers (e.g., f_T) that are lower than what would be expected for a real complementary SiGe HBT technology, but relative comparisons between npn and pnp devices are nonetheless valid, and the comparison methodology widely applicable.

MEDICI models of the devices were constructed using actual device layouts and measured SIMS data, and careful calibration of MEDICI simulations for both npn and pnp Si BJTs to measured complementary Si BJT hardware was performed. It was found that the default minority hole mobility modeling capability of MEDICI was deficient and tuning was required to obtain reasonable agreement between data and simulation, particularly under high-level injection. The SiGe model parameters determined from earlier calibrations of high-speed npn SiGe HBTs were used [3], and assumed to be the same for both npn and pnp transistors.

10.3 Profile Optimization Issues

A comparison of the equilibrium conduction and valence band edges for both npn and pnp devices without any Ge retrograding into the collector (i.e., an abrupt transition from the peak Ge content to zero Ge content in the CB junction) is shown in Figure 10.2 and Figure 10.3 for: (1) a Si BJT; (2) a triangular (linearly graded) Ge profile with a peak Ge content of 10%; and (3) a triangular Ge profile with a peak Ge content of 25%. Observe that while there is no visible conduction band barrier present in the npn HBT, there is an obvious valence band barrier in the pnp HBT, even for low Ge content. This is

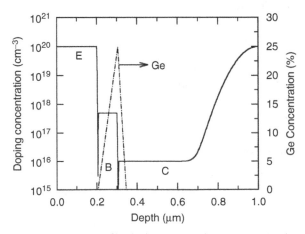

FIGURE 10.1 Hypothetical doping and Ge profiles for both pnp and npn SiGe HBTs. (From JD Cressler and G Niu. *Silicon–Germanium Heterojunction Bipolar Transistors.* Boston, MA: Artech House, 2003. With permission.)

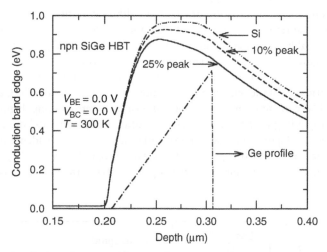

FIGURE 10.2 Valence band edge of an npn SiGe HBT for varying peak Ge content. (From G Zhang, JD Cressler, G Niu, and A Pinto. *Solid-State Electronics* 44:1949–1954, 2000. With permission.)

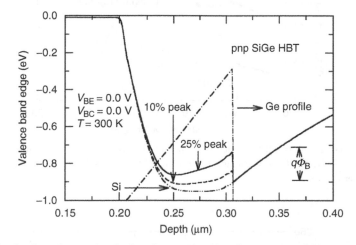

FIGURE 10.3 Simulated conduction band edge of a pnp SiGe HBT for varying peak Ge content. (From G Zhang, JD Cressler, G Niu, and A Pinto. *Solid-State Electronics* 44:1949–1954, 2000. With permission.)

consistent with the fact that there is a valence band offset in strained SiGe on Si (refer to Chapter 4), and clearly indicates that pnp SiGe HBT design is inherently more difficult than npn SiGe HBT design. In addition, due to the inherent minority carrier mobility differences between electrons and holes, it is also clear that npn devices will consistently out-perform pnp devices, everything else being equal.

Unlike for a well-designed npn SiGe HBT (i.e., Ge outside the neutral base edges), where conduction band barrier effects are uncovered only at high J_C under Kirk effect [3] (refer to Chapter 5), the valence band barrier in pnp SiGe HBTs is in play even at low injection, and acts to block minority holes transiting the base. This pileup of accumulated holes produces a retarding electric field in the base, which compensates the Ge-grading-induced drift field, dramatically decreasing both J_C, β, and f_T. This effect worsens as the current density increases, since more hole charge is stored in the base. In this case, the f_T of the pnp SiGe HBT is in fact significantly lower than that of the pnp Si BJT. As expected, however, retrograding of the Ge edge into the collector can "smooth" this valence band offset in the pnp

SiGe HBT, and thus improve this situation dramatically, although at the expense of film stability [4,5]. For an increase of the Ge retrograde from 0 to 40 nm, the pnp SiGe HBT performance is dramatically improved, yielding roughly a 2× increase in peak f_T over the pnp Si BJT performance at equal doping.

Figure 10.4 and Figure 10.5 show the variation in peak f_T and β as a function of peak Ge content for both npn and pnp SiGe HBTs for both a 0-nm Ge retrograde and 100-nm Ge retrograde. At 100-nm retrograde, the performance of the pnp SiGe HBT monotonically improves as the Ge content rises, while the maximum useful Ge content is limited to about 10% without retrograding. Figure 10.6 indicates that 40–50 nm of Ge retrograding in the pnp SiGe HBT is sufficient to "smooth" the valence band barrier, and this is reflected in Figure 10.7, which explicitly shows the dependence of pnp peak f_T on Ge retrograde distance, for both triangular and box Ge retrograde profile shapes. Observe that the box Ge retrograde is not effective in improving the pnp SiGe HBT performance, since it does not smooth

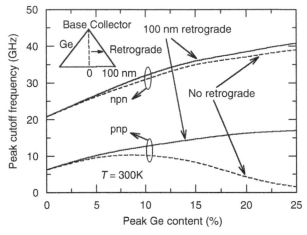

FIGURE 10.4 Simulated peak cutoff frequency as a function of peak Ge content for different Ge retrogrades for both npn and pnp SiGe HBTs. (From G Zhang, JD Cressler, G Niu, and A Pinto. *Solid-State Electronics* 44:1949–1954, 2000. With permission.)

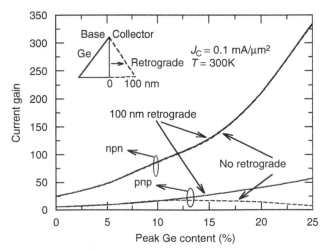

FIGURE 10.5 Simulated current gain as a function of peak Ge content for different Ge retrogrades for both npn and pnp SiGe HBTs. (From G Zhang, JD Cressler, G Niu, and A Pinto. *Solid-State Electronics* 44:1949–1954, 2000. With permission.)

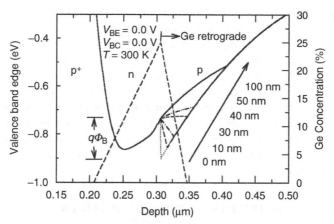

FIGURE 10.6   Valence band edge of a pnp SiGe HBT as a function of Ge retrograde distance. (From G Zhang, JD Cressler, G Niu, and A Pinto. *Solid-State Electronics* 44:1949–1954, 2000. With permission.)

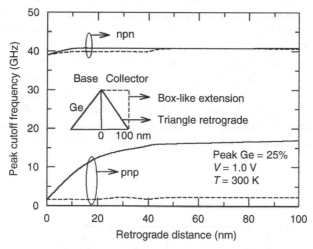

FIGURE 10.7   Simulated peak cutoff frequency as a function of both Ge retrograde distance and Ge profile shape for both npn and pnp SiGe HBTs. (From G Zhang, JD Cressler, G Niu, and A Pinto. *Solid-State Electronics* 44:1949–1954, 2000. With permission.)

the Ge barrier, but rather only pushes it deeper into the collector, where it is still felt at the high J_C needed to reach peak f_T. This box Ge retrograde is also clearly undesirable from a stability standpoint. The effects of Ge retrograding on the npn SiGe HBT performance, on the other hand, are minor, while the film stability is significantly worse due to the additional Ge content. This suggests that using one Ge profile design for both npn and pnp SiGe HBTs is not optimum for high-peak Ge content values. Note that while the peak f_T is unchanged with Ge retrograding in the npn SiGe HBT, the f_T response above peak f_T does not roll off as rapidly due to the high-injection-induced barrier, consistent with the results in Ref. [3] (refer to Chapter 5).

   An examination of the frequency response of the npn and pnp SiGe HBTs as a function of front-side Ge profile shape (in this case, triangle versus box Ge profile, with a fixed retrograde of 100 nm for both) and peak Ge content shows that for the npn SiGe HBT, the base transit time reduction from the Ge-grading-induced drift field of the triangle Ge profile shape gives a significant advantage above 10%

peak Ge, indicating that the npn SiGe HBT is base transit time limited. Interestingly, for the pnp SiGe HBT, however, the differences between the box and triangle Ge profiles are much less pronounced, everything else being equal. The box Ge profile gives a slight advantage at low Ge content due to the low β and hence importance of the emitter transit time ($\tau_E \propto 1/\beta$), but once the β is sufficiently high, the triangle Ge profile dominates at higher peak Ge content, where the base transit time limits the overall response. In both npn and pnp devices, the triangle Ge profile offers better performance and better stability (less-integrated Ge content), and thus can be considered an optimum shape for both devices. This is even more apparent if we examine the Early voltage of the devices, a key figure-of-merit for complementary analog circuits. In this case, the triangle Ge profile has a clear advantage due to its graded bandgap, as expected, and both npn and pnp transistors show a significant improvement in V_A with increasing Ge content.

10.4 Stability Constraints in pnp SiGe HBTs

The total amount of Ge that can be put into a given SiGe HBT is limited by the thermodynamic stability criterion. Above the critical thickness, the strain in the SiGe film relaxes, generating defects. In general, varying peak Ge content or retrograde distance (i.e., film thickness) moves the profile along different contours in stability space (Figure 10.8). Under the SiGe stability constraint, the peak Ge content must be traded off for the Ge retrograde distance in the collector–base junction. Figure 10.9 shows that a 11% peak Ge profile with a 25-nm retrograde gives the highest f_T for the pnp SiGe HBT at this design point. A similar exercise for the npn SiGe HBT shows that the ac performance is not sensitive to the SiGe profile shapes used, and, hence, without a significant loss of performance, the same Ge profile may in principle be used for both pnp and npn SiGe HBTs. This may be advantageous from a fabrication viewpoint. These results should be valid for current SiGe technology nodes with about 100-nm base width. If the base width is further reduced with technology scaling, the peak Ge content can be obviously increased, while maintaining film stability. The same optimization methodology employed here can be used in that case to determine the best SiGe profile for both devices.

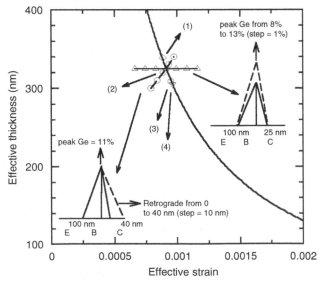

FIGURE 10.8 SiGe stability diagram illustrating the various pnp profile design tradeoffs. (From G Zhang, JD Cressler, G Niu, and A Pinto. *Solid-State Electronics* 44:1949–1954, 2000. With permission.)

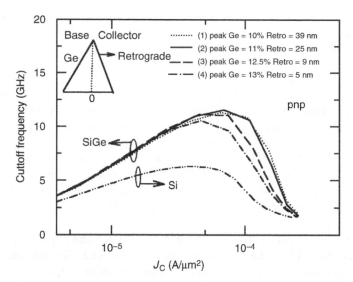

FIGURE 10.9 Simulated cutoff frequency as a function of collector current density for pnp SiGe HBTs with different Ge profiles. Refer to Figure 6.8 for the exact profile shapes. (From G Zhang, JD Cressler, G Niu, and A Pinto. *Solid-State Electronics* 44:1949–1954, 2000. With permission.)

10.5 Summary

In spite of the historical bias in favor of npn SiGe HBTs, complementary (npn + pnp) SiGe HBT technology has recently emerged as a viable mixed-signal technology. In this chapter, we have examined the fundamental profile design constraints associated with pnp SiGe HBTs, and importantly, how those constraints differ from those faced in conventional npn SiGe HBT design.

Acknowledgments

I am grateful to G. Zhang, G. Niu, and D. Harame for their contributions. This work was supported by the GEDC at Georgia Tech and IBM.

References

1. JD Cressler, J Warnock, DL Harame, JN Burghartz, KA Jenkins, and CT Chuang. A high-speed complementary silicon bipolar technology with 12 fJ power-delay product. *IEEE Electron Device Letters* 14:523–526, 1993.
2. CT Chuang, JD Cressler, and JD Warnock. ac-coupled complementary push-pull ECL circuits with 34 fJ power-delay product. *Electronics Letters* 29:1938–1939, 1993.
3. JD Cressler and G Niu. *Silicon–Germanium Heterojunction Bipolar Transistors*. Boston, MA: Artech House, 2003.
4. DL Harame, JMC Stork, BS Meyerson, EF Crabbé, GL Patton, GJ Scilla, E de Frésart, AA Bright, C Stanis, AC Megdanis, MP Manny, EJ Petrillo, M Dimeo, RC McIntosh, and KK Chan. SiGe-base pnp transistors fabrication with n-type UHV/CVD LTE in a "NO DT" process. Technical Digest of the IEEE Symposium on VLSI Technology, Honolulu, 1990, pp. 47–48.
5. DL Harame, BS Meyerson, EF Crabbé, JM Cotte, JMC Stork, AC Megdanis, GL Patton, SR Stiffler, JB Johnson, JD Warnock, JH Comfort, and JY-C Sun. 55 GHz polysilicon-emitter graded SiGe-base pnp transistors. Technical Digest of the IEEE Symposium on VLSI Technology, Oiso, Japan, 1991, pp. 71–72.

6. DL Harame, JH Comfort, EF Crabbé, JD Cressler, JD Warnock, BS Meyerson, KY-J Hsu, J Cotte, CL Stanis, JMC Stork, JY-C Sun, DA Danner, and PD Agnello. A SiGe-base pnp ECL circuit technology. Technical Digest of the IEEE Symposium on VLSI Technology, Honolulu, 1993, pp. 61–62.

7. B El-Kareh, S Balster, W Leitz, P Steinmann, H Yasuda, M Corsi, K Dawoodi, C Dirnecker, P Foglietti, A Haeusler, P Menz, M Ramin, T Scharnagl, M Schiekofer, M Schober, U Schultz, L Swanson, D Tatman, M Waitschull, J Weijtmans, and C Willis. A 5 V complementary-SiGe BiCMOS technology for high-speed precision circuits. Proceedings IEEE Bipolar/BiCMOS Circuits and Technology Meeting, Toulouse, France, 2003, pp. 211–214.

8. B Heinemann, R Barth, D Bolze, J Drews, P Formanek, O Fursenko, M Glante, K Glowatzki, A Gregor, U Haak, W Hoppner, D Knoll, R Kurps, S Marschmeyer, S Orlowski, H Rucker, P Schley, D Schmidt, R Scholz, W Winkler, and Y Yamamoto. A complementary BiCMOS technology with high-speed npn and pnp SiGe:C HBTs Technical Digest of the IEEE International Electron Devices Meeting, Washington, 2003, pp. 117–120.

9. G Zhang, JD Cressler, G Niu, and A Pinto. A comparison of npn and pnp profile design tradeoffs for complementary SiGe HBT technology. *Solid-State Electronics* 44:1949–1954, 2000.

10. MEDICI. 2-D Semiconductor Device Simulator, Avant!, Fremont, CA, 1997.

11

Temperature Effects

John D. Cressler
Georgia Institute of Technology

11.1 Introduction

Bandgap engineering generally has a positive influence on the low-temperature characteristics of bipolar transistors [1]. SiGe HBTs operate very well, in fact, in the cryogenic environment (e.g., liquid nitrogen temperature $= 77.3\,\text{K} = -320°\text{F} = -196°\text{C}$), an operational regime traditionally forbidden to Si BJTs. At present, cryogenic electronics represents a small but important niche market, with applications such as high-sensitivity cooled sensors and detectors, semiconductor–superconductor hybrid systems, space electronics, and eventually cryogenically cooled computers systems. While the large power dissipation associated with conventional bipolar digital circuit families such as emitter-coupled-logic (ECL) would likely preclude their widespread use in cooling-constrained cryogenic systems, the combination of cooled, low-power, scaled Si CMOS with SiGe HBTs offering excellent frequency response, low-noise performance, radiation hardness, and excellent analog properties represents a unique opportunity for the use of SiGe HBT BiCMOS technology in cryogenic systems. Furthermore, independent of the potential cryogenic applications that may exist for SiGe HBT BiCMOS technology, all electronic systems must successfully operate over an extended temperature range (e.g., -55 to $125°\text{C}$ to satisfy military specifications and 0 to $85°\text{C}$ for most commercial applications), and thus, understanding how Ge-induced bandgap engineering affects SiGe HBT device and circuit operation is important.

In this chapter, we address temperature effects in SiGe HBTs, by first reviewing the impact of temperature on bipolar transistor device and circuit operation. We then show how temperature couples to SiGe HBT dc and ac performance, how one optimizes SiGe HBTs specifically for cryogenic operation, and finally consider the operation of SiGe HBTs at elevated temperatures (to $300°\text{C}$).

11.2 The Impact of Temperature on Bipolar Transistors

The detrimental effects of cooling on homojunction bipolar transistor operation have been appreciated for many years [2–6]. While the precise dependence of Si BJT properties on cooling can be a strong function of technology generation and profile design, Si BJT device and circuit properties cooled to cryogenic temperatures typically exhibit [7–12]:

- A modest increase (degradation) in the junction turn-on voltage with decreasing temperature (monotonic).
- A strong increase (improvement) in the low-injection transconductance with cooling (monotonic).
- A strong increase (degradation) in the base resistance with cooling (typically, quasi-exponential below about 200 K).
- A mild decrease (improvement) in parasitic transistor depletion capacitances (monotonic).
- A strong decrease (degradation) in β with cooling (quasi-exponential).
- A modest decrease (degradation) in frequency response with cooling, with f_T typically degrading more rapidly than f_{max} with decreasing temperature (monotonic below about 200 K).
- An increase (degradation) in ECL circuit delay with cooling (monotonic below about 200 K).
- The noise margin of current-switch-based digital circuits (e.g., ECL) increases (improves) with cooling (monotonic), allowing reduced logic swing operation.

The impact of cooling of Si BJTs is typically largely one of serious device and circuit performance degradations, effectively precluding their use in cryogenic applications. The addition of SiGe to this problem can be used to change this situation dramatically.

11.3   Cryogenic Operation of SiGe HBTs

Intuitively, we expect that band-edge effects induced by bandgap engineering will generally couple strongly to bipolar transistor properties. This strong coupling is physically the consequence of the fact that the bipolar transistor is a minority carrier device, and hence the terminal currents are proportional to n_{i0}^2 via the Shockley boundary conditions, with n_{i0}^2 in turn proportional to the exponential of the bandgap. Hence, changes to the bandgap will couple exponentially to the currents. Furthermore, from very general statistical mechanical considerations, these bandgap changes will inevitably be divided by the thermal energy (kT), such that a reduction in temperature will greatly magnify any bandgap changes. Not surprisingly, then, even a cursory examination of the SiGe HBT device equations suggests that both the dc and ac properties of SiGe HBTs should be favorably affected by cooling [13,14]. In fact, the thermal energy (kT), *in every instance*, is arranged in the SiGe HBT equations such that it favorably affects the low-temperature properties of the particular performance metric in question, be it $\beta(T)$, $f_T(T)$, or $V_A(T)$.

The beneficial role of temperature in SiGe HBTs can be used to easily offset the inherent bandgap-narrowing-induced degradation in current gain of a Si BJT to achieve viable dc operation down to 77 K, even for a SiGe HBT that has not been optimized for the cryogenic environment. Figure 11.1 shows the evolution of peak current gain as a function of reciprocal temperature from early Si BJT technologies circa 1978 to SiGe technologies circa 1992. Clearly, the addition of Ge-induced bandgap engineering enables functional current gain down at least to 77 K with minimal effort. From a dynamic point of view, the Ge-grading-induced base drift field provides a means to offset the inherent τ_b degradation associated with cooled Si BJTs, yielding an f_T that does not degrade with cooling. Since the reduced thermal cycle nature of epitaxial growth techniques are generally more conducive to maintaining thinner, more heavily doped base profiles than conventional ion-implanted bases used in modern Si BJTs, it is fairly straightforward to control base freeze-out in SiGe HBTs, at least down to 77 K, and hence R_b at cryogenic temperatures can be more easily controlled. If f_T and R_b do not degrade significantly with cooling, then achieving respectable circuit performance down to 77 K becomes a reality unknown to Si BJT technologies.

Figure 11.2 shows the evolution of unloaded ECL gate delay as a function of publication date. As expected, optimized 300-K technology scaling successfully improved circuit speed over time. More surprising, perhaps, is that the rate of improvement in low-temperature performance was significantly faster. The 1991 and 1992 cryogenic data points are for SiGe HBT technologies, and clearly demonstrate that one can no longer out of hand dismiss Si-based bipolar technologies for

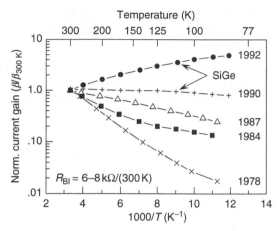

FIGURE 11.1 Evolution of current-gain temperature dependence with Si-based bipolar technology generation. The last two generations are SiGe HBT technologies. (From JD Cressler, JH Comfort, EF Crabbé, JMC Stork, and JY-C Sun. *IEEE Trans. Electron Dev.* 40:525–541, 1993. With permission.)

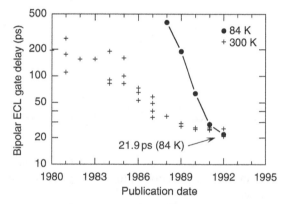

FIGURE 11.2 Evolution of unloaded ECL gate delay at 310 and 84 K with Si-based bipolar technology generation. The last two generations are SiGe HBT technologies. (From JD Cressler, JH Comfort, EF Crabbé, JMC Stork, and JY-C Sun. *IEEE Trans. Electron Dev.* 40:525–541, 1993. With permission.)

the cryogenic applications. SiGe can thus be viewed as an effective means to extend Si-based bipolar technology to the cryogenic environment (with little or no effort). This scenario is particularly appealing if we consider the state-of-the-art SiGe HBT BiCMOS technologies, since Si CMOS also performs well down to 77 K, and provides a major advantage in the reduction in power dissipation, an often serious constraint given the limited efficiency of cryocoolers. While it is unlikely that one would develop SiGe technology explicitly for cryogenic applications, if (as is the case) one could simply take a room temperature-optimized SiGe technology and operate it at low temperatures without serious modification, that prospect might prove cost-effective. With the present trend toward reduced-temperature operation of CMOS-based high-end servers as a performance and reliability enhancement vehicle (currently at 0 to −40°C and going lower), the appeal of SiGe HBT BiCMOS technologies for the cryogenic environment may naturally grow over time, since HBTs can provide numerous advantages over CMOS in analog, RF, heavily loaded digital, and high-speed driver or receiver applications.

We first examine the expected theoretical temperature dependence of the important SiGe HBT performance metrics. Compared to a comparably constructed Si BJT, $\beta(T)$ in a SiGe HBT should increase exponentially with decreasing temperature, since

$$\frac{\beta_{\text{SiGe}}}{\beta_{\text{Si}}}\bigg|_{V_{\text{BE}}} \simeq \left\{\frac{\tilde{\gamma}\tilde{\eta}\Delta E_{\text{g,Ge}}(\text{grade})/kT e^{\Delta E_{\text{g,Ge}}(0)/kT}}{1 - e^{-\Delta E_{\text{g,Ge}}(\text{grade})/kT}}\right\}. \tag{11.1}$$

As expected, a quasi-exponential increase in the SiGe-to-Si current gain ratio with decreasing temperature is typically observed experimentally. In addition, $V_A(T)$ and $\beta V_A(T)$ in a SiGe HBT should also increase exponentially with decreasing temperature compared to a comparably constructed Si BJT, since

$$\frac{V_{\text{A,SiGe}}}{V_{\text{A,Si}}}\bigg|_{V_{\text{BE}}} \simeq e^{\Delta E_{\text{g,Ge}}(\text{grade})/kT}\left[\frac{1 - e^{-\Delta E_{\text{g,Ge}}(\text{grade})/kT}}{\Delta E_{\text{g,Ge}}(\text{grade})/kT}\right] \tag{11.2}$$

and

$$\frac{\beta V_{\text{A,SiGe}}}{\beta V_{\text{A,Si}}} = \tilde{\gamma}\tilde{\eta}e^{\Delta E_{\text{g,Ge}}(0)/kT}e^{\Delta E_{\text{g,Ge}}(\text{grade})/kT}. \tag{11.3}$$

This is again confirmed experimentally. The anticipated temperature dependence of the frequency response of a SiGe HBT can be gleaned from the temperature dependence of the base and emitter transit times,

$$\frac{\tau_{\text{b,SiGe}}}{\tau_{\text{b,Si}}} = \frac{2}{\tilde{\eta}}\frac{kT}{\Delta E_{\text{g,Ge}}(\text{grade})}\left\{1 - \frac{kT}{\Delta E_{\text{g,Ge}}(\text{grade})}\left[1 - e^{-\Delta E_{\text{g,Ge}}(\text{grade})/kT}\right]\right\} \tag{11.4}$$

and

$$\frac{\tau_{\text{e,SiGe}}}{\tau_{\text{e,Si}}} \simeq \frac{J_{\text{C,Si}}}{J_{\text{C,SiGe}}} = \frac{1 - e^{-\Delta E_{\text{g,Ge}}(\text{grade})/kT}}{\tilde{\gamma}\tilde{\eta}\dfrac{\Delta E_{\text{g,Ge}}(\text{grade})}{kT}e^{\Delta E_{\text{g,Ge}}(0)/kT}}. \tag{11.5}$$

Both are favorably influenced by cooling, and thus, we expect that the influence of the graded SiGe base is sufficient to overcome the inherent electron diffusivity degradation on τ_b with cooling, and this is indeed the experimental case.

11.4   Design Constraints

While SiGe HBTs designed for room temperature operation function acceptably down to 77 K, second-order design constraints do, nonetheless, exist, and can impact profile optimization [15,16]. The first such constraint centers on the base current and its impact on the current gain at low injection. While conventional Shockley–Read–Hall (SRH) recombination exponentially decreases with cooling, thereby effectively eliminating reverse leakage in the collector–base junction, the same is not true of carrier tunneling processes, whether they are band-to-band or trap-assisted. Given that the EB junction of high-speed bipolar transistors (either Si or SiGe) are typically quite heavily doped (often in the vicinity of $1 \times 10^{18}\,\text{cm}^{-3}$), the doping-induced electric field is high, and can result in substantial parasitic tunneling leakage. While this is generally easily designed around in 300-K designs, it is more problematic at low temperatures, given that the collector and base currents decrease strongly at fixed V_{BE} as the temperature drops. In this case, as the base current decreases with cooling, any tunneling-induced leakage will remain roughly constant, hence uncovering a parasitic leakage "foot" on the base current (this effect can be clearly seen in Figure 11.3). This parasitic base leakage current can severely limit the current gain at low injection at cryogenic temperatures. Thus, as a rule of thumb, it can be safely stated that the ideality of the base current of a high-performance Si or SiGe bipolar transistor will never improve with cooling. If the base current is ideal (i.e., $e^{qV_{\text{BE}}/kT}$) down to a picoampere at 300 K, it may be ideal only to a nanoampere

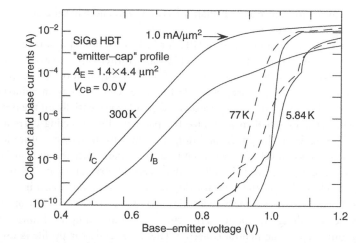

FIGURE 11.3 Gummel characteristics at 300, 77, and 5.84 K for a 77-K optimized emitter-cap SiGe HBT. (From AJ Joseph, JD Cressler, and DM Richey. *IEEE Electron Dev. Lett.* 16:268–270, 1995. With permission.)

at 77 K. If it is even modestly non-ideal at 300 K, it will be quite leaky at 77 K. How serious a limitation this leakage is depends strongly on the circuit application. In digital logic, for instance, it is not an issue, given that the devices are biased well out of the leakage regime, and β does not strongly couple to circuit speed. For more sensitive analog circuits, however, it can in principle require careful design consideration. As discussed below, one can optimize a SiGe HBT to reduce this leakage effect, a feat much more easily accomplished using epitaxial growth rather than ion-implantation for the base layer formation.

More worrisome than the base current at low temperatures, however, is the enhancement of high-injection, heterojunction barrier effects with cooling (refer to Chapter 5 for a detailed discussion of barrier effects in SiGe HBTs). Band-edge effects in bipolar transistors generally couple very strongly to the device properties, and barrier effects are no exception. In this case, given that barrier effects necessarily exist in all practical SiGe HBTs, cooling will make the situation decidedly worse. The consequences of barrier effects, as at room temperature, include a premature roll-off in both β and f_T at high J_C, and a limitation on maximum output current drive. What is different in the context of cryogenic operation, however, is that while a well-designed 300-K SiGe HBT may not show any clear evidence of barrier effect at 300 K, it will certainly show evidence of it at 77 K, and its impact on device performance will be correspondingly worse. That is, the design margin for 77-K operation is in essence narrower, always an undesirable situation. As discussed in Chapter 5, the device transconductance is a useful tool for assessing barrier effects in SiGe HBTs. A comparison of g_m between comparably designed i–p–i SiGe HBTs and i–p–i Si BJTs clearly shows that while g_m at low J_C increases with cooling as expected, a dramatic drop in g_m at a higher critical current density close to that of Kirk effect can be observed in the SiGe HBT. Fortunately, it is also true that this critical onset current density in fact increases with cooling, consistent with the fact that the saturation velocity rises at low temperatures, thus delaying Kirk effect until higher J_C. As discussed below, this result can be traded off to optimize SiGe HBTs for 77 K operation. One would also expect that barrier effect would have a serious impact on transistor dynamic response, given that enhanced charge storage in the base couples strongly to f_T. The approaches that can be used to design around barrier effects at cryogenic temperatures are the same as those outlined in Chapter 5, albeit with a narrower design margin than at 300 K.

11.5 Optimization of SiGe HBTs for 77 K

While conventional 300-K SiGe HBT designs will inherently function reasonably well down to 77 K, it remains to be seen whether a SiGe HBT designed *specifically* for 77 K operation can achieve significantly better device and circuit performance at 77 K than it has at 300 K, and what the design issues and trade-offs faced in achieving this goal would be.

To address the explicit optimization of a SiGe HBT for 77 K operation, a new profile design point and fabrication scheme is required [17]. In this case an epitaxial "emitter-cap" layer doped with phosphorus at about 1×10^{18} cm^{-3} was deposited *in situ* in a UHV/CVD deposition tool on top of the SiGe-base to form the EB junction. This 77-K optimized SiGe HBT will be referred to as an epitaxial "emitter-cap" SiGe HBT [18]. Because EB carrier tunneling processes depend exponentially on the peak junction field, the lightly doped emitter is expected to minimize the parasitic EB tunneling current compared to a conventional "i–p–i" SiGe HBT design. In addition, the increase in carrier saturation velocity with cooling, as well as the presence of velocity overshoot in the CB space–charge region at 77 K, results in an onset current density of base push-out (Kirk effect) that is about 50% larger at 77 K than at 300 K [15]. Thus, compared to a 300-K design, the collector doping level can be decreased in an optimized 77-K profile. In this case, the doping level at the metallurgical CB junction was lowered from 1×10^{17} cm^{-3} for the conventional SiGe HBT design to about 2×10^{16} cm^{-3}, and ramped upward toward the subcollector to minimize freeze-out deep in the neutral collector. This 77-K collector profile is used to reduce the parasitic CB capacitance under the constraint that the onset current density of the SiGe–Si heterojunction barrier be above the maximum operating current density of about 1.0 mA/μm^2.

To ensure a low emitter resistance, a 200-nm *in situ* doped polysilicon contact was deposited on top of the composite EB profile (n-cap/p-SiGe). Because the arsenic out-diffusion from the heavily doped polysilicon layer is used only to contact the epitaxial phosphorus emitter and does not determine the metallurgical EB junction, only a very short rapid thermal annealing (RTA) step is required to activate and redistribute the emitter dopants, allowing the maintenance of a thin, heavily doped base. A metallurgical emitter-cap thickness of about 10 nm was achieved at the end of processing (estimated by subtracting the arsenic out-diffusion of the emitter poly from the total EB junction depth). The boron doping of the base profile was increased over a more conventional i–p–i SiGe design to improve its base freeze-out properties, and was deposited as a box 10 nm wide by 2.5×10^{19} cm^{-3}. At the end of processing the metallurgical base was about 75 nm wide with a peak concentration of about 8×10^{18} cm^{-3}, well above the Mott transition for carrier freeze-out. To minimize minority carrier charge storage in the emitter-cap layer, a large 77 K β is also desirable ($\tau_e \propto 1/\beta_{ac}$). Therefore, a trapezoidal Ge profile with 3 to 4% Ge at the EB junction (compared to about 0 to 1% for the standard design) and ramping to 8.5% at the CB junction (compared to about 8.5% for the standard design) was used. The resultant emitter-cap Ge profile was about 65 nm thick, and satisfied the thermodynamic stability criteria for UHV/CVD blanket films.

This 77-K SiGe design point yields a transistor with reasonably ideal Gummel characteristics at low temperatures, with a maximum output current drive well above 1.0 mA/μm^2 at 84 K (Figure 11.3). The higher Ge concentration at the EB junction, the beneficial effects of the emitter high–low (n$^+$/n$^-$ cap) junction, and the bandgap narrowing of the heavily doped base, offset the bandgap narrowing of the heavily doped emitter region to yield a peak β that increases quasi-exponentially with cooling from 102 at 310 K to 498 at 84 K (Figure 11.4). This large β value at low temperatures serves to minimize the unwanted charge storage associated with the emitter-cap layer as well as to circumvent the degradation of β at medium injection levels due to bias-dependent Ge ramp effects (refer to Chapter 5), giving an ideal value of β of 99 at 84 K at a typical circuit operating point of 1.0-mA collector current [18]. An undesirable result of the high β at low temperature, however, is a decrease in the BV$_{CEO}$ from 3.1 V at 310 K to 2.3 V at 84 K, but it remains acceptable for most circuit applications. Depending on circuit requirements at 77 K, the low-temperature current gain can be easily tuned to a desired value.

The reduction in overall thermal cycle compared to a conventional design is key to maintaining the abrupt, as-deposited boron base profile, and thus providing immunity to carrier freeze-out at cryogenic temperatures (R_{bi} only increases from 7.7 to 11.0 kΩ/sq. between 310 and 84 K). Importantly, this immunity to base freeze-out does not come at the expense of increased EB leakage, as it does, for instance, in a spacer-free SiGe profile with a very heavily doped base [15]. The lower doping level of the emitter-cap layer results in a reverse EB leakage at 1.0 V at 84 K, which is more than 500 times smaller

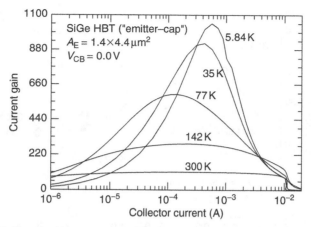

FIGURE 11.4 Current gain versus bias current as a function of temperature for a 77-K optimized emitter-cap SiGe HBT. (From AJ Joseph, JD Cressler, and DM Richey. *IEEE Electron Dev. Lett.* 16:268–270, 1995. With permission.)

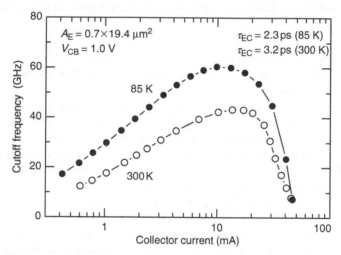

FIGURE 11.5 Cutoff frequency characteristics at 300 and 85 K for a 77-K optimized emitter-cap SiGe HBT. (From JD Cressler, EF Crabbé, JH Comfort, JY-C Sun, and JMC Stork. *IEEE Electron Dev. Lett.* 15:472–474, 1994. With permission.)

than for the conventional SiGe design. The consequence is a much smaller forward tunneling component in the base current (much larger low-current β), a smaller EB capacitance, and an expected improvement in hot-carrier reliability at cryogenic temperatures.

As shown in Figure 11.5, the transistor cutoff frequency (f_T) rises from 43 to 61 GHz with cooling to 85 K due to the beneficial effects of the Ge-grading-induced drift field. This improvement in f_T, coupled to the low total base resistance and slightly decreased CB capacitance, yields an increase in maximum oscillation frequency with cooling as well, from 40 GHz at 310 K to 50 GHz at 84 K. To assess the 77-K circuit capabilities of this technology, unloaded ECL ring oscillators were measured (Figure 11.6). High-power (12.45 mW) ECL circuits switch at a record 21.9 psec at 84 K, 3.5 psec faster than at 310 K. Circuits that were optimized for lower power operation achieve a minimum power-delay product of 61 fJ (41.3 psec at 1.47 mW) at 84 K, and are 9.6 psec faster than at 310 K. These 77-K optimized ECL circuits are expected to exhibit even more dramatic improvements in speed over room-temperature ECL circuits under heavy loading, due to the beneficial effects of cooling on metal interconnect resistance and

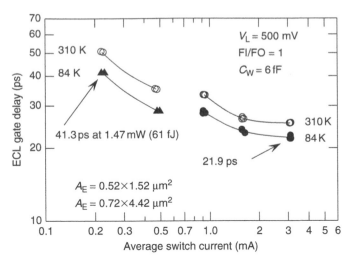

FIGURE 11.6 Unloaded ECL gate delay as a function of power at 310 and 84 K. (From JD Cressler, EF Crabbé, JH Comfort, JY-C Sun, and JMC Stork. *IEEE Electron Dev. Lett.* 15:472–474, 1994. With permission.)

circuit logic swing [16]. The delay improvement at long interconnect wire lengths can be dramatic (2.7× faster at 84 K than at 300 K at 10 mm wire length), and suggests that SiGe HBT based line-drivers might be attractive for 77-K applications.

Recent measurements on non-cryogenically optimized 200 GHz, third-generation SiGe HBTs, show even more impressive performance down to liquid nitrogen temperature [19]. Current–voltage measurements across the 300 to 85 K temperature range were made on SiGe HBTs with an emitter area of $0.12 \times 10.0 \, \mu m^2$. In spite of the high peak base and emitter doping levels associated with these aggressively scaled SiGe HBTs ($>10^{19} \, cm^{-3}$), the base current remains reasonably ideal at 85 K. This is the result of the lightly doped epitaxial spacer layer inserted between the base and emitter regions, and helps limit field-assisted tunneling and recombination at low temperatures. The base and emitter regions in this device are both doped above the Mott-transition, and ensure that carrier freeze-out does not negatively impact the base or emitter resistance below 100 K. This device is capable of very high current density operation ($>25 \, mA/\mu m^2$), and thus the high collector doping level effectively limits the impact of heterojunction barrier effects at low temperatures. The current gain increases monotonically with cooling, from 600 at 300 K to 3800 at 85 K (Figure 11.7). Two mechanisms are responsible for this improvement with cooling: (1) the (sizable) Ge-induced band offset in this device (exponentially) increases the current gain with cooling, and (2) the heavily doped base region partially offsets the doping-induced bandgap narrowing associated with the emitter region. There is a strong decrease in the current gain above its peak value at 85 K associated with the Ge-grading effect, but the current gain remains above 2000 at 85 K at the current density at which peak f_T is reached, effectively minimizing any emitter charge storage at low temperatures.

Figure 11.7 also shows the extracted peak cutoff frequency versus temperature for the $0.12 \times 10.0 \, \mu m^2$ SiGe HBT. An increase in peak f_T from 200 GHz at 300 K to 260 GHz at 85 K is observed. This increase in the peak f_T with cooling is proportionately smaller than has been reported in first-generation SiGe HBTs (Figure 11.5) operated at 85 K. This is because in the present case, the base and emitter transit times in this 200 GHz device, which are favorably affected by both the Ge-grading and cooling, are already small compared to the collector delay time, and thus their relative influence on the total transit time with cooling is smaller. The extrapolated total emitter-to-collector delay decreases from 0.7 psec at 300 K to 0.6 psec at 150 K and 0.5 psec at 85 K, and the total depletion capacitance of the device decreases with cooling, as expected, since the junction built-in voltages increase with cooling.

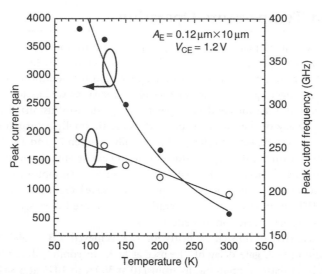

FIGURE 11.7 Peak current gain and peak cutoff frequency as a function of temperature for a third-generation SiGe HBT. (From B Banerjee, S Venkataraman, Y Lu, S Nuttinck, D Heo, E Chen, JD Cressler, J Laskar, G Freeman, and D Ahlgren. Proceedings of the IEEE Bipolar/BiCMOS Circuits and Technology Meeting, Toulouse, France, 2003, pp. 171–174. With permission.)

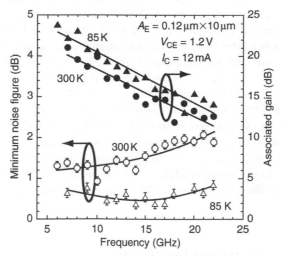

FIGURE 11.8 Minimum noise figure and associated gain as a function of frequency at 300 and 85 K for a third-generation SiGe HBT. (From B Banerjee, S Venkataraman, Y Lu, S Nuttinck, D Heo, E Chen, JD Cressler, J Laskar, G Freeman, and D Ahlgren. Proceedings of the IEEE Bipolar/BiCMOS Circuits and Technology Meeting, Toulouse, France, 2003, pp. 171–174. With permission.)

Figure 11.8 shows the measured minimum noise figure (NF_{min}) and associated gain (G_{assoc}) as a function of frequency at $I_C = 12$ mA (i.e., at peak f_T), for a 0.12×10.0 μm^2 SiGe HBT, at both 300 and 85 K. At 85 K, this device achieves a minimum NF_{min} of about 0.3 dB (with $G_{assoc} = 18$ dB) at 14 GHz, and a minimum NF_{min} of about 0.75 dB ($G_{assoc} = 15$ dB) at 20 GHz, record numbers for SiGe HBTs operating at cryogenic temperatures.

11.6   Helium Temperature Operation

Long-wavelength infrared focal-plane-arrays (FPA) and certain ultra-low-noise instrumentation amplifiers require transistors that operate down to liquid helium temperature (LHeT = 4.3 K). In addition to evaluating SiGe HBT performance at these potential application temperatures, the below 77-K regime is ideally suited for investigating new device physics phenomena, as well as for testing the validity of conventional theoretical formulations of device operation (e.g., drift-diffusion). This is particularly true for a SiGe HBT, since many of the transistor parameters are thermally activated functions of the Ge-induced band offsets, and are expected to change dramatically between 77 and 4 K. For instance, a simple calculation of the intrinsic carrier density, to which the terminal currents are proportional, shows that a n_{io} changes by a factor of e^{3056} between 77 and 4 K. Initial results on (unoptimized) Si BJTs to 10 K [20] showed transistor functionality but poor performance in the LHeT regime (<10 to 15 K). More recent work [21] on SiGe HBTs optimized for 77-K operation shows more impressive performance results as well as reveals interesting new device physics effects.

The emitter-cap SiGe HBT optimized explicitly for 77 K achieved a β of 500, f_T of 61 GHz, f_{max} of 50 GHz, and a minimum ECL gate delay of 21.9 psec at 84 K. In cooling this transistor from 77 K to LHeT, the current gain increases monotonically from 110 at 300 K to 1045 at 5.84 K, although parasitic base current leakage limits the useful operating current to above about 1.0 μA at 5.84 K. Figure 11.3 shows the Gummel characteristics of a 1.4×4.4 μm^2 emitter-cap SiGe HBT down to 5.84 K, and Figure 11.4 shows the current gain as a function of bias current down to 5.84 K.

The severity of the base current leakage at low injection, and the Ge-ramp effect at medium injection, limits the current range where one obtains the peak current gain. The aggressive base profile design in the emitter-cap SiGe HBT design (peak N_{ab}^- close to 8×10^{18} cm^{-3}) leads to an R_{bi} of <18 kΩ/sq. at 5.84 K, much lower than a more conventional SiGe HBT design. Base freeze-out below 77 K depends very strongly on peak base doping, and must be carefully optimized for LHeT applications. At temperatures as low as 5.84 K, this transistor has a maximum current drive in excess of 1.5 mA/μm^2 (limited by quasi-saturation and heterojunction barrier effects), with a peak transconductance of 190 mS. Theoretical calculations based on measured SIMS data were compared to the experimentally observed variation of peak current gain with temperature. Above 77 K, the temperature variation of peak current gain for the SiGe HBT is close to that theoretically expected, while at temperatures below 77 K, the exponential increase in current gain is primarily limited by parasitic base leakage due to field-enhanced tunneling. In contrast to this strong enhancement of current gain with cooling for the SiGe HBT, the current gain in a Si BJT fabricated with a comparable doping profile is significantly degraded at low temperatures, due to the strong bandgap narrowing in the emitter. A comprehensive discussion of other unique cryogenic phenomena in SiGe HBTs operating in the LHeT environment is presented in Ref. [1].

11.7   High-Temperature Operation

While it has been demonstrated that SiGe HBTs operate well down to deep cryogenic temperatures, there was historically early concern about their suitability for operation at elevated temperatures. Given that all electronic systems must successfully operate at temperatures considerably above 300 K (e.g., 125°C to satisfy military specifications and 85°C for many commercial applications), this is a potentially important issue. Given the narrow bandgap base region of the SiGe HBT compared to a Si BJT, and hence the expected negative temperature coefficient of the current gain (i.e., β decreases as temperature increases), it was often asked whether practical SiGe HBTs would have acceptable values of β at required high-end operational temperatures (e.g., 125°C). That this issue is not a valid concern for circuit designers is clearly demonstrated in Figure 11.9, which compares the percent change in peak current gain between 25 and 125°C for a Si BJT and a number or commercially relevant SiGe profiles. There are several important points to glean from these data:

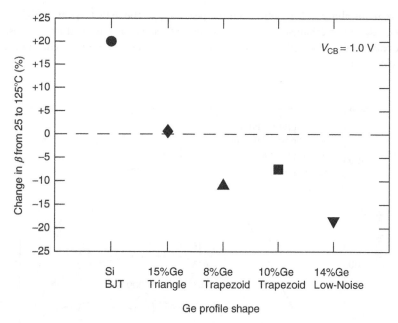

FIGURE 11.9 Percent change in peak current gain between 25 and 125°C for various Ge profiles. (From JD Cressler and G Niu. *Silicon–Germanium Heterojunction Bipolar Transistors.* Boston, MA: Artech House, 2003. With permission.)

- The current gain in SiGe HBTs does indeed have an opposite temperature dependence from that of a Si BJT, as expected from simple theory.
- These changes in β between 25 and 125°C, however, are modest at best (<25%), and clearly not cause for alarm for any realistic circuit.
- The negative temperature coefficient of β in SiGe HBTs is *tunable*, meaning that its temperature behavior between, say, 25 and 125°C can be trivially adjusted to its desired value by changing the Ge profile shape near the EB junction. In the case of the 15% Ge triangle profile, with 0% Ge at the EB junction, β is in fact temperature independent from 25 to 125°C. This points to a major advantage of bandgap engineering.
- Finally, it is well known that thermal-runaway in high-power Si BJTs is the result of the positive temperature coefficient of β (i.e., as the device heats up due to power dissipation, one gets more bias current since the β increases with temperature, leading to a positive feedback process, and hence thermal collapse). The fact that SiGe HBTs naturally have a negative temperature coefficient for β suggests that this might present interesting opportunities for power amplifiers, since emitter ballasting resistors (which degrade RF gain) could in principle be eliminated.

There is also an emerging interest in the operation of electronic devices *above* 125°C, for planetary space missions (e.g., Venus), or for on-engine electronics for both the automotive and aerospace sectors to support the "more-electric-vehicle" thrust of the military. In these cases, allowing the requisite electronic components to operate at relatively high temperatures (say 200 to 250°C) presents compelling cost-saving advantages, since the cooling system constraints can be dramatically relaxed. Conventional wisdom dictates that Si-based devices not be considered for these types of high-temperature applications, since Si is a fairly low-bandgap material, and thermal leakage (i.e., I_{on}/I_{off} ratios) depends exponentially on E_g. The fact that SiGe HBTs are capable of operation in such high-temperature environments can be easily demonstrated experimentally (as shown in Figure 11.10 and Figure 11.11). While performance degradation generally results at high temperatures, in these second-generation SiGe HBTs the peak current gain remains above 125 at 300°C and the peak f_T/f_{max} above 90 GHz at 200°C [22]. No serious reliability degradation mechanisms were identified at elevated temperatures. Thus, there is no fundamental reason why SiGe HBTs cannot satisfy this important emerging niche application of high-temperature electronics.

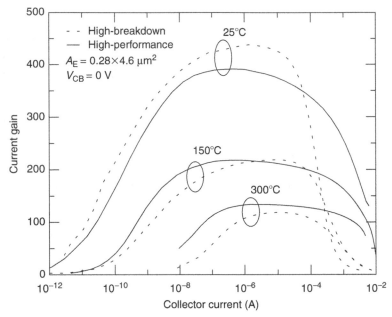

FIGURE 11.10   Current gain as a function of collector current at various temperatures for both high-performance (low-breakdown) and high-breakdown second-generation SiGe HBTs. (From T Chen, W-ML Kuo, E Zhao, Q Liang, Z Jin, JD Cressler, and AJ Joseph. *IEEE Trans. Electron Dev.* 51:1825–1832, 2004. With permission.)

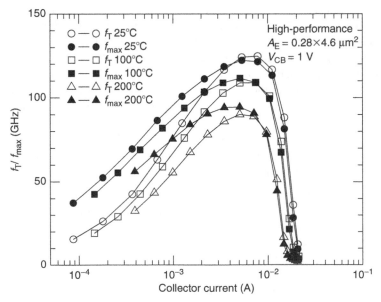

FIGURE 11.11   Cutoff frequency as a function of collector current at various temperatures for a second-generation SiGe HBT. (From T Chen, W-ML Kuo, E Zhao, Q Liang, Z Jin, JD Cressler, and AJ Joseph. *IEEE Trans. Electron Dev.* 51:1825–1832, 2004. With permission.)

11.8 Summary

Bandgap engineering has a positive influence on the low-temperature characteristics of bipolar transistors, enabling conventionally designed SiGe HBTs to operate very well in the cryogenic environment. We have addressed the effects of temperature on SiGe HBT device and circuit operation, by showing how temperature couples to SiGe HBT dc and ac performance, addressing how one optimizes SiGe HBTs specifically for cryogenic operation, and finally by considering the operation of SiGe HBTs at high temperatures. We conclude that the operation of SiGe HBTs at extreme temperatures (both low and high) is a viable path for commercial SiGe technology, and of potential importance for a growing number of niche applications.

Acknowledgments

I am grateful to A. Joseph, T. Chen, R. Krithivasan, Y. Lu, D. Richey, B. Banerjee, S. Nuttinck, G. Niu, D. Harame, G. Freeman, B. Meyerson, D. Herman, and the IBM SiGe team for their contributions. This work was supported by Hypres, the GEDC at Georgia Tech, and IBM.

References

1. JD Cressler and G Niu. *Silicon–Germanium Heterojunction Bipolar Transistors*. Boston, MA: Artech House, 2003.
2. WL Kauffman and AA Bergh. The temperature dependence of ideal gain in double diffused silicon transistors. *IEEE Trans. Electron Dev.* 15:732–735, 1968.
3. ES Schlig. Low-temperature operation of Ge picosecond logic circuits. *IEEE J. Solid-State Circuits* 3:271–276, 1968.
4. D Buhanan. Investigation of current-gain temperature dependence in silicon transistors. *IEEE Trans. Electron Dev.* 16:117–124, 1969.
5. WP Dumke. Effect of minority carrier trapping on the low-temperature characteristics of Si transistors. *IEEE Trans. Electron Dev.* 17:388–389, 1970.
6. WP Dumke. The effect of base doping on the performance of Si bipolar transistors at low temperatures. *IEEE Trans. Electron Dev.* 28:494–500, 1981.
7. JD Cressler, DD Tang, KA Jenkins, GP Li, and ES Yang. On the low-temperature static and dynamic properties of high-performance silicon bipolar transistors. *IEEE Trans. Electron Dev.* 36:1489–1502, 1989.
8. JD Cressler, DD Tang, KA Jenkins, and GP Li. Low temperature operation of silicon bipolar ECL circuits. Technical Digest of the IEEE International Solid-State Circuits Conference, New York, 1989, pp. 228–229.
9. JMC Stork, DL Harame, BS Meyerson, and TN Nguyen. High performance operation of silicon bipolar transistors at liquid nitrogen temperature. Technical Digest of the IEEE International Electron Devices Meeting, San Francisco, 1987, pp. 405–408.
10. JCS Woo and JD Plummer. Optimization of bipolar transistors for low temperature operation. Technical Digest of the IEEE International Electron Devices Meeting, San Francisco, 1987, pp. 401–404.
11. K Yano, K Nakazato, M Miyamoto, M Aoki, and K Shimohigashi. A high-current-gain low-temperature pseudo-HBT utilizing a sidewall base-contact structure (SICOS). *IEEE Electron Dev. Lett.* 10:452–454, 1989.
12. JD Cressler, TC Chen, JD Warnock, DD Tang, and ES Yang. Scaling the silicon bipolar transistor for sub-100-ps ECL circuit operation at liquid nitrogen temperature. *IEEE Trans. Electron Dev.* 37:680–691, 1990.
13. EF Crabbé, GL Patton, JMC Stork, JH Comfort, and BS Meyerson. The low-temperature operation of Si and SiGe bipolar transistors. Technical Digest of the IEEE International Electron Devices Meeting, Washington, 1990, pp. 17–20.

14. JD Cressler, JH Comfort, EF Crabbé, GL Patton, W Lee, JY-C Sun, JMC Stork, and BS Meyerson. Sub-30-ps ECL circuit operation at liquid-nitrogen temperature using self-aligned epitaxial SiGe-base bipolar transistors. *IEEE Electron Dev. Lett.* 12:166–168, 1991.

15. JD Cressler, JH Comfort, EF Crabbé, JMC Stork, and JY-C Sun. On the profile design and optimization of epitaxial Si- and SiGe-base bipolar technology for 77 K applications—Part I. Transistor dc design considerations. *IEEE Trans. Electron Dev.* 40:525–541, 1993.

16. JD Cressler, EF Crabbé, JH Comfort, JMC Stork, and JY-C Sun. On the profile design and optimization of epitaxial Si- and SiGe-base bipolar technology for 77 K applications—Part II. Circuit performance issues. *IEEE Trans. Electron Dev.* 40:542–556, 1993.

17. JH Comfort, EF Crabbé, JD Cressler, W Lee, JY-C Sun, J Malinowski, M D'Agostino, JN Burghartz, JMC Stork, and BS Meyerson. Single crystal emitter cap for epitaxial Si- and SiGe-base transistors. Technical Digest of the IEEE International Electron Devices Meeting, San Francisco, 1991, pp. 857–860.

18. JD Cressler, EF Crabbé, JH Comfort, JY-C Sun, and JMC Stork. An epitaxial emitter-cap SiGe-base bipolar technology optimized for liquid-nitrogen temperature operation. *IEEE Electron Dev. Lett.* 15:472–474, 1994.

19. B Banerjee, S Venkataraman, Y Lu, S Nuttinck, D Heo, E Chen, JD Cressler, J Laskar, G Freeman, and D Ahlgren. Cryogenic Performance of a 200 GHz SiGe HBT Technology. Proceedings of the IEEE Bipolar/BiCMOS Circuits and Technology Meeting, Toulouse, France, 2003, pp. 171–174.

20. AK Kapoor, HK Hingarh, and TS Jayadev. Operation of poly emitter bipolar npn and p-channel JFETs near liquid-helium (10 K) temperature. Proceedings of the IEEE Bipolar/BiCMOS Circuits and Technology Meeting, Minneapolis, 1988, pp. 210–214.

21. AJ Joseph, JD Cressler, and DM Richey. Operation of SiGe heterojunction bipolar transistors in the liquid-helium temperature regime. *IEEE Electron Dev. Lett.* 16:268–270, 1995.

22. T Chen, W-ML Kuo, E Zhao, Q Liang, Z Jin, JD Cressler, and AJ Joseph. On the high-temperature (to 300°C) characteristics of SiGe HBTs. *IEEE Trans. Electron Dev.* 51:1825–1832, 2004.

12

Radiation Effects

John D. Cressler
Georgia Institute of Technology

12.1 Introduction

There are currently two recent but rapidly growing thrusts within the space electronics community: (1) the use of commercial-off-the-shelf (COTS) parts whenever possible for space-borne systems as a cost-saving measure; and (2) the use of system-on-a-chip integration to lower chip counts and system costs, as well as simplify packaging and lower total system launch weight. The "holy-grail" in the realm of space electronics can thus be viewed as a conventional terrestrial IC technology with a system-on-a-chip capability, which is also radiation-hard as fabricated, without requiring any additional process modifications or layout changes. It is within this context that we discuss SiGe HBT BiCMOS technology as potentially such a "radiation-hard-as-fabricated" IC technology with possibly far-ranging implications for the space community.

Within the context of existing data for radiation exposure of SiGe HBTs, it is meaningful to distinguish between different SiGe HBT technology nodes, and is loosely defined by the ac performance of the SiGe HBT (e.g., peak f_T, which is a very strong function of the vertical profile and hence nicely reflects the degree of sophistication in structural design, lateral dimensional scaling, profile scaling, and net thermal cycle). We thus label a SiGe HBT technology node having a SiGe HBT with a peak f_T of roughly 50 GHz as "first-generation" (e.g., SiGe 5HP from IBM [1]), that with a peak f_T of roughly 100 GHz as "second-generation" (e.g., SiGe 7HP from IBM [2]), that with a peak f_T of roughly 200 GHz as "third-generation" (e.g., SiGe 8HP from IBM [3]), and that with a peak f_T of roughly 300 GHz as "fourth-generation" (e.g., SiGe 9T from IBM [4]). For brevity, here we only discuss radiation effects in SiGe HBT. For discussion on the impact of radiation on the Si CMOS devices found the SiGe HBT CMOS, the reader is referred to Ref. [5]. More recent results on other commercial SiGe HBT technology platforms (than the IBM results presented here) can be found in Refs. [6,7].

12.2 DC Effects

The response of SiGe HBTs to a variety of radiation types has been reported, including gamma rays, neutrons, and protons [8–14]. Since protons induce both ionization and displacement damage, they can be considered the worst case for radiation tolerance. For the following results, relevant proton energy of 63 MeV was used. At proton fluences of 1×10^{12} p/cm^2 and 5×10^{13} p/cm^2, the measured equivalent total ionizing dose (TID) was approximately 135 and 6759 krad(Si), respectively, the latter being far larger than most orbital missions require.

The typical response of a SiGe HBT to irradiation can be seen in Figure 12.1, which shows typical measured Gummel characteristics of a fourth-generation SiGe HBT, both before and after exposure to protons [8–14]. As expected, the base current increases after sufficiently high proton fluence due to the production of generation/recombination (G/R) trapping centers, and hence the current gain of the device degrades. There are two main physical origins of this degradation. The base current density is inversely proportional to the minority carrier lifetime in the emitter, so that a degradation of the hole lifetime will induce an increase in the base current. In addition, ionization damage due to the charged nature of the proton fluence produces interface states and oxide-trapped charges in the spacer layer at the emitter–base junction. These G/R centers also degrade I_B, particularly if they are placed inside the EB space–charge region, where they will yield an additional non-ideal base current component (non-kT/q exponential voltage dependence). By analyzing a variety of device geometries, it can be shown that the radiation-induced excess base current is primarily associated with the EB spacer oxide at the periphery of the transistor, as naively expected, and is hence the radiation response is dominated by ionization damage rather than displacement damage. The radiation-induced degradation of the base current and current gain for four generations of SiGe technology are shown in Figure 12.2 and Figure 12.3. Less than 30% degradation in peak current gain is observed across all four technology nodes, to 1.0 Mrad(Si) equivalent radiation levels, suggesting that SiGe HBTs are robust to TID for typical orbital proton fluences for realistic circuit operating currents above roughly 100 μA without any additional radiation hardening. These results are significantly better than for conventional diffused or even ion-implanted Si BJT technologies (even radiation-hardened ones).

Of particular interest is the inference of the spatial location of the proton-induced traps in these devices [10]. The existence of proton-induced traps in the EB space–charge region is clearly demonstrated by the G/R-induced increase in the non-ideal base current component shown in the Gummel

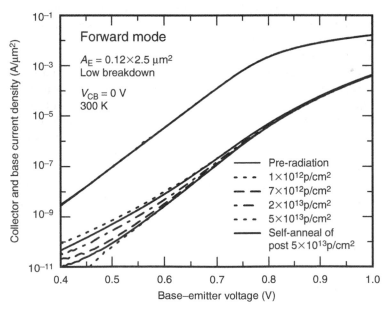

FIGURE 12.1 Forward-mode Gummel characteristics of a fourth-generation, 350 GHz SiGe HBT, during radiation exposure. (From A Sutton, B Haugerud, Y Lu, W-ML Kuo, JD Cressler, P Marshall, R Reed, J-S Rieh, G Freeman, and D Ahlgren. *IEEE Trans. Nuclear Sci.* 51:3736–3742, 2004. With permission.)

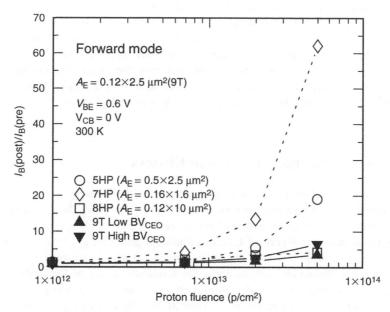

FIGURE 12.2 Normalized base current degradation for first-, second-, third-, and fourth-generation SiGe technology nodes. (From A Sutton, B Haugerud, Y Lu, W-ML Kuo, JD Cressler, P Marshall, R Reed, J-S Rieh, G Freeman, and D Ahlgren. *IEEE Trans. Nuclear Sci.* 51:3736–3742, 2004. With permission.)

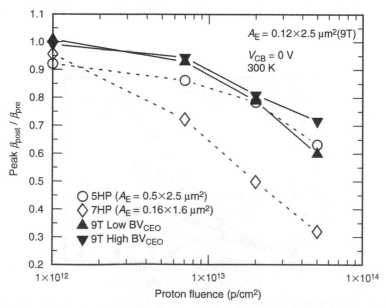

FIGURE 12.3 Current gain degradation for first-, second-, third-, and fourth-generation SiGe technology nodes. (From A Sutton, B Haugerud, Y Lu, W-ML Kuo, JD Cressler, P Marshall, R Reed, J-S Rieh, G Freeman, and D Ahlgren. *IEEE Trans. Nuclear Sci.* 51:3736–3742, 2004. With permission.)

characteristics. The existence of radiation-induced traps in the collector–base space–charge region was verified by measuring the inverse mode Gummel characteristics of the device (emitter and collector leads swapped). In this case the radiation-induced traps in the CB junction now act as G/R centers in the inverse EB junction, with a signature non-kT/q exponential slope. Two-dimensional simulations were calibrated to both measured data for the pre- and post-irradiated devices at a collector–base voltage of 0.0 V. In order to obtain quantitative agreement between the simulated and measured irradiated results, traps must be located uniformly throughout the device, and additional interface traps must be located around the emitter–base spacer oxide edge. Most of the radiation-induced recombination occurs inside the EB space–charge region, leading to a non-ideal base current, as expected.

12.3   AC Small-Signal and Noise Effects

To assess the impact of radiation on the ac performance of the transistors, the *S*-parameters were measured to 40 GHz both before and after proton exposure [15]. From the measured *S*-parameters, the transistor cutoff frequency as a function of bias current density can be extracted, and is shown for four technology generations in Figure 12.4. Only a slight degradation in f_T (and f_{max}) is observed, the latter expected from the minor increase of the base resistance with irradiation, due to either carrier removal, mobility and lifetime changes, or both. The broadband noise performance of SiGe HBTs is critical for space-borne transceivers and communications platforms. As shown in Figure 12.5, the minimum noise figure (NF$_{min}$) degrades only slightly at 2.0 GHz in a first-generation SiGe HBT after an extreme proton fluence of 5×10^{13} p/cm^2 (from 0.95 dB to a still-excellent value of 1.07 dB, a 12.6% degradation).

SiGe HBTs have the desirable feature of low $1/f$ noise commonly associated with Si bipolar transistors, which is of great importance because upconverted low-frequency noise (phase noise) typically limits the spectral purity of communication systems. Understanding the effects of radiation on $1/f$ noise in SiGe HBTs thus becomes a crucial issue for space-borne communications electronics. Physically, $1/f$ noise results from the presence of G/R center traps in the transistors, from which trapping–detrapping

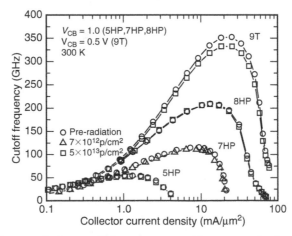

FIGURE 12.4 Pre- and post-radiation cutoff frequency versus bias current density for first-, second-, third-, and fourth-generation SiGe technology nodes. (From A Sutton, B Haugerud, Y Lu, W-ML Kuo, JD Cressler, P Marshall, R Reed, J-S Rieh, G Freeman, and D Ahlgren. *IEEE Trans. Nuclear Sci.* 51:3736–3742, 2004. With permission.)

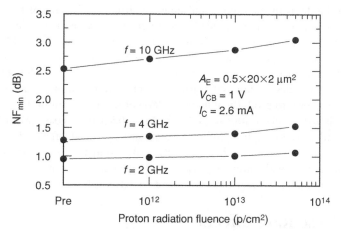

FIGURE 12.5 Extracted minimum noise figure as a function of proton fluence for multiple frequencies for first-generation SiGe HBTs. (From S Zhang, G Niu, SD Clark, JD Cressler, and M Palmer. *IEEE Trans. Nuclear Sci.* 46:1716–1721, 1999. With permission.)

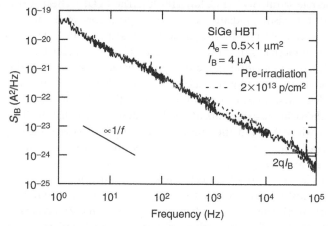

FIGURE 12.6 Input-referred base current low-frequency noise for a first-generation SiGe HBT, before and after irradiation. (From Z Jin, G Niu, JD Cressler, C Marshall, P Marshall, H Kim, R Reed, and D Harame. *IEEE Trans. Nuclear Sci.* 48:2244–2249, 2001. With permission.)

processes occur while carriers flow inside the device, thus modulating the number of carriers (and hence currents) to produce $1/f$ noise. The pre-irradiation low-frequency noise spectrum in these SiGe HBTs is typically $1/f$, with an I_B^2 dependence, while $S_{I_B} \times A_E$ is almost independent of A_E. The I_B^2 and $1/A_E$ dependencies of S_{I_B} are strong indicators of uniformly distributed noise sources over the entire emitter area. After 2×10^{13} p/cm^2 proton irradiation, the low-frequency noise spectrum in first-generation SiGe HBTs remains $1/f$ in frequency dependence, and free of G/R (burst) noise, and at roughly the same noise magnitude (i.e., no radiation-induced degradation) (as can be seen in Figure 12.6) [16].

12.4 Origin of Radiation Hardness

We note that careful comparisons between identically fabricated SiGe HBTs and Si BJTs (same device geometry and wafer lot, but without Ge in the base for the epitaxial-base Si BJT) show that the extreme level of total dose tolerance of SiGe HBTs is not per se due to the presence of Ge [10]. That is, the proton response of both the epitaxial base SiGe HBT and Si BJT is nearly identical. We thus attribute the observed radiation hardness to the unique and inherent structural features of the device itself, which from a radiation standpoint can be divided into three major aspects: (1) in these epitaxial base structures, the extrinsic base region is very heavily doped ($>5 \times 10^{19} cm^{-3}$) and located immediately below the emitter–base (EB) spacer oxide region, effectively confining any radiation-induced damage, and its effects on the EB junction; (2) the EB spacer, known to be the most vulnerable damage point in conventional BJT technologies, is thin ($<0.20 \mu m$ wide) and composed of an oxide–nitride composite, the latter of which is known to produce an increased level of radiation immunity; (3) the active volume of these transistors is very small (emitter stripe width $W_E = 0.5 \mu m$, and base width $W_b < 150 nm$), and the emitter, base, and collector doping profiles are quite heavily doped, effectively lessening the impact of displacement damage. We also note that these SiGe HBTs compare very favorably in both performance and radiation hardness with (more expensive) GaAs HBT technologies that are often employed in space applications requiring both very high speed and an extreme level of radiation immunity [17].

12.5 Low-Dose-Rate Effects

Within the past few years, a pronounced low-dose-rate sensitivity to gamma irradiation that is not screened by the current test methods for ionizing radiation has been observed in Si bipolar technologies. The enhancement in device and circuit degradation at low gamma dose rates has come to be known as "enhanced low-dose-rate sensitivity" (ELDRS) [18–20]. The ELDRS effect was first reported in 1991, which demonstrated that existing radiation hardness test assurance methodologies were not appropriately considering worst-case conditions. The physical origins underlying ELDRS have been hotly debated for years, and numerous mechanisms proposed. Recent attempts to understand ELDRS include a model suggesting that the lower net radiation-induced trapped charge density at high-dose-rates is a result of a space–charge phenomenon, caused by delocalized hole traps that occur in heavily damaged oxides such as bipolar base oxides. These traps can retain holes on a timescale of seconds to minutes, causing a buildup of positive charge in the oxide bulk during high-dose-rate irradiation. This is in contrast to low-dose-rate irradiation, where the irradiation time is much longer, effectively allowing the holes in the trap centers to be detrapped. Thus, in the high-dose-rate case, the larger total trapped hole density forces holes near the interface to be trapped closer to the interface, where they can be compensated by electrons from the silicon. This lowers the resultant net trapped charge density.

To assess ELDRS in first-generation SiGe technology, low-dose-rate (0.1 rad(Si)/sec) and high-dose-rate (300 rad(Si)/sec) experiments were conducted using Cobalt-60 [21]. As can be seen in Figure 12.7, low-dose-rate effects in these first-generation SiGe HBTs were found to be nearly non-existent, in striking contrast to reports of strong ELDRS in conventional Si bipolar technologies. We attribute this observed hardness to ELDRS to the same mechanisms responsible for the overall radiation hardness of the technology, and is likely more structural in nature than due to any unique advantage afforded by the SiGe base. Interestingly, an anomalous decrease in base current was also found in these devices at low-dose-rates, suggesting that a new physical phenomenon is present at low-dose-rates in these devices.

12.6 SiGe HBT Circuit Tolerance

For the successful deployment of SiGe technology into space-based systems, circuit-level radiation hardness is clearly more important than device-level hardness. As presented above, the TID device degradation is minor in the bias range of interest to most actual circuits (typically $I_C > 100 \mu A$). In order to assess the impact of radiation exposure on actual SiGe HBT circuits, we have compared two very important, yet very different circuit types, one heavily used in analog ICs (the bandgap reference circuit), and one heavily used in RFICs (the voltage-controlled oscillator) [22,23]. Each circuit represents a key building block for realistic SiGe ICs that might be flown in space. Each of these SiGe HBT circuits was designed using fully calibrated SPICE models, layed-out, and then fabricated on the same wafer to facilitate unambiguous comparisons. In addition, because any realistic RF IC must also

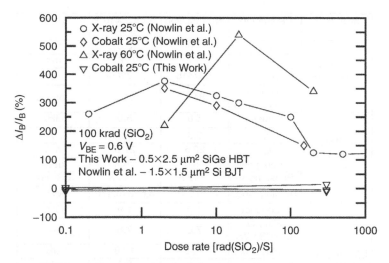

FIGURE 12.7 Normalized base current as a function of gamma radiation dose rate, for both Si BJTs and first-generation SiGe HBTs. (From G Banerjee, G Niu, JD Cressler, SD Clark, MJ Palmer, and DC Ahlgren. *IEEE Trans. Nuclear Sci.* 46:1620–1626, 1999. With permission.)

necessarily include passive elements such as monolithic inductors and capacitors, we have also investigated the effects of proton exposure on an RF LC bandpass filter. As can be seen from the data (Figure 12.8), the impact of even extreme proton fluences has minimal effect on either the output voltage or temperature sensitivity of BGRs, the phase noise or tuning range of VCOs or passive elements, and is indicative of the overall robustness of this SiGe technology for analog and RF circuit applications.

12.7 Single-Event Upset

Clearly, a space-qualified IC technology must demonstrate sufficient SEU immunity to support high-speed circuit applications as well as possess TID tolerance. It is well known that even III–V technologies that have significant TID tolerance often suffer from poor SEU immunity, particularly at high data rates. Recently, high-speed SiGe HBT digital logic circuits were found to be vulnerable to SEU at even low linear energy transfer (LET) values [24–26]. In addition, successfully employed III–V HBT circuit-level hardening schemes using the current-sharing hardening (CSH) technique were found to be ineffective for these SiGe HBT logic circuits (Figure 12.9). To understand single-event effects in SiGe HBTs, one must use calibrated two-dimensional or three-dimensional device simulation to assess the charge collection characteristics of SiGe HBTs. These device-level simulation results can then be coupled to circuit-level modeling to better understand circuit-level mitigation approaches. From a device perspective, it is important to first assess the transistor charge collection characteristics as a function of terminal bias, load condition, substrate doping, and ion strike depth [27,28]. Bias and loading conditions were chosen to mimic representative circuit conditions within an actual ECL/CML digital circuit. Figure 12.10 shows the charge collected by the collector versus time for different RC loads. The base and emitter terminals were grounded, the substrate bias was −5.2 V, the collector was connected to ground through an RC load, and the substrate doping was $5 \times 10^{15}\,cm^{-3}$. A uniform LET of 0.1 pC/μm (equivalent to 10 MeV cm^2/mg) over 10-μm depth was used, which generates a total charge of 1.0 pC. The results clearly show that charge collection is highly dependent on the transistor load condition (i.e., circuit topology). As the load resistance increases, the collector-collected charge decreases. Note, however, that the emitter-collected charge increases correspondingly. The underlying physics is that more electrons exit through the emitter, instead of the collector. A larger load resistance presents a higher impedance to the electrons at the collector, and thus more electrons exit through the emitter. The collector of the

SiGe HBT Circuit	Parameters	Pre-radiation	After 5×10^{13} p/cm^2	Units
Bandgap reference	V_{CC}	3.0	3.0	V
	I_{CC}	0.773	0.767	mA
	V_{out} at 300 K	1.37416	1.372096	V
	Stability (−55 to 85°C)	81.2	81.7	ppm/°C
Voltage controlled oscillator	Frequency	5.0	5.0	GHz
	V_{cc}	3.3	3.3	V
	I_{CC}	22.5	22.5	mA
	Outpout power	−5.0	−5.5	dBm
	Phase noise	−112.5	−111.8	dBc/Hz
	Tunning range	4,55–5,452	4,623–5,470	MHz
LC bandpass filter	Frequency	1.9	1.9	GHz
	Filter Q (at 3 dB BW)	7.6	7.6	—
	Insertion Loss	16.8	16.8	dB
	L	2.5	2.5	nH
	Inductor Q	7.4	7.4	—
	C	6.0	6.0	pF
	Capacitor Q	58	58	—

FIGURE 12.8 Summary of the measured radiation tolerance of some important first-generation SiGe HBT circuits and passives. (From JD Cressler, MC Hamilton, R Krithivasan, H Ainspan, R Groves, G Niu, S Zhang, Z Jin, CJ Marshall, PW Marshall, HS Kim, RA Reed, MJ Palmer, AJ Joseph, and DL Harame. *IEEE Trans. Nuclear Sci.* 48:2238–2243, 2001. With permission.)

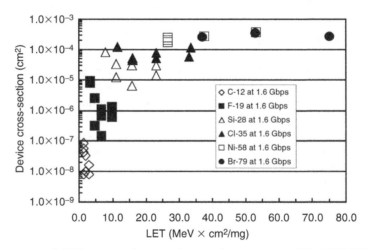

FIGURE 12.9 Experimental SEU cross-section test data on first-generation SiGe HBT shift registers. (From P Marshall, MA Carts, A Campbell, D McMorrow, S Buchner, R Stewart, B Randall, B Gilbert, and R Reed. *IEEE Trans. Nuclear Sci.* 47:2669–2674, 2000. With permission.)

adjacent device only collects a negligible amount of charge, despite the transient current spikes of the strike. Nearly all the electrons deposited are collected by the collector and the emitter, although the partition between emitter and collector collection varies with the load condition. The impact on the SiGe base layer on the charge collection properties is a secondary effect.

To better understand circuit-level SEU response, we combined these simulated charge–time profiles with circuit-level modeling in three different SiGe circuit architectures [29]. Circuit A is a

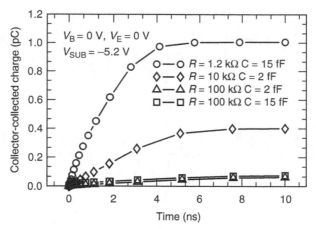

FIGURE 12.10 Simulated collector-collected charge versus time for different RC loads. (From G Niu, R Krithivasan, JD Cressler, P Marshall, C Marshall, R Reed, and D Harame. *IEEE Trans. Nuclear Sci.* 48:1849–1854, 2001. With permission.)

straightforward ECL implementation of the standard rising edge-triggered flip-flop logic. Circuit B is the unhardened version of the D flip-flop used in the shift register results shown in Figure 12.9. Circuit B uses fewer transistors and thus less power than circuit A, and is also faster than circuit A, allowing operation at higher clock rates. Because of these advantages, circuit B is very popular in high-speed bipolar digital circuit design. The circuit consists of a master stage and a slave stage. The master stage consists of a pass cell, a storage cell, a clocking stage, and a biasing control. The slave stage has a similar circuit configuration. Circuit C is the current-sharing hardened version of circuit B. (Refer to Ref. [5] for circuit-level schematics for A, B, and C.) The circuit was used as a basic building block of the 32-stage shift-register data shown in Figure 12.9. In this case, the current source transistor is divided into five paths, and these paths are maintained separately through the clocking stage and through the pass and storage cells. In essence, the input and output nodes of five copies of the switching circuits, including the controlling switch, clock, master, and storage cells, are connected in parallel. The load resistance is shared by all the current paths. The quasi-three-dimensional simulated SEU-induced transient currents were activated on one of the sensitive transistors in the respective circuits. The SEU currents were activated at 5.46 nsec (within the circuit hold time), immediately after the clock goes from low to high, a sensitive time instant for SEU-induced transient currents to produce an upset at the output. The input data are an alternating "0" and "1" series with a data rate of 2 Gbit/sec. Under these conditions, circuit A shows no upset at all, while circuits B and C show five and three continuous bits of data upset, respectively (Figure 12.11). These results suggest that circuit A has the best SEU tolerance, while circuit C, the CSH hardened version, has better SEU tolerance than its unhardened companion version, circuit B. Circuit A, which shows no data upset at a switching current of 1.5 mA, does in fact show an upset when the switching current is lowered to 0.6 mA. This is consistent with our earlier observation that increasing switching current is effective in improving SEU performance for circuit C.

The fundamental reason for the observed better SEU tolerance of circuit A than for circuits B and C is that only one of the two outputs of the emitter-coupled pair being hit is affected by the ion-strike SEU current transients. As long as the differential output is above the logic-switching threshold, the output remains unaffected, and no upset occurs. The collector voltage of the switching transistor decreases upon ion strike (compared to without SEU), however, and no upset is observed at the output, simply because the differential output remains above or below the relevant-switching threshold (Figure 12.12). These results suggest that circuit-level mitigation techniques can be used in SEU

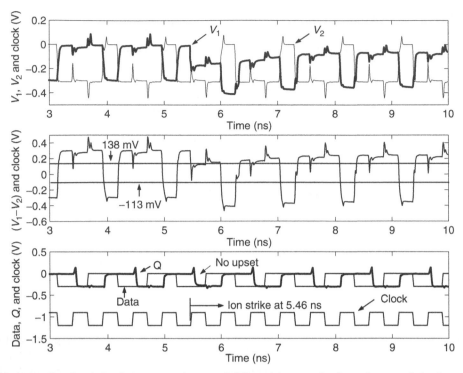

FIGURE 12.11 Simulated circuit A output voltages and differential output for the emitter-coupled pair struck by a heavy ion, with a switch current of 1.5 mA. (From G Niu, R Krithivasan, JD Cressler, PA Riggs, BA Randall, P Marshall, and R Reed. *IEEE Trans. Nuclear Sci.* 49:3107–3114, 2002. With permission.)

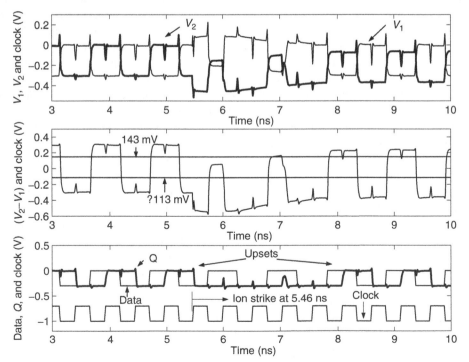

FIGURE 12.12 Simulated circuit B output voltages and differential output for the emitter-coupled pair struck by a heavy ion, with a switch current of 1.5 mA. (From G Niu, R Krithivasan, JD Cressler, PA Riggs, BA Randall, P Marshall, and R Reed. *IEEE Trans. Nuclear Sci.* 49:3107–3114, 2002. With permission.)

hardening of SiGe HBT logic, albeit at some level of additional power dissipation and circuit complexity. A potential SEU-hardening approach has been recently discussed in Ref. [30], but clearly more research is needed in the area of SEU mitigation before widespread deployment of SiGe circuitry in space is attempted.

12.8   Summary

While ionizing radiation degrades both the dc and ac performance of SiGe HBTs, this degradation is remarkably minor, and is far better than that observed in even radiation-hardened conventional Si BJT technologies. This fact is particularly significant given that no intentional radiation hardening is needed to ensure this level of both device-level and circuit-level tolerance (typically multi-Mrad TID). SEU effects are pronounced in SiGe HBT circuits, as expected, but circuit-level mitigation schemes will likely be suitable to ensure adequate tolerance for many orbital missions. While technology scaling can negatively impact the TID response of the SiGe HBT, it naturally improves the hardness of the CMOS devices, and thus 200 to 300 krad tolerance of the full BiCMOS technology can be achieved without radiation-hardening in second-generation SiGe HBT BiCMOS technology. Taken together, SiGe HBT BiCMOS technology offers many interesting possibilities for SoC applications of space-borne electronic systems.

Acknowledgments

I am grateful to A. Sutton, B. Haugerud, Y. Lu, W.-M. Kuo, G. Niu, P. Marshall, R. Reed, C. Marshall, J.-S. Rieh, G. Freeman, D. Ahlgren, L. Cohn, K. LaBel, H. Brandhorst, H. Ainspan, J. Lee, S. Nuttnick, H. Kim, C. Palor, S. Clark, D. Emily, B. Randall, K. Jobe, M. Palmer, A. Joseph, D. Harame, D. Herman, B. Meyerson, and the IBM SiGe team for their contributions. This work was supported by DTRA under the Radiation Hardened Microelectronics Program, NASA-GSFC under the Electronics Radiation Characterization Program, IBM, and the Georgia Electronic Design Center at Georgia Tech.

References

1. DC Ahlgren, G Freeman, S Subbanna, R Groves, D Greenberg, J Malinowski, D Nguyen-Ngoc, SJ Jeng, K Stein, K Schonenberg, D Kiesling, B Martin, S Wu, DL Harame, and BS Meyerson. A SiGe HBT BiCMOS technology for mixed-signal RF applications. Proceedings of the IEEE Bipolar/ BiCMOS Circuits and Technology Meeting, Minneapolis, 1997, pp. 195–198.
2. A Joseph, D Coolbaugh, M Zierak, R Wuthrich, P Geiss, Z He, X Liu, B Orner, J Johnson, G Freeman, D Ahlgren, B Jagannathan, L Lanzerotti, V Ramachandran, J Malinowski, H Chen, J Chu, M Gordon, P Gray, R Johnson, J Dunn, S Subbanna, K Schonenberg, D Harame, R Groves, K Watson, D Jadus, M Meghelli, and A Rylyakov. A 0.18 μm BiCMOS technology featuring 120/100 GHz (f_T/f_{max}) HBT and ASIC compatible CMOS using copper interconnect. Proceedings of the IEEE Bipolar/BiCMOS Circuits and Technology Meeting, Minneapolis, 2001, pp. 143–146.
3. AJ Joseph, D Coolbaugh, D Harame, G Freeman, S Subbanna, M Doherty, J Dunn, C Dickey, D Greenberg, R Groves, M Meghelli, A Rylyakov, M Sorna, O Schreiber, D Herman, and T Tanji. 0.13 μm 210 GHz f_T SiGe HBTs—expanding the horizons of SiGe BiCMOS. Technical Digest of the IEEE International Solid-State Circuits Conference, San Francisco, 2002, pp. 180–182.
4. J-S Rieh, B Jagannathan, H Chen, K Schonenberg, D Angell, A Chinthakindi, J Florkey, F Golan, D Greenberg, S-J Jeng, M Khater, F Pagette, C Schnabel, P Smith, A Stricker, K Vaed, R Volant, D Ahlgren, G Freeman, K Stein, and S Subbanna. SiGe HBTs with cut-off frequency near 300 GHz. Technical Digest of the IEEE International Electron Devices Meeting, Washington, DC, 2002, pp. 771–774.
5. JD Cressler and G Niu. *Silicon–Germanium Heterojunction Bipolar Transistors.* Boston, MA: Artech House, 2003.

6. J Comeau, A Sutton, B Haugerud, JD Cressler, W-ML Kuo, P Marshall, R Reed, A Karroy, and R Van Art. Proton tolerance of advanced SiGe HBTs fabricated on different substrate materials. 2004 IEEE Nuclear and Space Radiation Effects Conference, Atlanta, to be presented July 2004.

7. E Zhao, A Sutton, B Haugerud, JD Cressler, P Marshall, R Reed, B El-Kareh, and S Balster. The effects of radiation on $1/f$ noise in complementary (npn + pnp) SiGe HBTs. 2004 IEEE Nuclear and Space Radiation Effects Conference, Atlanta, to be presented July 2004.

8. JA Babcock, JD Cressler, LS Vempati, SD Clark, RC Jaeger, and DL Harame. Ionizing radiation tolerance of high performance SiGe HBTs grown by UHV/CVD. *IEEE Trans. Nuclear Sci.* 42:1558–1566, 1995.

9. J Roldan, WE Ansley, JD Cressler, SD Clark, and D Nguyen-Ngoc. Neutron radiation tolerance of advanced UHV/CVD SiGe HBTs. *IEEE Trans. Nuclear Sci.* 44:1965–1973, 1997.

10. J Roldan, G Niu, WE Ansley, JD Cressler, and SD Clark. An investigation of the spatial location of proton-induced traps in SiGe HBTs. *IEEE Trans. Nuclear Sci.* 45:2424–2430, 1998.

11. JD Cressler, M Hamilton, G Mullinax, Y Li, G Niu, C Marshall, P Marshall, H Kim, M Palmer, A Joseph, and G Freeman. The effects of proton irradiation on the lateral and vertical scaling of UHV/CVD SiGe HBT BiCMOS technology. *IEEE Trans. Nuclear Sci.* 47:2515–2520, 2000.

12. JD Cressler, R Krithivasan, G Zhang, G Niu, P Marshall, H Kim, R Reed, M Palmer, and A Joseph. An investigation of the origins of the variable proton tolerance in multiple SiGe HBT BiCMOS technology generations. *IEEE Trans. Nuclear Sci.* 49:3203–3207, 2002.

13. Y Lu, JD Cressler, R Krithivasan, Y Li, RA Reed, PW Marshall, C Polar, G Freeman, and D Ahlgren. Proton tolerance of third-generation, $0.12\,\mu m$ 185 GHz SiGe HBTs. *IEEE Trans. Nuclear Sci.* 50:1811–1815, 2003.

14. A Sutton, B Haugerud, Y Lu, W-ML Kuo, JD Cressler, P Marshall, R Reed, J-S Rieh, G Freeman, and D Ahlgren. Proton response of 4th-generation 350 GHz UHV/CVD SiGe HBTs. 2004. *IEEE Trans. Nuclear Sci.* 51:3736–3742, 2004.

15. S Zhang, G Niu, SD Clark, JD Cressler, and M Palmer. The effects of proton irradiation on the RF performance of SiGe HBTs. *IEEE Trans. Nuclear Sci.* 46:1716–1721, 1999.

16. Z Jin, G Niu, JD Cressler, C Marshall, P Marshall, H Kim, R Reed, and D Harame. $1/f$ noise in proton-irradiated SiGe HBTs. *IEEE Trans. Nuclear Sci.* 48:2244–2249, 2001.

17. S Zhang, G Niu, JD Cressler, SJ Mathew, SD Clark, P Zampardi, and RL Pierson. A comparison of the effects of gamma irradiation on SiGe HBT and GaAs HBT technologies. *IEEE Trans. Nuclear Sci.* 47:2521–2527, 2000.

18. RL Pease. Total-dose issues for microelectronics in space systems. *IEEE Trans. Nuclear Sci.* 43:442–452, 1996.

19. EW Enlow, RL Pease, W Combs, RD Schrimpf, and RN Nowlin. Response of advanced bipolar processes to ionizing radiation. *IEEE Trans. Nuclear Sci.* 38:1342–1351, 1991.

20. DM Fleetwood, SL Kosier, RN Nowlin, RD Schrimpf, RA Reber Jr, M Delaus, PS Winokur, A Wei, WE Combs, and RL Pease. Physical mechanisms contributing to enhanced bipolar gain degradation at low dose rates. *IEEE Trans. Nuclear Sci.* 41:1871–1883, 1994.

21. G Banerjee, G Niu, JD Cressler, SD Clark, MJ Palmer, and DC Ahlgren. Anomalous dose rate effects in gamma irradiated SiGe heterojunction bipolar transistors. *IEEE Trans. Nuclear Sci.* 46:1620–1626, 1999.

22. JD Cressler, MC Hamilton, R Krithivasan, H Ainspan, R Groves, G Niu, S Zhang, Z Jin, CJ Marshall, PW Marshall, HS Kim, RA Reed, MJ Palmer, AJ Joseph, and DL Harame. Proton radiation response of SiGe HBT analog and RF circuits and passives. *IEEE Trans. Nuclear Sci.* 48:2238–2243, 2001.

23. S Zhang, JD Cressler, G Niu, C Marshall, P Marshall, H Kim, R Reed, M Palmer, A Joseph, and D Harame. The effects of operating bias conditions on the proton tolerance of SiGe HBTs. *Solid-State Electron.* 47:1729–1734, 2003.

24. P Marshall, MA Carts, A Campbell, D McMorrow, S Buchner, R Stewart, B Randall, B Gilbert, and R Reed. Single event effects in circuit hardened SiGe HBT logic at gigabit per second data rates. *IEEE Trans. Nuclear Sci.* 47:2669–2674, 2000.

25. R Reed, P Marshall, H Ainspan, C Marshall, H Kim, JD Cressler, and G Niu. Single event upset test results on an IBM prescalar fabricated in IBM's 5HP germanium doped silicon process. Proceedings of the IEEE Nuclear and Space Radiation Effects Conference Data Workshop, Vancouver, 2001, pp. 172–176.

26. RA Reed, PW Marshall, J Pickel, MA Carts, T Irwin, G Niu, JD Cressler, R Krithivasan, K Fritz, P Riggs, J Prairie, B Randall, B Gilbert, G Vizkelethy, P Dodd, and K LaBel. Heavy-ion broad-beam and microprobe studies of single-event upsets in 0.20 μm silicon germanium heterojunction bipolar transistors and circuits. *IEEE Trans. Nuclear Sci.* 50:2184–2190, 2003.

27. G Niu, JD Cressler, M Shoga, K Jobe, P Chu, and DL Harame. Simulation of SEE-induced charge collection in UHV/CVD SiGe HBTs. *IEEE Trans. Nuclear Sci.* 47:2682–2689, 2000.

28. G Niu, R Krithivasan, JD Cressler, P Marshall, C Marshall, R Reed, and D Harame. Modeling of single event effects in circuit-hardened high-speed SiGe HBT logic. *IEEE Trans. Nuclear Sci.* 48:1849–1854, 2001.

29. G Niu, R Krithivasan, JD Cressler, PA Riggs, BA Randall, P Marshall, and R Reed. A comparison of SEU tolerance in high-speed SiGe HBT digital logic designed with multiple circuit architectures. *IEEE Trans. Nuclear Sci.* 49:3107–3114, 2002.

30. R Krithivasan, G Niu, JD Cressler, SM Currie, KE Fritz, RA Reed, PW Marshall, PA Riggs, BA Randall, and B Gilbert. An SEU hardening approach for high-speed SiGe HBT digital logic. *IEEE Trans. Nuclear Sci.* 50:2126–2134, 2003.

13

Reliability Issues

John D. Cressler
Georgia Institute of Technology

13.1 Introduction

Clearly, any new integrated circuit technology (SiGe or otherwise) must be proven to be "reliable." That is, under typical circuit-operating conditions, the circuits, and importantly, the systems constructed from those circuits, must not wear-out or degrade to a level at which they fail "in the field" over the functional life of the system. In integrated circuit circles, reliability of a given technology begins with assurance of the reliability of the underlying building block devices—the transistors certainly, but also the passive elements such as inductors or capacitors, and the interconnects linking the various elements. In this chapter, we will focus only on the reliability of the transistors; in this case SiGe HBTs.

From a transistor perspective, one ensures adequate reliability by subjecting the devices to extreme operating conditions for a given length of time, which, for a bipolar technology, historically encompasses two different operational scenarios: (1) hot carrier (hot electron or hot hole or both) stressing associated with reverse-biasing of the emitter–base (EB) junction, and (2) high forward collector current density (J_C) stressing. Both reliability "modes" will generally be conducted under "accelerated" conditions ("over stress") consisting of higher V_{EB} and J_C than the device would normally encounter during "typical use" circuit conditions, and will likely be performed at either elevated (e.g., 100°C for high J_C stressing) or reduced (e.g., −40°C for reverse EB stressing) temperatures to invoke worst-case stress conditions. One then defines the "reliability" of the transistors in terms of the measured change in a given defined device metric after a given amount of time under stress (e.g., the stress time it takes to produce decrease in current gain of 10%). From this time-dependent stress data, which is by definition limited in scope due to practical testing demands, one then typically projects an extrapolated "lifetime" of the technology (e.g., >10 years), a procedure which clearly invokes a number of assumptions. If the projected lifetime greatly exceeds the intended system life of the part, then all is well. In practice, during technology "qualification" various process splits and fabrication cycle variations are often changed to improve this or that reliability metric as needed, until the checkered flag is finally raised.

This is standard practice today for bipolar integrated circuit technologies. Interestingly, these basic bipolar reliability stress methodologies have been in place for well over 25 years in basically an unaltered form. Given the present reality that Si-based bipolar device performance has increased dramatically in recent years (largely due to the addition of Ge bandgap engineering), and classical bipolar circuit topologies have changed radically during this period from a high-speed digital ECL centric world to a wide variety of mixed-signal circuit types, it is logical to ask whether such reliability methodologies are in fact capturing all possible reliability degradation modes. As will be argued, they are not. For the purposes of this chapter, then, we define the concept of device "reliability" to be much broader than its standard usage in the industry, to include any possible degradation mechanism for any possible mixed-signal circuit application, in any of the various intended circuit application domains [1]. For instance, in addition to classical device reliability mechanisms associated with reverse EB and forward J_C stress, new reliability issues for SiGe HBTs include, for instance, the impact of Ge film stability on technology yield, impact-ionization induced "mixed-mode" stress, concerns associated with scaling-induced breakdown voltage compression and operating point instabilities, and geometrical scaling-induced low-frequency noise variations. The impact of ionizing radiation on device and circuit reliability is also important in this context, but is discussed in Chapter 12.

13.2 Technology-Driven Reliability and Yield Issues

It is obviously key to the viability of SiGe HBT technology that it has a clearly demonstrated reliability and yield that are comparable to or better than existing Si BJT technology. That is, any reliability or yield loss due to the incorporation of strained SiGe films are potential technology "showstoppers." Although published data on commercial SiGe technologies are sparse, there is no evidence to date that the use of thermodynamically stable SiGe films imposes any such reliability risk. Clearly, this is good news.

Interestingly, the reverse-bias EB stress response of SiGe HBTs can actually be substantially better than that for aggressively scaled ion-implanted Si BJTs. This is because the very shallow, low-energy base implants needed to realize high-performance implanted Si BJTs inevitably place the peak of the base doping at the metallurgical EB junction, and thus increase the EB electric field. In contrast, for an epitaxial base device (Si or SiGe), the boron can be placed inside the base region as a boron "box" profile, and while the finite thermal cycle spreads the boron during processing, a boron "retrograde" is naturally produced at the EB junction, thereby lowering the EB electric field. Since hot electron injection under reverse-bias EB stress conditions depends exponentially on the EB electric field [2], a transistor with an epitaxial base will have a fundamental and decided advantage over an implanted base device in terms of reliability. Importantly, however, this boron retrograde at the EB junction itself produces a doping-gradient-induced electric field that retards electron transport through the base under forward-bias, degrading f_T. In a SiGe HBT, this doping-induced retarding field is more than compensated by the Ge-induced accelerating field, but in an epi-base Si BJT, a performance penalty is inevitable. Thus, compositionally graded SiGe is clearly desirable for epi-base transistor design in Si technology.

High yield on large wafers is key to the cost advantage Si enjoys over its III–V competition, and as in reliability, the presence of a strained SiGe layer must not unfavorably impact device and circuit yield. It does not. CMOS yield in a SiGe HBT BiCMOS technology is typically evaluated using a large embedded static random-access-memory (SRAM) yield monitor (e.g., several hundred kbit). If any of the HBT films or residuals are not properly removed, then this will be reflected in the SRAM yield. Yield values can also be easily compared with CMOS-only processes to gauge the robustness of the CMOS section of the BiCMOS process. Typical yield numbers for the 154 k SRAM in first- and second-generation SiGe technology are well above 75% [3].

SiGe HBT yield is typically quantified using large chains of small transistors wired in parallel. A chain yield "failure" is defined as the intersection of emitter-to-collector shorts (pipes), high EB leakage, or high CB leakage (i.e., any of the three occurrences is defined as a "bad" or "dead" device chain). For instance, 4000 $0.42 \times 2.3 \ \mu m^2$ SiGe HBTs is used as a yield monitor in a first generation technology, and typically has greater than 85% to 90% yield. Choice of CMOS integration scheme does not appear to affect this result. Interestingly, the primary yield failure mechanism in both the CMOS and SiGe HBTs is the same, and can be traced to the shallow-trench isolation [3]. By assuming an ideal Poisson distribution relating defect density and emitter area, one can infer the net defect density associated with a given SiGe HBT BiCMOS technology, in this case yielding numbers in the range of 100 to 500 defects/cm^2. For orientation, a defect density of 426 defects/cm^2 would ideally produce a 60% yield on a integrated circuit containing 100,000 0.5 × 2.5 µm^2 SiGe HBTs, ample transistor count (and yield) to satisfy almost any imaginable mixed-signal application using SiGe technology [3].

The above yield considerations are clearly predicated on the use of thermodynamically stable SiGe films in the SiGe HBT BiCMOS technology. While stability considerations in the SiGe material system are not completely settled, and numerous "open issues" remain (see Ref. [4]), there do exist reasonably accurate simple theories for predicting film stability once film Ge content and dimensions are accurately known [5]. While not often openly discussed in the literature, there is a general consensus in industry that using stable SiGe films is a "good thing" from a yield perspective. The precise coupling of stability to local pattern density as well as local (added) strain associated with say the shallow- or deep-trench isolations, remain largely matters for conjecture, and should be quantified with dedicated research.

13.3 Conventional Device Degradation Mechanisms

As discussed above, reliability stress and "burn-in" of bipolar transistors historically proceed along two different paths [2–9]: (1) reverse emitter–base (EB) stress, which is used to inject hot electrons (or holes) into the EB spacer oxide, thereby introducing generation/recombination (G/R) center traps which lead to excess non-ideal base current (Figure 13.1) and hence current gain degradation (Figure 13.2) as well as increased low-frequency noise [7]; and (2) high forward-current density stress, which also results in current gain degradation, but is generally attributed to electromigration-induced changes in the emitter contact, resulting in a decrease in collector current with increasing stress time.

Accelerated lifetime testing of SiGe HBTs using reverse-bias EB stress is generally conducted under high reverse EB bias (e.g., 3.0 V) at reduced temperatures (e.g., −40°C), where carrier velocities are higher due to reduced scattering, whereas high forward-current density stress is conducted under a large J_C (e.g., three to four times of the J_C at peak f_T = 3.0–4.0 mA/µm^2 for a first-generation SiGe technology) at elevated temperatures (e.g., 100°C), where electromigration is inherently more severe.

Typical reverse-bias EB burn-in data from first-generation SiGe HBTs (Figure 13.3) show less than 5% change in the current gain after a 500-hour, −40°C reverse-bias EB stress at 2.7 V [3]. Comparison of reverse-bias EB stress data of SiGe HBTs having various Ge profile shapes with a comparably constructed epi-base Si BJT control (Figure 13.4) suggests that there is no enhanced reliability risk associated with the SiGe layer itself [9].

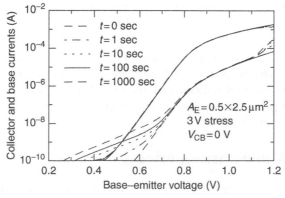

FIGURE 13.1 Gummel characteristics of a first-generation SiGe HBT showing the effects of reverse-bias emitter–base stress. Time (t) is measured in seconds. (From JD Cressler. *IEEE Trans. Device Mater. Reliab.*, 4:222–236, 2004. With permission.)

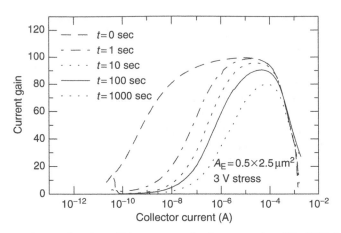

FIGURE 13.2 Current gain as a function of bias current of a first-generation SiGe HBT showing the effects of reverse-bias emitter–base stress. Time (t) is measured in seconds. (From U Gogineni, JD Cressler, G Niu, and DL Harame. *IEEE Trans. Electron Dev.* 47:1440–1448, 2000. With permission.)

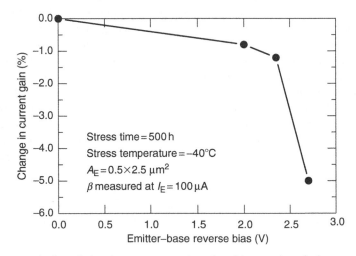

FIGURE 13.3 Current gain degradation due to reverse emitter–base bias stressing of a first-generation SiGe HBT as a function of stress voltage. (From DL Harame, DC Ahlgren, DD Coolbaugh, JS Dunn, G Freeman, JD Gillis, RA Groves, GN Hendersen, RA Johnson, AJ Joseph, S Subbanna, AM Victor, KM Watson, CS Webster, and PJ Zampardi. *IEEE Trans. Electron Dev.* 48:2575–2594, 2001. With permission.)

Typical high forward-current burn-in data from first-generation SiGe HBTs (Figure 13.5) show less than 5% change in the current gain after a 500-hour, 100°C forward-current stress at 1.3 mA/μm^2 [3]. Using empirically determined acceleration factors, this result is theoretically equivalent to a more-than-acceptable 10% current gain degradation after 100,000 power-on-hours (POH) under "normal use" conditions (1.25 mA/μm^2 at 100°C).

13.4 "Mixed-Mode" Stress Effects

Optimized transistor scaling leading to such rapid advances in SiGe HBT performance inevitably results in increased current density operation (i.e., the J_C at which peak f_T is achieved), in the presence of increased impact ionization due to the increased collector doping required to suppress both Kirk effect and high-injection heterojunction barrier effects [4].

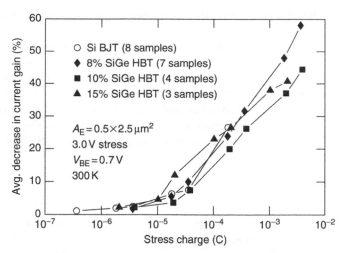

FIGURE 13.4 Percent current gain degradation as a function of injected stress charge for a variety of SiGe profiles and a comparably constructed epi-base Si BJT. (From U Gogineni, JD Cressler, G Niu, and DL Harame. *IEEE Trans. Electron Dev.* 47:1440–1448, 2000. With permission.)

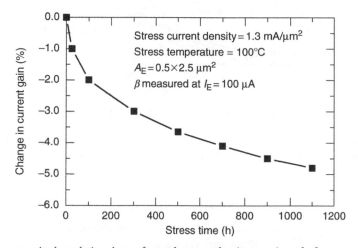

FIGURE 13.5 Current gain degradation due to forward current density stressing of a first-generation SiGe HBT as a function of stress time. (From DL Harame, DC Ahlgren, DD Coolbaugh, JS Dunn, G Freeman, JD Gillis, RA Groves, GN Hendersen, RA Johnson, AJ Joseph, S Subbanna, AM Victor, KM Watson, CS Webster, and PJ Zampardi. *IEEE Trans. Electron Dev.* 48:2575–2594, 2001. With permission.)

A new reliability damage mechanism in SiGe HBTs was recently reported [10,11], which was termed "mixed-mode" degradation, since it results from the simultaneous application of high J_C and high V_{CB}, and which differs fundamentally from conventional bipolar device reliability damage mechanisms associated with either reverse emitter–base stress [2,9], or high forward current density stress [13,14]. (We note parenthetically that the 120 GHz, second-generation SiGe HBTs used in this mixed-mode stress study [12] showed negligible (acceptable) degradation for conventional reverse EB and forward J_C stressing.)

To carefully control the total injected charge during mixed-mode stressing, a robust time-dependent stress methodology was used which operates the transistor in common-base mode under variable forced I_E and V_{CB} conditions. The stress times ranged from 1 msec to 1000 sec, with excellent repeatability. Both forward-mode and inverse-mode (emitter and collector swapped) Gummel characteristics were measured at specific (adjustable) time intervals, and the base current degradation determined at $V_{BE} = 0.5\,V$ ($V_{CB} = 0\,V$).

Typical forward-mode and inverse-mode Gummel characteristics as a function of cumulative stress time are shown in Figure 13.6 and Figure 13.7 for a $J_E = 40\,mA/\mu m^2$ and $V_{CB} = 3.0\,V$ mixed-mode stress condition. The mixed-mode stressing produces interface traps and subsequent G/R base current leakage at both the emitter–base spacer (forward-mode), and the shallow-trench edge (inverse-mode), consistent with Ref. [10]. The latter effect is new and unexpected compared to conventional reliability stress modes. The specific J_E and V_{CB} dependence of the damage process is shown in Figure 13.8 and Figure 13.9. Damage thresholds can be observed in second-generation SiGe HBTs at about $25\,mA/\mu m^2$ ($3.0\,V$), and $V_{CB} = 1.0\,V$ ($35\,mA/\mu m^2$).

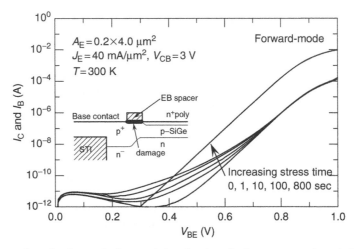

FIGURE 13.6 Forward-mode Gummel characteristics showing the base current degradation with increasing mixed-mode stress time ($J_E = 40\,mA/\mu m^2$ and $V_{CB} = 3.0\,V$). (From C Zhu, Q Liang, R Al-Huq, JD Cressler, A Joseph, J Johansen, T Chen, G Niu, G Freeman, J-S Rieh, and D Ahlgren. Technical Digest IEEE International Electron Devices Meeting, Washington, DC, 2003, pp. 185–188. With permission.)

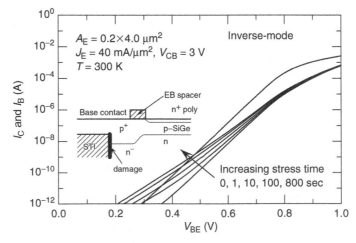

FIGURE 13.7 Inverse-mode Gummel characteristics showing the base current degradation with increasing mixed-mode stress time ($J_E = 40\,mA/\mu m^2$ and $V_{CB} = 3.0\,V$). (From C Zhu, Q Liang, R Al-Huq, JD Cressler, A Joseph, J Johansen, T Chen, G Niu, G Freeman, J-S Rieh, and D Ahlgren. Technical Digest IEEE International Electron Devices Meeting, Washington, DC, 2003, pp. 185–188. With permission.)

We consistently observed random fluctuations in the base current during stress (both within a single device and device-to-device), which we believe are due to simultaneous creation and annealing of stress-created interface traps. As argued in Ref. [13], base current fluctuations in the 1 to 100 pA range are quite consistent with reported hot-carrier-generated trap capture cross sections of 10^{-13} to 10^{-15} cm^{-2} for Si–SiO$_2$ interface traps. We have performed poststress annealing studies (at 400°C for 30 min in forming gas), which demonstrate that most of the mixed-mode-induced damage can be annealed, consistent with the known behavior of interface traps.

To gain deeper insight into the mixed-mode damage physics, the hot-carrier injection current was simulated under mixed-mode conditions (35 mA/μm^2, 3.0 V V_{CB}) using fully calibrated (using SIMS, device layout, and dc and ac data), isothermal two-dimensional MEDICI simulations (with the "gate

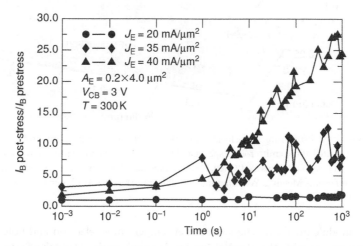

FIGURE 13.8 Base current damage ratio versus stress time for different emitter current densities ($V_{CB} = 3.0$ V). Poststress data are measured at $V_{BE} = 0.5$ V and $V_{CB} = 0.0$ V. (From From C Zhu, Q Liang, R Al-Huq, JD Cressler, A Joseph, J Johansen, T Chen, G Niu, G Freeman, J-S Rieh, and D Ahlgren. Technical Digest IEEE International Electron Devices Meeting, Washington, DC, 2003, pp. 185–188. With permission.)

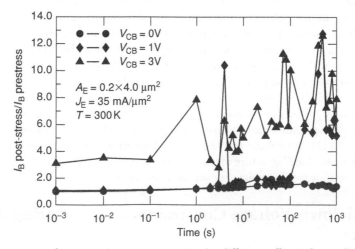

FIGURE 13.9 Base current damage ratio versus stress time for different collector–base voltages ($J_E = 35$ mA/ μm^2). Poststress data are measured at $V_{BE} = 0.5$ V and $V_{CB} = 0.0$ V. (From From C Zhu, Q Liang, R Al-Huq, JD Cressler, A Joseph, J Johansen, T Chen, G Niu, G Freeman, J-S Rieh, and D Ahlgren. Technical Digest IEEE International Electron Devices Meeting, Washington, DC, 2003, pp. 185–188. With permission.)

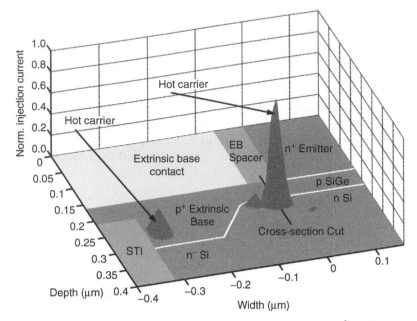

FIGURE 13.10   Simulated distribution of the local hot-carrier current ($J_E = 35\,\text{mA}/\mu\text{m}^2$ and $V_{CB} = 3.0\,\text{V}$). The peak injection currents are located at the emitter–base spacer and the shallow-trench isolation edge. (From C Zhu, Q Liang, R Al-Huq, JD Cressler, A Joseph, J Johansen, T Chen, G Niu, G Freeman, J-S Rieh, and D Ahlgren. Technical Digest IEEE International Electron Devices Meeting, Washington, DC, 2003, pp. 185–188. With permission.)

current analysis" module invoked). The local carrier temperatures (electron and hole) were calculated using energy balance. As with the classical "lucky-electron" model [15], the Si–SiO$_2$ interface trap production is correlated with the hot-carrier injection current density [16], and is the product of the local electron current density and the probability that these electrons reach the oxide interface with kinetic energy higher than the interface trap creation activation energy (taken here to be 2.3 eV [16]). Note that the emitter–base spacer (and the shallow-trench isolation edge) are well within a mean-free path length of the (randomly moving) hot carriers generated in the CB junction by impact ionization. Figure 13.10 shows the normalized distribution of the simulated local hot-carrier injection current density. For both the emitter–base spacer and shallow-trench damage regions, we find that injection current density is clearly present and dominated by hot electrons (hot holes exist but in smaller numbers), consistent with the data shown in Figure 13.6 and Figure 13.7.

A comparison of these second-generation, 120 GHz SiGe HBT mixed-mode stress results with stress results on more aggressively scaled third-generation 200 GHz SiGe HBTs (peak f_T in the 200 GHz devices occurs at a J_C of nearly 20 mA/μm^2) shows that transistor lateral and vertical scaling appears to improve the mixed-mode damage thresholds at fixed J_E and V_{CB} conditions, consistent with observations reported in Refs. [10,17]. We believe that the observed improvement is likely to be a consequence of the new "raised extrinsic base" structure of the 200 GHz device, which has a reduced level of the impact ionization at the device edge due to its very shallow extrinsic base formation.

13.5   Breakdown Voltage Constraints and Operating Point Instabilities

Optimized scaling of SiGe HBTs necessarily results in the decrease of maximum operating voltages as the technology evolves. For CMOS technology, the reduction in V_{DD} is driven by hot-carrier reliability constraints. For SiGe HBTs, this voltage reduction has a different physical origin; namely, the increase in impact ionization as the collector doping is increased to support increasingly higher operating current densities. In the case of SiGe HBTs, BV_{CEO} decreases, for instance, from 3.3 to 2.5 to 1.7 V for first-, second-, and third-generation SiGe technology, respectively. Operating voltage compression is rarely a good thing from a circuit design and system performance perspective, except for helping maintain power dissipation as operating frequencies rise. Particularly in high-speed analog and RFIC design, and even in cascoded digital circuits, all application arenas for which bipolar technology is more naturally suited than CMOS, voltage compression can present serious problems for maintaining adequate signal-to-noise ratio, voltage headroom for transistor "stacking," and loss of efficiency.

This voltage compression issue in SiGe HBTs is far more interesting in many ways than it is for CMOS, since the actual maximum operating voltages that the transistor can sustain depends very strongly on how the transistor is driven (i.e., its local circuit environment), and the static as well as dynamic bias currents which it sees as it is operated. This makes for a seriously complicated situation, particularly with regard to predictive modeling and robust reliability testing. The ubiquitous BV_{CEO} of SiGe HBT technologies, for instance, represents the worst-case bias configuration, since it electrically opens the base, thereby providing a (bad) positive feedback path for the impact ionization induced currents originating in the CB junction, leading to premature breakdown. Even in such cases, however, BV_{CEO} is of questionable relevance to real mixed-signal circuit design since all circuits will present a dynamically varying finite impedance between the base and emitter terminals (i.e., if the base is truly open the circuit cannot do very much!). Thus, the maximum sustainable operating voltage on a SiGe HBT generally lies between that of BV_{CEO} (worst case) and the open-emitter collector–base breakdown voltage BV_{CBO} (best case). For instance, in a first-generation SiGe HBT, BV_{CEO} might be 3.3 V, while BV_{CBO} might be 10 V, three times higher. Figure 13.11 shows the measured maximum sustainable collector-to-emitter voltage as a function of operating current density for different input bias configurations for a second-generation SiGe HBT with a 2.2 V BV_{CEO} and 7.5 V BV_{CBO}. Observe the substantial (worrisome) structure in these curves. This should give any reliability engineer food for thought; since in principle the operational bias configuration of the transistor is application driven, and can even vary by

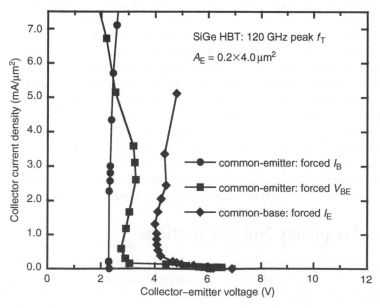

FIGURE 13.11 Collector current density as a function of applied collector to emitter voltage for three different transistor drive conditions. The data are for a second-generation 120 GHz peak f_T SiGe HBT. (From JD Cressler. *IEEE Trans. Device Mater. Reliab.*, 4:222–236, 2004. With permission.)

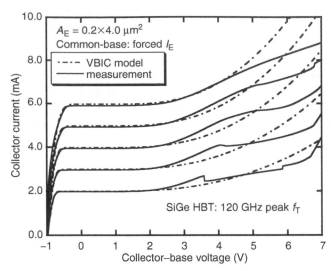

FIGURE 13.12 Common-base output characteristics for forced emitter-current drive showing bias induced instabilities, and a comparison with results from a "calibrated" compact model. (From JD Cressler. *IEEE Trans. Device Mater. Reliab.*, 4:222–236, 2004. With permission.)

architecture within the same basic application. Said another way, ensuring reliability of the transistor building block is no solid guarantee of overall circuit and system reliability.

Perhaps even more troubling is the fact that very complex operating bias point "instabilities" exist in SiGe HBTs (in all bipolar devices, actually) [18,19]. Such instabilities are generally believed to be triggered by impact ionization, resulting in so-called "pinch-in" current constriction phenomena, but the simple truth is that they are both highly complex in nature and extremely difficult to both predict and test for. The simple and tempting proclamation of "do not design your circuits to operate the transistors anywhere near such instabilities," may seem like a safe and reasonable approach, but this stance is increasingly problematic given the ever-shrinking voltage supplies of scaled IC technologies, and the ever-increasing need for circuit designers to maximize both performance and efficiency. In addition, even circuits that are well behaved with respect to such instabilities at dc may be inadvertently forced to dynamically switch through such unstable regimes, with unpredictable consequences. This is particularly true for certain of the myriad classes of amplifiers, for instance. Figure 13.12 shows a typical result for a second-generation SiGe HBT operating in forced I_E mode. At V_{CE} in the range of 3 V, the output characteristics develop a chaotic-like behavior, the potential circuit implications of which will be frightening to most designers. More alarming perhaps, even well-calibrated compact models cannot capture such instabilities in a robust manner, and hence such effects are effectively not modeled in even mature technology design kits. The open questions from a reliability perspective are: (1) how do we meaningfully test our devices for exposure to such instabilities, and (2) can we predict the results of such behavior on our circuits. These questions remain largely unanswered at present.

13.6 Low-Frequency Noise Variations

Low-frequency noise (LFN) is up-converted to phase noise (noisy sidebands on the carrier) through the nonlinearities of transistors, producing a fundamental limit on the achievable spectral purity of communications systems. While LFN is not traditionally considered to be a reliability issue, per se, its importance in mixed-signal circuit design makes it worthy of fresh consideration, within both the context of aggressive geometrical scaling, as well as the addition of SiGe to the problem. One of the unique merits of SiGe HBTs is that they can simultaneously provide very small broadband noise and low

$1/f$ noise, giving them a decided advantage over scaled CMOS and III–V devices for high-frequency wireless building blocks limited by phase noise (e.g., oscillators and mixers) [20,21]. SiGe HBTs are in fact capable of extremely impressive levels of LFN, even when they are aggressively scaled in geometry, vertical profile, and thermal budget to improve the broadband performance. For instance, a $1/f$ noise corner frequency of 220 Hz ($I_B = 1\,\mu A$) was achieved in a $0.12 \times 0.50\,\mu m^2$ (drawn) third-generation SiGe HBT with a peak f_T of greater than 200 GHz and minimum NF_{min} of less than 0.5 dB at 10 GHz. This combined low-frequency plus broadband noise performance is superior to *any* semiconductor device technology, even InP pHEMT technology. This impressive performance noted, we must also point out that an unusual statistical variation (in effect, a device-to-device statistical "scatter") in the LFN spectra of small geometry SiGe HBTs has been recently reported [22]. We view such variations to be inherent reliability concerns, with largely unknown circuit implications, and are generally underappreciated in the reliability community.

This LFN statistical variation with size has also been observed in MOSFETs, JFETs, and BJTs in small-sized devices [23–25]. Fundamentally, the noise-generating mechanism inside transistors is generally regarded as a superposition of individual trapping or detrapping processes due to the presence of G/R centers in the device. Each G/R center contributes a Lorentzian-type ($1/f^2$) noise signature, and given a sufficient number of traps (a statistical ensemble), these Lorentzian processes combine to produce the observed $1/f$ noise behavior. At sufficiently small device size, however, the total number of traps is small enough that non-$1/f$ behavior, and hence large statistical variations, can be easily observed.

Figure 13.13 compares a typical family of noise spectra measured on a "small" SiGe HBT, with those measured on a "large" SiGe HBT, at fixed base current density (to allow easy comparison). In these SiGe HBTs, the noise magnitude as a function of bias current (at 10 Hz) for the devices exhibiting "clean" $1/f$ behavior exhibit a classical I_B^2 plus $1/A_E$ dependence across the useful bias range, independent of the technology generation, consistent with classical number fluctuation theory [26,27].

The noise variation in these SiGe HBTs can be quantified using a classical standard deviation approach [22]. Cross-generational noise variation data in SiGe HBT technology are shown in Figure 13.14 (the effect is negligible in the first-generation 50 GHz SiGe HBTs due to their larger emitter size)

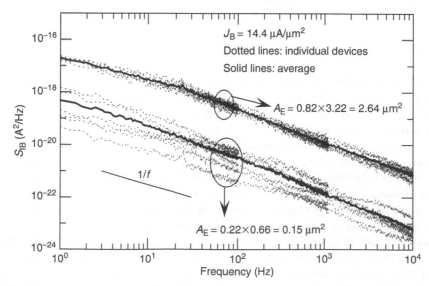

FIGURE 13.13 Base current noise spectra for a "large" area SiGe HBT and a "small" area SiGe HBT showing the large statistical variation from sample-to-sample for small geometry transistors. (From J Johansen, Z Jin, JD Cressler, Y Cui, G Niu, Q Liang, J-S Rieh, G Freeman, D Ahlgren, and A Joseph. Proceedings of the IEEE International Semiconductor Device Research Symposium, Washington, DC, 2003, pp. 12–13.)

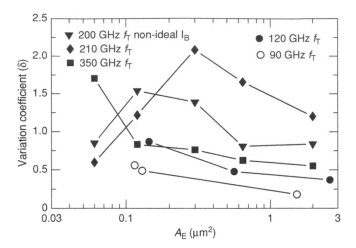

FIGURE 13.14   Noise variation coefficient versus emitter area for multiple SiGe HBT technology generations. The base bias current is 1.0 μA. (From J Johansen, Z Jin, JD Cressler, Y Cui, G Niu, Q Liang, J-S Rieh, G Freeman, D Ahlgren, and A Joseph. Proceedings of the IEEE International Semiconductor Device Research Symposium, Washington DC, 2003, pp. 12–13.)

[28]. Interestingly, observe that the noise variation in the 200 GHz SiGe technology generation shows anomalous scaling behavior below about 0.2 to 0.3 μm^2 emitter geometry, below which the noise variation rapidly decreases. Such LFN variations in very small geometry SiGe HBTs can be qualitatively explained by calculations based on the superposition of Lorentzian ($1/f^2$) G/R traps [22,23]. Both simple calculations and more sophisticated microscopic noise simulations indicate that as the emitter geometry scales, the device-to-device LFN variation becomes larger due to the decreased number of G/R traps participating in the noise process. Conventional reverse-bias EB stressing, as well as exposure to ionizing radiation, can be used to further probe and understand the origins of this unique scaling-induced reliability issue [22,29].

13.7   Summary

In this chapter, I have attempted to give a new (and hopefully refreshing) perspective on the reliability issues and concerns associated with emerging SiGe HBT technologies, particularly as they are increasingly used in a wide variety of mixed-signal circuit applications. As any honest reliability engineer will admit, the scariest scenarios are those reliability failure mechanisms that may loom beyond the horizon and remain unseen at present, even to the trained eye. While I have managed to address in this chapter the conventional reliability failure mechanisms, as well introduce a number of additional nonstandard reliability issues in SiGe HBTs, which are becoming increasingly important in the emerging mixed-signal application domain, it is impossible, by definition, to anticipate them all. Thermal effects, for instance, which are unavoidable in today's high-performance technologies operating at very high current levels, will eventually come back to plague us in a major way (they already do). Such thermal issues couple in strong ways to virtually all failure mechanisms in devices, and importantly, are both difficult to measure, and even harder to predictively model. Impact ionization induced bias point instabilities are another concern of increasing importance, which demands attention. Operating voltages necessarily compress with scaling for optimal performance, and what effect such instabilities have on circuit and system-level reliability remains unclear, and hence is worthy of increased focus. We will likely soon reach a point in certain mixed-signal circuits when simply avoiding such operational bias regimes will not be a tractable solution. All of this said, I do not view any of the reliability issues addressed here as "show-stopping" in nature, and as with all reliability concerns, they must be understood, quantified, and then carefully but relentlessly "designed around."

Acknowledgments

I am grateful to G. Niu, C. Zhu, Q. Liang, R. Al-Huq, J. Johansen, Z. Jin, T. Chen, U. Gogineni, C. Grens, J. Babcock, A. Joseph, D. Harame, G. Freeman, J.-S. Rieh, D. Ahlgren, F. Guarin, J. Dunn, B. Meyerson, D. Herman, and the IBM SiGe team for their contributions. This work was supported by the Semiconductor Research Corporation, the GEDC at Georgia Tech, and IBM.

References

1. JD Cressler. Emerging SiGe HBT reliability issues for mixed-signal circuit applications. *IEEE Trans. Device Mater. Reliab.*, 4:222–236, 2004.
2. DD Tang and E Hackbarth. Junction degradation in bipolar transistors and the reliability imposed constraints to scaling and design. *IEEE Trans. Electron Dev.* 35:2101–2107, 1988.
3. DL Harame, DC Ahlgren, DD Coolbaugh, JS Dunn, G Freeman, JD. Gillis, RA Groves, GN Hendersen, RA Johnson, AJ Joseph, S Subbanna, AM Victor, KM Watson, CS Webster, and PJ Zampardi. Current status and future trends of SiGe BiCMOS technology. *IEEE Trans. Electron Dev.* 48:2575–2594, 2001.
4. JD Cressler and G Niu. *Silicon–Germanium Heterojunction Bipolar Transistors*. Boston, MA: Artech House, 2003.
5. A Fischer, H-J Osten, and H Richter. An equilibrium model for buried SiGe strained layers. *Solid-State Electron.* 44:869–873, 2000.
6. A Neugroschel, CT Sah, and MS Carroll. Current–acceleration for rapid time-to-failure determination of bipolar junction transistors under emitter–base reverse-bias stress. *IEEE Trans. Electron Dev.* 42:1380–1383, 1995.
7. A Neugroschel and CT Sah. Comparison of time-to-failure of GeSi and Si bipolar transistors. *IEEE Electron Dev. Lett.* 17:211–213, 1996.
8. JA Babcock, JD Cressler, LS Vempati, AJ Joseph, and DL Harame. Correlation of low-frequency noise and emitter–base reverse-bias stress in epitaxial Si- and SiGe-base bipolar transistors. Technical Digest IEEE International Electron Devices Meeting, San Francisco, 1995, pp. 357–360.
9. U Gogineni, JD Cressler, G Niu, and DL Harame. Hot electron and hot hole degradation of SiGe heterojunction bipolar transistors. *IEEE Trans. Electron Dev.* 47:1440–1448, 2000.
10. G Zhang, JD Cressler, G Niu, and A Joseph. A new "mixed-mode" reliability degradation mechanism in advanced Si and SiGe bipolar transistors. *IEEE Trans. Electron Dev.* 49:2151–2156, 2002.
11. C Zhu, Q Liang, R Al-Huq, JD Cressler, A Joseph, J Johansen, T Chen, G Niu, G Freeman, J-S Rieh, and D Ahlgren. An investigation of the damage mechanisms in impact ionization-induced "mixed-mode" reliability stressing of scaled SiGe HBTs. Technical Digest IEEE International Electron Devices Meeting, Washington, DC, 2003, pp. 185–188.
12. A Joseph, D Coolbaugh, M Zierak, R Wuthrich, P Geiss, Z He, X Liu, B Orner, J Johnson, G Freeman, D Ahlgren, B Jagannathan, L Lanzerotti, V Ramachandran, J Malinowski, H Chen, J Chu, M Gordon, P Gray, R Johnson, J Dunn, S Subbanna, K Schonenberg, D Harame, R Groves, K Watson, D Jadus, M Meghelli, and A Rylyakov. A 0.18 μm BiCMOS technology featuring 120/100 GHz (f_T/f_{max}) HBT and ASIC compatible CMOS using copper interconnect. Proceedings of the IEEE Bipolar/BiCMOS Circuits and Technology Meeting, Minneapolis, 2001, pp. 143–146.
13. RA Wachnik, TJ Bucelot, and GP Li. Degradation of bipolar transistors under high current stress at 300 K. *J. Appl. Phys.* 63:4734, 1988.
14. MS Carroll, A Neugroschel, and C-T Sah. Degradation of silicon bipolar junction transistors at high forward current density. *IEEE Trans. Electron Dev.* 44:110, 1997.
15. S Tam, PK Ko, and C Hu. Lucky-electron model of channel hot-electron injection in MOSFET's. *IEEE Trans. Electron Dev.* 31:1116, 1984.

16. D DiMaria and JW Stasiak. Trap creation in silicon dioxide produced by hot electrons. *J. Appl. Phys.* 65:2342, 1989.

17. Z Yang, F Guarin, E Hostetter, and G Freeman. Avalanche current induced hot carrier degradation in 200 GHz SiGe heterojunction bipolar transistors. Proceedings of the IEEE International Reliability Physics Symposium, 2003, pp. 339–343.

18. M Rickelt, HM Rein, and E Rose. Influence of impact-ionization-induced instabilities on the maximum usable output voltage of silicon bipolar transistors. *IEEE Trans. Electron Dev.* 48:774–783, 2001.

19. G Freeman, B Jagannathan, S-J Jeng, J-S Rieh, A Stricker, D Ahlgren, and S Subbanna. Transistor design and application considerations for >200 GHz SiGe HBTs. *IEEE Trans. Electron Dev.* 50:645–655, 2003.

20. LS Vempati, JD Cressler, JA Babcock, RC Jaeger, and DL Harame. Low-frequency noise in UHV/CVD epitaxial Si and SiGe bipolar transistors. *IEEE J. Solid-State Circ.* 31:1458–1467, 1996.

21. R Plana, L Escotte, JP Roux, J Graffeuil, A Gruhle, and H Kibbel. 1/f noise in self-aligned Si/SiGe heterojunction bipolar transistors. *IEEE Electron Dev. Lett.* 16:58–60, 1995.

22. Z Jin, JD Cressler, G Niu, and A Joseph. Impact of geometrical scaling on low-frequency noise in SiGe HBTs. *IEEE Trans. Electron Dev.* 50:676–682, 2003.

23. M Sanden, O Marinov, and MJ Deen. A new model for the low-frequency noise and the noise level variation in polysilicon emitter BJTs. *IEEE Trans. Electron Dev.* 49:514–520, 2002.

24. P Llinares, D Celi, and O Roux-dit-Buisson. Dimension scaling of 1/f noise in the base current of quasi-self-aligned polysilicon emitter bipolar junction transistors. *J. Appl. Phys.* 82:2671–2675, 1997.

25. MJ Deen, S Rumysantesev, R Bashir, and R Taylor. Measurements and comparison of low frequency noise in npn and pnp polysilicon emitter bipolar transistors. *J. Appl. Phys.* 84:625–633, 1998.

26. HAW Markus and TGM Kleinpenning. Low-frequency noise in polysilicon emitter bipolar transistors. *IEEE Trans. Electron Dev.* 42:720–727, 1995.

27. A Mounib, F Balestra, N Mathieu, J Brini, G Ghibaudo, A Chovet, A Chantre, and A Nouailhat. Low-frequency noise sources in polysilicon emitter BJTs: influence of hot-electron-induced degradation and post-stress recovery. *IEEE Trans. Electron Dev.* 42:1647–1652, 1995.

28. J Johansen, Z Jin, JD Cressler, Y Cui, G Niu, Q Liang, J-S Rieh, G Freeman, D Ahlgren, and A Joseph. On the scaling limits of low-frequency noise in SiGe HBTs. Proceedings of the IEEE International Semiconductor Device Research Symposium, Washington DC, 2003, pp. 12–13.

29. Z Jin, JA Johansen, JD Cressler, RA Reed, PW Marshall, and AJ Joseph. Using proton irradiation to probe the origins of low-frequency noise variations in SiGe HBTs. *IEEE Trans. Nuclear Science*, 50:1816–1820, 2003.

14

Self-Heating and Thermal Effects

Jae-Sung Rieh
Korea University

14.1 Introduction

Within semiconductor devices under external bias, the carriers experience successive scatterings while traveling under the influence of an electric field. In most scattering events, they give up kinetic or potential energy they retained, which is converted into various other forms of energy as a result of generating phonons, photons, or electron–hole pairs. Among these energy-conversion modes, the generation of phonons, or the increase of lattice vibration level, is the most frequently encountered mode and results in the generation of heat in the devices. The consequent junction temperature rise, or *self-heating*, significantly influences device behaviors. It modulates the device operation condition since the characteristics of semiconductors are inherently given as a function of temperature. Device reliability is also affected since the raised temperature promotes most of the long-term degradation mechanisms as well as the catastrophic failures of the devices. Therefore, an accurate characterization of self-heating is critical for the precise prediction of device operation and degradation as well as the prevention of device failures. Historically, self-heating has been a concern mainly for high-power devices in which extensive power consumption results in a great heat generation. However, recent aggressive scalings intended for speed enhancement, which usually accompany a considerable increase in operation current level, have made the self-heating in high-speed devices a major concern. Hence, self-heating has become a generic issue for semiconductor systems in general, and their thermal properties need to be properly analyzed along with the electrical properties.

This chapter provides an overview of the issues related to self-heating and thermal effects in SiGe HBTs. As SiGe HBTs are nearly identical to conventional Si BJTs from the thermal point of view, except for a fractional amount of Ge included in the base, the analyses and discussions made in this review are for Si BJTs in general. Although SiGe in the base has a poorer thermal conductivity than Si [1], its impact will not be pronounced since the principal heat source in bipolar transistors is the base–collector space–charge region and most of the generated heat is dissipated downward through the substrate. Section 14.2 overviews the modeling of self-heating and its impact on device operation, followed by

Section 14.3 in which thermal resistance measurement methods are introduced along with the trend of the thermal resistance for various SiGe HBT dimensions and structures. In Section 14.4, selected reliability issues related to self-heating are discussed.

14.2 Modeling of Self-Heating

Thermal Resistance R_{th}

The heat conduction in a homogeneous isotropic solid is described by following time-dependent equation [2]:

$$\nabla^2 T = \frac{\rho c}{\kappa}\frac{\partial T}{\partial t},\tag{14.1}$$

where T is temperature, ρ is density, c is specific heat, κ is thermal conductivity, and t is time. With appropriate initial and boundary conditions, the temperature is determined as a function of time and position within the conducting media. The thermal resistance R_{th} is defined as the ratio of the final value of the temperature at a given position $\vec{r}$ and the dissipated power (P_{diss}) from the heat source:

$$R_{th}(\vec{r}) = \frac{T(\vec{r},\, t=\infty) - T(\vec{r},\, t=0)}{P_{diss}} = \frac{\Delta T(\vec{r})}{P_{diss}}.\tag{14.2}$$

Although R_{th} is a function of position in general, R_{th} *of a device* conventionally refers to the thermal resistance at the region inside the device with the peak temperature, which is called the junction temperature T_j. Hence, following position-independent formula is widely accepted:

$$T_j = R_{th}P_{diss} + T_0,\tag{14.3}$$

where R_{th} is the thermal resistance of the given device, and T_0 is the temperature in the absence of power dissipation, or the ambient temperature.

The simplest practical boundary condition for the heat dissipation in a device fabricated on a semiconductor substrate is a hemisphere with an adiabatic surface just above the heat source. Although the introduction of an image heat source above the surface simplifies such boundary condition [3], there exists no closed-form solution for this apparently simple geometry, implying the level of complexity involved in the proper modeling of R_{th} in practical devices. Joy and Schlig [3] proposed an approximate expression, based on numerical calculations, to best represent R_{th} for such boundary condition:

$$R_{th} = \frac{1}{2\kappa\sqrt{LW}}f(L,\, W,\, H,\, D)\tag{14.4a}$$

$$\simeq \frac{1}{4\kappa\sqrt{LW}},\tag{14.4b}$$

where $f(L, W, H, D)$ is a function of the dimensions for the heat source: length (L), width (W), height (H), and the distance from the surface (D). For bipolar transistors, in which most of heat is generated within the base–collector space–charge region, the dimensions of this region suitably serve for the estimation when substituted. Equation 14.4a can be further reduced to Equation 14.4b based on the fact that $f(L, W, H, D)$ typically falls in the proximity of 0.5 for most practical devices [3], which is valid even for the aggressively scaled devices of today. Therefore, R_{th} of plain bulk devices can be estimated from Equation 14.4b with a reasonable accuracy.

While an ideal homogeneous medium was assumed for the substrate of the device in the analysis above, actual bipolar transistors typically employ oxide-based structures for improved isolation, such as deep trenches or buried oxides with SOI substrates. Since the thermal conductivity of silicon dioxide is only one hundredth of that of silicon (see Table 14.1 [1,4,5]), such oxide-based isolations severely impede heat dissipation, resulting in R_{th} significantly greater than predicted by Equation 14.4b [6–11]. Therefore, modified approaches are necessary to accurately model the devices with oxide isolations, which is challenging as the oxide-based structures impose complicated boundary conditions. Despite the difficulty, there have been efforts to tackle the problem, which are briefly reviewed below.

Deep trench is the predominant isolation scheme for modern bipolar transistors and the modeling of R_{th} for deep trench-isolated devices has been a subject of numerous studies. Walkey et al. [7,12] treated the vertical trench walls as adiabatic boundaries and the trench bottom plane as a boundary with a constant temperature, and then applied the image source method similar to Ref. [3]. Pacelli et al. [10] introduced an additional R_{th} term to Equation 14.4b to account for the thermal resistance increase that results from the blocking of radial heat propagation by the trench. Rieh et al. [11,13] treated the trench as a perfect heat-insulator that limits the lateral extent of heat flux, and assumed that the flux below the trench is confined within a cone-shaped boundary [14]. In this work, the thermal resistance of the device was estimated based on the following general expression for R_{th}, applied to the geometrical heat flux boundary assumed:

$$R_{th} = \int \frac{1}{\kappa(z)A(z)} dz, \qquad (14.5)$$

where κ and A are the thermal conductivity and the cross section of the heat flux, respectively, both given as a function of z. Equation 14.5 implies that the increase in R_{th} for trench-isolated devices is more pronounced with deeper trenches and narrower trench-enclosed areas. This geometrical approach leads to a simple analytic expression of R_{th} in terms of the device dimensions, which can be readily implemented into device models. Despite the different approaches, all these thermal models predict a substantial increase in R_{th} for the deep trench-isolated devices compared to plain bulk devices (50% to 100% for typical trench dimensions), manifesting the adverse effects of the oxide trench isolations on heat dissipation.

SOI-based bipolar transistors have gained increasing attention recently [15–17], owing to improved isolation or compatibility with SOI CMOS, or both. From the thermal point of view, however, such structure is detrimental because the buried oxide, often combined with oxide trenches and BEOL dielectric layers, severely impedes the heat dissipation. A few modeling approaches [8,10] and measurements [6,17] have been reported regarding the thermal characteristics of bipolar transistors on SOI substrates, consistently showing that SOI devices exhibit R_{th} values far larger (50% to 300%) than those of bulk devices of a similar dimension. It is noted that the increase of R_{th} in SOI structures is far more significant when combined with deep trenches, as verified by simulations in Ref. [17].

Thermal Capacitance C_{th}

For steady-state conditions, the thermal resistance R_{th} alone is sufficient to describe the relationship between temperature and dissipated power. However, when the *transient* behavior of self-heating is important, the concept of the thermal capacitance C_{th} needs to be introduced, which will eventually constitute the thermal impedance Z_{th} along with R_{th}. A general expression for the thermal capacity (heat capacity) in a solid is given by

TABLE 14.1 Thermal Conductivity of Selected Semiconductors and Insulators at $T = 300\,\text{K}$ [1,4,5]

	Si	GaAs	InP	SiC(6H)	GaN	$Si_{0.7}Ge_{0.3}$	SiO_2	Si_3N_4	Diamond
κ (W/cm K)	1.41	0.46	0.68	4.6	1.3	0.08	0.014	0.19	2000

$$C_{th} = \rho c V, \tag{14.6}$$

where V is the volume to be heated. It is assumed in Equation 14.6 that the temperature is uniform throughout the volume V. For practical device structures, however, thermal gradients always exist inside the device, and the volume V is not clearly defined since the substrate around the device is partially heated as well. Hence, Equation 14.6 is seldom employed to model the thermal capacitance for practical purposes. Instead, C_{th} is usually estimated from the measured time-dependent response of the junction temperature to the power dissipation, typically applied as a step function [18–20]. Together with R_{th} values, which can be readily measured as described in the next section, C_{th} is obtained from the estimated time delay $\tau = R_{th}C_{th}$ of the transient response.

Reported values of τ for typical bipolar transistors range from ∼50 ns to ∼1 μs [3,20]. It is noted that this is the time delay associated with the *local* heating of an individual device. Another time delay related to the *global* heating of a chip, which arises from the averaged heat dissipation of all the devices embedded in the chip, is much larger and ranges from milliseconds to minutes depending on the packaging and air flow design around the package [21]. It is the global heating that causes the overall chip temperature rise, which typically ranges around 80 to 120°C. The local heating is superimposed onto the global heating, causing the junction temperature to rise above the chip temperature. The thermal modeling of individual devices generally pertains to the local heating and the chip temperature is considered as a fixed ambient temperature.

Thermal Impedance Z_{th}

The thermal impedance Z_{th} is a generalized form of R_{th} to include the time dependence of the junction temperature rise. With a simple electrical circuit analogy, Z_{th} can be represented as a parallel combination of R_{th} and C_{th}, with the dissipated power P_{diss} replacing the current I as shown in Figure 14.1a. Then the frequency domain expression of thermal impedance $Z_{th}(s)$ is given by

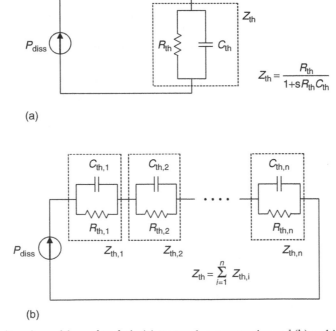

FIGURE 14.1   Circuit analogy of thermal analysis: (a) monopole representation and (b) multipole representation.

$$Z_{th}(s) = \left(\frac{1}{R_{th}} + sC_{th}\right)^{-1} = \frac{R_{th}}{1 + sR_{th}C_{th}}. \tag{14.7}$$

The corresponding time domain expression of thermal impedance $Z_{th}(t)$, assuming P_{diss} is applied as a unit step function, can be obtained by taking the inverse Laplace transform of Equation 14.7 multiplied by $1/s$:

$$Z_{th}(t) = R_{th}\left(1 - \exp\left(-\frac{t}{R_{th}C_{th}}\right)\right). \tag{14.8}$$

This is a *monopole* approach with a single time constant ($\tau = R_{th}C_{th}$), which is widely adopted by most bipolar models, including VBIC, largely owing to its simplicity. When a higher level of accuracy is required, *multipole* approaches with more than one time constant can be employed. The circuit representation for the multipole approximation is shown in Figure 14.1b and the corresponding Z_{th} is given by

$$Z_{th}(t) = \sum_i R_{th,i}\left(1 - \exp\left(-\frac{t}{R_{th,i}C_{th,i}}\right)\right). \tag{14.9}$$

It was shown that even two-pole approximations provide a significantly improved level of accuracy, in terms of match to measurement, compared to monopole approaches [8,20].

For bipolar transistors, the dissipated power is given by $P_{diss} = I_B V_{BE} + I_C V_{CE}$ and the final junction temperature T_j is expressed in terms of Z_{th} as

$$T_j = Z_{th}(I_B V_{BE} + I_C V_{CE}) + T_{amb}. \tag{14.10}$$

An accurate prediction of transistor operation in the presence of self-heating is then obtained by simultaneously solving Equation 14.10 and relevant electrical equations in which T_j is substituted for the temperature in electrical parameter expressions. Electrothermally self-consistent circuit simulators can thus be developed based on such relationship linking electrical and thermal properties of bipolar transistors [21–25].

14.3 Thermal Resistance Measurement and Trends

Measurement of R_{th}

The thermal resistance of a device can be extracted from the relation between the power dissipation and the junction temperature. Such relation is usually obtained by exploiting the temperature dependence of an electrical parameter of the device, in which the electrical parameter is measured for various power dissipation levels and then translated into temperature variations by a careful calibration. The most widely used such temperature-sensitive electrical parameters (TSEPs) for R_{th} extraction in bipolar transistors are the base–emitter voltage V_{BE} [13,26–30] and current gain β [30–32]. The temperature dependence of current levels can also be utilized for the extraction [33–35]. Here, an approach utilizing V_{BE} as a TSEP is briefly introduced following the description in Ref. [13].

As a first step, the device is biased with a fixed emitter current I_E and collector–base voltage V_{CB}, and then V_{BE} is measured for substrate temperature T_S which is swept for the range of interest. The resultant $V_{BE} - T_S$ correlation, such as shown in Figure 14.2a, is called the calibration curve. As a second step, the device is biased with the same I_E as used in step 1 at a fixed substrate temperature (denoted as ambient temperature T_{amb}), and V_{BE} is measured for different dissipated power ($P_{diss} = I_C V_{CE} + I_B V_{BE}$) with varying V_{CB} (Figure 14.2b). A moderate V_{CB} range is suggested to avoid avalanche effects. Now, by

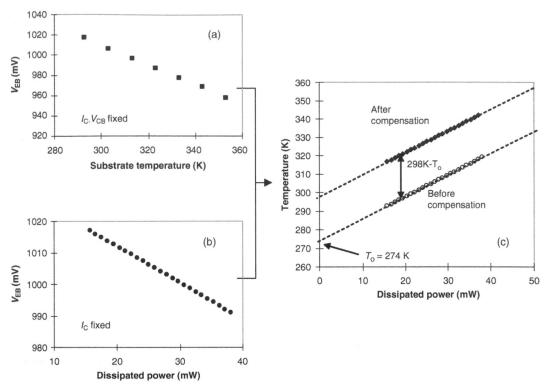

FIGURE 14.2   The extraction procedure for R_{th}: (a) step 1: I_E and V_{CB} is fixed, and V_{BE} is measured for various substrate temperature T_S. (b) Step 2: I_E is fixed and V_{BE} is measured for a range of P_{diss} with varying V_{CB}. (c) Step 3: P_{diss}–T_j relation is obtained by eliminating V_{BE} from step 1 and step 2, and compensation is made to account for self-heating in step 1. The slope of the curve is R_{th}.

eliminating V_{BE} from the two measurements obtained from step 1 and step 2, the relationship between temperature and power is obtained (as shown in Figure 14.2c). As a final step, a compensation is made in order to account for the self-heating in step 1, since it related V_{BE} to the substrate temperature T_S, not to the junction temperature T_j. This can be done by taking the y-axis intercept point of the obtained temperature–power relation, denoted by T_o in Figure 14.2c, and shifting the entire curve upward by the difference between the ambient temperature T_{amb} and T_o [31]. This final curve presents the junction temperature as a function of power dissipation, and R_{th} can be extracted from its slope. If the self-heating in step 1 is significant enough to cause a considerable power dissipation variation due to the small V_{BE} change over the T_S variation (note that any increase in T_S would slightly reduce V_{BE}, resulting in a finite increase in the total power dissipation, which was assumed negligible above), an additional compensation is needed which involves a correction in the slope of the temperature–power curve [29].

R_{th} Trend in SiGe HBTs

Several studies have been published which report the measured thermal resistance of SiGe HBTs [11,13,17,28,29,34]. Here, results [11] obtained from the IBM's deep trench-isolated 200 GHz SiGe HBTs [36] are introduced as an example, which were extracted based on the method described above. Figure 14.3 shows the measured R_{th} for different emitter lengths and a fixed emitter width of 0.12 μm. Also included as a solid line is a prediction from the analytical model in Ref. [13], in which the heat dissipation through the metal lines is treated as a fitting parameter. As is clear from Figure 14.3, shorter

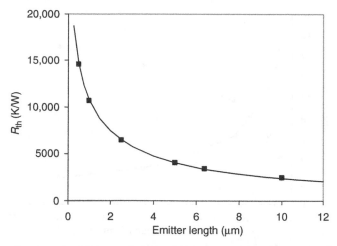

FIGURE 14.3 Thermal resistance of deep trench-isolated SiGe HBTs with various emitter lengths [11]. Measurement (symbols) is compared with model prediction (solid line) where the heat dissipation through metal line is treated as a fitting parameter.

devices (with a smaller emitter length) exhibit larger R_{th} due to a smaller cross-sectional area enclosed by the deep trench, which effectively pinches the heat flux toward substrate. However, the deep trench-enclosed area *per emitter area* is larger for shorter devices, which results in a smaller junction temperature rise for shorter devices when a *fixed power density* is assumed over the various emitter lengths (Figure 14.4). An opposite trend is exhibited by the junction temperature rise for a *fixed power*, which closely follows the trend of thermal resistance (recall $\Delta T_j = P_{diss}R_{th}$). However, the fixed power density assumption is more realistic since it is a similar current density, rather than current, that is shared for devices with different sizes. This indicates that shorter devices, despite their larger thermal resistances, tend to cause less self-heating effects and are favored from the thermal standpoint. A similar trend is observed for the variation of R_{th} with emitter width. With emitter length fixed, narrower devices exhibit smaller junction temperature rise for a fixed power density, despite the increasing trend of R_{th} with decreasing emitter width [13].

Device Layout for Reduced R_{th}

Due to the considerable self-heating effects, the layout of today's scaled high-speed devices needs to be optimized for both thermal and electrical performance. This section provides a couple of examples for such considerations. As the deep trenches significantly suppress the heat dissipation, R_{th} is expected to decrease with increasing deep trench-enclosed area. In order to verify such tendency experimentally, R_{th} was measured and compared for SiGe HBTs with various deep trench-enclosed areas, which was achieved by adjusting the distance from emitter finger to trench [11]. As shown schematically in the inset of Figure 14.5, the device used in the experiment has an emitter finger located off the center of the deep trench-enclosed area to allow for the collector contact, and two parameters are defined for the finger-to-trench distance: S_1 (shorter distance) and S_2 (longer distance). R_{th} was measured for various S_1/S_2 ratios for which S_2 is fixed. Figure 14.5 clearly depicts a decreasing trend of R_{th} with increasing S_1: 17% reduction when S_1/S_2 ratio is increased from 0.23 to 1. Also compared is a device without deep trench, which exhibits a 32% reduction in R_{th} by eliminating the trench.

Another approach to improve the thermal resistance is to partition the emitter finger into segments, which effectively increases the cross section of the heat flux beneath the emitter fingers. To verify the effect, devices with segmented emitter fingers with various spacings between the segments and different

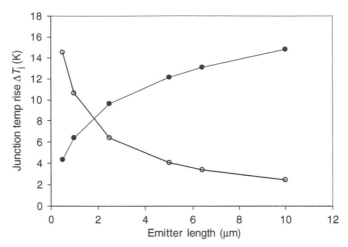

FIGURE 14.4 Junction temperature rise of deep trench-isolated SiGe HBTs with various emitter lengths [11]. Two difference conditions are assumed: fixed power (open symbols) and fixed power density (solid symbols).

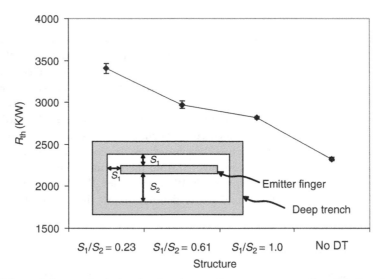

FIGURE 14.5 Schematic layout of a device showing the two distance parameters S_1 and S_2 (inset), and measured R_{th} with various S_1/S_2 ratios for which S_1 is fixed. Also shown is the R_{th} for the device without deep trench. The error bars indicate minimum and maximum of the data acquired across the wafer.

numbers of the segments were fabricated [11]. First, the emitter finger was divided into two, three, and four segments with fixed total spacing and total emitter length, for which the deep trench-enclosed area remained unchanged. Then, for the same set of device structures, the total spacing was varied from 0 to 3 μm. Figure 14.6 shows that R_{th} decreases with increasing total spacing, which is a combined effect of increased deep trench-enclosed area and segmented emitter finger. More interestingly, when the total spacing is fixed, the devices with a larger number of segments (with smaller individual segment length) exhibit smaller R_{th}, an effect solely due to the segmented emitter finger (the deep trench-isolated area is fixed and its effect is isolated). Such reduction in R_{th} can be ascribed to the fact that the heat source is more evenly spread with more segments, although the total heat source area and deep trench-enclosed area are fixed.

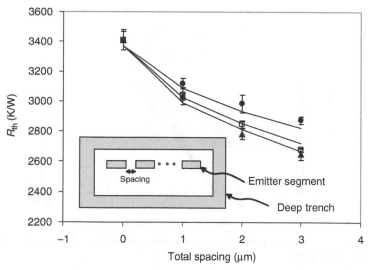

FIGURE 14.6 Schematic layout of a device with segmented emitter finger (inset), and measured R_{th} for various total spacings and number of segments: two (solid circles), three (open squares), and four segments (solid triangles). Solid lines are model prediction, and the error bars indicate minimum and maximum of the data acquired across the wafer.

14.4 Thermal Effects and Device Reliability

Thermal Runaway

Thermal runaway is a phenomenon caused by the electrothermal positive feedback widely observed in bipolar transistors with an excessive junction temperature rise. The origin of the thermal runaway (or thermal instability) is closely related to the positive temperature dependence of collector current I_C, which increases with increasing temperature for a fixed V_{BE}. Consider a typical I_C–V_{BE} relation of a bipolar transistor in the presence of strong self-heating (as shown in Figure 14.7 [37]), and assume the device is biased near a critical regression point on the curves. Now, if V_{BE} is forced to increase by a small amount of ΔV_{BE}, then an increase in I_C will follow, leading to an increase in power dissipation and thus temperature. The raised temperature, due to the positive temperature dependence of I_C, will further increase I_C, which results in yet another temperature rise. This mutual interaction would build up a positive feedback between temperature and I_C, which may eventually lead to an instantaneous burn-out of the device. Since ambient electrical noise may cause fluctuations on V_{BE} large enough to trigger the thermal runaway when a device is biased near the critical point, it is strongly suggested to keep devices away from such bias point with an enough margin. Such instability can be triggered by a perturbation in the spatial distribution of the current over the device also. If a certain location over a device develops a current density higher than surrounding area, the region will be selectively heated up and the local temperature will rise sharply, creating a "hot spot." Then, this spot would further attract current from neighboring regions due to the aforementioned positive temperature dependence of I_C, triggering a positive feedback similar to the one described above. In fact, this is a more commonly observed triggering mode of thermal runaway in practical devices.

The thermal instability is also generally believed to be a direct cause of the second breakdown [38–40] (especially for forward mode second breakdown [41,42]), in which an abrupt V_{CE} reduction is observed with I_C raised beyond a critical point. When the thermal instability results in an excessive level of local temperature rise in a device, an intrinsic zone may develop at the hot spot (which happens at $T_j \gtrsim$ 1300 K) [43]. At this intrinsic zone, the carrier concentration is now determined by the intrinsic carrier

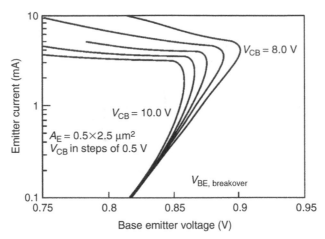

FIGURE 14.7 Typical I_C–V_{BE} curves in the presence of a strong self-heating, obtained from IBM's 50 GHz SiGe HBT [37]. (Copyright 2000 IEEE.)

concentration at the given temperature rather than the doping concentration. As the intrinsic carrier concentrations at such high temperatures are much larger than typically available doping concentrations, a highly conductive local region is created in the device, triggering an abrupt reduction in the voltages across junctions, notably V_{CE} of bipolar transistors. This phenomenon is generally called the second breakdown.

As a greater temperature increase is expected with a larger power dissipation, thermal runaway is in general more likely to take place with higher V_{CB} and I_C, and a safe operation boundary needs to be accordingly defined. As the junction temperature decreases with decreasing R_{th} for a given power dissipation, a reduction in R_{th}, either by structural modification or employing a heat sink, is favored to relax the safe operation boundary and lower the chance for thermal runaway. Alternatively, an emitter ballasting resistor can be employed [44], which is probably the most practical and widely accepted approach to suppress thermal runaway. With an extra resistance component inserted in series with the emitter, any increase in I_C will cause a voltage drop across the inserted emitter resistance, leading to a reduction in the *intrinsic* V_{BE}. The reduced intrinsic V_{BE} will suppress any further increase of I_C, thus providing a negative feedback that counterbalances the electrothermal positive feedback.

Long-Term Reliability

Most of the device degradation mechanisms are accelerated with temperature, and self-heating imposes negative impacts on the long-term reliability of devices. In general, accelerated degradations with temperature follow an Arrhenius relation, and the mean time to failure (MTTF) can be estimated in terms of activation energy E_0 and junction temperature T_j as following:

$$\text{MTTF} = C \exp\left(\frac{E_0}{kT_j}\right) = C \exp\left(\frac{E_0}{k(T_{j0} + \Delta T_j)}\right), \tag{14.11}$$

where T_{j0} is the junction temperature without self-heating, ΔT_j is the junction temperature rise due to self-heating, k is the Boltzmann constant, and C is a coefficient. It is clear from Equation 14.11 that any junction temperature rise will lead to a reduction in MTTF. Such effects are illustrated in Figure 14.8 in which the normalized MTTF is plotted as a function of ΔT_j up to 200 K, for various E_0 values within the practical range. The chip temperature was fixed at 100°C (373 K), a typical value for commercial chips, and C is assumed constant implying MTTF is dominated by temperature. The plot shows a rapid

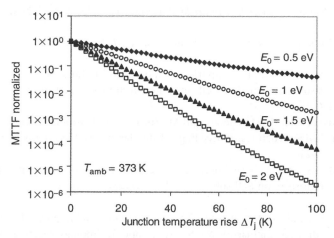

FIGURE 14.8 Calculated MTTF plotted as a function of junction temperature rise ΔT_j based on Equation (14.11), with various activation energies in the practical range.

reduction in MTTF with increasing T_j, which is more pronounced with larger activation energies. According to the plot, ΔT_j of a few tens of kelvins, which is realistic for modern high-speed SiGe HBT operation, may lead to a reduction in MTTF by multiple orders of magnitude depending on the activation energy. It is obvious from this simple calculation that self-heating has a significant impact on the long-term device degradation and any effort for R_{th} reduction will lead to a substantial improvement in the device lifetime.

14.5 Summary

In this chapter, the general issues regarding the self-heating and thermal effects in Si-based bipolar transistors, particularly for SiGe HBTs, were reviewed, which covered the modeling of self-heating, the measurement of the thermal resistance, the trend of R_{th} for various device structures, and the effect of self-heating on device reliability. As stressed in this chapter, the proper consideration and analysis of the self-heating effects are critical for an accurate prediction of device operation and understanding of device reliability. Although extensive knowledge on this field has been accumulated owing to decades-long efforts, there still exist unexplored territories to be investigated for better understanding and control of the thermal effects.

Acknowledgments

The author would like to thank Andreas Stricker, Ping-Chuan Wang, and Kim Watson for helpful discussions and is also grateful to Thomas Adam and Greg Freeman for the careful review of the manuscript.

References

1. KL Wang and X Zheng. Thermal properties of SiGe. In *Properties of Strained and Relaxed Silicon Germanium*, E. Kasper, Ed. London: INSPEC, 1995.
2. HS Carslaw and JC Jaeger. *Conduction of Heat in Solids*, 2nd ed. Oxford: Oxford University Press, 1959.
3. R Joy and E Schlig. Thermal properties of very fast transistors. *IEEE Transactions on Electron Devices* 17:586–594, 1970.

4. R Muller and T Kamins. *Device Electronics for Integrated Circuits*, 2nd ed. New York: John Wiley & Sons, 1986.

5. O Madelung. *Semiconductors: Group IV Elements and III–V Compounds*. Berlin: Springer-Verlag, 1991.

6. PR Ganci, J-JJ Hajjar, T Clark, P Humphries, J Lapham, and D Buss. Self-heating in high performance bipolar transistors fabricated on SOI substrates. Technical Digest of Electron Devices Meeting, 1992, pp. 417–420.

7. D Walkey, T Smy, H Tran, D Marchesan, and M Schröter. Prediction of thermal resistance in trench isolated bipolar device structures. Proceedings of Bipolar/BiCMOS Circuits and Technology Meeting, 1998, pp. 207–210.

8. JS Brodsky, RM Fox, and DT Zweidinger. A physics-based dynamic thermal impedance model for vertical bipolar transistors on SOI substrates. *IEEE Transactions on Electron Devices* 46:2333–2339, 1999.

9. P Palestri, A Pacelli, and M Mastrapasqua. Thermal resistance in $Si_{1-x}Ge_x$ HBTs on bulk-Si and SOI substrates. Proceedings of Bipolar/BiCMOS Circuits and Technology Meeting, 2001, pp. 98–101.

10. A Pacelli, P Palestri, and M Mastrapasqua. Compact modeling of thermal resistance in bipolar transistors on bulk and SOI substrates. *IEEE Transactions on Electron Devices* 49:1027–1033, 2002.

11. J-S Rieh, J Johnson, S Furkay, D Greenberg, G Freeman, and S Subbanna. Structural dependence of the thermal resistance of trench-isolated bipolar transistors. Proceedings of Bipolar/BiCMOS Circuits and Technology Meeting, 2002, pp. 100–103.

12. D Walkey, T Smy, C Reimer, M Schröter, H Tran, and D Marchesan. Modeling thermal resistance in trench-isolated bipolar technologies including trench heat flow. *Solid-State Electronics* 46:7–17, 2002.

13. J-S Rieh, D Greenberg, B Jagannathan, G Freeman, and S Subbanna. Measurement and modeling of thermal resistance of high speed SiGe heterojunction bipolar transistors. Proceedings of Topical Meeting on Silicon Monolithic Integrated Circuits in RF Systems, 2001, pp. 110–113.

14. D Kennedy. Spreading resistance in cylindrical semiconductor devices. *Journal of Applied Physics* 31:1490–1497, 1960.

15. K Washio, E Ohue, H Shimamoto, K Oda, R Hayami, Y Kiyota, M Tanabe, M Kondo, T Hashimoto, and T Harada. A 0.2-μm 180-GHz-f_{max} 6.7-ps-ECL SOI/HRS self aligned SEG SiGe HBT/CMOS technology for microwave and high-speed digital applications. Technical Digest of International Electron Devices Meeting, 2000, pp. 741–744.

16. J Cai, A Ajmera, C Ouyang, P Oldiges, M Steigerwalt, K Stein, K Jenkins, G Shahidi, and T Ning. Fully-depleted-collector polysilicon-emitter SiGe-base vertical bipolar transistor on SOI. Digest of Symposium on VLSI Technology, 2002, pp. 172–173.

17. M Mastrapasqua, P Palestri, A Pacelli, GK Celler, MR Frei, PR Smith, RW Johnson, L Bizzarro, W Lin, TG Ivanov, MS Carroll, IC Kizilyalli, and CA King. Minimizing thermal resistance and collector-to-substrate capacitance in SiGe BiCMOS on SOI. *IEEE Electron Device Letters* 23:145–147, 2002.

18. RT Dennison and KM Walter. Local thermal effects in high performance bipolar devices/circuits. Proceedings of Bipolar Circuits and Technology Meeting, 1989, pp. 164–167.

19. DT Zweidinger, RM Fox, JS Brodsky, T Jung, and S-G Lee. Thermal impedance extraction for bipolar transistors. *IEEE Transactions on Electron Devices* 43:342–346, 1996.

20. DJ Walkey, TJ Smy, D Marchesan, H Tran, C Reimer, TC Kleckner, MK Jackson, M Schroter, and JR Long. Extraction and modelling of thermal behavior in trench isolated bipolar structures. Proceedings of Bipolar/BiCMOS Circuits and Technology Meeting, 1999, pp. 97–100.

21. RM Fox, S-G Lee, and DT Zweidinger. The effects of BJT self-heating on circuit behavior. *IEEE Journal of Solid-State Circuits* 28:678–685, 1993.

22. RM Fox and S-G Lee. Scalable small-signal model for BJT self-heating. *IEEE Electron Device Letters* 12:649–651, 1991.

23. DT Zweidinger, S-G Lee, and RM Fox. Compact modeling of BJT self-heating in SPICE. *IEEE Transactions on Computer-Aided Design of Integrated Circuits and Systems* 12:1368–1375, 1993.

24. CC McAndrew. A complete and consistent electrical/thermal HBT model. Proceedings of Bipolar/ BiCMOS Circuits and Technology Meeting, 1992, pp. 200–203.

25. RS Vogelsong and C Brzezinski. Extending SPICE for electro-thermal simulation. Proceedings of Custom Integrated Circuits Conference, 1989, pp. 21.4/1–21.4/4.

26. MG Adlerstein and MP Zaitlin. Thermal resistance measurements for AlGaAs/GaAs heterojunction bipolar transistors. *IEEE Transactions on Electron Devices* 38:1553–1554, 1991.

27. BM Cain, PA Goud, and CG Englefield. Electrical measurement of the junction temperature of an RF power transistor. *IEEE Transactions on Instrumentation and Measurement* 41:663–665, 1992.

28. M Pfost, V Kubrak, and P Brenner. A practical method to extract the thermal resistance for heterojunction bipolar transistors. Conference on European Solid-State Device Research, 2003, pp. 335–338.

29. T Vanhoucke, HMJ Boots, and WD van Noort. Revised method for extraction of the thermal resistance applied to bulk and SOI SiGe HBTs. *IEEE Electron Device Letters* 25:150–152, 2004.

30. DE Dawson, AK Gupta, and ML Salib. CW measurement of HBT thermal resistance. *IEEE Transactions on Electron Devices* 39:2235–2239, 1992.

31. JR Waldrop, KC Wang, and PM Asbeck. Determination of junction temperature in AlGaAs/GaAs heterojunction bipolar transistors by electrical measurement. *IEEE Transactions on Electron Devices* 39:1248–1250, 1992.

32. N Bovolon, P Baureis, J-E Muller, P Zwicknagl, R Schultheis, and E Zanoni. A simple method for the thermal resistance measurement of AlGaAs/GaAs heterojunction bipolar transistors. *IEEE Transactions on Electron Devices* 45:1846–1848, 1998.

33. W Liu and A Yuksel. Measurement of junction temperature of an AlGaAs/GaAs heterojunction bipolar transistor operating at large power densities. *IEEE Transactions on Electron Devices* 42:358–360, 1995.

34. H Tran, M Schroter, DJ Walkey, D Marchesan, and TJ Smy. Simultaneous extraction of thermal and emitter series resistances in bipolar transistors. Proceedings of Bipolar/BiCMOS Circuits and Technology Meeting, 1997, pp. 170–173.

35. SP Marsh. Direct extraction technique to derive the junction temperature of HBTs under high self-heating bias conditions. *IEEE Transactions on Electron Devices* 47:288–291, 2000.

36. B Jagannathan, M Khater, F Pagette, J-S Rieh, D Angell, H Chen, J Florkey, F Golan, DR Greenberg, R Groves, SJ Jeng, J Johnson, E Mengistu, KT Schonenberg, CM Schnabel, P Smith, A Stricker, D Ahlgren, G Freeman, K Stein, and S Subbanna. Self-aligned SiGe NPN transistors with 285 GHz f_{max} and 207 GHz f_T in a manufacturable technology. *IEEE Electron Device Letters* 23:258–260, 2002.

37. J Dunn, D Harame, SS Onge, A Joseph, N Feilchenfeld, K Watson, S Subbanna, G Freeman, S Voldman, D Ahlgren, and R Johnson. Trends in silicon germanium BiCMOS integration and reliability. Proceedings of International Reliability Physics Symposium, 2000, pp. 237–242.

38. CG Thornton and CD Simmons. A new high current mode of transistor operation. *IRE Transactions on Electron Devices* 5:6–10, 1958.

39. HA Schafft and JC French. A survey of second breakdown. *IEEE Transactions on Electron Devices* 13:613–618, 1966.

40. HA Schafft. Second breakdown—a comprehensive review. *Proceedings of the IEEE* 55:1272–1288, 1967.

41. PL Hower and VGK Reddi. Avalanche injection and second breakdown in transistors. *IEEE Transactions on Electron Devices* 17:320–335, 1970.

42. M Jovanovic. A transistor model for numerical computation of forward-bias second-breakdown boundary. *IEEE Transactions on Power Electronics* 6:199–207, 1991.

43. F Bergmann and D Gerstner. Some new aspects of thermal instability of the current distribution in power transistors. *IEEE Transactions on Electron Devices* 13:630–634, 1966.

44. G-B Gao, MS Unlu, H Morkoc, and DL Blackburn. Emitter ballasting resistor design for, and current handling capability of AlGaAs/GaAs power heterojunction bipolar transistors. *IEEE Transactions on Electron Devices* 38:185–196, 1991.

15

Device-Level Simulation

Guofu Niu

Auburn University

15.1 Introduction

Device simulation is now an integral part of SiGe technology development, and is routinely used for understanding SiGe HBT operation and device optimization. All the major commercial device simulators support SiGe devices, including MEDICI from Avant (now Synopsys), DESSIS from ISE, and ATLAS from Silvaco. They are typically part of a technology computer-aided-design (TCAD) package, which includes process simulation, device simulation, and parameter extraction programs. Fortunately or unfortunately, these device simulators were historically developed as general semiconductor equation solvers, and the user must choose his or her model of physics, such as mobility, carrier statistics (Fermi–Dirac or Boltzmann), bandgap narrowing (BGN), that best suit the device in question. The default physical models are usually the simplest ones, and often give inaccurate results, particularly for advanced device technologies such as SiGe. Users are also responsible for the "meshing" of the device structure, which can affect the simulation results significantly. This chapter addresses these practical issues of device-level simulation for SiGe HBTs, and presents techniques of simulation results analysis.

15.2 Semiconductor Equations

The basic set of equations solved in device simulation are Poisson's equation and the current continuity equations for electrons and holes:

$$\nabla \cdot \varepsilon \nabla \phi = -q(p - n + C) \qquad (15.1)$$

$$\frac{1}{q}\nabla \cdot \overrightarrow{J_n} - R = \frac{\partial n}{\partial t}, \tag{15.2}$$

$$-\frac{1}{q}\nabla \cdot \overrightarrow{J_p} - R = \frac{\partial p}{\partial t}, \tag{15.3}$$

where ϕ is potential, n and p are electron and hole concentrations, J_n and J_p are the electron and hole current densities, C is the net concentration of ionized dopants and charged traps, and R is the net rate of recombination (including impact ionization). The fundamental variables are ϕ, n, and p. All the other variables are functions of ϕ, n, and p that need to be modeled based on semiconductor physics. For simple drift-diffusion model of carrier transport, J_n and J_p are given by

$$\overrightarrow{J_n} = qn\mu_n \overrightarrow{E_n} + qD_n\nabla n, \tag{15.4}$$

$$\overrightarrow{J_p} = qp\mu_p \overrightarrow{E_p} - qD_p\nabla p, \tag{15.5}$$

where μ_n and μ_p are electron and hole mobilities, D_n and D_p are diffusivities which are related to μ_n and μ_p by Einstein relation, and $\overrightarrow{E_n}$ and $\overrightarrow{E_p}$ are the effective fields for electron and hole drift:

$$\overrightarrow{E_n} = \nabla \frac{E_C}{q} + \frac{kT}{q}\nabla \ln \frac{N_C}{T^{3/2}} = -\nabla \left(\phi + \frac{\chi}{q} - \frac{kT}{q} \ln \frac{N_C}{T^{3/2}} \right), \tag{15.6}$$

$$\overrightarrow{E_p} = \nabla \frac{E_V}{q} - \frac{kT}{q}\nabla \ln \frac{N_V}{T^{3/2}} = -\nabla \left(\phi + \frac{\chi}{q} + \frac{E_g}{q} + \frac{kT}{q} \ln \frac{N_V}{T^{3/2}} \right), \tag{15.7}$$

where χ is electron affinity, and N_C and N_V are the effective conduction and valence band density of states, and E_g is the bandgap. The band edges E_C and E_V are determined by ϕ, χ, and E_g:

$$E_C = -q\phi - \chi + \Delta, \tag{15.8}$$

$$E_V = -q\phi - \chi + \Delta - E_g, \tag{15.9}$$

where Δ is a constant depending on the choice of energy reference. In a SiGe HBT, the Ge mole fraction is a function of position, and thus both χ and E_g vary with position.

Boundary conditions are required for solving the equations described above. Two types of boundary conditions are of particular importance:

1. The "Dirichlet" boundary condition at ohmic contacts, such as the base, emitter, and collector contacts. The values of ϕ, n, and p are fixed at their equilibrium values, which are then determined by the applied voltages and doping, as well as the carrier statistics chosen.
2. The "Neumann" boundary condition at other edges of simulation domain (except for ohmic contacts). The fluxes of electric field and currents are assumed to be zero. The user needs to make sure the simulation domain is large enough so that the "Neumann" boundary condition implemented in the simulator is consistent with reality.

15.3 Physical Model Selection

A number of physical parameters are required in the semiconductor equations, including N_C and N_V, χ, E_g, μ_n, and μ_p, and R. In commercial simulators such as MEDICI, only χ and E_g are modeled as a function of Ge mole fraction. For parameters such as the mobilities, a number of models are available from which the user must choose. The model equations can be found in the user manuals, but the relevant question is which parameter models to select, and sometimes, which model parameters can be tuned in a meaningful way if needed.

N_C, N_V, and E_g

The Ge dependence of E_g is relatively well understood and accounted for in device simulators, at least at low doping levels. The Ge dependence of N_C and N_V, however, are not well understood, particularly at heavy doping. Strictly speaking, the use of a single effective N_C or N_V is only meaningful for Boltzmann statistics in case of multiple conduction band minima or valence band maxima at different energy levels, like in strained SiGe. Due to strained, induced band splitting, N_C and N_V decreases with increasing Ge mole fraction first, and then "saturates" when the band split exceeds a couple of kT. This assumes no change of the effective mass for each band minima. At heavy doping, which is of practical interest, the situation becomes quite complicated, due to the complicated interaction of N_C/N_V change, BGN, and Fermi–Dirac statistics. For practical purposes, one may modify N_C and N_V so that the $N_C N_V$ product in SiGe is about 40% of that in Si (or other numbers necessary to fit measured I–V).

Mobility and Velocity Saturation

For bipolar transistor simulations, the so-called "Philips unified mobility model" [1] should be chosen, as this model distinguishes majority and minority carrier mobilities. Velocity saturation, which is not accounted for by default, should be turned on, as it is important in determining at what current density the peak f_T is reached.

Incomplete Ionization

Complete ionization of dopants in Si and SiGe is typically assumed. At heavy doping levels found in the base and emitter of SiGe HBTs, the simulation results with and without incomplete ionization can be quite different if the simplest models of incomplete ionization are used. This difference is not truly physical because the dopants should be completely ionized at all temperatures for concentrations above a certain doping level known as the "Mott" or "metal–insulator" transition. If one continues to use the incomplete ionization relations for such heavy doping levels, the majority carrier concentration is significantly underestimated, and the minority carrier concentration is equally significantly overestimated. Significant shifts of both I_C and I_B are then observed on the Gummel characteristics for a typical SiGe HBT. This situation has been corrected in later versions of MEDICI by applying incomplete ionization relations for doping levels below a defined low-valued threshold, and applying complete ionization for doping levels above a defined high-valued threshold, and then interpolating between the two thresholds. This option is chosen by specifying "`high.dop`" together with "`incomplete`" in the MEDICI model statement.

BGN, Statistics, and Mobility

It is well known that the bandgap E_g narrows at heavy doping, which increases the pn product at equilibrium. This is often referred to as heavy doping induced BGN. Another heavy doping effect is that Boltzmann statistics is no longer accurate, and Fermi–Dirac statistics is needed instead. Naturally, one may attempt to select both BGN and Fermi–Dirac statistics for SiGe HBT simulations, as the doping levels are heavy. This, however, is not necessarily correct, depending on the BGN model chosen and the device simulator chosen, as detailed below.

Perhaps the most widely used BGN model is the Slotboom BGN model. The idea is to artificially decrease the apparent electrical bandgap so that one can continue to apply Boltzmann statistics to describe the equilibrium pn product at heavy doping [2],

$$pn = n_{i0}^2 e^{\Delta G/kT},$$ (15.10)

where n_{i0} is the intrinsic carrier concentration at low doping levels. The pn product changes due to a combination of degeneracy (Fermi–Dirac statistics), doping-induced rigid BGN, and density-of-states perturbations, which can all be lumped into a single parameter ΔG, commonly called the "apparent BGN." ΔG is modeled as a function of doping N [2]:

$$\Delta G = \Delta G_0 \left[\ln \frac{N}{N_0} + \sqrt{\left[\ln \frac{N}{N_0} \right]^2 + C} \right].$$ (15.11)

The model parameters were first reported in Ref. [2], and later updated in Ref. [3] by reinterpreting the same data using the Philips unified mobility model: $\Delta G_0 = 6.92\,\mathrm{meV}$, $N_0 = 1.3 \times 10^{17}\,\mathrm{cm}^{-3}$, and $C = 0.5$ for Si. One should, therefore, always use this set of model parameters when the Philips unified mobility model is used. The underlying reason is that the pn product at equilibrium is not directly measured, and is instead inferred from the Gummel characteristics. For instance, the base current is given by

$$J_B = kT \mu_n \frac{n_{i0}^2}{N_{de}^+ W_e} e^{\Delta G/kT} e^{qV_{BE}/kT},$$ (15.12)

where N_{de}^+ and W_e are the emitter doping level and the emitter depth. The effects of degeneracy (i.e., Fermi–Dirac statistics), rigid BGN, and the $N_C N_V$ changes are all lumped into the ΔG term. To determine ΔG from J_B, μ_n, and N_{de}^+ are needed. Rigorous derivation of the above familiar transport equation including the effects of degeneracy, rigid BGN, and $N_C N_V$ changes can be performed, as was reviewed in Ref. [4], and the same analysis can be applied to derive the collector current in SiGe HBTs [5]. The equations derived by including advanced physics share the same functional *form* as the older equations derived using simplified physics, but differ in *substance*.

The Ge dependence of ΔG is largely unknown. Experimental determination of ΔG involves the N_C and N_V of SiGe, as well as minority carrier mobilities in SiGe, whose Ge dependences are not well understood yet. Experimental results in Ref. [6] suggest that the apparent BGN is different for SiGe and Si, and the true BGN (after Fermi–Dirac correction) in SiGe and Si are close. The later, however, could potentially cause negative apparent BGN at heavy doping and large Ge mole fraction in the experience of this author. Before systematic measurement and modeling of BGN in SiGe becomes available, one may assume that the apparent BGN for SiGe is the same as for Si.

Since Boltzmann statistics is used in obtaining the apparent BGN expression, one should also use Boltzmann statistics in device simulation if the apparent BGN parameters are used "as is." Otherwise, the effect of degeneracy on the pn product is effectively accounted for twice. For a doping level of $10^{20}\,\mathrm{cm}^{-3}$, the ΔG due to degeneracy is $-31.2916\,\mathrm{meV}$, and is thus significant for SiGe HBTs. Therefore, Boltzmann statistics should be used as opposed to the more accurate Fermi–Dirac statistics for SiGe HBTs when the default Boltzmann-statistics-based BGN model is used. This approach, however, may potentially cause other problems in cases where Fermi–Dirac statistics is necessary, either in another region of the device where the doping level is moderate, or at low temperatures. Another potential problem is that the Einstein relations depend on the carrier statistics, which can affect minority carrier diffusivity and hence f_T. One solution is to automatically adjust the value of ΔG based on the user's choice of the statistics, as was done in DESSIS for the Slotboom BGN model.

Figure 15.1 shows the apparent BGN ΔG as a function of n-type doping using the Slotboom model. We show two curves; the dashed curve is calculated "as is" (i.e., as found in most simulators), while the solid curve is calculated by applying a correction factor to remove the degeneracy effect [7]. If Boltzmann statistics is used for device simulation, the dashed curve should be used, since the degeneracy

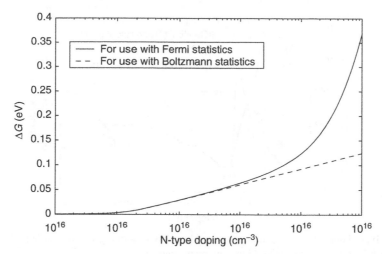

FIGURE 15.1 Apparent bandgap narrowing ΔG for n-type doping. For device simulation using Fermi–Dirac statistics, the solid curve should be used. For device simulation using Boltzmann statistics, the dashed curve should be used.

effect is already lumped into the ΔG term. If Fermi–Dirac statistics is used for device simulation, however, the solid curve should be used, since it does not contain the degeneracy effect. The amount of correction needed to account for Fermi statistics is in principle dependent on N_C and N_V, for n- and p-type dopants, which are in turn dependent on the Ge mole fraction.

The distribution of the heavy-doping-induced BGN between the conduction and valence bands is also important, as it affects the high-injection potential barrier effects [7]. This issue becomes increasingly important in scaled devices with heavy base doping. In simulators, an equal split between conduction and valence bands is assumed by default.

15.4 Application Issues

Device Structure Specification

The basic input to a device simulator is the doping and Ge profiles, which can be obtained either from process simulation or SIMS measurement. Figure 15.2 shows an example of doping and Ge profiles measured by SIMS. The polysilicon–silicon interface can be identified by the As segregation peak. The measured As doping "tail" into the Si is apparently higher than the base doping across the entire base, which is not real. In fact it is simply the result of the finite resolution limit of SIMS in following very rapidly changing doping profiles. The true metallurgical EB junction can be determined from the "dip" in the B SIMS profile of the base.

The dopant activation percentage in the polysilicon emitter is quite low for As due to As clustering. A 5 to 10% activation rate is often assumed. The As profile in the single crystalline silicon emitter is Gaussian-like and falls from the polysilicon–silicon interface toward the base. The SIMS measured Ge profile has limited accuracy, and should be compared with the intended Ge profile during SiGe epitaxial growth. An example of the net doping profile and Ge profile used for simulation is shown in Figure 15.3.

For two-dimensional and three-dimensional simulations, the vertical doping profile in the extrinsic base region can be obtained using SIMS in a similar manner. The lateral doping transition between extrinsic and intrinsic device, however, can only be estimated from device layout and fabrication details, since two-dimensional doping profile information is typically unavailable. For vertical profile design, one-dimensional simulations can be used first because of the low simulation overhead, since one-dimensional

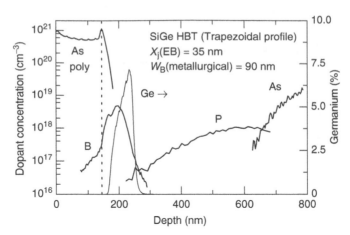

FIGURE 15.2 Typical doping and Ge profiles measured by SIMS for a first-generation SiGe HBT. The true emitter As profile is much less abrupt than the SIMS measurement suggests.

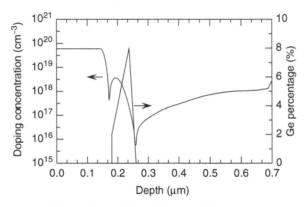

FIGURE 15.3 Typical doping and Ge profiles used for device simulation.

simulation involves much easier gridding, easier transit time analysis, much shorter simulation time, and easier debugging. The resulting design can then be refined using two-dimensional simulation, which is necessary when accurate f_{max} or noise analysis is desired. Full three-dimensional simulation becomes necessary for problems which are inherently three-dimensional in nature, such as in single-event upset.

Meshing Guidelines

The next step is to define the coordinate values of the points (nodes) at which the semiconductor equations are discretized. Even though commercial simulators all provide some means of regridding (e.g., based on the doping gradient), taking extra time to specify a reasonably good initial mesh usually pays off in the end. Regridding, if not well controlled, can easily generate a large number of obtuse triangle elements, which can cause numerical problems. A popular meshing method is to use a rectangular grid (e.g., in MEDICI and ATLAS).

The optimum grid in a given problem depends strongly on the device metric of most interest. To simulate the forward-mode SiGe HBT operation, for instance, the EB spacer oxide corner mesh does not need to be fine. However, to simulate the reverse emitter–base junction band-to-band-tunneling current for an EB reliability study, the grid at the oxide corner needs to be very fine in order to accurately locate the peak electric field.

A number of empirical criteria can be applied in meshing. In general, fine meshing is necessary where the space-charge density and its spatial gradient are large, e.g., in the depletion layers of the EB and CB junctions. Placing nodes along the physical junction interfaces is important for accurate simulation. In addition, grid lines must be placed at the critical points defining the SiGe profile in order to avoid creating artificial SiGe profiles that inadvertently differ from what one has in mind. When a simulator such as MEDICI is used, for instance, the initial grid lines must be placed with the SiGe and doping profiles in mind. Figure 15.4 shows an example of *bad* mesh line specification, while Figure 15.5 shows an example of mesh lines placed properly with the intended Ge profile in mind.

Initial coarse meshes are often refined based on where the physical properties of the device structure dictate it. That is, the mesh must be refined where a given variable or change in that variable across an element exceeds a given defined tolerance. If breakdown voltage is the concern, for instance, the impact ionization rate can be used. Because of the strong nonlinearities in semiconductor problems, the doping concentration at the newly generated nodes should be determined from the original doping profile specification, instead of interpolation from the existing mesh.

Theoretically speaking, the potential difference or quasi-Fermi potential difference between two adjacent nodes should generally be kept less than the thermal voltage kT/q in order to minimize discretization error. In practice, this requirement is often relaxed to about 10 to 15 kT/q between adjacent nodes. The doping concentration change between adjacent nodes should be less than two to three orders of magnitude. In high-level injection, very fine meshing is often required where the minority carrier concentration exceeds the doping concentration (e.g., in the CB space–charge region of a SiGe HBT).

Mesh Quality Assurance

For assurance of mesh validity, the electrical parameters of interest (e.g., f_T–I_C for a SiGe HBT) should *always* be resimulated using a finer mesh to check for grid sensitivity effects. Identical results using

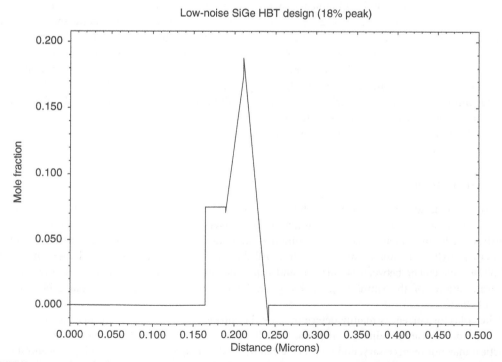

FIGURE 15.4 Ge mole fraction from a mesh *improperly* specified without the Ge profile in mind.

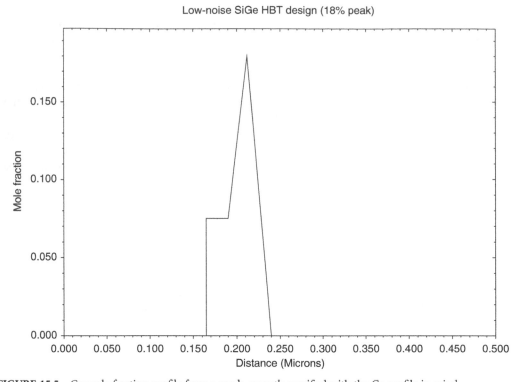

FIGURE 15.5   Ge mole fraction profile from a mesh *correctly* specified with the Ge profile in mind.

different gridding can generally be taken to imply that a robust mesh has been achieved. In MEDICI, an overall finer mesh can be obtained conveniently using the statement "Regrid potential factor = 1.5 smooth = 1." The "factor" parameter requests an automatic increase of the number of nodes by a factor of 1.5×. The "potential" parameter indicates that the refinement is performed where the potential change between adjacent nodes is large. One can have more confidence in the mesh used if the various simulated metrics no longer change with further mesh refining. This technique can also be applied to determine the acceptable coarse meshing limit for a particular problem before running extensive parametric analysis, and can dramatically minimize overall simulation time. Since one-dimensional simulation is quite fast, very fine (finer than necessary) mesh can be used, which adds little extra simulation time, but may save time in the end spent on generating an accurate mesh with fewer nodes.

I–V Simulation

DC simulations are used to capture the transistor *I–V* behavior, which for a SiGe HBT usually means the Gummel characteristics. A number of parameters can affect the simulated Gummel characteristics, including carrier statistics, recombination parameters, BGN model parameters, mobility models, as well as the doping and Ge profiles. All these models must be considered when attempting to obtain agreement between simulation and experimental data (we refer to this (iterative) process as "calibration" of the simulator). A few general guidelines for simulator-to-data calibration are given below.

The collector current is mainly determined by the intrinsic carrier concentrations and base doping. The N_C and N_V can be adjusted, and made smaller than in Si. The detailed dependence on Ge mole fraction may not be necessary, and an average can be used. The majority carrier (hole) concentration in the base at equilibrium, however, can be inferred from the intrinsic base sheet resistance R_{bi} data, which is readily available from simple measurements on "ring-dot" test structures. Slight changes to the SIMS measured doping profile can be made to match measured I_C, at least to the range of accuracy of the SIMS data. This calibration technique is particularly useful when no "dip" exists in the base boron profile to indicate the precise EB junction location, as in a SiGe HBT with a phosphorous-doped emitter or a pnp SiGe HBT.

The base current is primarily determined by the emitter structure in a SiGe HBT. Practically all viable SiGe HBT technologies have polysilicon emitter contacts. The polysilicon region can be either modeled as a Schottky contact or simply as an extension of the crystalline silicon emitter (the so-called "extended emitter" structure). The default model parameters for polysilicon are the same as those for silicon, and need to be modified by the user. The work function and surface recombination velocities can be adjusted as fitting parameters in order to calibrate the I_B if using a Schottky contact. For the extended emitter approach, the hole lifetime parameters can be adjusted to obtain agreement. The two approaches, however, can result in different emitter charge storage. For highly scaled HBTs where the emitter transit time is significant, the extended emitter approach is recommended, as it accounts for charge storage in polysilicon.

Figure 15.6 shows a calibration example for the SiGe HBT Gummel characteristics using the techniques described above. The model parameters were calibrated to 200 K data and then used to reproduce the 300 K data as is (i.e., no further tuning of parameters). Accurate simulation of the Gummel characteristics can be quite challenging, particularly for the high V_{BE} range when high injection occurs, and the impact of emitter and base resistance is not negligible.

High-Frequency Simulation

High-frequency two-port parameters can be simulated using small-signal ac analysis. Here, f_T, f_{max}, as well as the various noise parameters can all be extracted from the simulated two-port parameters [8,9]. Although there are many parameters that one can adjust, determining a single set of simulation parameters for a SiGe HBT that can reproduce the four complex network parameters at all biases of interest for frequencies up to f_T requires substantial effort. An in-depth understanding of the interaction between the physics underlying the simulation models and the device operation is important for achieving sensible results. For instance, at low currents, the total transit time is dominated by the time constants related to the EB space–charge region capacitance rather than the diffusion capacitance. Therefore the adjustment of extrinsic device structure as well as the intrinsic EB junction is needed to match the measured f_T at low J_C. Even though mobility model parameters (including parameters for both the low field mobility and the velocity saturation models) can be modified for f_T calibration at high

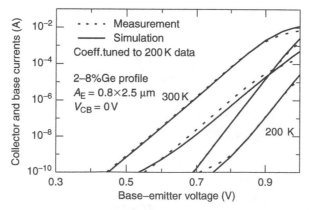

FIGURE 15.6 A calibration example for the Gummel characteristics of a SiGe HBT using the described calibration strategy. The same set of model parameters was used for both the 300 and the 200 K simulations.

J_C near the f_T peak, it should be used as a last resort. Instead, adjustments to the two-dimensional structure and lateral doping profile transitions should first be attempted. Because exact matching of the Gummel characteristics is difficult, high-frequency calibration of ac parameters such as f_T should always be made at fixed J_C, and not at fixed V_{BE}.

For efficient calibration of the Y-parameters across a wide frequency range and a large J_C range, one can first calibrate the f_T–J_C and f_{max}–J_C curves. For state-of-the-art SiGe HBTs with narrow emitters, the shallow-trench isolation and extrinsic CB capacitances can often be comparable to the intrinsic CB capacitance, and are therefore nonnegligible. For accurate Y-parameter simulation, all the two-dimensional lateral structure must be included. The extrinsic base and collector structures (geometric overlap as well as lateral doping profile transition) can then be modified to calibrate f_{max}–J_C. Typically, once f_T–J_C and f_{max}–J_C are calibrated, the simulated Y-parameters will match the measured Y-parameters reasonably well. For accurate separation of the intrinsic and extrinsic base resistances and CB capacitances, transistors with different emitter widths (if available on the test die) can be measured. By simulating and measuring devices with different emitter widths, the contribution of the extrinsic and intrinsic elements can be accurately separated.

A useful technique for high-frequency SiGe HBT calibration is to extract the equivalent circuit parameters such as C_{BE}, C_{BC}, and r_b as a function of J_C. Analytical extraction methods, which use only single frequency data, are highly desirable because they are efficient, since we only need a rough picture to guide us on the appropriate changes to make in our device structure or model coefficients. The parameter extraction method proposed in Ref. [10], for instance, can be used.

By comparing the simulated and measured C_{BE}, C_{BC}, r_b, r_e, and r_c, one can readily identify the dominant factors for any simulation-to-measurement discrepancy, and adjust the lateral doping extension accordingly. The diffusion capacitance component of C_{BE} is proportional to J_C at relatively low current densities (i.e., before the f_T roll-off), and can thus be distinguished from the depletion capacitance component. Figure 15.7 shows an example of f_T–J_C calibration for a typical first-generation SiGe HBT obtained using the techniques described above. This calibration was successfully achieved using two-dimensional MEDICI simulations without modifying the model parameters of the Philips unified mobility model and the velocity saturation model. The intrinsic base and collector doping profiles from SIMS were also used as measured. Most of the required adjustments were instead made in the extrinsic device regions.

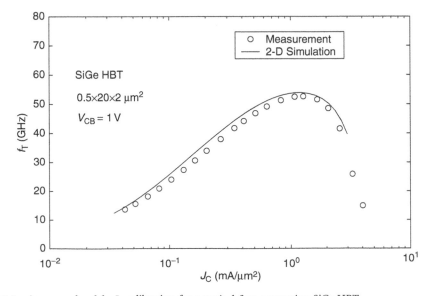

FIGURE 15.7 An example of f_T–J_C calibration for a typical first-generation SiGe HBT.

Qualitative versus Quantitative Simulations

Obviously, qualitative simulation is much easier and quicker than quantitative simulation. Doing a rough relative comparison between two device structures is much easier than simulating a single device structure to high accuracy. An advantage of qualitative simulation is that fewer grid points can be used and hence simulation time can be dramatically reduced. For instance, the comparison of current gain and cutoff frequency between Si BJT and SiGe HBT can be made using a coarse grid. An "exact" simulation, however, is obviously quite involved.

15.5 Probing Internal Device Operation

The fundamental reason for the degradation of transistor performance at higher frequency is charge storage. The resulting capacitive current, typically at the transistor input, increases with frequency, leading to the degradation of current gain and power gain. The most effective way to examine the details of charge storage is to perform a small-signal ac simulation in the frequency domain. The small signal electron concentration (n_{ac}) profile contains information on the spatial distribution of the total transit time. n_{ac} is in general a complex number, but reduces to a real number at low frequency. Using n_{ac} and the small-signal collector current density $J_{C,ac}$, the "effective transit time velocity" (v_τ) can be defined as

$$v_\tau = \frac{J_{C,ac}}{q n_{ac}}. \tag{15.13}$$

The "accumulated transit time" can then be defined for a given position along the path of electron transport according to

$$\tau_{acc}(x) = \int_0^x \frac{1}{v_\tau} dx = \frac{1}{J_{C,ac}} \int_0^x q n_{ac} \, dx. \tag{15.14}$$

Figure 15.8 shows τ_{acc} versus depth at the peak f_T point calculated using 1 MHz ac simulation results. The f_T estimated from $\tau_{ec} = \tau_{acc}(x = x_{cc})$ (i.e., the transit time defined from quasi-static analysis) is 42.3 GHz, with x_{cc} being the location of the collector contact. The f_T extrapolated from h_{21}, however, is 45 GHz. In general, there is a good correlation between the f_T determined from h_{21} and the f_T

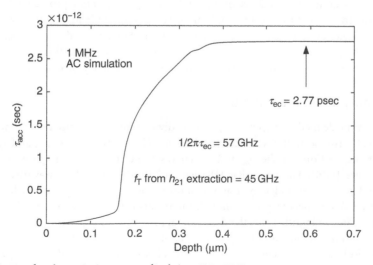

FIGURE 15.8 Accumulated transit time versus depth in a SiGe HBT.

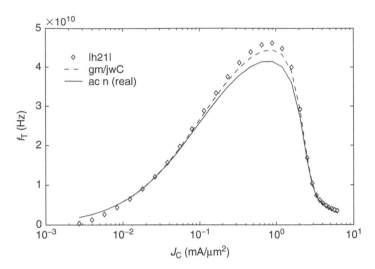

FIGURE 15.9 Comparison of three methods of determining f_T using numerical simulation.

determined from $1/2\pi\tau_{ec}$. A comparison of the f_T extracted using the above two methods is shown in Figure 15.9, together with that extracted using

$$f_T = \frac{g_{cb}}{2\pi C_{bb}}, \tag{15.15}$$

where g_{cb} is the real part of Y_{21}, C_{bb} is defined by the imaginary part of Y_{11}, and $Y_{11} = g_{11} + j\omega C_{bb}$. Here, C_{bb} was evaluated at 1 MHz in the above example, and is nearly independent of the frequency used in the simulation as long as the frequency chosen is well below f_T.

For comparison with experimental f_T data, the value from $|h_{21}|$ extrapolation should be used, as opposed to that from the accumulated transit time, which uses the quasistatic approximation. The practical reason for this is that experimental f_T data are all obtained from h_{21} extrapolation.

Despite the fact that the resulting f_T value may be off compared to the value obtained from h_{21} extrapolation, the transit time analysis of n_{ac} and the $\tau_{acc}(x)$ profiles provide information of the local contribution to the total transit time, and can be very useful in identifying the transit time limiting factor in a given device design (i.e., for profile optimization). Since n_{ac} and $J_{C,ac}$ are nearly independent of frequency up to f_T, we can evaluate $\tau_{acc}(x)$ at any frequency below f_T. In this example, the results are nearly the same from 1 MHz to 60 GHz. This insensitivity to frequency proves useful in practice.

Regional Analysis of Transit Time

The total transit time defined by $\tau_{acc}(x = x_{cc})$ can be divided into five components to facilitate physical interpretation [11]. Two boundaries, the electrical EB and CB junction depths x_{eb}^* and x_{cb}^* are defined to be the in-most intersections of the n_{ac} and p_{ac} curves inside the junction space–charge regions (as illustrated in Figure 15.10). The same SiGe HBT with a 2 to 8% graded base was used in this case. The peaks of n_{ac} and p_{ac} can be understood as the approximate space–charge region boundaries on the emitter and base sides, respectively, even though the results clearly show that no abrupt space–charge region boundary can be identified (i.e, the depletion approximation is invalid). The "neutral base" that corresponds to traditional bipolar theory can be approximately identified as where $n_{ac} \approx p_{ac}$. The "neutral" base width is clearly smaller than the electrical base width defined by $x_{cb}^* - x_{eb}^*$. With scaling of base width into the nanometer regime, the region where $n_{ac} \approx p_{ac}$ eventually disappears, and the

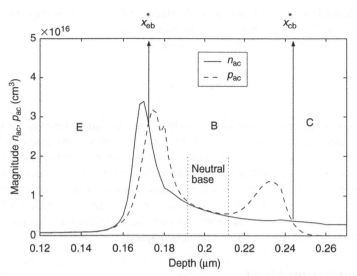

FIGURE 15.10 Definition of the electrical EB and CB junctions from the simulated n_{ac} and p_{ac} profiles. The bias is chosen at the peak f_T point.

electrical base width should be used instead for device analysis. The electrical EB and CB junction locations (x_{eb}^* and x_{cb}^*) are in general different from the metallurgical junctions (x_{eb} and x_{cb}), as expected. The total τ_{ec} can be divided into five components with the help of p_{ac} [11]:

1. The emitter transit time due to minority carrier storage in the emitter

$$\tau_e^* = \frac{q}{J_{C,ac}} \int_0^{x_{eb}^*} p_{ac} \, dx.$$ (15.16)

2. The EB depletion charging time due to the storage of uncompensated mobile carriers

$$\tau_{eb}^* = \frac{q}{J_{C,ac}} \int_0^{x_{eb}^*} (n_{ac} - p_{ac}) dx.$$ (15.17)

3. The base transit time due to electron charge storage in the electrical base (which includes the traditional "quasineutral base")

$$\tau_b^* = \frac{q}{J_{C,ac}} \int_{x_{eb}^*}^{x_{cb}^*} n_{ac} \, dx.$$ (15.18)

4. The CB depletion charging time

$$\tau_{cb}^* = \frac{q}{J_{C,ac}} \int_{x_{eb}^*}^{x_{cc}} (n_{ac} - p_{ac}) dx.$$ (15.19)

5. The collector transit time

$$\tau_c^* = \frac{q}{J_{C,ac}} \int_{x_{eb}^*}^{x_{cc}} p_{ac} \, dx.$$ (15.20)

Note that τ_c^*, as defined above, is different from the traditional τ_c, and τ_c^* is important only when holes are injected into the collector after the onset of high injection.

The sum of all the transit time components is equal to τ_{ec}

$$\tau_{ec} = \frac{q}{J_{C,ac}} \int_0^{x_{cc}} n_{ac} \, dx = \tau_e^* + \tau_{eb}^* + \tau_b^* + \tau_{cb}^* + \tau_c^* \qquad (15.21)$$

The transit time due to electron charge storage in the EB space–charge region is not treated separately, but is instead included in the modified transit times of the emitter and base (the "*" transit times above). Under high injection, however, the whole transistor from emitter to collector is flooded with a high concentration of electrons and holes, and hence no clear boundaries can be identified. Strictly speaking, the concepts of base, emitter, and collector consequently lose their conventional meanings, and thus the concept of regional transit times is no longer meaningful. In SiGe HBTs, the SiGe-to-Si transition at the CB junction causes additional electron charge storage at high injection. In this case, the x_{eb}^* and x_{cb}^* definitions discussed above cannot be applied.

High-Injection Barrier Effect

We now examine the evolution of the small-signal $qn_{ac}/J_{C,ac}$ and $qp_{ac}/J_{C,ac}$ profiles with increasing current density J_C from well below the peak f_T current density to slightly above the peak f_T current density. The simulated $qn_{ac}/J_{C,ac}$ and $qp_{ac}/J_{C,ac}$ profiles at three current densities representing low to high injection levels are shown in Figure 15.11a–c. The small-signal magnitude of the V_{BE} increase is 2.6 mV. At a typical low-injection J_C of 0.127 mA/μm^2, well below the peak f_T point, n_{ac} is positive across most of the device. Most of the charge modulation occurs in the EB space–charge region. The transit time related to this component of the charge storage decreases with increasing J_C because of increasing $J_{C,ac}$, which can be seen by comparing the magnitude of the first peak on the curves for the electrons in Figure 15.11a and b. Note that different scales are used on the y-axis for different injection levels to help visualize the details of the profiles.

At $J_C = 1$ mA/μm^2, near peak f_T, the base and collector transit time contributions become dominant compared to the EB space–charge region contribution (as shown in Figure 15.11b), mainly due to a decrease of the EB space–charge region transit time. One consequence of high-level injection is that the CB space–charge region pushes toward the collector n^+ buried layer much more obviously than at lower J_C, despite a decrease of V_{CB}. This is manifested as a large *negative* n_{ac} and hence negative $n_{ac}/J_{C,ac}$ around 0.37 μm. Physically, this corresponds to the *extension* of the CB space–charge region towards the n^+ buried layer, which causes a *decrease* of electron concentration at the front of the space–charge region. In the simulation, the base voltage is increased while the collector and emitter voltages are fixed. Because of the existence of negative n_{ac}, the real part of the simulated n_{ac}, as opposed to the absolute value of the simulated n_{ac}, should be used for calculation of the total transit time. A significant error can be introduced under high injection when the integral over the negative n_{ac} portion becomes significant to the total integral. We note that this negative-going n_{ac} component under high-injection is generally not treated properly in the literature.

Figure 15.11c shows the $qn_{ac}/J_{C,ac}$ and $qp_{ac}/J_{C,ac}$ profiles at a slightly higher J_C of 1.76 mA/μm^2, just past the peak f_T. The SiGe-Si interface, which was buried in the CB space–charge region under low injection, is now exposed to the large density of electrons and holes. The valence band potential barrier to holes induces a conduction band potential barrier to electrons as well. The most important consequence is increased dynamic charge storage, as seen from the high $qn_{ac}/J_{C,ac}$ and $qp_{ac}/J_{C,ac}$ peaks near the SiGe–Si transition in Figure 15.11c. This additional charge storage results in a significant increase of the total transit time and hence a strong decrease of f_T to 29 GHz, even though the current density is just above the value needed to reach the peak f_T (1.0 mA/μm^2).

At an even higher J_C of 3.56 mA/μm^2, both $qn_{ac}/J_{C,ac}$ and $qp_{ac}/J_{C,ac}$ are very large, and nearly equal to each other (as shown in Figure 15.12). No clear space–charge regions can be identified from the $qn_{ac}/$

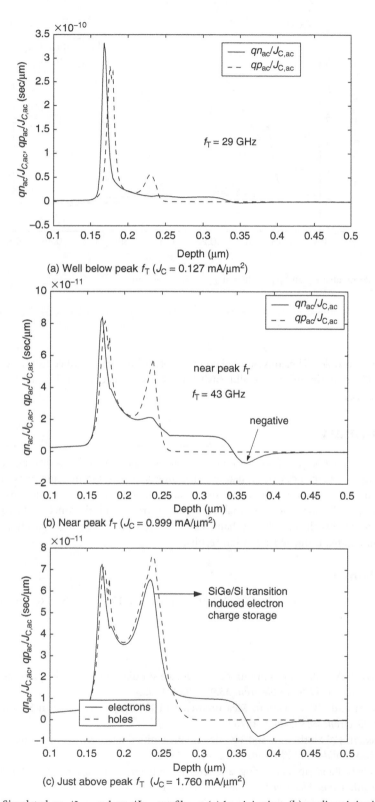

FIGURE 15.11 Simulated $qn_{ac}/J_{C,ac}$ and $qp_{ac}/J_{C,ac}$ profiles at (a) low injection, (b) medium injection, and (c) high injection.

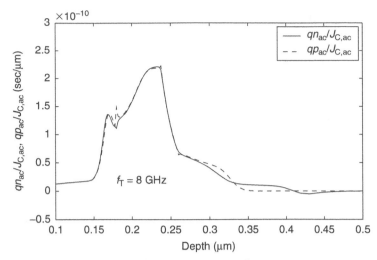

FIGURE 15.12 Simulated n_{ac} and p_{ac} profiles at $J_C = 3.56\,\text{mA}/\mu\text{m}^2$.

$J_{C,ac}$ and $qp_{ac}/J_{C,ac}$ profiles. The conventional concepts of emitter, base, and collector no longer apply in this situation. The majority of the overall transit time, however, is contained inside the SiGe "base," as intuitively expected.

15.6 Summary

We have introduced the basics of semiconductor device simulation, and practical aspects of SiGe HBT simulation using commercial device simulators, including structure specification, meshing, and physical model selection. Strategies for calibration of dc and RF device characteristics are presented and illustrated. The spatial distributions of small signal ac electron and hole concentrations and the transit time velocity provide insight into the mechanisms of charge storage. The high-injection barrier effect in SiGe HBTs is illustrated using ac simulation results.

Acknowledgment

This work is supported by NSF under ECS-0112923 and ECS-0119623 and by SRC under SRC-2001-NJ-937 and SRC-2003-NJ-1133.

References

1. DBM Klaassen. A unified mobility model for device simulation I. Model equations and concentration dependence. *Solid-State Electron.* 35:953–959, 1992.
2. JW Slotboom and HC de Graaff. Measurements of bandgap narrowing in Si bipolar transistors. *Solid-State Electron.* 19:857–862, 1976.
3. DBM Klaassen, JW Slotboom, and HC de Graaff. Unified apparent bandgap narrowing in n- and p-type silicon. *Solid-State Electron.* 35:125–129, 1992.
4. CM van Vliet. Bandgap narrowing and emitter efficiency in heavily doped emitter structures revisited. *IEEE Trans. Electron Dev.* 40:1141–1147, 1993.
5. GF Niu. Modeling and Simulation of SiGe Microelectronic Devices. M.S. Thesis, Fudan University, 1994.

6. Z Matutinovic-Krstelj, V Venkataraman, EJ Prinz, JC Sturm, and CW Magee. Base resistance and effective bandgap reduction in npn Si/SiGe/Si HBTs with heavy base doping. *IEEE Trans. Electron Dev.* 43:457–466, 1996.

7. Y Shi, GF Niu, and JD Cressler, On the modeling of BGN for consistent device simulation of highly scaled SiGe HBTs. *IEEE Trans. Electron Dev.* 50:1370–1377, 2003.

8. G Niu, S Zhang, JD Cressler, AJ Joseph, JS Fairbanks, LE Larson, CS Webster, WE Ansley, and DL Harame. Noise modeling and SiGe profile design tradeoffs for RF applications. *IEEE Trans. Electron Dev.* 47:2037–2044, 2000.

9. GF Niu, WE Ansley, S Zhang, JD Cressler, C Webster, and R Groves. Noise parameter optimization of UHV/CVD SiGe HBTs for RF and microwave applications. *IEEE Trans. Electron Dev.* 46:1347–1354, 1999.

10. P Baureis and D Seitzer. Parameter extraction for HBTs temperature dependent large signal equivalent circuit model. Technical Digest of IEEE GaAs IC Symposium, 1993, pp. 263–266.

11. JJH van den Biesen. A simple regional analysis of transit times in bipolar transistors. *Solid-State Electron.* 29:529–534, 1986.

16

SiGe HBT Performance Limits

Greg Freeman and
Andreas Stricker
IBM Microelectronics

David R. Greenberg
*IBM Thomas J. Watson
Research Center*

Jae-Sung Rieh
Korea University

16.1 Performance

How high a performance can be achieved in silicon-based bipolar transistors? An answer to such a question surely considers many assumptions. New discoveries continue to affect the critical aspects of device operation such as charge storage, carrier transport, and parasitics. Other discoveries affect the processing of the device, leading to even better ways to make the device structurally ideal. One example of a historic discontinuity in device fabrication and operation is the development of production-ready SiGe epitaxy. Before the advent of SiGe epitaxy, predictions toward device limits would likely have made certain assumptions regarding emitter charge storage or minority carrier diffusion and this would clearly be off the mark due to the significant advancement in SiGe band engineering. More recently, the incorporation of carbon has provided a boost, strongly affecting the diffusion of dopants and thus providing a greater control over the device structure. Similar innovations are expected to continue to provide a boost to the device operation, and so continually change the assumptions that may go into predicting limits of device operation.

It is common to think principally of the f_T figure-of-merit in discussion of performance limits, yet this figure-of-merit in itself is a poor predictor of most circuit performance. Depending on the application, other device figures-of-merit such as collector–emitter breakdown voltage (e.g., BV_{CES}), linearity, power added efficiency, or f_{MAX}, may be preferred for predicting circuit performance. Most broadly applicable is the f_{MAX} figure-of-merit, which more strongly takes into account key parasitic elements and better predicts the power capability and digital switching delay, and is to first order related to f_T through the following relation:

$$f_{MAX} \cong \sqrt{\frac{f_T}{8\pi R_B C_{CB}}} \qquad (16.1)$$

where R_B and C_{CB} are the total base resistance and collector–base capacitance, respectively. f_T then is not only a figure-of-merit, but also a key component of f_{MAX}, and so it is important to understand f_T limitations since they are also limitations to other figures-of-merit. A simplified expression for the f_T delay components of a SiGe bipolar transistor is

$$\frac{1}{2\pi f_T} = \tau_{EC} = \tau_E + \tau_C + \tau_B + \tau_{CSCL} \cong \frac{kT}{qI_C}C_{EB} + \left(\frac{kT}{qI_C} + R_C + R_E\right)C_{CB} + \frac{W_B^2}{\gamma D_n} + \frac{W_{CSCL}}{2v_{SAT}} \quad (16.2)$$

where C_{EB} and C_{CB} are emitter–base and base–collector capacitance, R_C and R_E are collector and emitter resistance, W_B and W_{CSCL} are neutral base and base–collector space–charge layer width, respectively, k is the Boltzmann constant, T is temperature, q is unit electron charge, γ is field factor, and v_{SAT} is the electron saturation velocity. This clearly illustrates the complex nature of the transit time, which includes neutral base and collector space–charge layer transit times τ_B and τ_{CSCL}, as well as R * C delays. These R * C delay terms are improved by transconductance (qI_C/kT) improvements and improvements in parasitic resistances.

Limitations to performance come from the physical reality in achieving narrow base width, narrow collector–base space–charge layers, low resistances, low capacitances, and the ability to achieve high current densities with acceptable device self-heating and reliability. We assert that performance advancement has taken place uniformly across the various limiting parameters and device structures over time, and so the performance continues to be limited by the same effects as prior generation devices. This means for instance that unwanted diffusion still has a major impact on transit times and parasitic capacitances. Also, base widths are much larger than deposited and collector pedestals are wider than implanted due to dopant diffusion. Yet dopant diffusion is not always bad, since diffusion is needed to define required junction depths and to improve device properties such as base resistance and junction leakage [1]. With reduced dimensions, the constraints in device design only become tighter with increased sensitivity to dopant profile details, such as implant tails and two-dimensional effects. Therefore, the engineering of a device becomes a complex tradeoff of such effects, and is highly constrained by the device structures chosen and process steps available.

One should expect improvements along the same line as previously established—e.g., that the base will become more narrow and more highly doped, that the collector will become more narrow (vertically and laterally) and self-aligned for lower parasitic capacitance as well as lower resistance, and that the emitter and base will become lower resistance. Numerous process steps not implemented into the SiGe HBT processes are already available for such improvements, including anneals, silicides, self-alignment techniques [2,3]. This chapter takes the approach of considering a nearly ideal device structure, which all these improvements will continue to approach, and provide discussion of the remaining effects limiting performance, in order to provide some insight into the eventual device limitations and effects to be overcome for continued improvements. By this approach, the more practical effects are considered and eventual limits may be better understood.

16.2 Intrinsic and Extrinsic Partitioning

To understand the constraints to device performance, consider the "intrinsic" versus the "extrinsic" portion of the device. By understanding to what extent the performance of the complete device is impacted by the extrinsic portion of the device, we can better understand what structural improvements may influence the device performance.

The intrinsic portion is commonly considered the region of the device defined vertically between the neutral emitter to the neutral collector. The lateral dimension of this intrinsic region, approximately 100 nm, is considerably smaller than the complete device, which is typically over 1 μm in width. The exact partitions between the intrinsic and extrinsic regions are somewhat arbitrary, and for the purposes of the study to follow, we wish to define these partitions such that the more fundamental aspects of the device operation are contained in the intrinsic portion of the device. This means that the intrinsic portion contains as little as possible of the resistive elements in the neutral emitter and collector, since one may improve these by structural changes in the device such as with silicides or improved dopant levels. The shape of this intrinsic region is generally shown to be rectangular, as a sort of expanded one-dimensional device operation. As shown by the current flow lines in Figure 16.1, actual intrinsic device operation takes place in a very much two-dimensional fashion and so the shape of the intrinsic region is not necessarily a rectangle. The electron flow from the emitter spreads laterally from the actual emitter–base junction due to the lateral potential drops in the base and the complex injection from the edges of the emitter–base junction. This flow becomes vertical and more one-dimensional through the neutral base, and then spreads to a wider region of the device as it traverses the collector–base space–charge region because the low resistance portion of the collector is generally wider compared to the emitter opening.

Structural improvements may be made to either the intrinsic or the extrinsic device. The intrinsic device structure changes result in relatively predictable performance benefits and tradeoffs. The extrinsic device provides the performance-enhancing opportunity without significant tradeoff, since this portion of the device is not fundamental to the device operation and so parasitic reductions are not generally accompanied with other performance losses. Consider the trend in R_B and C_{CB} versus the advancements in f_T as shown in Figure 16.2 [4–11]. The C_{CB} values continue to increase with increasing f_T because the intrinsic portion is a majority portion of the total C_{CB} for the device (where the pedestal is defined). The R_B has been dominated by the extrinsic portion of the device and so is not fundamental to achieving the higher f_T and is shown to be more suited to improvement with structural modifications [6,9].

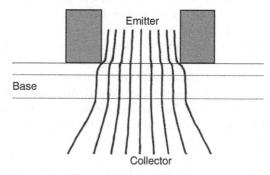

FIGURE 16.1 Current flow lines from TCAD simulations illustrating strong two-dimensional behavior of the SiGe HBT device.

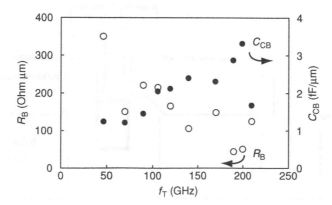

FIGURE 16.2 Base resistance R_B (open symbols) and collector–base capacitance C_{CB} (solid symbols) trend with f_T. Values are normalized to emitter length. Values reported here are limited to those published results where $f_{MAX} > f_T$ and where R_B and C_{CB} are both reported [4–11].

We have performed process and device (TCAD) simulations in order to estimate the performance of the intrinsic-only device without the extrinsic device [12,13]. In order to estimate the two-dimensional nature of device operation described above, yet stay within the straightforward implementation of the simulator, we define the intrinsic region of the device as a rectangle, and this is approximately twice the width of the emitter opening. This allows for the realistic injection of the carriers from the emitter junction edge, and the spreading of these carriers to the pedestal portion of the collector. The region is also bounded at the emitter side approximately 2 nm into the neutral emitter and at the collector side approximately 40 nm into the neutral collector. Figure 16.3 shows the cross section of the intrinsic device within the complete device. Note that the intrinsic device contains ohmic contacts to the base, emitter, and collector as if the rectangle were in free space connected by nonresistive contacts at the four access points. These four contacts also provide a thermal sink to reduce the intrinsic-only device heating, and we will discuss the thermal issues in real devices in the last section of this chapter.

The simulator attempts to mimic the process steps and the electrical characteristics of the $f_T = 350$ GHz device reported in Ref. [14], and as such the simulation parameters have been calibrated to achieve good predictability in the fabrication of this device and its electrical behavior. The simulators and methods used are similar to that used in prior generation devices [12]. Figure 16.4 shows the results of the simulation comparison (dark arrows). Compared to the parasitic values that would be present in an intrinsic-only device, we observe that the extrinsic device contributes significantly to the overall parasitics, at an additional 35%, 180%, 124%, and 42% in C_{BE}, C_{CB}, R_C, and R_B, respectively. This extrinsic device strongly reduces the device performance. f_T and f_{MAX} comparisons also shown in Figure 16.4 indicate that values of 557 and 630 GHz respectively become 332 and 224 GHz, which correspond approximately to measured values reported in Ref. [14].

From this analysis, it is clear that contributed parasitic resistances and capacitance from the extrinsic device degrades performance as measured in f_T and f_{MAX}. Other device designs will show different partitioning of performance. Namely, devices with lower intrinsic base sheet resistance will exhibit a lesser portion of the total base resistance from the intrinsic device. Due to the thermal contacts at the base, emitter, and collector of the intrinsic-only device, we observe similar self-heating characteristics between the simulations. Without the thermal sinks, the performance would significantly degrade. The thermal issue will only become more acute with increasing device performance (due to the increasing current density), with increasing need for effective thermal conductivity away from the device. The intrinsic device scenario depicted here, where the electrical contacts provide a thermal sink, may be realistic in some small-integration device configurations (e.g., power amplification devices), but in a more general configuration of highly integrated chips, a thermal sink through the substrate will be

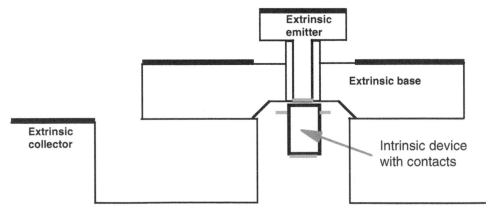

FIGURE 16.3 Intrinsic device portion of full extrinsic device represented in TCAD simulations. The intrinsic dimension is approximately twice the emitter-opening dimension in order to capture the two-dimensional nature of the device operation.

required. For such highly integrated chips, the extrinsic device has the significant function of conducting heat away from the intrinsic device, and this aspect cannot be ignored in device optimization.

Thus, in consideration of the device limits, the intrinsic device cannot be considered independent of the extrinsic device. One should consider both what can practically be done to improve performance reducing effects of the extrinsic device, and the thermal benefits provided by the extrinsic device. In the next two sections, we consider the scaling aspects of the intrinsic and extrinsic devices separately.

16.3 Intrinsic Device Scaling

Classical scaling, with the reduction in delay elements captured in Equation 16.2, is expected to continue to improve SiGe HBT device performance. As mentioned previously, SiGe HBT vertical profiles, while in the range of 10 nm, are still dominated by practical processing effects such as diffusion and growth constraints. As in the past, improved device structures, process integration techniques, and tooling are expected to open up new performance territory through continued device scaling. We explore issues and unknowns related to intrinsic device scaling above 1 THz in this section.

Experience has shown that, in the graded-base SiGe HBT, the carriers travel in the range of saturation velocity through most of the neutral base and collector space–charge region. Therefore, transit time is largely a function of carrier velocity and dimensions. Clearly, one goal is the reduction in base dimension W_B and collector–base space–charge layer W_{CSCL}. However, the ever-reduced dimensions begin to put the commonly assumed drift–diffusion physics, which determine the effective v_{SAT}, into question. In transistors with large critical dimensions (e.g., base widths), the acceleration of carriers in an electric field is counteracted by impurity and phonon scattering, which act toward randomizing the momentum and bringing the carriers into thermal equilibrium with the lattice. This balance results in a steady-state velocity for any given field, characterized by the mobility and saturation velocity of the sample. Since scattering takes place at a finite rate, however, a carrier may not scatter at all over a sufficiently short interval of time. As a result, carriers crossing a sufficiently small base may be able to do so with very little scattering and thus with very little change in their initial velocity. In silicon, acoustic phonons scatter

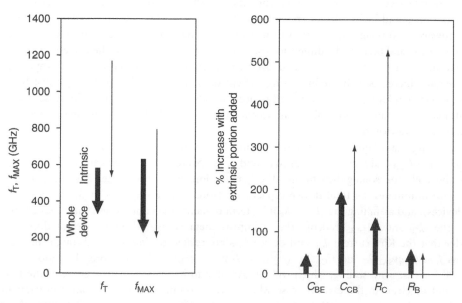

FIGURE 16.4 Electrical parameter shifts with addition of the device extrinsic portion as predicted by TCAD simulation. Thick arrows are the 350 GHz device as reported in Ref. [14]; thinner arrows are simulated 1.16 THz (intrinsic) device.

electrons between sixfold-generate conduction band minima with a time constant on the order of 10^{-13} to 10^{-14} sec. Based on an electron velocity in the range of 10^6 to 10^7 cm/sec, ballistic transport across a HBT base becomes possible as base widths shrink below $\sim$10 nm.

To maximize the benefits of ballistic transport, a HBT design must provide a method for electrons to attain and maintain a high velocity for a large fraction of their trip across the base. One such method is to introduce a very high electric field at the emitter side of the base, rapidly accelerating electrons to "overshoot" velocities greater than predicted by the scattering-driven steady-state velocity versus field curve. Since accelerating electrons from low velocity can consume an appreciable portion of the base transit and reduce the overall effective velocity, however, an improved method involves the inclusion of a "launcher" capable of injecting electrons from the emitter into the base at high initial velocities.

Studies of source velocity in nanoscale NFETs illustrate that injection over a simple energy barrier, such as in the emitter–base junction of a homojunction BJT or of a graded SiGe HBT with no SiGe mole fraction at the junction, limits the injection velocity to the average thermal velocity of electrons moving in the desired direction. Such velocities may exceed 1.2×10^7 cm/sec, reducing the transit time across a 10 nm base to less than 0.08 psec [15].

Bandgap engineering, commonly used in III–V device design, can be used to increase injected carrier velocity still further. One possible design is an abrupt conduction band discontinuity at the emitter–base junction, with a lower conduction band energy in the base. Such a heterojunction can be realized with a strained Si emitter grown atop an unstrained SiGe base, for example. As an electron crosses this junction, the conduction band "floor" drops out from under it, converting potential energy into a large kinetic energy.

Since electrons injected in this manner will cover a wide range of velocities, the average velocity is reduced by the slowest of the population. A tunneling barrier, such as a very thin layer of SiO_2, can be inserted between the emitter and base to filter the carrier population and pass only the highest energy members of the population. A quantum well can be used for the same purpose, with the energy levels of the well tuned by well width to form a "pass band" for the desired electron energies. Using these structures, and with the ever-reduced dimensions of the SiGe HBT, transit time benefits may be obtained through increased effective carrier velocity.

In consideration of the capacitance charging time reduction with scaling, the principal benefit is higher Kirk-effect current, which results from the collector design scaling. This translates to higher current densities and higher device transconductance qI_C/kT before the base-push-out effect reduces the microwave performance of the transistor. Lateral scaling typically accompanies the device current density increase, and because the dimensions scale down at a similar rate to the current density increase, a similar net device current is obtained between generations. Unit collector–base capacitance also fundamentally increases with the increasing current density, since collector doping must be increased or the collector epi layer must be decreased to accommodate the current density increase. Like with the current density increase, lateral scaling can decrease the effect of the higher unit capacitance by reducing the effect at the device level.

By reviewing again Equation 16.1 and Equation 16.2, a fundamental tradeoff between increasing I_C and increasing C_{CB} and their impact to f_{MAX} and f_T is apparent. In f_T, one can see that a C_{CB} increase offsets the collector scaling benefit in W_{CSCL} reduction and I_C increase. Assuming that the C_{CB} capacitance dominates the total device capacitance (today surpassing C_{EB} in the highest speed SiGe HBT devices), and assuming the $(R_E + R_C)C_{CB}$ term remains small due to resistance reduction, we note that in the C_{CB} and W_{CSCL} tradeoff, the f_T improvement is favored with collector doping increase. Consider that the Kirk current J_{CP} and device C_{CB} are related to doping N_C through the well-known relations $J_{CP} = qv_{SAT}N_C$ and $C_{CB} = \sqrt{qN_C\varepsilon_s/2(V_{BI} + V_{CB})}$ [16]. Therefore, the ratio of $J_{CP}/C_{CB} \propto \sqrt{N_C}$ will increase with scaling, which will be favorable to increasing f_T. The effect of the tradeoff on f_{MAX} is not as clear. In the extreme case wherein C_{CB} dominates the capacitance portion in the f_T expression (such as in many III–V devices), f_{MAX} sees diminishing returns with vertical scaling as approximated by $f_{MAX}^2 \propto f_T/C_{CB} \propto (\sqrt{N_C} + K)/\sqrt{N_C}$, where K represents the remaining terms of the f_T expression unaffected by the collector scaling.

Other authors have performed simulations of intrinsic-only devices above 1 THz [17]. While that report does provide useful insight into the operation of the device, we wish to include certain phenomena not included in that simulation. For instance, as previously mentioned, we wish to utilize a calibrated simulation deck to a high-speed device, as well as capture the two-dimensional device behavior with the two-dimensional current injection from the emitter and the lateral spreading of the current into the collector. We also wish to understand the effects of self-heating and of the extrinsic device. We still rely on drift–diffusion device simulations, and inaccuracies are likely in this regime as a result. However, we have found surprising predictability over many generations of devices, and so believe that these same simulation techniques should (to first order) continue to provide insights into the device scaling.

Following this approach, we have simulated the fabrication and operation of a complete device, and when the intrinsic portion of this device is separated and measured in the device simulator, an f_T of 1165 GHz is obtained. The device geometry corresponds to the two-dimensional intrinsic device cutout and contact configuration shown in Figure 16.3 and discussed previously. Note that, as before, the full two-dimensional device construction is defined in the process simulator environment, and the cutout is made and contact made to represent the intrinsic device. Vertical scaling was used to increase f_T, and realistic fabrication steps were utilized in the process simulator to define the dopant profiles and diffusion. Accompanying the higher f_T and reduced transit times are approximately three times higher peak f_T current density, three times higher C_{CB}, and 1/3 times R_C compared to the previously discussed device.

Thus, when considering the intrinsic portion of the device only, we may expect that over 1 THz f_T operation may be obtained. Very importantly, we have neglected two very critical elements to the device, which are device reliability and extrinsic parasitics. In the next two sections, we build on the results from this section and describe the impact of these effects on device performance and the challenges to device scaling imposed.

16.4 Extrinsic Device RC Delays

To gauge the effects of the extrinsic device, we simulate the THz intrinsic device of the last section and add its extrinsic portion, which is similar to the device of the previous section. The impact of the extrinsic device is shown in Figure 16.4 (as the more narrow arrows), adjacent to the same analysis of the 350 GHz device. Clearly, the extrinsic device has a greater impact on the THz device than on the lower performance device. The largest impact is on the parasitic R_C, where the intrinsic device has a reduced value from the increased doping concentration, and the extrinsic device has not been reengineered to commensurately reduce the extrinsic resistance. Also impacted severely is the C_{CB} parasitic, because the high performance is achieved in part from a reduced vertical spacing between the heavily doped subcollector and the base region of the device. This increases the extrinsic capacitance as well as the intrinsic capacitance. The net result is a more severe reduction in both f_T and f_{MAX}, when the extrinsic region is added to the THz intrinsic device. Compared to the parasitic values that would be present in an intrinsic-only device, we again observe that the extrinsic device contributes significantly to the overall parasitics, at an additional 55%, 300%, 525%, and 44% in C_{BE}, C_{CB}, R_C, and R_B. f_T and f_{MAX} comparisons shown in Figure 16.4 indicate that values of 1165 and 525 GHz become 798 and 193 GHz, respectively. f_T becomes 45% of the 1165 intrinsic device value, and f_{MAX} becomes 24% of its intrinsic device value.

With the higher performance, the device operation is more sensitive to the parasitic resistance and capacitance compared to the lower performance intrinsic device. This is a result of the need to reduce all the terms in Equation 16.2, and with the reduction in the intrinsic portions, the extrinsic portions become more significant. As scaling of the intrinsic device is critical to achieving higher performance, improvements in the extrinsic device need to be commensurate. Moreover, like the advancements in the intrinsic device follow material and structure advances, so do the advancements in the extrinsic device, but often with a different set of material and structure advances.

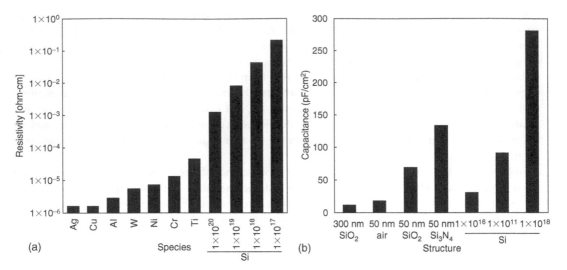

FIGURE 16.5   Resistance and capacitance of different materials and structures.

Challenges for extrinsic device improvement are found in many places. Resistive parasitics are found in the conducting layers leading to the device. Where the emitter, base, and collector connecting layers into the device are relatively low resistivity at a distance from the device (i.e., the metal interconnect layers) these layers increase in resistivity approaching the intrinsic device (i.e., silicides, then heavily doped semiconductors, and then less heavily doped semiconductors). Shown in Figure 16.5a are the resistivities of various layers typical in semiconductor devices. The device designers and integration engineers are challenged to choose lower resistance layers and to fabricate them as close as possible to the intrinsic device. This is typically achieved through lithography advances and through self-aligned processes, such as described in an earlier chapter.

Other resistive elements that challenge the device engineers are the interfaces. These interfaces are often found between the polysilicon and the single-crystal emitter, between polysilicon and epitaxy base layers, and in work function differences between layers. Reducing the impact of these interfaces involves careful process integration [18–20] or material advances such as silicides with reduced contact resistance [3].

Capacitive parasitics are also reduced. As discussed in the previous section, the intrinsic device design involves performance versus capacitance tradeoffs, both in the emitter and the collector side junctions. Regarding the extrinsic device, shrinking lateral device dimensions, especially related to the active area and the pedestal dimensions, is probably the most significant technical advancement to provide capacitance reduction. However, to approach the intrinsic device performance, the extrinsic capacitive elements need to continue to reduce. The shrinking emitter lateral dimensions make this particularly difficult due to the increased device perimeter to area ratio that results, and because the extrinsic capacitances are scaled with the device perimeter. For instance, the emitter–base perimeter junction and fringing capacitances become ever more significant. In addition, the collector capacitance perimeter becomes more significant, because the polysilicon base contact area does not shrink at the same rate as the emitter dimension. One solution is to employ increasing amounts of (preferably self-aligned) low dielectric constant materials, such as replacing portions of the active area with silicon dioxide or even air or vacuum. Shown in Figure 16.5b is a comparison of unit capacitances of typical device insulating structures. Typical films are for instance a 300 nm SiO_2 for the shallow trench material and thickness. Fifty nanometers is a typical spacer dimension, which can be made of different materials as shown. Note that the choice of different materials may result in significantly different capacitances in the device. New processing techniques and materials will permit incorporation of such improvements.

While techniques such as described will take the device toward the intrinsic-only device, there will always be practical limitations to the implementation. For instance, a relatively radical approach to reducing collector resistance and capacitance components has been tried by different groups. This approach, called "transferred substrate," has been applied with significant performance enhancement in III–V devices [21] and with limited improvement to silicon devices [22]. The intent of this approach is to address the resistive element and the excess capacitance in the collector, because the collector typically needs to be contacted through a large relatively resistive and capacitive region from the bottom of the device. By removing the substrate and providing a contact to the collector from the device bottom, performance advancements may be achieved. However, thermal issues begin to dominate as shown in Ref. [22], and practical solutions would require replacement of the substrate function as a thermal sink.

16.5 Practical Limitations

Performance capability is not the only concern in understanding the limits. One must also consider avalanche and thermal properties, both may degrade device or circuit performance, and may affect the device reliability. Understanding of these effects has recently been improved with the advent and characterization of devices with f_T over 200 GHz [23].

Avalanche current, which is responsible for the device breakdown, or BV_{CEO} and BV_{CES}, is a result of the higher electric fields in the scaled collector–base space–charge region (i.e., reduced W_{CSCL} of Equation 16.2). BV_{CEO} (collector–emitter breakdown with the base terminal open) is often cited as a limit to device biasing voltage. This parameter is now generally below 2 V for devices with $f_T > 100$ GHz, to a value of 1.4 V for the 350 GHz device [14]. However, a typical transistor within a circuit has relatively low impedance connected to the base terminal, such that the breakdown is more closely related to BV_{CES}, which is with the base terminal tied to the emitter. An evaluation of the $f_T * BV_{CES}$ product scaling has been performed. This parameter is shown to be collector-doping dependent and significantly greater than the $f_T * BV_{CEO}$ product [24].

Reliability in SiGe HBTs appears to be robust to degradation resulting from avalanche current. This contrasts to some III–V devices, which exhibit crystal degradation through carrier recombination and generation. For example, 200 GHz SiGe HBTs have been shown to exhibit increased base current nonideality as a result of operation above BV_{CEO}, yet this degradation is expected to remain negligible for most applications over typical product lifetimes [25]. Furthermore, the avalanche current is expected to continue to increase with increased device performance, yet the voltage causing the same avalanche multiplication factor is found to reduce only a small amount looking to future generations of devices [23]. This indicates that the avalanche will not provide a significant limitation to the performance of the SiGe HBT.

Voltage and current limits are also related to device self-heating. We have already discussed this aspect of performance scaling with relation to its impact to performance degradation. However, as with most semiconductor devices and integrated circuits, reliability is impacted with greater temperatures, which can be caused by increased power density in a device [23]. In particular, electromigration in the metal interconnects is highly sensitive to increased temperatures. Unlike in CMOS devices, which have exhibited linear currents in the range of 700 μA/μm, with recent maximum voltages in the range of 1 V, the SiGe HBTs with $f_T > 100$ GHz have called for currents in the range of 1500 μA/μm achieved at voltages in the range of 1.5 V. For small devices, this does not present much of an issue due to their relatively smaller thermal resistance × power product, but designs with larger devices require attention to self-heating and robust wiring for reliable operation. Looking ahead, device designers must continue to focus on linear current and thus the power density reduction, through such techniques as dimension reduction. Accordingly, ever-higher current densities are anticipated to be acceptable and reliable with respect to device self-heating.

Another issue with respect to higher current densities is the well-established base current degradation accelerated by higher current densities [26–28]. Due to the requirement for higher current densities for performance enhancement, these current densities once again challenge the device designer and integration engineers, since they must find materials and processes that decrease base current degradation. Low hydrogen-containing materials have been shown to improve the base current degradation [28], and it has been shown that advanced devices may be fabricated with degradation substantially less than the prior generation devices at similar current densities [23]. New solutions through materials and experimentation must be established to propel the SiGe HBT to continued reliable high performances.

16.6 Summary

The industry will continue to demonstrate SiGe HBT performance improvements. In achieving these further advancements, the intrinsic performance improvements will be a relatively straightforward continuation of recent advancements in materials and processes. Techniques made available through CMOS processing will be leveraged for such progress, and ballistic effects will start to be seen in devices with sub-10 nm transit dimensions. The extrinsic portion of the device is perhaps the most challenging, due to the practical availability of construction methods and materials in semiconductor fabrication facilities. Thus, limits are mainly due to practical ability to implement the structures needed for reduced parasitics and thermal conduction within a generalized application technology.

References

1. D Terpstra and WB de Boer. Anomalous collector base leakage in selectively grown SiGe base heterojunction bipolar transistors. Proceedings of the 29th European Solid State Device Research Conference, Leuven, 1999, pp. 720–723.
2. SK Earles, ME Law, J Kevin, S Talwar, and S Corcoran. Nonmelt laser annealing of 5 keV and 1 keV boron implanted silicon. *IEEE Transactions on Electron Devices* 49:1118–1123, 2002.
3. MC Öztürk, J Liu, and H Mo. Low resistivity nickel germanosilicide contacts to ultra-shallow $Si_{1-x}Ge_x$ source/drain junctions for nanoscale CMOS. Proceedings of the International Electron Devices Meeting, Washington, DC, 2003, pp. 497–500.
4. G Freeman, D Greenberg, K Walter, and S Subbanna. SiGe HBT performance improvements from lateral scaling. Proceedings of the European Solid-State Device Research Conference, Leuven, 1999.
5. A Joseph, D Coolbaugh, M Zierak, R Wuthrich, P Geiss, Z He, X Liu, B Orner, J Johnson, G Freeman, D Ahlgren, B Jagannathan, L Lanzerotti, V Ramachandran, J Malinowski, H Chen, J Chu, P Gray, R Johnson, J Dunn, and S Subbanna. A 0.18 μm BiCMOS technology featuring 120/100 GHz (f_T/f_{MAX}) HBT and ASIC-compatible CMOS using copper interconnect. Proceedings of the 2001 Bipolar/BiCMOS Circuits and Technology Meeting, pp. 143–146.
6. B Jagannathan, M Khater, F Pagette, J-S Rieh, D Angell, H Chen, J Florkey, F Golan, D R Greenberg, R Groves, S J Jeng, J Johnson, E Mengistu, KT Schonenberg, CM Schnabel, P Smith, A Stricker, D Ahlgren, G Freeman, K Stein, and S Subbanna. Self-aligned SiGe NPN transistors with 285 GHz f_{MAX} and 207 GHz f_T in a manufacturable technology. *IEEE Electron Device Letters* 23:258–260, 2002.
7. J Bock, H Schafer, H Knapp, D Zoschg, K Aufinger, M Wurzer, S Boguth, R Stengl, R Schreiter, and TF Meister. High-speed SiGe:C bipolar technology. Proceedings of the International Electron Devices Meeting, Washington, DC, 2001, pp. 344–347.
8. K Washio, M Kondo, E Ohue, K Oda, and R Hayami. A 0.2 μm self-aligned SiGe HBT featuring 107-GHz f_{MAX} and 6.7-ps ECL. Proceedings of the International Electron Devices Meeting, Washington, DC, 1999, pp. 557–560.
9. H Rücker, B Heinemann, R Barth, D Bolze, J Drews, U Haak, W Höppner, D Knoll, K Köpke, S Marschmeyer, HH Richter, P Schley, D Schmidt, R Scholz, B Tillack, W Winkler, H-E Wulf, and Y Yamamoto. SiGe:C BiCMOS technology with 3.6 ps gate delay. Proceedings of the International Electron Devices Meeting, Washington, DC, 2003, pp. 121–124.
10. P Deixler, R Colclaser, D Bower, N Bell, W De Boer, D Szmyd, S Bardy, W Wilbanks, P Barre, Mv Houdt , JCJ Paasschens , H Veenstra , Evd Heijden, JJTM Donkers, JW Slotboom. QUBiC4G:

a f_T/f_{MAX} = 70/100 GHz 0.25 μm low power SiGe-BiCMOS production technology with high quality passives for 12.5 Gb/s optical networking and emerging wireless applications up to 20 GHz. Proceedings of the Bipolar and BiCMOS Circuits and Technology Meeting, 2002, pp. 201–204.

11. S Wada, Y Nonaka, T Saito, T Tominari, K Koyu, K Ikeda, K Sakai, K Sasahara, K Watanabe, H Fujiwara, F Murata, E Ohue, Y Kiyota, H Shimamoto, K Washio, R Takeyari, H Hosoe, and T Hashimoto. A manufacturable 0.18-μm SiGe BiCMOS technology for 40-Gb/s optical communication LSIs. Proceedings of the Bipolar and BiCMOS Circuits and Technology Meeting, 2002, pp. 84–87.

12. J Dunn, DC Ahlgren, DD Coolbaugh, NB Feilchenfeld, G Freeman, DRGreenberg, RA Groves, FJ Guarin, A Joseph, LD Lanzerotti, SA St Onge, BA Orner, J-S Rieh, KJ Stein, S Voldman, P-C Wang, MJ Zierak, S Subbanna, DL Harame, DA Herman Jr, BS Meyerson, and Y Hammad. Foundation of RF CMOS and SiGe BICMOS technologies. *IBM Journal of Research and Development* 47:101–138, 2003.

13. M Ieong and P Oldigies. Technology modeling for emerging SOI devices. International Conference on Simulation of Semiconductor Processes and Devices, September 2002, pp. 225–230.

14. J-S Rieh, B Jagannathan, H Chen, KT Schonenberg, D Angell, A Chinthakindi, J Florkey, F Golan, D Greenberg, S-J Jeng, M Khater, F Pagette, C Schnabel, P Smith, A Stricker, K Vaed, R Volant, D Ahlgren, G Freeman, K Stein, and S Subbanna. SiGe HBTs with cut-off frequency of 350 GHz. Proceedings of the International Electron Devices Meeting, 2002, pp. 771–774.

15. M Lundstrom and Z Ren. Essential physics of carrier transport in nanoscale MOSFETs. *IEEE Trans. Electron Devices* 49:133–141, 2002.

16. SM Sze. *Physics of Semiconductor Devices*. John Wiley & Sons, New York, 1981.

17. Y Shi and G Niu. Vertical profile design and transit time analysis of nano-scale SiGe HBTs for terahertz f_T. Proceedings Bipolar and BiCMOS Circuits and Technology Meeting, 2004, pp. 213–216.

18. M Kondo, T. Kobayashi, and Y Tamaki. Hetero-emitter-like characteristics of phosphorus doped polysilicon emitter transistors—Part I. Band structure in the polysilicon emitter obtained from electrical measurements. *IEEE Transactions on Electron Devices* 42:419–426, 1995.

19. S Jouan, R Planche, H Baudry, P Ribot, JA Chroboczek, D Dutartre, D Gloria, M Laurens, P Llinares, M Marty, A Monroy, C Morin, R Pantel, A Perrotin, J de Pontcharro, JL Regolini, G Vincent, and A Chantre. A high-speed low 1/f noise SiGe HBT technology using epitaxially-aligned polysilicon emitters. *IEEE Transactions on Electron Devices* 46:1525–1531, 1999.

20. K Oda, E Ohue, T Onai, and K Washio. Heterojunction Bipolar Transistor. U.S. Patent 5,962,880, October 5, 1999.

21. MJW Rodwell, M Urteaga, T Mathew, D Scott, D Mensa, Q Lee, J Guthrie, Y Betser, SC Martin, RP Smith, S Jaganathan, S Krishnan, SI Long, R Pullela, B Agarwal, U Bhattacharya, L Samoska, and M Dahlstrom. Submicro scaling of HBTs. *IEEE Transactions on Electron Devices* 48:2606–2624, 2001.

22. LK Nanver, N Nenadovic, V d'Alessandro, H Schellevis, HW van Zeijl, R Dekker, DB de Mooij, V Zieren, and JW Slotboom. A back-wafer contacted silicon-on-glass integrated bipolar process—Part I. The conflict electrical versus thermal isolation. *IEEE Transactions on Electron Devices* 51:42–50, 2004.

23. G Freeman, J-S Rieh, Z Yang, and F Guarin. Reliability and performance scaling of very high speed SiGe HBTs. *Microelectronics Reliability* 44:397–410, 2004.

24. J-S Rieh, B Jagannathan, D Greenberg, G Freeman, and S Subbanna. A doping concentration-dependent upper limit of the breakdown voltage–cutoff frequency product in Si bipolar transistors. *Solid-State Electronics* 48:339–343, 2004.

25. Z Yang, F Guarin, E Hostetter, and G Freeman. Avalanche current induced hot carrier degradation in 200 GHz SiGe heterojunction bipolar transistors. Proceedings of International Reliablity Physics Symposium, March 2003, pp. 339–343.

26. J-S Rieh, K Watson, F Guarin, Z Yang, P-C Wang, A Joseph, G Freeman, and S Subbanna. Reliability of high-speed SiGe heterojunction bipolar transistors under very high forward current density. *IEEE Transactions on Device and Materials Reliability* 3:31–38, 2003.

27. MS Carroll, A Neugroschel, and C-T Sah. Degradation of silicon bipolar junction transistors at high forward current densities. *IEEE Transactions on Electron Devices* 44:110–117, 1997.

28. K Hoffmann, G Bruegmann, and M Seck. Impact of inter-metal dielectric on the reliability of SiGe NPN HBTs after high temperature electrical operation. IEEE Topical Meeting on Si Monolithic Integrated Circuits in RF Systems, April 2003, pp. 126–129.

17

Overview: Heterostructure FETs

John D. Cressler
Georgia Institute of Technology

Ironically, despite the fact that SiGe HBTs at present dominate the commercial silicon heterostructure world, the first Si-based heterostructure field effect transistor was demonstrated in 1986, predating the first SiGe HBT by over 1 year. These earliest FETs were Schottky-gated, III–V-like n- and p-channel modulation doped devices, which rapidly gave rise to a variety SiGe-based MOSFET topologies. More recently, the field has centered on strained Si MOSFETs, because of its better compatibility with mainstream CMOS, and the impressive mobility enhancements that can be realized in that system at aggressively scaled gate lengths. Two fundamentally different ways of producing strained Si CMOS exist, utilizing both biaxial and uniaxial strain techniques. In Chapter 18, K. Rim of IBM Research discusses "Biaxial Strained Si CMOS," while in Chapter 19, by S. Thompson of the University of Florida gives an overview of "Uniaxial Strained Si CMOS." More conventional SiGe-channel FETs, of various flavors, are presented in Chapter 20, "SiGe-Channel HFETs," by S. Banerjee of the University of Texas at Austin. Finally, in Chapter 21, "Industry Examples at the State-of-the-Art: Intel's 90 nm Logic Technologies," by S. Thompson of the University of Florida, an overview of the world's first commercially available strained Si CMOS technology is presented. In addition to this substantial collection of material, and the numerous references contained in each chapter, a number of review articles and books detailing the operation and modeling of SiGe and strained Si FETs exist, including Refs. [1–8].

References

1. R People. Physics and applications of Ge_xSi_{1-x}/Si strained layer heterostructures. *IEEE Journal of Quantum Electronics* 22:1696–1710, 1986.
2. B Meyerson. UHV/CVD growth of Si and SiGe alloys: Chemistry, physics, and device applications. *Proceedings of the IEEE* 80:1592–1608, 1992.
3. U König and H Daembkes. SiGe HBTs and HFETs. *Solid-State Electronics* 38:1595–1602, 1995.
4. CK Maiti, LK Bera, and S Chattopadhyay. Strained-Si heterostructure field-effect transistors. *Semiconductor Science and Technology* 13:1225–1246, 1998.
5. U König, M Gluck, and G Hock. Si/SiGe field-effect transistors. *Journal of Vacuum Science and Technology B* 16:2609–1614, 1998.
6. CK Maiti and GA Armstrong. *Applications of Silicon–Germanium Heterostructure Devices*. London: Institute of Physics Publishing, 2001.

7. CK Maiti, NB Chakrabarti, and SK Ray. *Strained Silicon Heterostructures: Materials and Devices.* London: The Institution of Electrical Engineers, 2001.

8. JL Hoyt, HM Nayfeh, S Eguchi, I Aber, G Xia, T Drake, EA Fitzgerald, and DA Antoniadis. Strained silicon MOSFET technology. Technical Digest of the IEEE International Electron Devices Meeting, Washington, 2002, pp. 23–26.

18

Biaxial Strained Si CMOS

Kern (Ken) Rim

IBM Thomas J. Watson Research Center

18.1 Introduction

Scaling of Si CMOS devices has fueled the exponential growth in the electronics industry. Along with continued increase in density of integration (e.g., number of devices per area), CMOS scaling has also enabled circuit speed enhancements at the rate of 1.2 times per year or higher.

CMOS logic circuit speed is determined by the current drive of MOSFETs and the load capacitance, and is often described by a simple expression for circuit delay, $\tau = C_L(V/I)$, where C_L is the load capacitance, V the voltage swing, and I the MOSFET current drive. The load capacitance has been reduced by miniaturization of device dimension and innovations such as silicon-on-insulator (SOI), while the intrinsic current drive capability of a MOSFET is determined by the channel carrier density and carrier velocity. As the channel length has scaled to the deep submicron regime, impact of channel length scaling on the carrier velocity and current drive has diminished, and the industry has heavily relied on the increase of gate capacitance, which is achieved through aggressive reduction of gate oxide thickness, in order to maximize the channel carrier density and MOSFET current drive. However, gate oxide scaling has recently neared the physical limit where the direct tunneling leakage current is now a major component of the total leakage in a transistor.

The current drive of MOSFET can also be enhanced by modifying the carrier transport properties of silicon. Current drive enhancement obtained by such material property change is in addition to that induced by the geometric scaling of CMOS, and can be combined with the advantage obtained by continued scaling of CMOS. For applications where power consumption is a concern, the current drive increase can be traded off to reduce the standby leakage current by allowing higher threshold voltage V_T while maintaining equivalent current drive. Alternatively, the enhancement can be traded off to control active power consumption by enabling lower supply voltage V_{DD} [1].

Strain is an effective mechanism to modify the carrier transport properties of silicon. Strained Si/SiGe MOSFETs take advantage of strain-induced changes of carrier transport in silicon and obtain current drive enhancements [2]. Figure 18.1a illustrates the structure of MOSFETs fabricated on strained Si on relaxed SiGe. When a thin layer of Si is pseudomorphically grown on a thick, relaxed SiGe layer (Figure 18.1b), the lattice mismatch leads to biaxial tensile strain in the Si layer. If the SiGe layer is fully relaxed and the Si layer fully strained (i.e., the in-plane lattice constants of Si conform to those of the underlying relaxed SiGe layer), the amount of strain in Si is approximately $(4.2x)\%$ where x is the Ge mole fraction

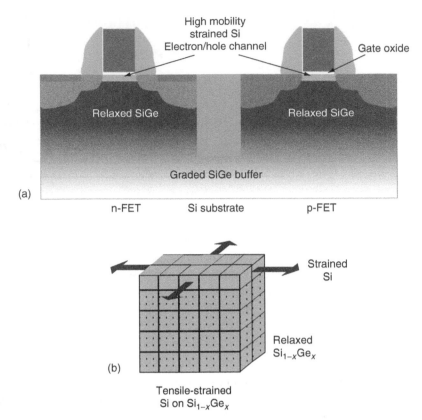

FIGURE 18.1 Typical structures of (a) strained Si–relaxed SiGe bulk MOSFETs and (b) strained Si–relaxed SiGe heterolayers.

in the SiGe layer. So, for instance, a pseudomorphic Si layer grown on fully relaxed SiGe with 25% [Ge] would be under ~1% biaxial tensile strain.

In a strained Si MOSFET, a surface channel is formed at the oxide–Si interface. This is in contrast to a buried channel structure formed in a SiGe/Si/SiGe FET [3,4], or a SiGe surface channel in a SiGe MOSFET where gate oxide is formed on SiGe [5]. Compared to the buried channel or the SiGe surface channel alternatives, strained Si *surface channel* MOSFETs have the following advantages: a single epi layer can potentially enhance both electron and hole mobilities, the surface channel structure leads to better scaling behavior in deep submicron channel lengths, and advanced gate oxides can be thermally grown on pure Si as opposed to on SiGe. Obtaining high-quality oxide interface through thermal oxidation of SiGe is difficult. As can be seen in Figure 18.1, the essential structure and operation principle of a strained Si/SiGe MOSFET is identical to typical Si MOSFETs, except that the inversion channel is formed in the thin Si layer under biaxial tension, and much of the MOSFET body and source–drain junctions are located within the underlying SiGe layer.

18.2 Mobility Characteristics

Theoretical calculations [6–11] have predicted electron and hole mobility enhancements in strained Si. In the conduction band of silicon (Figure 18.2a), biaxial tensile strain splits the six-fold degeneracy in the Δ-valleys, and lowers the two-fold degenerate perpendicular Δ-valleys with respect to the four-fold in-plane Δ-valleys in energy space. Such energy splitting suppresses intervalley carrier scattering between the two-fold and four-fold degenerate valleys, and causes preferential occupation of the two-fold valleys

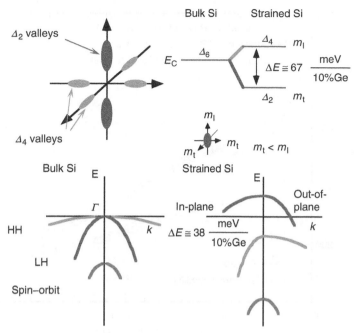

FIGURE 18.2 In-plane biaxial tension-induced changes in (a) conduction and (b) valence bands in strained Si.

where the in-plane conduction mass is lower. These two effects combine and lead to increased electron mobility in strained Si. Similarly, strain splits the valence band degeneracy (Figure 18.2(b)) between the heavy and light hole bands (HH and LH) at the Γ-point and shifts the spin–orbit band. The resulting band deformation effectively lowers the in-plane conduction mass, and the splitting suppresses inter-band scattering between the two bands, improving the in-plane hole mobility.

Observation of high electron mobility in two-dimensional electron gas formed in strained Si was reported in late 1980s and early 1990s [12,13]. Application of strained Si in MOSFET channel was proposed by Keyes as early as in 1986 [14], and the first demonstration of strained Si MOSFET was reported by Welser et al. using relaxed SiGe graded buffer [15,16].

Electron mobility measured in NFET is plotted against vertical effective field in Figure 18.3 [17]. Effective inversion mobility is extracted using *I–V* and split *C–V* measurements on long-channel MOSFETs, and effective field is calculated by counting depletion and inversion charges as described in Ref. [18]. Amount of strain increases with the Ge mole fraction in the relaxed SiGe layer increases, and strained Si NFET mobility is enhanced by as much as 110% over the mobility of the unstrained control device. Even with the modest amount of strain ($\sim$0.5% strain, 13% [Ge]), over 50% electron mobility enhancement is observed.

Figure 18.4 [17] shows the typical PFET mobility characteristics as a function of tensile strain. Similar PFET mobility behaviors have been reported by various publications [19–21]. Compared to electron mobility, the amount of hole mobility enhancement is modest, and shows a strong vertical effective field dependence. Figure 18.5 compares the strain dependence of electron and hole mobilities in strained Si MOSFETs. With $\sim$1% strain, NFET mobility is enhanced by more than two times. On the other hand, hole mobility is slightly degraded with small strain before it begins to increase at larger amounts of strain.

Both electron and hole mobility characteristics exhibit interesting deviations from the theoretical predictions. For electrons in NFETs, quantum mechanical confinements in the inversion layer result in Δ_2–Δ_4 energy splitting even in unstrained Si devices. Strong strain dependence of electron mobility suggests that the intervalley phonon scattering has a strong influence on the mobility of electron inversion layer [7], and the additional energy splitting caused by strain further reduces the intervalley

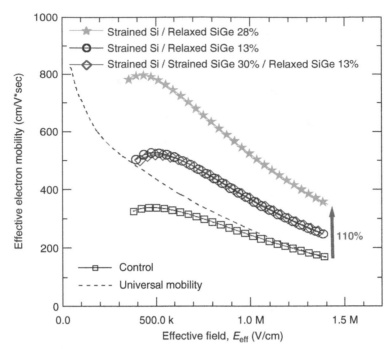

FIGURE 18.3 Effective electron inversion mobility versus vertical effective field as a function of strain in strained Si NMOSFETs. (From K Rim, J Chu, H Chen, KA Jenkins, T Kanarsky, K Lee, A Mocuta, H Zhu, R Roy, J Newbury, J Ott, K Petrarca, P Mooney, D Lacey, S Koester, K Chan, D Boyd, M Ieong, and H-S Wong. Digest of Symposium on VLSI Technology, Honolulu, HI, 2002, pp. 98–99. With permission.)

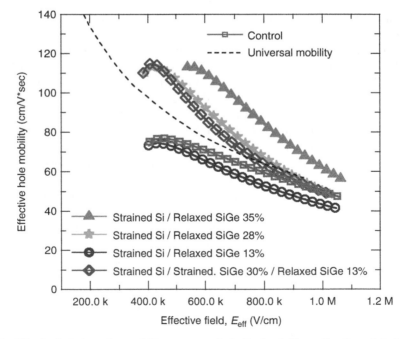

FIGURE 18.4 Effective hole inversion mobility versus vertical effective field as a function of strain in strained Si NMOSFETs. (From K Rim, J Chu, H Chen, KA Jenkins, T Kanarsky, K Lee, A Mocuta, H Zhu, R Roy, J Newbury, J Ott, K Petrarca, P Mooney, D Lacey, S Koester, K Chan, D Boyd, M Ieong, and H-S Wong. Digest of Symposium on VLSI Technology, Honolulu, HI, 2002, pp. 98–99. With permission.)

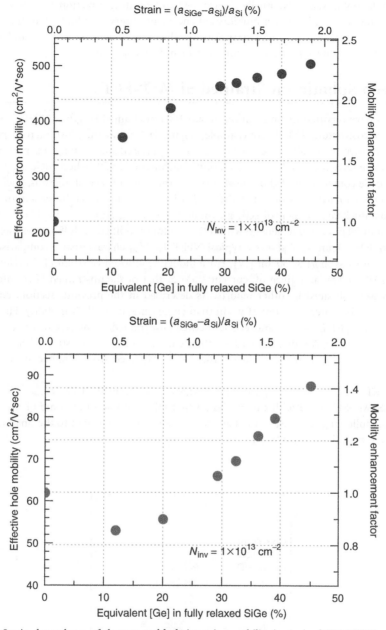

FIGURE 18.5 Strain-dependence of electron and hole inversion mobility in strained SiMOSFETs.

scattering. On the other hand, surface scattering, which dominates MOSFET mobility at high vertical fields, is expected to be similar in strained and unstrained Si devices, and so the observed strong electron mobility enhancement even at high vertical field is an unexpected trend in this regard.

Strain-dependence of hole mobility can also be described by the interplay of energy splittings induced by strain and confinement effects. Quantum-mechanical confinement in the hole inversion layer splits the HH–LH band degeneracy, but in the opposite direction compared to the splitting induced by tensile strain, moving the HH-associated band toward the midgap. As tensile strain increases in a strained Si PFET, strain-induced changes in LH and HH bands first compensate the confinement-induced splitting and degrade hole mobility, and eventually reverse the band-splitting direction and increase the amount of splitting and mobility [22]. Smaller mobility enhancement observed in high vertical fields can also be explained by this mechanism—at high vertical fields, the confinement is stronger, and strain-induced splitting needs to overcome larger initial confinement-induced splitting.

18.3 Deep Submicron Strained Si MOSFETs

Various groups have reported demonstration of sub-100 nm channel length MOSFETs on strained Si–relaxed SiGe heterostructures [23–27, for example]. Figure 18.6 shows the TEM micrograph of a strained Si MOSFET [17] with cobalt silicide formed on selective epi raised source drain. In most of the reported cases, relaxed SiGe layer is created by a graded buffer technique [13,28] where SiGe is epitaxially grown on Si wafer and Ge content is graded up slowly over the thickness of several microns. Such graded buffer structure is known to reduce dislocation defect density in the top layer by orders of magnitude in comparison to a thick layer of SiGe directly grown on Si without grading. Strained Si channel layer with thickness of 10 to 20 nm is epitaxially grown on SiGe before modified CMOS process steps are used to fabricate MOSFETs. Figure 18.7 shows a typical NFET $I_{on}-I_{off}$ characteristics comparison for devices with channel lengths as short as 50 nm [17]. For a moderate amount of strain (13% [Ge]), NFET I_{on} is enhanced by 15% to 20% at a given off current I_{off} over that of the unstrained Si counterpart. Similar characteristics were obtained by other reports. As described in the previous section, enhancement of PFET mobility requires larger amount of strain than enhancement of NFET mobility. Figure 18.8 shows the comparison of PFET $I_{on}-I_{off}$ characteristics between unstrained Si and strained Si devices with 28% [Ge] (i.e., ~1.2% strain). About 7% to 10% enhancement is observed in strained Si PFETs. At lower strain, minimal or no performance gain is observed as expected from strained dependence of hole mobility.

In both NFET and PFETs, the electrostatic design of the strained Si device is very similar to the conventional Si MOSFETs. So with careful design that takes into account some of the issues that are described in the following sections, short channel effects that are equivalent to the unstrained Si devices can be obtained in strained Si MOSFETs.

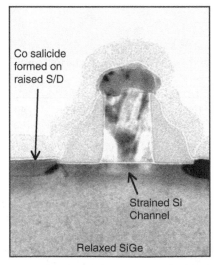

FIGURE 18.6 TEM micrograph of strained Si MOSFET with selective epi raised source–drain structure. (From K Rim, J Chu, H Chen, KA Jenkins, T Kanarsky, K Lee, A Mocuta, H Zhu, R Roy, J Newbury, J Ott, K Petrarca, P Mooney, D Lacey, S Koester, K Chan, D Boyd, M Ieong, and H-S Wong. Digest of Symposium on VLSI Technology, Honolulu, HI, 2002, pp. 98–99. With permission.)

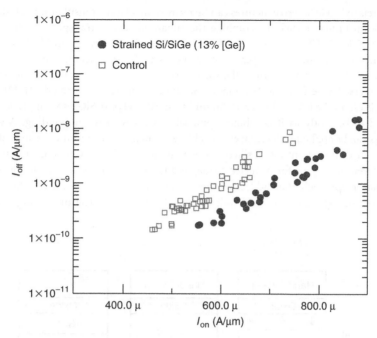

FIGURE 18.7 I_{on}–I_{off} comparison between strained Si and unstrained Si NFETs. (From K Rim, J Chu, H Chen, KA Jenkins, T Kanarsky, K Lee, A Mocuta, H Zhu, R Roy, J Newbury, J Ott, K Petrarca, P Mooney, D Lacey, S Koester, K Chan, D Boyd, M Ieong, and H-S Wong. Digest of Symposium on VLSI Technology, Honolulu, HI, 2002, pp. 98–99. With permission.)

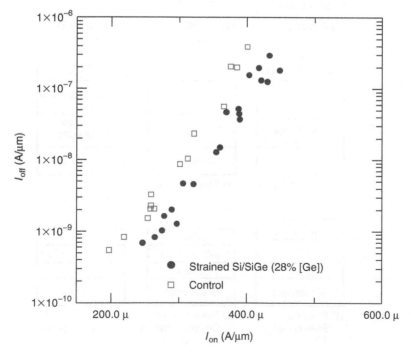

FIGURE 18.8 I_{on}–I_{off} comparison between strained Si and unstrained Si PFETs. (From K Rim, J Chu, H Chen, KA Jenkins, T Kanarsky, K Lee, A Mocuta, H Zhu, R Roy, J Newbury, J Ott, K Petrarca, P Mooney, D Lacey, S Koester, K Chan, D Boyd, M Ieong, and H-S Wong. Digest of Symposium on VLSI Technology, Honolulu, HI, 2002, pp. 98–99. With permission.)

Strained Si-relaxed SiGe heterostructures can be formed on a layer of insulator. Such structures, which are called SiGe-on-insulator (SGOI), combine the enhanced carrier transport in strained Si with the advantages of SOI MOSFETs. One major difference between SGOI devices and bulk strained Si devices on graded buffer is that the SiGe layer in SGOI needs to be significantly thinner (a few thousand angstroms or less). Figure 18.9 describes the various techniques to create SGOI structures. By transferring a layer of relaxed SiGe from buffer, bonded SGOI (BSGOI) can be formed [21,29]. This approach utilizes the well-established SiGe buffer technique to create relaxed SiGe to supply layers to transfer, but is expected to be costly as it combines epitaxial process with wafer bonding. A variation of the SIMOX process can be applied to SiGe layers to achieve a potentially economic method of forming SGOI [20], but Ge out-diffusion into the silicon substrate can be difficult to control during the high-temperature oxidation step to form buried oxide, and the quality of buried oxide formed in SiGe is a concern. A bilayer approach that addresses these concerns has been proposed [30]. Finally, the Ge condensation or mixing technique [31] utilizes Ge diffusion and rejection during high-temperature

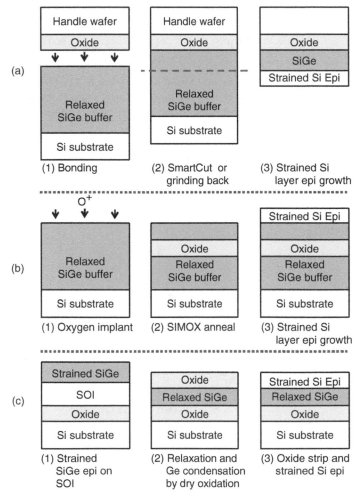

FIGURE 18.9 Various techniques to fabricate SiGe-on-insulator (SGOI). (From K Rim, R Anderson, D Boyd, F Cardone, K Chan, H Chen, S Christansen, J Chu, K Jenkins, T Kanarsky, S Koester, B Lee, K Lee, V Mazzeo, A Mocuta, D Mocuta, PM Mooney, P Oldiges, J Ott, P Ronsheim, R Roy, A Steegen, M Yang, H Zhu, M Ieong, and H-SP Wong. *Solid-State Electronics* 47:1133–1139, 2003. With permission.)

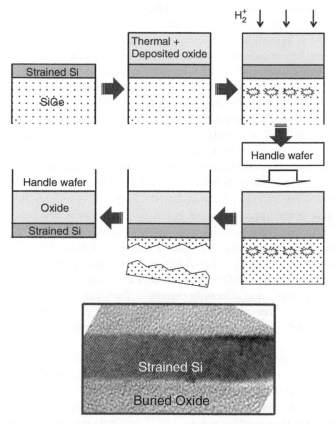

FIGURE 18.10 Fabrication of strained-Si directly on insulator (SSDOI) by layer transfer and selective etch of SiGe. (From K Rim, K Chan, L Shi, D Boyd, J Ott, N Klymko, F Cardone, L Tai, S Koester, M Cobb, D Canaperi, B To, E Duch, I Babich, R Carruthers, P Saunders, G Walker, Y Zhang, M Steen, and M Ieong. Technical Digest — IEEE International Electron Devices Meeting, Washington, DC, 2003, pp. 49–52. With permission.)

oxidation to form ultrathin SGOI. This is a very promising approach, but the mechanism that governs the relaxation of SiGe lattice is not well understood at this point.

Sub-100 nm MOSFETs have been demonstrated on SGOI substrates fabricated using thermal mixing technique [32], which is similar to the Ge condensation technique. Excellent MOSFET characteristics and NFET current drive enhancement have been reported.

The extreme case of ultrathin SGOI is strained Si directly on insulator (SSDOI) [29,33,34]. Tensile strained Si layer can be directly formed on buried oxide by a combination of layer transfer and selective etch removal of relaxed SiGe (Figure 18.10). Retention of strain and electron and hole mobility enhancements have been demonstrated on SSDOI MOSFETs [34].

18.4 Device Physics and Design Issues

In order to realize the strain-enhanced CMOS performance to the fullest extent in ULSI applications, both fundamental and technological challenges need to be addressed. As discussed above, biaxial tension-induced hole mobility enhancement requires large amounts of strain (i.e., [Ge]), and the amount of enhancement diminishes at high vertical field, consistent with the theoretical calculation in Ref. [10]. High strain and high Ge content increase the difficulty of various process integration issues, and due to the small hole mobility enhancement at high vertical fields, strained Si is expected to provide only limited enhancement in the future bulk and PDSOI PFETs with high channel doping and vertical field. On the other hand, for device structures such as symmetric double gate or FDSOI pMOSFETs in which the vertical field is inherently low during operation, strained Si can provide a significant hole mobility enhancement. Recent experimental results [35] suggest that a significant hole mobility enhancement at high vertical field might be possible with an optimized epi layer preparation technique, but further investigation is needed for verification. The electron mobility of strained Si FET, on the other hand, exhibits unexpectedly large enhancements even at high vertical fields as described in the previous section. Presently, few theoretical explanations have been proposed to account for this observation.

Although low field mobility is a property that can be measured and compared with relatively little ambiguity and is an essential parameter for device and circuit modeling, it is inadequate in quantitatively predicting the actual saturation current drive of the deep submicron devices. Velocity saturation effect strongly influences device characteristics in short channel lengths, and the observed impact by the changes in low field mobility is limited. Modern day NFETs operate at roughly 50% of the ballistic limit, and so only about half of the impact observed in low field mobility is observed in saturation current [36]. In addition, accurately extracting the intrinsic enhancement in device performance is difficult because short-channel FET characteristics are sensitive functions of many device design parameters such as the junction- and channel-doping profiles as well as the extrinsic resistance.

Even so, strain-induced energy splitting may influence the transport of "warm" carriers in addition to the low-field mobility, improving saturation current in short-channel devices [1]. Smaller density of states in strained Si should contribute to the reduction of scattering rates even at carrier energies several times larger than kT, contributing to enhanced nonequilibrium transport and deep submicron NMOSFET current drive.

In order to maximize the impact of strain-induced enhancement of intrinsic device performance, device scaling and design have to be achieved without compromise to the optimization of the parasitics in the extrinsic parts of device. This point is illustrated with a simple estimation of expected enhancements in the device on-resistance R_{ON} (which is inversely proportional to the drain current in the linear region) and the saturation transconductance $g_{m,sat}$ [37]. For instance, assuming $R_{ON} = 300\,\Omega\,\mu m$ for the control device and parasitic resistance $R_{ext} = 200\,\Omega\,\mu m$ for both the control and SS devices, a 70% increase in mobility will be diluted to provide only 16% improvement in R_{ON}. Similarly, assuming $g_{m,sat} = 1000\,\mu S/\mu m$ for the control device, R_{ext} of $200\,\Omega\,\mu m$ (equally divided between source and drain) dilutes a 30% enhancement in intrinsic g_m to 26% even in a simplified (neglecting the impact of a finite drain conductance in saturation) and optimistic estimation.

Due to the low thermal conductivity of the thick SiGe layer, the SS MOSFETs exhibit significant self-heating [38], analogous that observed in SOI MOSFETs. In typical CMOS digital logic circuits, device duty cycle is expected to be much shorter than the thermal time constant of self-heating. However, in analog applications, self-heating will in effect induce a significant rise in device operation temperature, affecting the performance. Techniques to reduce the SiGe layer thickness in the structure can improve the self-heating characteristics.

Band offsets at the Si/SiGe interface that lower the conduction band edge in strained Si, along with the narrower energy gap in SiGe, lower V_T of strained Si NFETs. V_T lowering can be close to 100 mV in strained Si NFETs on SiGe with 20% [Ge]. Additional well and halo doping can offset the V_T lowering, but the extra doping can diminish the mobility gain [39].

18.5 Material and Integration Issues

The foremost critical challenge in the SS CMOS technology is the control of dislocation defects in the epitaxial layers. A finite density of misfit and threading dislocations are present in the SiGe buffers grown by the graded buffer growth techniques. Such dislocations can cause increase in junction leakage and device OFF current I_{off} (as in Refs. [25,26]). For ULSI implementation, innovations and optimizations are required to minimize the propagation of dislocations in strained Si-relaxed SiGe structures.

Thermal processing during CMOS fabrication steps can cause relaxation of the strain in the Si layer, or out-diffusion of Ge. In thermal stability experiments [40], where Raman spectroscopy was used to measure the changes in strained Si-relaxed SiGe structure before and after annealing steps, the position of the strained Si peak did not shift distinctly, while the signal strength decreased with increasing amounts of thermal annealing. This indicated that while thermal annealing at 1000°C does not cause measurable strain relaxation, it effectively reduces the Si thickness by Ge out-diffusion. The thermal budget during device fabrication has to be optimized carefully to avoid strain relaxation and Ge out-diffusion into the channel layer. Optimal thickness of strained Si channel for a given amount of strain needs to be thick enough to allow such process window against Si consumption and Ge out-diffusion, and thin enough to prevent random nucleation and propagation of dislocation defects. In addition, geometric effects and interaction with process-related stress have to be understood and controlled. Understanding in this area is very limited at this time.

Doping diffusion in SiGe affects device design. While the well-known suppression of boron diffusion in SiGe is beneficial to the formation of abrupt junction in strained Si PFETs, arsenic diffusion in SiGe is significantly enhanced in comparison to that in Si [23,41]. In device design to control short-channel effects, such difference must be taken into account for the proper design of junction depth, overlap capacitance, and V_T. SiGe also interacts with silicide formation. For the cobalt salicide process, Ge hinders transition to the low-resistivity disilicide phase. When a typical cobalt silicide process is used on SiGe, a rough cobalt germano-silicide film can result with ten times higher sheet resistance. Alternative material or integration schemes, such as raised source drain as in Figure 18.3 [17], are required to achieve acceptable salicide properties.

18.6 Summary

Strained Si MOSFET is a case example of how Si–SiGe heterostructure can make direct and significant impact on today's Si device technology, following the success of SiGe HBTs. Significant challenges remain before the realization of a manufacturable strained Si CMOS technology, but the potential of geometric scaling-independent performance enhancement provides a strong motivation. One interesting application is a combination of strained Si and high-k dielectrics. Strain can be used to recover the mobility degraded in devices with high-k dielectrics [42,43], combining the advantage of strained Si with gate leakage reduction in high-k gate dielectrics. As CMOS scaling continues on, such material innovations that can complement geometric scaling and enable performance boost will play an increasingly important role in Si technology.

References

1. K Rim, JL Hoyt, and JF Gibbons. Fabrication and analysis of deep submicron strained Si n MOSFETs. *IEEE Transactions on Electron Devices* 47: 1406–1413, 1998.
2. JL Hoyt, HM Nayfeh, S Eguchi, I Aberg, G Xia, T Drake, EA Fitzgerald, and DA Antoniadis. Strained Si MOSFET technology. Technical Digest IEEE International Electron Devices Meeting, San Francisco, CA, 2002, pp. 23–26.
3. K Ismail. Si/SiGe high speed field effect transistors. Technical Digest IEEE International Electron Devices Meeting, Washington, DC, 1995, pp. 509–512.
4. VP Kesan, S Subbana, PJ Restle, MJ Tejwani, JM Aitken, SS Iyer, and JA Ott. High performance 0.25 μm p-MOSFETs with silicon–germanium channels for 300 K and 77 K operation. Technical Digest IEEE International Electron Devices Meeting, Washington, DC, 1991, pp. 977–980.
5. T Tezuka, N Sugiyama, T Mizuno, and S Takagi. Novel fully-depleted SiGe-on-insulator pMOSFETs with high-mobility SiGe surface channels. Technical Digest IEEE International Electron Devices Meeting, Washington, DC, 1991, pp. 946–948.
6. MV Fischetti and SE Laux. Band structure, deformation potentials, and carrier mobility in strained Si, Ge, and SiGe alloys. *Journal of Applied Physics* 80:2234–2252, 1994.

7. S Takagi, JL Hoyt, JJ Welser, and JF Gibbons. Comparative study of phonon limited mobility of two dimensional electrons in strained and unstrained Si metal oxide semiconductor field effect transistors. *Journal of Applied Physics* 80:1567–1577, 1994.

8. JB Roldan, F Gamiz, JA Lopez-Villanueva, and JE Carceller. A Monte Carlo study on the electron transport properties of high performance strained Si on relaxed SiGe channel MOSFETs. *Journal of Applied Physics* 80:5121–5128, 1994.

9. GF Formicone, D Vasileska, and DK Ferry. Transport in the surface channel of strained Si on a relaxed $Si_{1-x}Ge_x$ substrate. *Solid State Electronics* 41:879–885, 1997.

10. R Oberhuber, G Zandler, and P Vogl. Subband structure and mobility of two dimensional holes in strained Si/SiGe MOSFET's. *Physical Review B* 58:9941–9948, 1998.

11. D Nayak, JCW Woo, JS Park, L Wang, and KP MacWilliams. High-mobility p-channel metal-oxide-semiconductor field-effect transistor on strained Si. *Applied Physics Letters* 62:2853–2855, 1993.

12. SF Nelson, K Ismail, JO Chu, and BS Meyerson. Room-temperature electron mobility in strained Si/SiGe heterostructures. *Applied Physics Letters* 18:435–437, 1997.

13. EA Fitzgerald, YH Xie, D Monroe, PJ Silverman, JM Kuo, AR Kortan, FA Thiel, and BE Weir. Relaxed GeSi structures for III–V integration with Si and high mobility two-dimensional electron gases in Si. *Journal of Vacuum Science & Technology B* 10:1807, 1992.

14. RW Keyes. High-mobility FET in strained silicon. *IEEE Transactions on Electron Devices* 33:863, 1986.

15. J Welser, JL Hoyt, and JF Gibbons. NMOS and PMOS transistors fabricated in strained silicon/relaxed silicon–germanium structures. Technical Digest IEEE International Electron Devices Meeting, San Francisco, CA, 1992, pp. 1000–1002.

16. J Welser, JL Hoyt, S Takagi, and JF Gibbons. Strain dependence of the performance enhancement in strained-Si n-MOSFETs. Technical Digest IEEE International Electron Devices Meeting, San Francisco, CA, 1994, pp. 947–950.

17. K Rim, J Chu, H Chen, KA Jenkins, T Kanarsky, K Lee, A Mocuta, H Zhu, R Roy, J Newbury, J Ott, K Petrarca, P Mooney, D Lacey, S Koester, K Chan, D Boyd, M Ieong, and H-S Wong. Characteristics and device design of sub-100 nm strained Si N- and pMOSFETs. Digest of Symposium on VLSI Technology, Honolulu, HI, 2002, pp. 98–99.

18. S Takagi, A Toriumi, M Iwase, and H Tango. On the universality of inversion layer mobility in Si MOSFETs: Part I—Effects of substrate impurity concentration. *IEEE Transactions on Electron Devices* 41:2357–2362, 1994.

19. K Rim, J Welser, JL Hoyt, and JF Gibbons. Enhanced hole mobilities in surface channel strained Si p MOSFETs. Technical Digest IEEE International Electron Devices Meeting, Washington, DC, 1995, pp. 517–520.

20. T Mizuno, S Takagi, N Sugiyama, J Koga, T Tezuka, K Usuda, T Hatakeyama, A Kurobe, and A Toriumi. High performance strained Si p MOSFETs on SiGe on insulator substrates fabricated by SIMOX technology. Technical Digest IEEE International Electron Devices Meeting, Washington, DC, 1999, pp. 934–937.

21. LJ Huang, JO Chu, S Goma, CP D'Emic, SJ Koester, DF Canaperi, PM Mooney, SA Cordes, JL Speidell, RM Anderson, and H-S P Wong. Carrier mobility enhancement in strained Si on insulator fabricated by wafer bonding. Digest Symposium on VLSI Technology, Kyoto, Japan, 2001, pp. 57–58.

22. K Rim, K Chan, L Shi, D Boyd, J Ott, N Klymko, F Cardone, L Tai, S Koester, M Cobb, D Canaperi, B To, E Duch, I Babich, R Carruthers, P Saunders, G Walker, Y Zhang, M Steen, and M Ieong. Fabrication and mobility characteristics of ultra-thin strained Si directly on insulator (SSDOI) MOSFETs. Technical Digest—IEEE International Electron Devices Meeting, Washington, DC, 2003, pp. 49–52.

23. K Rim, JL Hoyt, and JF Gibbons. Transconductance enhancement in deep submicron strained Si nMOSFETs. Technical Digest IEEE International Electron Devices Meeting, San Francisco, CA, 1998, pp. 707–710.

24. K Rim, S Koester, M Hargrove, JO Chu, PM Mooney, J Ott, T Kanarsky, P Ronsheim, M Ieong, A Grill, and H-SP Wong, Strained Si NMOSFETs for high performance CMOS technology. Digest of Symposium on VLSI Technology, Kyoto, Japan, 2001, pp. 59–60.

25. Q Xiang, J-S Goo, J Pan, B Yu, S Ahmed, J Zhang, and M-R Lin. Strained silicon NMOS with nickel–silicide metal gate. Digest of Technical Papers—Symposium on VLSI Technology, Kyoto, Japan, 2003, pp. 101–102.

26. HC-H Wang, Y-P Wang, S-J Chen, C-H Ge, SM Ting, J-Y Kung, R-L Hwang, C-K Chiu, LC Sheu, P-Y Tsai, L-G Yao, S-C Chen, H-J Tao, Y-C Yeo, W-C Lee, and C Hu. Substrate-strained silicon technology: process integration. Technical Digest—IEEE International Electron Devices Meeting, Washington, DC, 2003, pp. 61–64.

27. T Sanuki, A Oishi, Y Morimasa, S Aota, T Kinoshita, R Hasumi, T Takegawa, K Isobe, H Yoshimura, M Iwai, K Sunouchi, and T Noguchi. Scalability of strained silicon CMOSFET and high drive current enhancement in the 40 nm gate length technology. Technical Digest—IEEE International Electron Devices Meeting, Washington, DC, 2003, pp. 65–68.

28. FK LeGoues, BS Meyerson, JF Morar, and PD Kirchner. Mechanism and conditions for anomalous strain relaxation in graded thin films and superlattices. *Journal of Applied Physics* 71:4230–4243, 1992.

29. K Rim, L Shi, K Chan, J Chu, D Boyd, K Jenkins, J Ott, D Lacey, P Mooney, M Cobb, N Klymko, F Jamin, S Koester, and T Kanarsky. Strained Si MOSFETs on bulk and SiGe-on-insulator (SGOI) substrates. Program and Abstracts of First International SiGe Technology and Device Meeting (ISTDM 2003), Nagoya, Japan, 2003, pp. 9–10.

30. T Mizuno, N Sugiyama, T Tezuka, and S Takagi. Novel SOI p-channel MOSFETs with higher strain in Si channel using double SiGe heterostructures *IEEE Transactions on Electron Devices* 49:7–14, 2002.

31. T Tezuka, N Sugiyama, T Mizuno, M Suzuki, and S Takagi. A novel fabrication technique of ultrathin and relaxed SiGe buffer layers with high Ge fraction for sub 100 nm strained silicon on insulator MOSFETs. *Japanese Journal of Applied Physics* 40:2866–2875, 2001.

32. BH Lee, A Mocuta, S Bedell, H Chen, D Sadana, K Rim, P O'Neil, R Mo, K Chan, C Cabral, C Lavoie, D Mocuta, A Chakravarti, RM Mitchell, J Mezzapelle, F Jamin, M Sendelbach, H Kermel, M Gribelyuk, A Domenicucci, KA Jenkins, S Narasimha, SH Ku, M Ieong, IY Yang, E Leobandung, P Agnello, W Haensch, and J Welser. Performance enhancement on sub-70 nm strained silicon SOI MOSFETs on ultra-thin thermally mixed strained silicon/SiGe on insulator (TM-SGOI) substrate with raised S/D. Technical Digest—IEEE International Electron Devices Meeting, Washington, DC, 2002, pp. 946–948.

33. TA Langdo, A Lochtefeld, MT Currie, R Hammond, VK Yang, JA Carlin, CJ Vineis, G Braithwaite, H Badawi, MT Bulsara, and EA Fitzgerald. Preparation of novel SiGe-free strained Si on insulator substrates. IEEE International SOI Conference, 2002, pp. 211–212.

34. TS Drake, N Chleirigh, ML Lee, AJ Pitera, EA Fitzgerald, DA Antoniadis, DH Anjum, JL Hoyt, DH Anjum, J Li, R Rull, N Klymko, and JL Hoyt. Fabrication of ultra-thin strained silicon on insulator. *Journal of Electronic Materials* 32:972–975, 2003.

35. N Sugii, D Hisamoto, K Washio, N Yokoyama, and S Kimura. Enhanced performance of strained-Si MOSFETs on CMP SiGe virtual substrate. Technical Digest—IEEE International Electron Devices Meeting, Washington, DC, 2001, pp. 737–740.

36. DA Antoniadis. MOSFET scalability limits and "new frontier" devices. Digest of Symposium on VLSI Technology, Honolulu, HI, 2002, pp. 2–5.

37. K Rim, R Anderson, D Boyd, F Cardone, K Chan, H Chen, S Christansen, J Chu, K Jenkins, T Kanarsky, S Koester, B Lee, K Lee, V Mazzeo, A Mocuta, D Mocuta, PM Mooney, P Oldiges, J Ott, P Ronsheim, R Roy, A Steegen, M Yang, H Zhu, M Ieong, and H-SP Wong. Strained Si CMOS (SS CMOS) technology: opportunities and challenges. *Solid-State Electronics* 47:1133–1139, 2003.

38. KA Jenkins and K Rim. Measurement of the effect of self-heating in strained-silicon MOSFETs. *IEEE Electron Device Letters* 23:360–362, 2002.

39. HM Nayfeh, C Leitz, A Pitera, EA Fitzgerald, JL Hoyt, and DA Antoniadis. Influence of high channel doping on the inversion layer electron mobility in strained silicon n-MOSFETs. *IEEE Electron Device Letters* 24:248–250, 2003.

40. SJ Koester, K Rim, JO Chu, PM Mooney, JA Ott, and MA Hargrove. Effect of thermal processing on strain relaxation and interdiffusion in Si/SiGe heterostructures studied using Raman spectroscopy. *Applied Physics Letters* 79:2148–2150, 2001.

41. S Eguchi, JL Hoyt, CW Leitz, and EA Fitzgerald. Comparison of arsenic and phosphorus diffusion behavior in silicon-germanium alloys. *Applied Physics Letters* 80:1743, 2002.

42. K Rim, EP Gusev, C D'Emic, T Kanarsky, H Chen, K Chu, J Ott, K Chan, D Boyd, V Mazzeo, BH Lee, A Mocuta, J Welser, SL Cohen, M Ieong, and H-S Wong. Mobility enhancement in strained Si NMOSFETs with HfO_2 gate dielectrics. Digest of Symposium on VLSI Technology, Honolulu, HI, 2002, pp. 12–13.

43. S Datta, G Dewey, M Doczy, BS Doyle, B Jin, J Kavalieros, R Kotlyar, M Metz, N Zelick, and R Chau. High mobility Si/SiGe strained channel MOS transistors with HfO_2/TiN gate stack. Technical Digest—IEEE International Electron Devices Meeting, Washington, DC, 2003, pp. 653–656.

19

Uniaxial Stressed Si MOSFET

Scott E. Thompson
University of Florida

19.1 Introduction

Over the past 40 years [1–7], to improve MOSFET performance, strain introduced via biaxial tensile stress using a Si–SiGe heterostructure has received substantial attention. Little attention, however, has been paid to uniaxial stress created via heterostructures [8]. Both biaxial and uniaxial stress offer large enhanced electron and hole mobility and great potential to continue Moore's law when conventional scaling slows. Since biaxial tensile stress introduces advantageous strain for both n- and p-type MOSFETs, it has potential importance to CMOS logic technologies. However, biaxial stress has not yet been adopted into high-volume manufacturing due to cost and integration complexity. The use of uniaxial stress for CMOS logic is not without its own complexities, as will be described in Chapter 21, but has been adopted at the 90 nm technology generation [8–10] including heteroepitaxy introduced for the first time in commercial CMOS chips.

This chapter summarizes and quantitatively explains some advantageous electrical characteristics for uniaxial (as compared to biaxial) in-plane stressed MOSFETs fabricated on conventional (0 0 1) wafers. These advantages originate because of (i) favorable uniaxial stress-induced valence band warping and (ii) the often unrecognized benefit of straining both the gate and Si channel with process strain versus just the Si channel with wafer substrate strain. Using the Luttinger–Kohn [11,12] and Bir–Pikus [13,14] strain Hamiltonian, and Herring and Vogt deformation potential theory [15], three key advantages observed in the uniaxial stress experimental data are explained: large hole mobility enhancement at low stress and high vertical electric field and a small n-channel threshold voltage shift.

The hole mobility enhancement is experimentally observed to reduce to near zero at high vertical fields for biaxial stress but is maintained for uniaxial stress. Band curvature calculations show this can be explained by the relative magnitude of the out-of-plane light and heavy hole conductivity effective masses. For biaxial stress, smaller light than heavy hole mass reduces the strain-induced band splitting at high vertical field (lattice strain is "canceled" by surface confinement band splitting [16]). For uniaxial strain, the larger out-of-plane light than heavy hole mass increases the band splitting at high vertical fields (strain and surface confinement splitting is "additive"). Second, the technologically important question of why uniaxial stress hole mobility enhancement is present at low strain (and not present for biaxial) is shown to result from more favorable band warping. In-plane effective mass calculations show that the light hole mass is 50% smaller for uniaxial than biaxial stress. Finally, for biaxial stress, the undesirably large n-channel threshold voltage shift [2,17,18] is shown to be due to two effects. The first effect is differences in the gate and Si channel electron affinity. The difference arises since uniaxial process strain is present in both the n$^+$ poly-Si gate and Si channel while biaxial stress introduced via the substrate only strains the Si channel. The second effect is strain-induced bandgap narrowing [19], which is larger for biaxial than uniaxial tensile stress.

This chapter is organized as follows: Section 19.2 briefly covers how uniaxial stress is created in the CMOS transistor structure. Section 21 quantifies the key electrical differences for uniaxial versus biaxial stress. Chapter 21 describes the industry examples for uniaxial stressed Si in nanoscale CMOS logic technologies.

19.2   Uniaxial Strained Silicon

There are several ways to introduce uniaxial stress into the Si lattice: a tensile or compressive capping layer over the device [20,21], high-stress shallow-trench isolation fill [22], or heteroepitaxy in the source and drain of nanoscale transistors [1,6–10,23]. Heteroepitaxy offers the greatest potential due to the ability to create large strain. A schematic diagram of heteroepitaxy is shown in Figure 19.1 for the recently proposed uniaxial stress (a) and conventional biaxial stress (b) [8–10] device structures. For the case shown in Figure 19.1a, $Si_{1-x}Ge_x$ in the source–drain creates uniaxial compressive stress in the silicon channel in the direction of current flow. Alternately, $Si_{1-y}C_y$ in the source and drain would create tensile stress in the silicon channel. The mismatch strain is given by

$$\varepsilon_{mismatch} = \frac{a_{sub} - a_{film}}{a_{film}} \qquad (19.1)$$

where a_{sub} and a_{film} are the lattice constants of the substrate and heteroepitaxy film, respectively the stress in the layer is given by

$$\sigma_o = -2\gamma \frac{\nu + 1}{\nu - 1} \varepsilon_{mismatch} \qquad (19.2)$$

where γ is the shear modulus of elasticity and ν is Poisson's ratio.

A key technology question is how much strain can be introduced into the silicon channel by heteroepitaxy in the source and drain. Direct physical measurement of the side-induced uniaxial channel stress is difficult since the strain depends on channel length and is only present in the small nanoscale devices. At present, three-dimensional finite element simulations are the best-known method for

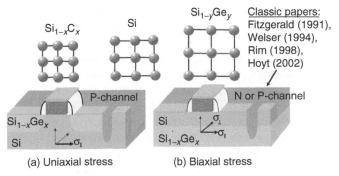

FIGURE 19.1   Device structures for uniaxial and biaxial stress.

quantifying the channel stress in small devices. Using FLOOPs [24,25] three-dimensional finite element analysis, for $Si_{0.83}Ge_{0.17}$ on a Si substrate [8–10], 1.4 GPa of compressive stress is introduced into the $Si_{0.83}Ge_{0.17}$ layer. This creates ~500 MPa of uniaxial compressive stress in the 45-nm gate length MOSFET channel [26]. A TEM micrograph of a 45-nm p-channel MOSFET is shown in Figure 19.2, which will be described in Chapter 21 [10].

19.3 Electrical Properties: Uniaxial versus Biaxial Stressed Silicon

To understand the electrical differences between uniaxial and biaxial stressed Si MOSFETs, it is required to formulate how strain changes the position and shape of the energy bands. The classic papers on this subject can be found in the references [11–15,27–31]. The topic of strain on electronic states is broad so this discussion will be restricted to the technologically important biaxial and uniaxial stressed Si MOSFETs on (0 0 1) wafers and ⟨1 1 0⟩ channel orientation (the dominant orientation used by the microelectronics industry). Figure 19.3 lists the key properties for one and two axis stresses. As a note, stress and strain are often confused. Stress is the force per unit area. Strain is the resulting deformation of the lattice. For uniaxial tensile stress, strain is present in all three directions (see Figure 19.3), with the sideways contraction (direction perpendicular to the stress) given by Poisson's ratio (ν). For a good review of these concepts, see Ref. [32]. Due to space considerations, derivations for most statements cannot be given but are cited in references.

The change in electronic states due to strain starts with a strain Hamiltonian proposed by Luttinger–Kohn [11,12] and Bir–Pikus [13,14]. To evaluate the strain Hamiltonian, the strain tensor is equivalently decomposed into three matrices ($\varepsilon_{ij} = \varepsilon_{ij}^{\text{trace}} + \varepsilon_{ij}^{\text{shear}-100} + \varepsilon_{ij}^{\text{shear}-111}$): a diagonal hydrostatic strain matrix with trace $\text{Tr}(e) = \varepsilon_x + \varepsilon_y + \varepsilon_z$ which is identical to the fractional change of the volume, $\delta V/V = \varepsilon_x + \varepsilon_y + \varepsilon_z$, and two traceless matrices that represent shear strain created by stress along ⟨1 0 0⟩ and ⟨1 1 1⟩, respectively:

$$\varepsilon_{ij}^{\text{trace}} = \frac{1}{3}\begin{bmatrix} \varepsilon_{xx} + \varepsilon_{yy} + \varepsilon_{zz} & 0 & 0 \\ 0 & \varepsilon_{xx} + \varepsilon_{yy} + \varepsilon_{zz} & 0 \\ 0 & 0 & \varepsilon_{xx} + \varepsilon_{yy} + \varepsilon_{zz} \end{bmatrix} \quad (19.3)$$

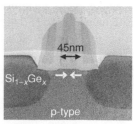

45nm

$Si_{1-x}Ge_x$ →←

p-type

Thompson, Elec. Dev. Lett. 1994.

FIGURE 19.2 TEM micrograph of 45 nm transistors with uniaxial stress.

Type of Stress	Principal Stresses	Principal Strain	$\Delta V/V$ (Silicon)
Uniaxial	$\sigma_1 = -E\varepsilon_1$ $\sigma_2 = 0$ $\sigma_3 = 0$	$\varepsilon_1 = \dfrac{\sigma_1}{E}$ $\varepsilon_2 = -\nu\varepsilon_1$ $\varepsilon_3 = -\nu\varepsilon_1$	$\varepsilon = \varepsilon_1$ $(1-2\nu)\varepsilon_1$ $0.44\varepsilon_1$
Biaxial	$\sigma_1 = \dfrac{E(\varepsilon_1 + \nu\varepsilon_2)}{1-\nu^2}$ $\sigma_2 = \dfrac{E(\varepsilon_2 + \nu\varepsilon_1)}{1-\nu^2}$ $\sigma_3 = 0$	$\varepsilon_1 = \dfrac{\sigma_1 - \nu\sigma_2}{E}$ $\varepsilon_2 = \dfrac{\sigma_2 - \nu\sigma_1}{E}$ $\varepsilon_3 = \dfrac{-\nu\sigma_1 - \nu\sigma_2}{E}$	$\varepsilon = \varepsilon_1 = \varepsilon_2$ $\dfrac{2(1-2\nu)}{1-\nu}\varepsilon$ 1.22ε

FIGURE 19.3 Relationships for uniaxial and biaxial stress.

$$\varepsilon_{ij}^{\text{shear}-100} + \varepsilon_{ij}^{\text{shear}-111} = \frac{1}{3}\begin{bmatrix} 2\varepsilon_{xx} - (\varepsilon_{yy} + \varepsilon_{zz}) & 0 & 0 \\ 0 & 2\varepsilon_{yy} - (\varepsilon_{zz} + \varepsilon_{xx}) & 0 \\ 0 & 0 & 2\varepsilon_{zz} - (\varepsilon_{xx} + \varepsilon_{yy}) \end{bmatrix}$$
$$+ \begin{bmatrix} 0 & \varepsilon_{xy} & \varepsilon_{xz} \\ \varepsilon_{xy} & 0 & \varepsilon_{yz} \\ \varepsilon_{xz} & \varepsilon_{yz} & 0 \end{bmatrix} \tag{19.4}$$

This decomposition simplifies the evaluation of the Hamiltonian. The effect of hydrostatic and shear strains on the electronic states is different [33]. From simple symmetry arguments, hydrostatic strain maintains the same symmetry, therefore, only shifts the energy position of the bands (changes the bandgap and electron affinity but does not split the degeneracy). Shear strain removes symmetry, and thus, splits degenerate states in the conduction and valence bands. Also, as pointed out by Herring and Vogt in 1955 [15] and still valid today [31], the experimental error in determining the hydrostatic energy shifts is large while the error in determining the shear splitting is small.

Deformation potential theory (first proposed by Bardeen and Shockley [34] and then modified for indirect many-valley semiconductors (Si,Ge) by Herring and Vogt [15]) expresses the conduction band edge shifts and splitting as a function of strain. For the valence band, Pikus and Bir developed analytical expressions for energy levels and effective mass near $\mathbf{k} = 0$ using perturbation theory and the strain Hamiltonian [13,14]. The band edge shifts are given by

$$E_{\text{hh}}, E_{\text{lh}} = a(\varepsilon_x + \varepsilon_y + \varepsilon_z) \pm \sqrt{b^2(\varepsilon_x - \varepsilon_z)^2 + d^2\varepsilon_{xy}^2} \tag{19.5}$$

$$\Delta E_c = \begin{cases} \Delta_2 = (\Xi_d + \Xi_u/3)(\varepsilon_x + \varepsilon_y + \varepsilon_z) + (2/3)\Xi_u(\varepsilon_z - \varepsilon_x) & \tag{19.6} \\ \Delta_4 = (\Xi_d + \Xi_u/3)(\varepsilon_x + \varepsilon_y + \varepsilon_z) - (1/3)\Xi_u(\varepsilon_z - \varepsilon_x) & \tag{19.7} \end{cases}$$

where E_{hh}, E_{lh}, Δ_2, Δ_4 are the heavy and light hole valence bands and two-fold and four-fold conduction band edge states, and the relationship $\varepsilon_x = \varepsilon_y \neq \varepsilon_z$ is used for simplification. The terms $\Xi_d + (1/3)\Xi_u$ in Equation 19.6 and Equation 19.7 (called E_1 by Bardeen and Shockley [34]) and a are the hydrostatic deformation potentials for the conduction and valence bands, respectively. Ξ_u, b, and d (needed for $\langle 1\,1\,0 \rangle$ stress since ε_{xy} is nonzero) are shear deformation potentials.

As seen from Equation 19.5 to Equation 19.7, the hydrostatic term shifts while the shear term splits the valence and conduction band states. It is the shear term that modulates conductivity and mobility. The shear deformation potentials are well known with good accuracy from piezoresistance measurements ($\Xi_u = 9.16$, $b = -2.35$, and $d = 5.0$) [15]. The hydrostatic deformation potential affects important electronic properties such as energy gap and electron affinity. However, the hydrostatic deformation cannot be directly measured and a very wide range of values from ~ 2 to ~ -10.7 have been reported [31]. Some optical experimental techniques can directly measure differences in the conduction and valence band deformation potential, which reduces the uncertainty for strain-induced bandgap narrowing. However, at present, there is still large uncertainty in the electron affinity of strained Si. This uncertainty (as will be shown in section "Strain-induced n-Channel MOSFET Threshold Voltage Shift") makes it difficult to calculate the threshold voltage shift for biaxial strained Si MOSFETs.

The shape of the energy bands determines the effective mass. Neglecting spin–orbit coupling, analytical expression for in-plane and out-of-plane effective mass is given by Hensel and Feher [28] and Chao and Chuang [27] at $\mathbf{k} = 0$,

$$\frac{m_{\text{hh},\perp}^*}{m_0} = \frac{1}{\gamma_1 - 2\gamma_2}, \quad \frac{m_{\text{lh},\perp}^*}{m_0} = \frac{1}{\gamma_1 + 2\gamma_2}, \quad \frac{m_{\text{hh},\parallel}^*}{m_0} = \frac{1}{\gamma_1 + \gamma_2}, \quad \frac{m_{\text{lh},\parallel}^*}{m_0} = \frac{1}{\gamma_1 - \gamma_2} \tag{19.8}$$

where γ_1 and γ_2 are material band parameters in the Kohn–Luttinger strain Hamiltonian. The above results are for biaxial stress but similar results are obtained by Hensel and Feher for uniaxial stress along $\langle 1\ 1\ 0 \rangle$ (by swapping $\parallel$ with $\perp$ in Equation 19.8 and replacing the A, B, and C band parameters with the Kohn–Luttinger parameters [28]). Including spin–orbit coupling, Chao and Chuang [27] have derived expressions at $\mathbf{k} = 0$. As a note of caution, the effective mass at $\mathbf{k} = 0$ should only be used as a guide. The valence band is not a parabolic function of $\mathbf{k}$, hence the effective mass is not constant. The effective mass can be determined at higher $\mathbf{k}$ by diagonalizing the 6×6 strain Hamiltonian. Within this band shift and effective mass description, we now review the current understanding of uniaxial and biaxial stress on MOSFETs.

Strain-Induced Hole Mobility Enhancement versus Vertical Electric Field

One of the biggest differences for biaxial tensile and uniaxial compressive stressed p-channel MOSFETs is the field dependence of the mobility enhancement. The hole mobility enhancement at large vertical fields is important in nanoscale CMOS since aggressive gate oxide scaling has dramatically increased the oxide field to about 5 MV/cm and silicon inversion effective vertical field (E_{EFF}) to greater than 1 MV/cm [35]. For uniaxial stress introduced by $Si_{1-x}Ge_x$ in the source and drain, the hole mobility for a 45 nm gate length transistor increases 50% as shown in Figure 19.4 [8]. An important observation in Figure 19.4 is that the uniaxial stress hole mobility enhancement is present at large vertical fields unlike that of biaxial stress [8,26]. This important feature will now be discussed in more detail.

Figure 19.5 summarizes what is known about the hole band structure for unstrained and strained Si. The valence bands are plotted in the in-plane direction of the MOSFET channel [36]. For the unstrained lattice, the valence bands are degenerate and are the primary reason for the low bulk hole mobility. The unstrained valence band structure consists of a degenerate heavy hole and light hole band at $\mathbf{k} = 0$ and a slightly offset spin–orbit split-off band. The large valence band degeneracy is undesirable and creates an opportunity for larger hole than electron mobility enhancement [7].

An important factor in the field dependence of the mobility is how the stress-induced light to heavy hole band separation changes with surface confinement (triangular surface potential setup by the MOSFET vertical field). Uniaxial compressive and biaxial tensile stress cause a similar magnitude separation in the light to heavy hole band at $\mathbf{k} = 0$ as calculated from Equation 19.5 to Equation 19.7 [37]. Hole mobility enhancement can result when additional holes populate the "light hole like" band [13,14,16,38] since it potentially has lower conductivity effective mass. However, the situation, as will be shown in the next section, is more complex due to band warping.

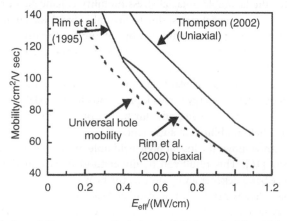

FIGURE 19.4 Enhanced hole mobility versus vertical electric field for uniaxial and biaxial stress.

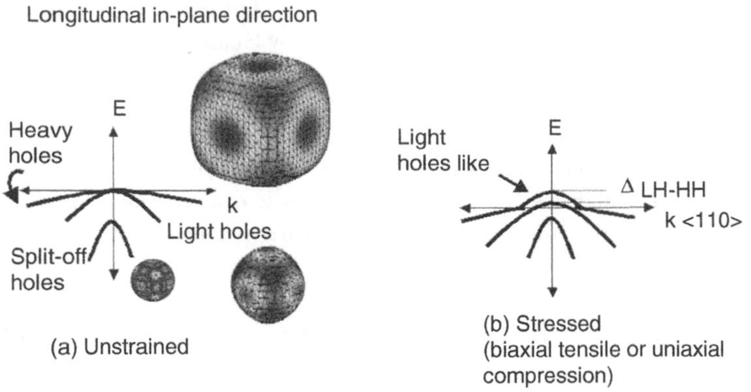

Longitudinal in-plane direction

(a) Unstrained

(b) Stressed
(biaxial tensile or uniaxial
compression)

FIGURE 19.5 Simplified hole valence band structure for longitudinal in-plane directions: (a) unstrained and (b) strained Si.

For biaxial tensile stress, the reduction in hole mobility enhancement with vertical electric field is caused by surface confinement [16]. Fischetti showed numerically that the strain-induced separation between the light and heavy hole bands decreases at high vertical field. The reduced splitting can be understood from the difference in the light and heavy hole out-of-plane effective mass (Equation 19.8). The out-of-plane mass along with surface confinement shifts the energy levels as shown by Stern [39] and qualitatively in Figure 19.6. The energy level shift for the light and heavy hole bands can be approximated by

$$E_j = \left[\frac{2hqE_s}{4\sqrt{2m_z}} \left(j + \frac{3}{4} \right) \right]^{2/3}, \quad j = 0, 1, 2, \ldots \tag{19.9}$$

where E_s is the vertical electric field in the silicon. The important observation from Equation 19.9 is a band with a low out-of-plane effective mass ($m_\perp$) will shift and become depopulated with increasing vertical field. Biaxial stress is known to have a low out-of-plane effective mass for the light hole band [31,40]. Thus, the biaxial strain-induced band splitting (Δ_{LH-HH} in Figure 19.5) will decrease ("be canceled") with vertical field.

The physical origin for the favorable uniaxial stress mobility enhancement at high field results from the same mechanism [41]. For uniaxial stress, due to favorable band warping, the surface confinement increases the band splitting. This can be quantified by solving for the out-of-plane conductivity mass using the Hensel–Feher formulation [28,42] and band parameters extracted from cyclotron resonance data [28,42]. The results show that the out-of-plane light hole mass is larger than the heavy hole mass (opposite to the biaxial case). The out-of-plane effective masses are quantitatively shown in Figure 19.6 near $k = 0$ along with a qualitative energy versus k plot. Thus for uniaxial strain, the out-of-plane mass relation $m_{lh} > m_{hh}$ increases the splitting and further populates the light hole band, which helps maintain the high vertical field mobility enhancement.

Finally, not all biaxial stress shows the loss of high-field hole mobility enhancement, as is the case with biaxial compressive stress [16]. This result can be understood from Equation 19.5 since compressive strain removes the valence band degeneracy by shifting the heavy hole band to higher energy (as opposed to light hole). Since holes already primarily populate the heavy hole band, the vertical electric field does not cause a light to heavy hole repopulation due to surface confinement. However, biaxial compressive stress is less interesting since the maximum hole mobility enhancement is less [16].

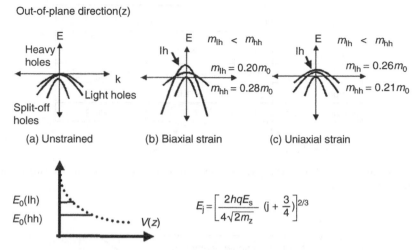

FIGURE 19.6 Simplified hole valence band structure for out-of-plane directions: (a) unstrained and (b) biaxial strained Si, and (c) uniaxial strained Si.

Hole Mobility Enhancement at Low Strain

The next important question is why biaxial tensile versus uniaxial compressive stress hole mobility enhancement is different at low strain. At low strain, biaxial stress shows little if any mobility enhancement [2] versus large uniaxial stress-induced mobility improvement [8]. Band warping is again responsible for this effect. Band warping leads to more favorable curvature [37] and population in **k** space [41] for uniaxial than biaxial stress. In the low-strain region, the well-studied piezoresistance effect in Si can be used to determine how bands warp and repopulate with strain. Piezoresistance coefficients are valid for strain less than approximately 250 to 500 MPa, where the piezoresistance varies linearly with strain. Yamada et al. found the nonlinearity in the piezoresistance to be small ($\sim$1% for longitudinal compression) up to 250 MPa [43]. For this discussion, we assume industry standard Si wafers with (0 0 1) surface and wafer notch on the [1 1 0] axis. The effect of mechanical stress on the mobility can then be expressed as follows:

$$\frac{\Delta\mu}{\mu} \approx \left| \pi_\parallel T_\parallel + \pi_\perp T_\perp \right| \tag{19.10}$$

where the subscripts $\parallel$ and $\perp$ refer to the directions parallel and transverse to the current flow in the plane of the MOSFETs. $T_\parallel$ and $T_\perp$ are the longitudinal and transverse stresses, and $\pi_\parallel$ and $\pi_\perp$ are the piezoresistance coefficients expressed in Pa^{-1}. The $\pi_\parallel$ and $\pi_\perp$ can be expressed in terms of the three fundamental cubic piezoresistance coefficients π_{11}, π_{12}, and π_{44}.

For the case of the technologically important (0 0 1) wafer, the longitudinal and transverse piezoresistance coefficients for the standard layouts are given in Figure 19.7. For simplicity, we use the bulk values for π_{11}, π_{12}, and π_{44} first measured 50 years ago by Smith [44], though technically piezoresistance coefficients should take into account the two-dimensional nature of transport in the MOSFET and depend on temperature, gate voltage, and doping [45,46]. Using the bulk coefficients, $\pi_\parallel$ and $\pi_\perp$ are calculated in Figure 19.7 and the following can be concluded.

Both biaxial tensile and uniaxial compressive stress split and repopulate the light and heavy hole bands by similar amounts [37]. However, the enhanced mobility as calculated from Equation 19.10 and plotted in Figure 19.8 shows large differences for biaxial versus uniaxial stress. For biaxial stress,

Polarity	<100>		<110>	
	$\pi_{\parallel}$	$\pi_{\perp}$	$\pi_{\parallel}$	$\pi_{\perp}$
N or P	π_{11}	π_{12}	$(\pi_{11}+\pi_{12}+\pi_{44})/2$	$(\pi_{11}+\pi_{12}-\pi_{44})/2$
N-type	−102	53.4	−31.6	−17.6
P-type	6.6	−1.1	71.8	−66.3

Biaxial stress   Uniaxial stress

FIGURE 19.7 Longitudinal and transverse piezoresistance coefficients evaluated for standard layout and wafer orientation (units of 10^{-12} dyn^{-1}).

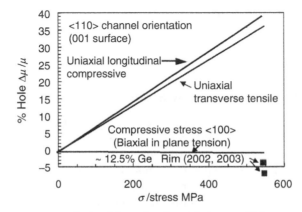

FIGURE 19.8 Mobility versus stress calculated from the piezoresistance coefficients.

the negligible mobility improvement in the low-strain regime results from the low π_{12} coefficients −1.1 (compared to 71.8×10^{-12} cm^2 dyn^{-1} for uniaxial stress).

The results in Figure 19.8 can be understood by realizing how biaxial and uniaxial stresses warp the bands causing changes in the in-plane conductivity mass. The strain-altered in-plane effective masses are calculated from the Luttinger–Kohn and Bir–Pikus 6×6 strain Hamiltonian for the $p_{3/2}$ [27]. The results for biaxial and uniaxial stress near $\mathbf{k} = 0$ are summarized in Figure 19.9. In Figure 19.9, the uniaxial stress-altered light hole in-plane conductivity mass is 50% smaller than for biaxial stress [37].

In summary, similar to the out-of-plane effective mass, biaxial stress band warping causes less desirable in-plane mass. For biaxial stress, the "light" hole mass is actually slightly larger than the heavy hole mass. Thus, repopulation creates a slight negative resistance which is in good agreement with the slight negative piezoresistance coefficient and the biaxial MOSFET data at low Ge concentration (see Figure 19.8 and Ref. [2]). Biaxial stress can enhance the hole mobility at high Ge concentration and strain since large band splitting reduces interband scattering [16]. However, high strain is more difficult to integrate since it requires additional reduction to the midsection thermal cycles to avoid strain relaxation.

Strain-Induced n-Channel MOSFET Threshold Voltage Shift

Biaxial and uniaxial stress-enhanced electron mobility is better understood and results from conductivity effective mass improvement due to increased electron concentration in the Δ_2 valleys (Equation 19.6) and reduced intervalley scattering [47]. However, less attention has been paid to strain-induced

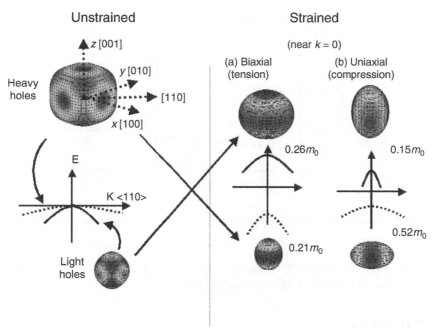

FIGURE 19.9 Effective mass calculated from the Luttinger–Kohn and Bir–Pikus 6 × 6 strain Hamiltonian to illustrate quantitatively how uniaxial compressive and biaxial tensile stress alter the valence band structure.

threshold voltage shifts. Recent literature [2,17–19] suggests that large threshold voltage shifts occur for biaxial stressed n-channel MOSFETs, while much smaller shifts are observed for uniaxial tensile stress [8,10,26]. Large threshold voltage shifts are undesirable for high-performance logic transistors since it increases off-state leakage or if retargeted by increasing the well doping degrades mobility and subthreshold slope [19].

There is some confusion over the exact origin of the threshold voltage shift but changes in the electron affinity and strain-induced bandgap narrowing are key contributors [48]. These two effects can be quantified using Equation 19.5 to Equation 19.7 following an approach used by Van de Walle and Martin [49,50] and People [30] for biaxial stressed heterostructures. Figure 19.10 plots the conduction and valence band edge versus strain for longitudinal uniaxial and biaxial tensile stress. For simplicity, a MOSFET with ⟨1 0 0⟩ channel orientation is chosen but the results are similar for ⟨1 1 0⟩ devices. The strain-induced bandgap narrowing is ∼3× less for uniaxial stress (see Figure 19.10). In the bandgap narrowing calculations, it is necessary to include spin–orbit coupling [51], which causes the light hole band shift to be approximately twice as large as the heavy hole shift.

The conduction band edge shift also gives the change in the electron affinity, which is simply the shift in the Δ_2 valleys. The decrease in Δ_2 valleys produces a large negative (lower) threshold voltage shift directly proportional to the shift in Δ_2 [48] (providing no electron affinity change in the gate which is the case for biaxial stress). Using the range of reported hydrostatic deformation potential, 2 to −10.7 eV, and Equation 19.6, approximately −100 to −400 mV threshold voltage shifts occur for 1% biaxial strain, respectively, which negatively affects performance.

As a final note in this uniaxial versus biaxial comparison, longitudinal uniaxial tensile stress is chosen since this type of stress is widely adopted in 90-nm logic technologies [8,26] (as will be discussed in Chapter 21). Since the implementation of uniaxial stress with a capping layer also strains the n^+ gate, the electron affinity change has no effect on the threshold voltage. Thus, the threshold voltage shift for uniaxial stress is smaller and dominated by bandgap narrowing, which is also smaller compare to biaxial stress. However, the significantly less bandgap narrowing for uniaxial tensile stress in the n-channel MOSFET comes at a price since it results from the heavy hole band shifting up. This is undesirable for

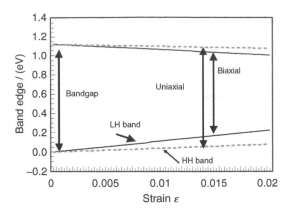

FIGURE 19.10   Conduction and valence bands shifts and bandgap narrowing due to uniaxial and biaxial strain.

hole transport; p-channel MOSFETs would be significantly degraded if nothing was changed to compensate for the tensile stress in the p-channel transistor.

19.4   Summary

Both biaxial and uniaxial stress provide significant enhanced mobility to improve MOSFET perform-ance. Using the strain Hamiltonian, equations are given for the conduction and valence band edge shifts, which explain the differences observed in biaxial and uniaxial stressed nanoscale MOSFETs. The key advantage for biaxial stress is that both hole and electron mobilities can be enhanced for the same strain. The key advantages of uniaxial stress are larger hole mobility enhancement at low strain, mobility enhancement at high vertical electric field, and less n-channel threshold voltage shift due to less bandgap narrowing and the n^{+} gate also strained.

Acknowledgments

The author would like to thank the efforts of his former colleagues in the Portland Technology Development, Technology Computer Aided Design, and in the Corporate Quality and Reliability Groups of Intel. The author also acknowledges the support and encouragement from Mark Bohr, Robert Chau, and William Holt and many helpful discussions with Professor C.T. Sah, Philippe Matagne, Borna Obradovic, Lucian Shifren, Toshi Nishida, and Kehuey Wu.

References

1. K. Rim, S. Koester, M. Hargrove, J. Chu, P. Mooney, J. Ott, T. Kanarsky, P. Ronsheim, M. Ieong, A. Grill, H.-S.P. Wong. Strained Si NMOSFETs for high performance CMOS technology. Symp. VLSI Tech. Dig., Kyoto, Japan, 2001, pp. 59–60.
2. K. Rim, J. Chu, H. Chen, K.A. Jenkins, T. Kanarsky, K. Lee, A. Mocuta, H. Zhu, R. Roy, J. Newbury, J. Ott, K. Petrarca, P. Mooney; D. Lacey, S. Koester, K. Chan, D. Boyd, M. Ieong and H.-S Wong. Characteristics and device design of sub-100 nm strained Si n- and p-MOSFETs. Symp. VLSI Tech. Dig., Hawaii, 2002, pp. 98–99.
3. K. Rim, K. Chan, L. Shi, D. Boyd, J. Ott, N. Klymko, F. Cardone, L. Tai, S. Koester, M. Cobb, D. Canaperi, B. To, E. Duch, I. Babich, R. Carruthers, P. Saunders, G. Walker, Y. Zhang, M. Steen, and M. Ieong. Fabrication and mobility characteristics of ultra-thin strained Si directly on insulator (SSDOI) MOSFETs. Technical Digest of the IEEE International Electron Devices Meeting, San Francisco, 2003, pp. 49–52.

4. K. Rim, J. Welser, J.L. Hoyt, and J.F. Gibbons. Enhanced hole mobilities in surface-channel strained-Si p-MOSFETs. Technical Digest of the IEEE International Electron Devices Meeting, San Francisco, 1995, pp. 517–520.

5. C.W. Leitz, M.T. Currie, M.L. Lee, C. Zhi-Yuan, D.A. Antoniadis, and E.A. Fitzgerald. Channel engineering of SiGe-based heterostructures for high mobility MOSFETs. Materials Issues in Novel Si-Based Technology. Symposium (Materials Research Society Symposium Proceedings, Vol. 686), 2002, pp. 113–118.

6. M.L. Lee, C.W. Leitz, C. Zhiyuan, A.J. Pitera, G. Taraschi, D.A. Antoniadis, and E.A. Fitzgerald. Strained Ge channel p-type MOSFETs fabricated on $Si_{1-x}Ge_x/Si$ virtual substrates. Materials Issues in Novel Si-Based Technology. Symposium (Materials Research Society Symposium Proceedings, Vol. 686), 2002, pp. 39–43.

7. J.L. Hoyt, H.M. Nayfeh, S. Eguchi, I. Aberg, G. Xia, T. Drake, E.A. Fitzgerald, and D.A. Antoniadis. Strained silicon MOSFET technology. Technical Digest of the IEEE International Electron Devices Meeting, Washington, 2002, pp. 23–26.

8. S.E. Thompson, M. Armstrong, C. Auth, S. Cea, R. Chau, G. Glass, T. Hoffman, J. Klaus, Zhiyong Ma, B. Mcintyre, A. Murthy, B. Obradovic, L. Shifren, S. Sivakumar, S. Tyagi, T. Ghani, K. Mistry, M. Bohr and Y. El-Mansy. A logic nanotechnology featuring strained silicon. *IEEE Electron Dev. Lett.* 25, 191–193, 2004.

9. S. Thompson, N. Anand, M. Armstrong, C. Auth, B. Arcot, M. Alavi, P. Bai, J. Bielefeld, R. Bigwood, J. Brandenburg, M. Buehler, S. Cea, V. Chikarmane, C. Choi, R. Frankovic, T. Ghani, G. Glass, W. Han, T. Hoffmann, M. Hussein, P. Jacob, A. Jain, C. Jan, S. Joshi, C. Kenyon, J. Klaus, S. Klopcic, J. Luce, Z. Ma, B. Mcintyre, K. Mistry, A. Murthy, P. Nguyen, H. Pearson, T. Sandford, R. Schweinfurth, R. Shaheed, S. Sivakumar, M. Taylor, B. Tufts, C. Wallace, P. Wang, C. Weber and M. Bohr. A 90 nm logic technology featuring 50nm strained silicon channel transistors, 7 layers of Cu interconnects, low k ILD, and 1mm² SRAM cell. Technical Digest of the IEEE International Electron Devices Meeting, Washington, 2002, pp. 61–64.

10. T. Ghani, M. Armstrong, C. Auth, M. Bost, P. Charvat, G. Glass, T. Hoffmann, K. Johnson, C. Kenyon, J. Klaus, B. McIntyre, K. Mistry, A. Murthy, J. Sandford, M. Silberstein, S. Sivakumar, P. Smith, K. Zawadzki, S. Thompson and M. Bohr. A 90 nm high volume manufacturing logic technology featuring novel 45 nm gate length strained silicon CMOS transistors. Technical Digest of the IEEE International Electron Devices Meeting, San Francisco, 2003, pp. 978–980.

11. J.M. Luttinger. Quantum theory of cyclotron resonance in semiconductors: general theory. *Phys. Rev.* 102, 1030–1041, 1956.

12. J.M. Luttinger and W. Kohn. Motion of electrons and holes in perturbed periodic fields. *Phys. Rev.* 97, 869–883, 1955.

13. G.E. Pikus and G.L. Bir. Effect of deformation on the energy spectrum and the electrical properties of imperfect germanium and silicon. *Sov. Phys. Solid-State* 1, 136–138, 1959.

14. G.E. Pikus and G.L. Bir. Cyclotron and paramagnetic resonance in strained crystals. *Phys. Rev. Lett.* 6, 103–105, 1961.

15. C. Herring and E. Vogt. Transport and deformation-potential theory for many-valley semiconductors with anisotropic scattering. *Phys. Rev.* 101, 944–961, 1956.

16. M.V. Fischetti, Z. Ren, P. M. Solomon, M. Yang, and K. Rim. Six-band **k. p** calculation of the hole mobility in silicon inversion layers: dependence on surface orientation, strain, and silicon thickness. *J. Appl. Phys.* 94, 1079–1095, 2003.

17. J. Goo, X. Qi, Y. Takamura, F. Arasnia, E.N. Paton, P. Besser, J. Pan, and L. Ming-Ren. Band offset induced threshold variation in strained-Si nMOSFETs. *IEEE Electron Dev. Lett.* 24, 568–570, 2003.

18. J. Goo, Qi Xiang, Y. Takamura, Haihong Wang, J. Pan, F. Arasnia, E.N. Paton, P. Besser, M.V. Sidorov, E. Adem, A. Lochtefeld, G. Braithwaite, M.T. Currie, R. Hammond, M.T. Bulsara and Ming-Ren Lin. Scalability of strained-Si nMOSFETs down to 25nm gate length. *IEEE Electron Dev. Lett.* 24, 351–353, 2003.

19. J.G. Fossum and W. Zhang. Performance projections of scaled CMOS devices and circuits with strained Si-on-SiGe channels. *IEEE Trans. Electron Dev.* 50, 1042–1048, 2003.

20. A. Shimizu, K. Hachimine, N. Ohki, H. Ohta, M. Koguchi, Y. Nonaka, H. Sato and F. Ootsuka. Local mechanical-stress control (LMC): a new technique for CMOS-performance enhancement. Technical Digest of the IEEE International Electron Devices Meeting, San Francisco, 2001, pp. 433–436.

21. S. Ito, H. Namba, K. Yamaguchi, T. Hirata, K. Ando, S. Koyama, S. Kuroki, N. Ikezawa, T. Suzuki, T. Saitoh and T. Horiuchi. Mechanical stress effect of etch-stop nitride and its impact on deep submicron transistor design. Technical Digest of the IEEE International Electron Devices Meeting, Washington, 2000, pp. 247–250.

22. H. Cheng-Liang, H. Soleimani, G. Grula, N.D. Arora, and D. Antoniadis. Isolation process dependence of channel mobility in thin-film SOI devices. *IEEE Electron Dev. Lett.* 17, 291–293, 1996.

23. C.K. Maiti, L.K. Bera, S.S. Dey, D.K. Nayak, and N.B. Chakrabarti. Hole mobility enhancement in strained-Si p-MOSFETs under high vertical field. *Solid-State Electron.* 41, 1863–1869, 1997.

24. M.E. Law and S.M. Cea. Continuum based modeling of silicon integrated circuit processing: an object oriented approach. *Comput. Mater. Sci.* 12, 289–308, 1998.

25. S. Cea and M. Law. Multidimensional nonlinear viscoelastic oxidation modeling. Simulation of Semiconductor Devices and Processes, H. Ryssel and P. Pichler, Eds., Erlangen, Germany, Vol. 6, 1995, pp. 139–142.

26. S. E. Thompson, M. Armstrong, C. Auth, M. Alavi, M. Buehler, R. Chau, S. Cea, T. Ghani, G. Glass, T. Hoffman, C. H. Jan, C. Kenyon, J. Klaus, k. Kuhn, Z. Ma, B. Mcintyre, K. Mistry, A. Murthy, B. Obradovic, R. Nagisetty, P. Nguyen, R. Shaheed, L. Shifren, S. Sivakumar, B. Tuffs, S. Tyagi, M. Bohr and Y. El-Mansy, "A 90 nm logic technology featuring strained silicon," *IEEE Trans. Elec. Devices*, vol. 51, pp. 1790–1797, 2004.

27. C.Y.-P. Chao and S.L. Chuang, Spin-orbit-coupling effects on the valence-band structure of strained semiconductor quantum wells. *Phys. Rev. B* 46, 4110–4122, 1992.

28. J.C. Hensel and G. Feher. Valence band parameters in silicon from cyclotron resonances in crystals subjected to uniaxial stress. *Phys. Rev. Lett.* 5, 307–309, 1960.

29. F. Herman. The electronic energy band structure of silicon and germanium. *Proc. IRE* 40, 1703–1732, 1955.

30. R. People. Indirect band gap of coherently strained Ge_xSi_{1-x} bulk alloys on $(0\ 0\ 1)$ silicon substrates. *Phys. Rev. B: Cond. Matter* 32, 1405–1408, 1985.

31. M.V. Fischetti and S.E. Laux. Band structure, deformation potentials, and carrier mobility in strained Si, Ge, and SiGe alloys. *J. Appl. Phys.* 80, 2234–2252, 1996.

32. W.A. Brantley. Calculated elastic constants for stress problems associated with semiconductor devices. *J. Appl. Phys.* 44, 534–535, 1973.

33. C.-T. Sah. Fundamentals of Solid-State Electronics and Fundamentals of Solid-State Electronics, World Scientific Publishing, 1995.

34. J. Bardeen and W. Shockley. Deformation potentials and mobilities in non-polar crystals. *Phys. Rev.* 80, 72–80, 1950.

35. S. Thompson. Technology performance: trends and challenges. IEDM short cource, Washington, DC, 1999.

36. G. Dresselhaus, A.F. Kip, and C. Kittel. Cyclotron resonance of electrons and holes in silicon and germanium crystals. *Phys. Rev.* 98, 368–384, 1955.

37. K. Wu. Comparison of hole mobility in uniaxial and biaxial strained pMOSFETs. *IEEE Electron Dev. Lett.* submitted.

38. R. Oberhuber, G. Zandler, and P. Vogl. Subband structure and mobility of two-dimensional holes in strained Si/SiGe MOSFETs. *Phys. Rev. B* 58, 9941–9948, 1998.

39. F. Stern. Self-consistent results for n-type Si inversion layers. *Phys. Rev. B: Solid State* 5, 4891–4899, 1972.

40. M.L. Lee and E.A. Fitzgerald. Hole mobility enhancements in nanometer-scale strained-silicon heterostructures grown on Ge-rich relaxed $Si_{1-x}Ge_x$. *J. Appl. Phys.* 94, 2590–2596, 2003.

41. M.D. Giles, M. Armstrong, C. Auth, S.M. Cea, T. Ghani, T. Hoffmann, R. Kotlyar, P. Matagne, K. Mistry, R. Nagisetty, B. Obradovic, R. Shaheed, L. Shifren, M. Stettler, S. Tyagi, X. Wang, C. Weber and K. Zawadzki. Understanding stress enhanced performance in Intel 90nm CMOS technology. Symp. VLSI Tech. Dig., Hawaii, 2004.

42. J.C. Hensel and G. Feher. Cyclotron resonance experiments in uniaxially stressed silicon: valence band inverse mass parameters and deformation potentials. *Phys. Rev.* 129, 1041–1062, 1963.

43. K. Yamada. Temperature dependence of the piezoresistance effects of p-type silicon diffused layers. *Trans. IEE Japan* 103A, 555–562, 1983.

44. C.S. Smith. Piezoresistance effect in germanium and silicon. *Phys. Rev.* 94, 42–49, 1954.

45. G. Dorda. Effective mass change of electrons in silicon inversion layers observed by piezoresistance. *Appl. Phys. Lett.* 17, 406–408, 1970.

46. G. Dorda. Piezoresistance in quantized conduction bands in silicon inversion layers. *J. Appl. Phys.* 42, 2053–2060, 1971.

47. S.-i. Takagi, J.L. Hoyt, J.J. Welser, and J.F. Gibbons. Comparative study, of phonon-limited mobility of two-dimensional electrons in strained and unstrained Si metal-oxide-semiconductor field-effect transistors. *J. Appl. Phys.* 80, 1567–1577, 1996.

48. W. Zhang and J.G. Fossum. On the threshold voltage of strained-Si/SiGe MOSFETs. *IEEE Trans. Electron Dev.* 52(2): 263–268, 2005.

49. C.G. Van de Walle. Band lineups and deformation potentials in the model-solid theory. *Phys. Rev. B* 39, 1871–1883, 1989.

50. C.G. Van de Walle and R.M. Martin. Theoretical calculations of heterojunction discontinuities in the Si/Ge system. *Phys. Rev. B* 34, 5621–5634, 1986.

51. H. Hasegawa. Theory of cyclotron resonance in strained silicon crystals. *Phys. Rev.* 129, 1029–1040, 1963.

20

SiGe-Channel HFETs

Sanjay Banerjee
University of Texas at Austin

20.1 Introduction

Since the earlier chapters have provided excellent overviews of the materials and bandstructure issues of SiGe, in this chapter we will focus on their application in heterostructure FETs (HFETs). We will briefly discuss the transport issues of SiGe and SiGeC alloys that are relevant to HFETs. Then will discuss their applications in buried- and surface-channel HFETs, with an emphasis on pHFETs. We will conclude by briefly describing the vertical HFETs and implications for n-channel devices and CMOS.

20.2 Bandstructure and Transport

The main motivation for grafting Ge and C onto Si technology is that the use of such heterostructures enables one to do "bandgap" and "strain" engineering to achieve enhanced transport properties [1–5]. Seminal work in this area was done by Meyerson and his group at IBM [1]. There is initially a gradual decrease of the bandgap with increasing Ge mole fraction x as the valleys in unstrained Si$_{1-x}$Ge$_x$ are lowered, but the overall bandstructure remains Si-like. For higher Ge mole fractions above ∼0.85, there is a more rapid decrease of the energy gap with x, corresponding to the L-valleys decreasing more rapidly with x, and the bandstructure of Si$_{1-x}$Ge$_x$ becomes Ge-like. In the presence of biaxial compressive strain of a Si$_{1-x}$Ge$_x$ layer grown on a Si substrate, the six-fold degeneracy of the conduction band (for low x, where the bands are still Si-like) is broken into lower energy four "in-plane" valleys, and higher energy valleys in the growth direction. Similarly, the degeneracy in the valence band is also split, and the heavy hole band is lifted relative to the light hole band (Figure 20.1) [6]. The bandgap in compressively strained Si$_{1-x}$Ge$_x$ grown on an Si substrate is reduced compared to the bulk, unstrained system, and is determined by the transition from the heavy hole (HH) to the fourfold in-plane degenerate conduction bands Δ [4,7]. For this case, one has a Type I band alignment where most of the band discontinuity is in the valence band, and $\Delta E_c \cong 0$. This results in effective hole confinement in the Si$_{1-x}$Ge$_x$ layer, but no electron confinement [8].

Because of the warped nature of the spheres in the valence band, according to the Luttinger–Kohn parameters, the heavy hole band actually presents a lower in-plane effective mass than the light hole

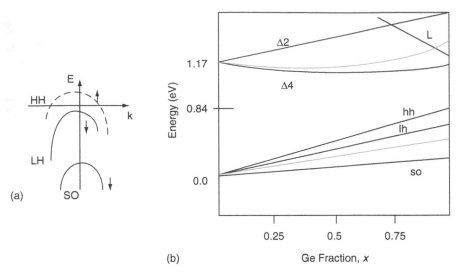

FIGURE 20.1 (a) Valence bandstructure in $Si_{1-x}Ge_x$ with compressive strain. The heavy hole band is occupied. The deformation increases the curvature of the heavy hole (HH) band, and increases hole mobility. (b) With increasing Ge mole fraction (compressive strain), the degeneracies in the conduction and valence band are broken, as shown.

band. There is, thus, an increase of the in-plane hole mobility, which is attractive for pHFETs (Figure 20.2) [6,9]. There are additional strain-related factors such as the lifting of the degeneracy between heavy and light holes, which reduces inter-subband scattering. On the other hand, several factors reduce the mobility. These include alloy scattering in random $Si_{1-x}Ge_x$ alloys. If there are misfit dislocations generated in these strained heterolayers, there can be dislocation scattering. Finally, for in-plane transport, there can be surface roughness scattering at the heterointerfaces, where strain and surface free energies play a role in the morphology of the heterointerface.

Obviously, for ultrasmall HFETs the mobility alone (especially low field μ) is not the only important parameter, which determines speed and drive current. The saturation drift velocity, V_{SAT}, and velocity overshoot can have a big impact on performance [10,11]. Although the V_{SAT} is slightly higher in strained $Si_{1-x}Ge_x$ than in bulk Si the enhancement is not very dramatic. In spite of this, there are significant advantages from a device point of view by increasing mobility. That is because V_{SAT} is achieved at a lower field in $Si_{1-x}Ge_x$ than in Si, i.e., carriers would travel at V_{SAT} over a longer portion of the channel in $Si_{1-x}Ge_x$ devices than in Si. In such heterostructures, it is also possible to do "modulation doping" to increase carrier mobilities further in MODFETs. For example, if the acceptors are introduced only in Si, the holes would spill over to the adjacent $Si_{1-x}Ge_x$ layers if the layers are sufficiently thin. Then for in-plane transport, the holes would not undergo ionized impurity scattering.

For strained $Si_{1-x}Ge_x$ channel HFETs or MODFETs [12,13], the critical device parameters can be identified by first examining the drain current expression in a *long-channel* MOSFET,

$$I_{DSAT} = \frac{W}{L}\frac{\mu C_{OX}}{2}[V_G - V_T]^2 \tag{20.1}$$

where W and L are the width and length, μ is the effective channel mobility, C_{OX}, the gate dielectric capacitance per unit area, is the ratio of the dielectric constant over the gate oxide thickness ($= \varepsilon_{OX}/t_{OX}$), V_T is the threshold voltage, and V_G and V_D are the gate and drain biases, respectively. As discussed earlier, the attraction of introducing strained $Si_{1-x-y}Ge_xC_y$ channels in HFETs or modulation-doped layers in MODFETs is the increase of μ [10–13]. In a short-channel MOSFET, because of the high

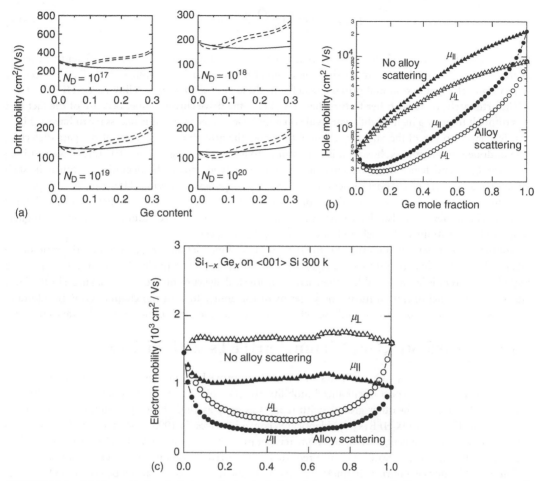

FIGURE 20.2 (a) Monte Carlo calculations of minority hole mobilities of $Si_{1-x}Ge_x$ for four doping levels (in cm^{-3}) at 300 K: dot-dashed line is the vertical mobility of strained $Si_{1-x}Ge_x$, solid line is the mobility of unstrained $Si_{1-x}Ge_x$, dashed line is the planar mobility of strained $Si_{1-x}Ge_x$ [9]. (b) Hole mobility with or without alloy scattering in-plane and out-of-plane. (c) Electron mobility with or without alloy scattering in-plane and out-of-plane [6].

longitudinal electric field along the channel, the carriers tend to travel at the saturation drift velocity, V_{SAT}, over a large fraction of the channel. Here the drain current is given by

$$I_{DSAT} = WC_{OX}[V_G - V_T]V_{SAT} \tag{20.2}$$

rather than the expression in Equation 20.1. In that case the benefits of using Si–Ge alloys in the channel are a little less clear because V_{SAT} is very similar in Si and these alloys. However, a higher μ still helps somewhat because the carriers then attain V_{SAT} earlier on in the channel where the channel is not yet pinched off.

More interesting is another viewpoint of drain current limitation in *extremely* short channel, "quasi-ballistic" MOSFETs [14]. In this "scattering" picture, based on the so-called Landauer–Buttiker formalism for transport in mesoscopic systems, the drain current in such devices is limited by carrier injection from the source across the source–channel potential barrier, rather than velocity saturation in the pinch-off region near the drain.

$$I_{DSAT} = WC_{OX}V_{th}\left(\frac{1-r}{1+r}\right)[V_G - V_T] \tag{20.3}$$

where V_{th} is the thermal velocity of carriers in the source, and r is the reflection coefficient of carriers at the source–channel barrier. Increasing the channel mobility in *long-channel* MOSFETs reduces the reflection coefficient and should provide benefits in these *very-short-channel* alloy-based HFETs.

Unfortunately, there are other factors that complicate the drive current issue. Because of the fact that it is difficult to grow a high-quality gate oxide on a high Ge content layer, it is necessary to have a thin Si "buffer" layer on top of the $Si_{1-x}Ge_x$ channel to enable the growth or deposition of the gate dielectric. The undoped Si buffer layer, however, is tantamount to adding to the gate dielectric thickness, resulting in lower C_{OX}, and hence the drive current. Of course, μ tends not to be degraded as much in such structures as in surface channel devices due to surface roughness scattering. However, it is important that the reduction of C_{OX} is more than compensated by the increase of μ in such strained channel HFETs. The buffer layer also determines the gate bias "window" where conduction is in the higher-mobility buried strained channel, and not in the top Si cap layer.

Another design factor that is critical is the subthreshold slope that, in turn, determines the ratio of the ON (or drive) current to the OFF (or leakage) current. It turns out that often for deep submicron devices, a high drive current is less of a problem than achieving low leakage current. Here also, having a buffer layer and buried-channel operation hurts one in terms of the ability to turn the channel OFF. In addition, defects such as misfit dislocations in these heterolayers would also tend to increase the leakage current.

20.3 Buried $Si_{1-x}Ge_x$ Channel pHFETs with Si Cap

Compressively strained $Si_{1-x}Ge_x$ provides an avenue to improve hole mobility and thus increase pHFETs drive current. The first report of enhanced mobility $Si_{1-x}Ge_x$ pHFET was by Nayak et al. [12]. This was subsequently validated by other groups, with reports as early as 1993 showing 90% mobility enhancement for $Si_{0.7}Ge_{0.3}$ SIMOX HFETs over identical Si control devices. In 1995, Widener et al. [15] reported 70% mobility enhancement over Si at room temperature for $Si_{0.8}Ge_{0.2}$ pMOSFETs using a standard 0.6 μm technology process flow. They also reported a 20% drive current enhancement for these devices.

The $Si_{1-x}Ge_x$ device is buried channel because a high-quality gate oxide cannot be grown on $Si_{1-x}Ge_x$; thus a Si cap has to be grown on top of the $Si_{1-x}Ge_x$ to enable the growth of a thermal gate oxide (Figure 20.3) [16,17]. In order to maximize performance in buried-channel $Si_{1-x}Ge_x$ pHFETs, the Si cap needs to be as thin as possible, though there are trade-offs. Thicker Si caps degrade gate capacitance, but very thin caps lead to increased surface roughness scattering [13].

$Si_{0.9}Ge_{0.1}$ pHFETs show that mobility-enhanced drive current as well as improved short-channel effects can be achieved in buried-channel $Si_{1-x}Ge_x$ pHFETs with an optimized Si cap and relatively modest amounts of Ge (up to ~20%) (Figure 20.4) [18,19]. The gate-to-channel inversion capacitance difference for the $Si_{0.9}Ge_{0.1}$ and Si control devices results from the unconsumed Si cap layer in $Si_{0.9}Ge_{0.1}$ HFETs. The inversion equivalent oxide thickness (EOT) is 4.6 and 5.0 nm for Si control and $Si_{0.9}Ge_{0.1}$, respectively, including polydepletion and quantum-mechanical effects. The slightly higher inverse subthreshold slope (SS) in the long-channel $Si_{1-x}Ge_x$ HFET is due to buried-channel operation, but when $L_G < 0.2$ μm, the SS for Si and $Si_{1-x}Ge_x$ is about the same due to the improved short-channel effects (SCE) in the $Si_{1-x}Ge_x$ devices. The improved SCE is due to the reduction of B diffusivity in $Si_{1-x}Ge_x$, from the source–drain (S–D) regions. If the Si cap is completely consumed, then the gate–oxide interface reaches the $Si_{0.9}Ge_{0.1}$ layer, and a much higher SS is observed due to the poor gate–oxide interface quality.

The transistor I_{DS}–V_{GS} characteristics for $L_G = 70$ nm show that the turn-off characteristics are good for both Si and $Si_{1-x}Ge_x$ pHFETs (Figure 20.4a), with $I_{off} = $ ~25 nA/μm, and SS for Si and $Si_{1-x}Ge_x$ pHFETs are both ~107 mV/dec. $Si_{0.9}Ge_{0.1}$ shows a slightly smaller DIBL, and this could be due to shallower S–D junctions because of reduced B diffusivities in SiGe, that would lead to an increased effective channel length. The p–n junction leakage level for $Si_{0.9}Ge_{0.1}$ device is only slightly higher than the Si control, and more importantly, the OFF current at normal device operation condition

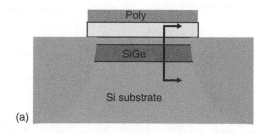

(a)

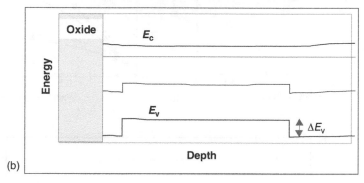

(b)

FIGURE 20.3 (a) pHFET structure with compressive $Si_{1-x}Ge_x$ or partially strain compensated $Si_{1-x-y}Ge_xC_y$ channel, Si cap and p^+ source–drains. (b) The band diagram as a function of depth indicates that the holes will be confined in the $Si_{1-x}Ge_x$ or $Si_{1-x-y}Ge_xC_y$ layer due to the valence band offset, as long as we operate in a suitable gate voltage operating window.

($V_{GS} = 0\,V$, $V_{DS} = -V_{DD}$) for $Si_{0.9}Ge_{0.1}$ device is about the same, or can be slightly lower than for the Si control device if V_T differences are normalized. The I_{DS}–V_{DS} characteristics for the $L_G = 70\,nm$, as seen in Figure 20.5b, show that $Si_{0.9}Ge_{0.1}$ has 17% higher drive current than Si even for such short-channel lengths and modest Ge mole fractions, indicating that the improved long-channel mobility translates to better source injection of carriers into the channel.

The linear V_T (defined by I_{DS}–V_{GS} curve at $V_{DS} = -100\,mV$, extrapolated to zero from maximum transconductance, G_m, point) as a function of channel length shows minimal V_T roll-off for both Si and $Si_{0.9}Ge_{0.1}$ devices. The V_T for $Si_{0.9}Ge_{0.1}$ is about 90 mV lower than for Si. This is due to the valence band offset between Si and $Si_{0.9}Ge_{0.1}$ [20,21]. X-ray diffraction (XRD) scans for $Si_{0.9}Ge_{0.1}$ before and after processing show no shift of the Ge peak. This suggests that the $Si_{0.9}Ge_{0.1}$ channel is still under compressive strain after processing. This is critical for strained $Si_{1-x}Ge_x$ HFET integration since for low Ge mole fractions the benefits of strained $Si_{1-x}Ge_x$ channel HFET are lost once the $Si_{1-x}Ge_x$ layer is relaxed. These results show that modest amounts of Ge (10% to 20%) in $Si_{1-x}Ge_x$ films can be used to fabricate high-performance buried-channel pHFETs. By carefully engineering a triangular Ge profile in $Si_{1-x}Ge_x$ [22], and the Si cap in these devices, it is possible to fabricate buried-channel $Si_{1-x}Ge_x$ pHFETs that have higher drive currents than surface channel Si pMOSFETs [23]. Since the mobility of holes in Si is lower than electrons, by using this approach, it is possible to obtain a more balanced CMOS process than is currently achieved in conventional Si CMOS processes.

20.4 Strain-Compensated $Si_{1-x-y}Ge_xC_y$ Buried-Channel pMOSFET

The first application of a *partially* strain-compensated ternary $Si_{1-x-y}Ge_xC_y$ alloy in a pHFET was by Ray et al. [24]. Since C is smaller than both Si and Ge, it can introduce local tensile strain and compensate

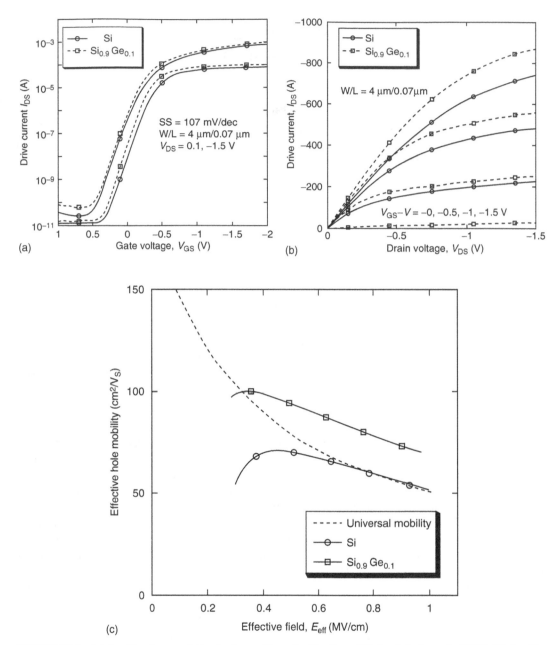

FIGURE 20.4 (a) I_{DS}–V_{GS} characteristics for $L_G = 70\,nm$ for $Si_{0.9}Ge_{0.1}$–SiO_2 buried-channel pHFET. (b) I_{DS}–V_{DS} characteristics for $L_G = 70\,nm$ for $Si_{0.9}Ge_{0.1}$–SiO_2 buried-channel pHFET. (c) Effective hole mobility versus field for $Si_{0.9}Ge_{0.1}$–SiO_2 buried-channel pHFET.

the compressive strain introduced by Ge, leading to complete strain compensation at a Ge-to-C ratio of 8:1 [25,26]. Carbon seems to have a lesser impact on band structure than on strain. Thus, with ternary $Si_{1-x-y}Ge_xC_y$ alloys, another degree of freedom is possible where C can, somewhat *independently*, adjust the bandgap and strain to some extent, which is not possible with $Si_{1-x}Ge_x$ alone. It also makes the $Si_{1-x-y}Ge_xC_y$ layers more robust in terms of the allowable thermal budget during processing, and allows higher levels of Ge incorporation.

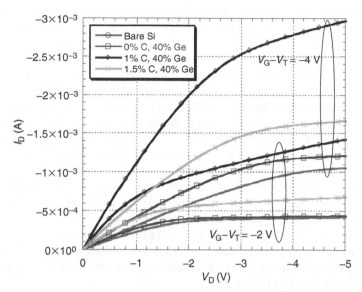

FIGURE 20.5 $Si_{0.585}Ge_{0.4}C_{0.015}$ shows 55% drive current enhancement over bulk Si and 42% enhancement over $Si_{0.6}Ge_{0.4}$; $W/L = 10/0.5\,\mu m$; $t_{ox} = 6\,nm$.

Results for pHFETs with high (40%) Ge show that incorporation of dilute levels of C (1.5%) allows the compressive strain to be retained much better than for the binary alloys, leading to higher drive currents. The $Si_{1-x-y}Ge_xC_y$ pHFETs have higher drive current than both control Si and $Si_{1-x}Ge_x$ devices (Figure 20.5). This is believed to be due to the fact that for these high levels of Ge, C helps maintain the compressive strain even after high-temperature device processing better than $Si_{1-x}Ge_x$ alone. Carbon also yields a smoother heterointerface as seen by atomic force microscopy, which improves channel mobility compared to that in binary $Si_{1-x}Ge_x$ alloys. The results showed that by partially compensating the strain, it is possible to enhance mobility and drive currents for $Si_{1-x-y}Ge_xC_y$ pHFETS with high Ge mole fractions. A key point is that for the range of Ge mole fractions studied, the optimal amount of C depends on the amount of Ge in the $Si_{1-x-y}Ge_xC_y$ pMOSFETs, as well as the channel length. In particular, for $x = 0.15$ and $y = 0.006$, there is mobility degradation, while for $x = 0.2$ and $y = 0.007$ there is enhanced drive current. Other results have also shown for higher amounts of Ge, an increased amount of C can be used to enhance mobility and drive current. The conclusion that can be drawn from these studies is that while C can be used to compensate strain, full strain-compensation is undesirable; instead C should be used to relax thermal budget constraints while maintaining sufficient compressive strain for mobility enhancement. Alloy scattering is especially critical for these ternary alloys because of the high deformation potential for C in Si [27–29].

Obviously, there are trade-offs in terms of the Ge and C mole fractions and the strained channel layer thickness in terms of how it impacts the allowable thermal budget [22,23]. An obvious extension is to go to much higher Ge mole fractions up to 100% Ge where Fitzgerald's group [30] has shown dramatic enhancements of hole mobility. Perhaps Ge:C layers can be grown directly on Si, without having to grow thick relaxed SiGe buffer layers, which present their own manufacturing and device challenges.

20.5 Surface-Channel Si and $Si_{1-x}Ge_x$/High-k pHFETs

We have seen that by careful engineering of the sacrificial Si cap on $Si_{1-x}Ge_x$, which is required for a high-quality thermal gate oxide, it is possible to obtain device performance enhancement in nanometer scale buried-channel $Si_{1-x}Ge_x$ and $Si_{1-x-y}Ge_xC_y$ pHFETs. An alternate approach is to remove the need for the Si cap by using a deposited gate oxide, such as a high dielectric constant (high-k) gate dielectric, thereby rendering a surface channel device [31,32]. This is particularly attractive for a $Si_{1-x}Ge_x$ or pure Ge channel because it avoids the problems of Ge segregation during thermal gate oxidation.

HfO_2 is considered to be one of the most promising high-k gate dielectrics to replace SiO_2 and achieve lower leakage currents at a comparable EOT. With a deposited HfO_2 gate dielectric replacing thermally grown SiO_2, SCE associated with buried-channel MOSFETs, and Si cap layer control challenges would no longer exist. In addition, one could harness the benefits associated with having a high-k gate dielectric, namely higher drive currents and lower off-state leakage currents. One drawback with HfO_2 is that it causes channel mobility degradation. The use of a higher mobility channel layer could recover some of this mobility and drive current degradation.

Compared to a Si–SiO_2 pMOSFET, a Si–HfO_2 pMOSFET exhibits a *peak* mobility degradation of about 25%, and 12% degradation at $E_{eff} = 1\,MV/cm$ (Figure 20.6a). However, the $Si_{0.8}Ge_{0.2}$ pHFET channel mobility with high-k is significantly higher than the Si–SiO_2 pMOSFET control sample, maintaining a 30% mobility enhancement at $E_{eff} = 1\,MV/cm$. These results show that the use of a compressively strained $Si_{1-x}Ge_x$ channel can provide a means to overcome the mobility degradation on Si caused by using HfO_2 as a gate dielectric. This mobility enhancement enables a higher drive current, even at channel lengths down to 180 nm.

Unfortunately, the subthreshold characteristics of such 180 nm devices (Figure 20.6b) show that while the drain-induced barrier lowering (DIBL) is comparable for both $Si_{0.8}Ge_{0.2}$ and Si control devices, the SS and junction leakage is significantly worse for $Si_{0.8}Ge_{0.2}$ pHFETs. The higher junction leakage may be caused by the narrower bandgap in $Si_{0.8}Ge_{0.2}$ or defects such as misfit dislocations. The degraded SS, 105 mV/dec for the $Si_{0.8}Ge_{0.2}$ device versus 85 mV/dec for the Si device is indicative of a higher interface trap density (D_{IT}). This is probably due to the $Si_{1-x}Ge_x$–HfO_2 having higher dangling bond density than the Si–HfO_2 interface. Another possibility is that Ge segregates out of the strained lattice and accumulates at the $Si_{1-x}Ge_x$–HfO_2 interface in a manner similar to the $Si_{1-x}Ge_x$–(Si/Ge)O_2 case causing increased D_{IT}.

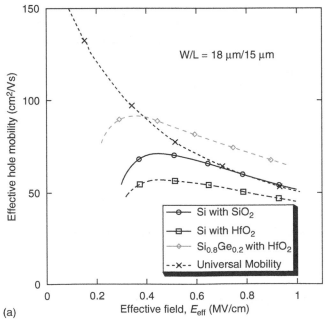

(a)

FIGURE 20.6 (a) I_{DS}–V_{GS} characteristics for $L_G = 180$ nm for $Si_{0.8}Ge_{0.2}$–HfO_2 pHFET.

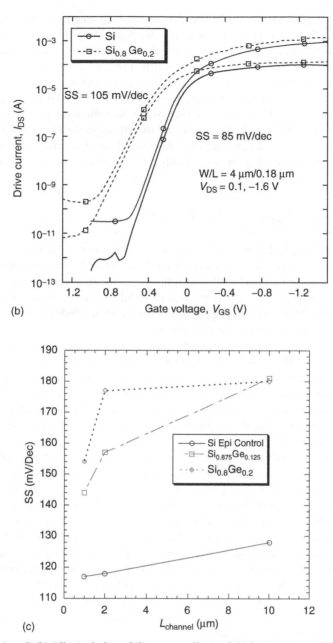

FIGURE 20.6 (*continued*) (b) Effective hole mobility versus effective field for $Si_{0.8}Ge_{0.2}$–HfO_2 pHFET. (c) SS versus channel length for the $Si_{0.875}Ge_{0.125}$ and $Si_{0.8}Ge_{0.2}$ samples compared to an epitaxial Si control device.

The significant difference in V_T between the $Si_{0.8}Ge_{0.2}$ and Si devices (210 mV) can be attributed to the bandgap difference between $Si_{0.8}Ge_{0.2}$ and Si. The bandgap difference is manifested mostly in the valence band and for Ge mole fraction of 0.2 represents a 168 mV valence band offset. Finally, if Ge is segregating to the surface during processing, as in the $Si_{1-x}Ge_x$–$(Si/Ge)O_2$ case, this could represent an increase in fixed negative oxide charge that would correspond to a further reduction in V_T for the $Si_{0.8}Ge_{0.2}$ device.

There is an anomalous dependence of V_T and SS on channel length, L, with HfO$_2$ gate dielectrics (Figure 20.6c). While V_T becomes more positive with decreasing L, SS *decreases* with reduction in L. SS changes from 144 to 105 mV/dec as the channel length decreased from 8 to 0.18 μm for the Si$_{1-x}$Ge$_x$ pHFETs, and 128 to 85 mV/dec for the Si pMOSFETs. These observations indicate that the longer channel length devices must have an increased D_{IT} and fixed oxide charges. There are reports that HfO$_2$ may be impervious to H$_2$ at the 400°C forming gas anneal temperature that was used for sintering. It is possible that the diffusion of H$_2$ during sintering occurs laterally from the contact holes along the channel, rather than through the polysilicon gate and then through the HfO$_2$, which can explain these anomalous results in terms of SS dependence on L. For long-channel devices, a significant portion of the channel would remain unsintered, as H$_2$ would only diffuse into a small portion of the channel, thus rendering devices with high SS.

20.6   n-Channel Devices for CMOS

We have so far focused on pHFETs in Si$_{1-x}$Ge$_x$ channels. The rationale used here is that in CMOS, since hole channel mobility is roughly 2.5 times lower than electron mobility, improving the PFET performance would enable a more balanced CMOS layout. However, while this is valid for static logic, in dynamic CMOS, one uses a PFET to precharge the nodes, and the speed depends more on the pull-down NFETs. Even for static CMOS, for many applications, one is often interested in improving the ratio of ($I_{Dsat,n-ch} + I_{Dsat,p-ch}$) over the sum of the OFF currents.

Unfortunately, the in-plane electron mobility *decreases* with compressive strain in Si$_{1-x}$Ge$_x$, for modest Ge mole fractions; thus it would be expected that the NFET drive current would also decrease. However, Yeo et al. [33] have shown that while this is the case for long-channel devices, as the channel length is scaled below 0.4 μm, enhanced drive current is obtained in Si$_{1-x}$Ge$_x$ NHFETs compared to similarly processed Si control devices. They attributed this to the reduced scattering caused by conduction band splitting. They postulated that this reduced scattering would yield higher optical phonon-limited carrier saturation velocity.

As the Ge mole fraction is increased, all the way to pure Ge, clearly the electron mobility in Si$_{1-x}$Ge$_x$ or Ge is higher than in Si [30]. Hence, it should lead to improvement of both p- and n-channel devices. However, such high Ge mole fraction or pure Ge layers can currently only be grown defect-free on thick relaxed SiGe buffer layers. However, the problems with such thick SiGe buffer layers (as with tensile strained Si on SiGe relaxed buffers) include cost, manufacturability challenges, isolation issues with shallow-trench isolation along the SiGe layers, lower thermal conductivity of the SiGe leading to self-heating effects in the FETs, and threading dislocation propagation into the channel regions during device fabrication.

Perhaps a more palatable solution may be to avoid growing thick SiGe buffer layers, and instead use the compressively strained thin Si$_{1-x}$Ge$_x$-on-Si channels only in the pHFET active regions (unfortunately requiring selective epitaxy or etching). Clever ideas include a *dual-channel* CMOS concept proposed by O'Neil and Antoniadis [34], where in the n-channel devices, conduction is in a Si channel while in the PFETs it is in the Si$_{1-x}$Ge$_x$ layer.

20.7   Vertical HFETs

Yet another solution that may prove to be attractive is to grow a thin, compressively strained Si$_{1-x}$Ge$_x$ channel directly on a Si substrate without requiring a thick SiGe relaxed buffer, etch mesas, and fabricate vertical HFETs on the *sidewalls* of the Si$_{1-x}$Ge$_x$ islands [35–39]. As shown in Figure 20.7, one can achieve significant enhancement of drive currents over control Si PFETs. As seen, it is important for current enhancement that the compressive strain be maintained which causes the splitting of the heavy and light hole bands, leading to lower interband scattering. These FETs were on large mesas, leading to partially depleted FETs with nonoptimized short-channel effects. However, it should be possible to achieve much better DIBL using fully depleted FETs.

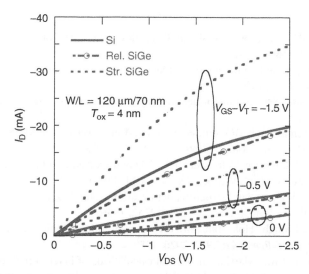

FIGURE 20.7 Vertical Si, relaxed and strained $Si_{1-x}Ge_x$ pHFETs showing the enhancement of drive current over control Si MOSFETs only in the presence of strain.

In this case, it also leads to an enhancement of the n-channel drive current [38,39]. This is because, as mentioned in Section 20.2, the four in-plane conduction band valleys are lowered in energy. While that leads to an *increase* of *in-plane* electron effective mass, obviously it leads to a *lower* (*transverse*) *out-of-plane* effective mass. This, along with a reduction of the f-type intervalley electron scattering, leads to an improvement of electron mobility and NFET drive currents. Of course, vertical FETs pose their own process integration challenges compared to planar devices.

20.8 Summary

As challenges to scaling continue to grow, it is prudent to examine nontraditional methods of improving CMOS performance. Because hole mobility is lower than electron mobility in silicon, it would be attractive to use materials that could enhance hole over electron mobility, and thus, provide a balanced CMOS process that reduces the total area used for a circuit as well as improves its performance. This chapter has presented results and discussion of buried-channel $Si_{1-x}Ge_x$, and $Si_{1-x-y}Ge_xC_y$ pHFETs. It has been shown that with even modest amounts of Ge (10% to 20%), buried-channel $Si_{1-x}Ge_x$ pHFETs can be used in deeply scaled devices and provide performance enhancement over Si. For much higher Ge mole fraction channels grown on Si *without* a SiGe buffer layer, partially strain-compensated $Si_{1-x-y}Ge_xC_y$ may be a viable option. It may be possible that for higher amounts of Ge, up to pure Ge, the increased hole and electron mobilities may enable improved pHFETs as well as NHFETs. Experimental results presented in this chapter show that $Si_{1-x}Ge_x$–HfO_2 surface channel pHFETs provide a means to recover the mobility degradation that occurs in Si–HfO_2 pMOSFETs. Vertical HFETs are another intriguing device structure in our repertoire that should be considered. Unfortunately, all these options have their pros and cons.

Acknowledgments

The author would like to acknowledge his former doctoral students, D. Onsongo, S. John, E. Quinones, Z. Shi, K. Jayanarayan, X. Chen, and post-docs, Dr. Samit Ray and Dr. Freek Prins, for their invaluable contributions. The work has been supported over the years by SRC, NSF, DARPA, Texas Advanced Technology/Research Program, Intel, TI, Micron Foundation, and Applied Materials.

References

1. BS Meyerson. UHV/CVD growth of Si and Si:Ge alloys: chemistry, physics and device applications. *Proc. IEEE* 80: 1592–1608, 1992.

2. SK Banerjee. Bandgap and strain engineered SiGeC vertical and planar MOSFETs. *Microelectr. Eng.* 69(2–4): 106–117, 2003 (invited paper).

3. JC Bean. Silicon-based semiconductor heterostructures: column IV bandgap engineering. *Proc. IEEE* 80: 571–587, 1992.

4. KL Wang, SG Thomas, and MO Tanner. SiGe band engineering for MOS, CMOS and quantum effect devices. *J. Mater. Sci.: Mater. Electron.* 6: 311–324, 1995.

5. C Li, S John, and SK Banerjee. Cold-wall UHVCVD of doped and undoped Si and $Si_{1-x}Ge_x$ epitaxial films using SiH_4 and Si_2H_6. *J. Vac. Sci. Technol. A* 14: 170–183, 1996.

6. MV Fischetti and SE Laux. Band structure, deformation potentials, and carrier mobility in strained Si, Ge and SiGe alloys. *J. Appl. Phys.* 80: 2236–2252, 1996.

7. JM Hinckley and J Singh. Hole transport theory in pseudomorphic $Si_{1-x}Ge_x$ alloys grown on Si (0 0 1) substrates. *Phys. Rev. B* 41: 2912–2926, 1989.

8. CG Van de Walle and RM Martin. Theoretical calculations of heterojunction discontinuities in the Si/Ge system. *Phys. Rev. B* 34: 5621–5634, 1986.

9. FM Bufler, P Graf, B Meinerzhagen, G Fischer, and H Kibbel. Hole transport investigation in unstrained and strained SiGe. *J. Vac. Sci. Technol. B* 16(3), 1667–1669, 1998.

10. A Sadek and K Ismail. Si/SiGe CMOS possibilities. *Solid State Electron.* 38: 1731–1736, 1995.

11. A Sadek, K Ismail, MA Armstrong, D Antoniadis, and F Stern. Design of Si/SiGe heterojunction complementary metal-oxide-semiconductor transistors. *IEEE Trans. Electron Dev.* 43: 1224–1232, 1996.

12. D Nayak, JCS Woo, JS Park, KL Wang, and KP MacWilliams. Enhancement-mode quantum-well GeSi MOSFET. *IEEE Electron Dev. Lett.* 12: 154–156, 1991.

13. S Verdonck-Vandebroek, EF Crabbe, BS Meyerson, DL Harame, PJ Restle, JMC Stork, and JB Johnson. SiGe-channel heterojunction p-MOSFETs. *IEEE Trans. Electron Dev.* 41: 90–101, 1994.

14. M Lundstrom. Elementary scattering theory of the Si MOSFET. *IEEE Electron Dev. Lett.* 18, 361–363, 1997.

15. JO Widener, KR Hofmann, F Hormann, and L Risch. Characterization of SiGe quantum-well p-channel MOSFETs. *J. Mater. Sci.—Mater. Electron.* 6(5): 325, 1995.

16. SK Ray, S Maikap, SK Samanta, and S Banerjee. Charge trapping characteristics of ultrathin oxynitrides on $Si/Si_{1-x-y}Ge_xC_y/Si$ heterolayers. *Solid State Electron.* 45(11): 1951–1955, 2001.

17. S Maikap, SK Ray, and SK Banerjee. Electrical properties of O_2/NO-plasma grown oxynitride films on partially strain compensated $Si/Si_{1-x-y}Ge_xC_y/Si$ heterolayers. *Semicond. Sci. Technol.* 16(3): 160–163, 2001.

18. ZH Shi, D Onsongo, and SK Banerjee. Mobility and performance enhancement in compressively strained SiGe channel pMOSFETs. *Appl. Surf. Sci.* 224(1–4): 248–253, 2004.

19. GS Kar, SK Ray, T Kim, and SK Banerjee. Estimation of hole mobility in strained $Si_{1-x}Ge_x$ buried channel heterostructure pMOSFET. *Solid State Electron.* 45(5): 669–676, 2001.

20. CG Van de Walle. Band lineups and deformation potentials in the model-solid theory. *Phys. Rev. B* 3: 1871–1883, 1989.

21. X Wang, DL Kencke, KC Liu, F Register, and SK Banerjee. Band alignments in sidewall strained Si/strained SiGe heterostructures. *Solid State Electron.* 46(12): 2021–2025, 2002.

22. SP Voingescu, CAT Salama, JP Noel, and TI Kamins. Optimized Ge channel profiles for VLSI compatible Si/SiGe p-MOSFETs. *Proc. IEDM* 369–372, 1994.

23. Z Shi, X Chen, D Onsongo, E Quinones, and SK Banerjee. Simulation and optimization of strained $Si_{1-x}Ge_x$ buried channel p-MOSFETs. *Solid State Electron.* 44(7): 1223–1228, 2000.

24. SK Ray, S John, E Quinones, SK Oswal, and SK Banerjee. Heterostructure p-channel metal-oxide-semiconductor transistor utilizing a $Si_{1-x}Ge_xC_y$ channel. *IEEE IEDM Tech. Dig.* 261–264, 1996; also *Appl. Phys. Lett.* 74: 847, 1999.

25. SS Iyer, K Ebert, AR Powell, and BA Ek. SiCGe ternary alloys-extending Si-based heterostructures. *Microelectron. Eng.* 19: 351–354, 1992.

26. K Ebert, SS Iyer, S Zollner, JC Tsang, and FK LeGoues. Growth and strain compensation effects in the ternary $Si_{1-x-y}Ge_xC_y$ alloy system. *Appl. Phys. Lett.* 60: 3033–3035, 1992.

27. GS Kar, A Dhar, LK Bera, and SK Banerjee. Effect of carbon on lattice strain and hole mobility in $Si_{1-x}Ge_x$ alloys. *J. Mater. Sci.—Mater. Electron.* 13(1): 49–55, 2002.

28. GS Kar, S Maikap, SK Ray, and SK Banerjee. Effective mobility and alloy scattering in the strain compensated SiGeC inversion layer. *Semicond. Sci. Technol.* 17(5): 471–475, 2002.

29. GS Kar, S Maikap, SK Banerjee, and SK Ray. Series resistance and mobility degradation factor in C-incorporated SiGe heterostructure p-type metal-oxide semiconductor field-effect transistors. *Semicond. Sci. Technol.* 17, 938–941, 2002.

30. ML Lee and EA Fitzgerald. Electron mobility characteristics of n-channel metal-oxide-semiconductor field-effect transistors fabricated on Ge-rich single- and dual-channel SiGe heterostructures. *J. Appl. Phys.* 95(3): 1550–1555, 2004.

31. ZH Shi, D Onsongo, K Onishi, J Lee, and SK Banerjee. Mobility enhancement in surface channel SiGe pMOSFETs with HfO_2 gate dielectrics. *IEEE Electron Dev. Lett.* 24(1): 34–36, 2003.

32. T Ngai, WJ Qi, R Sharma, and SK Banerjee. Transconductance improvement in surface-channel SiGe p-metal-oxide-silicon field-effect transistors using a ZrO_2 gate dielectric. *Appl. Phys. Lett.* 78(20): 3085–3087, 2001.

33. YC Yeo, Q Lu, T King, C Hu, T Kawashima, M Oishi, S Mashiro, and J Sakai. Enhanced performance in sub-100 nm CMOSFETs using strained epitaxial Si–Ge. *IEEE IEDM Tech. Dig.* 753–756, 2000.

34. AG O'Neill and DA Antoniadis. Deep submicron CMOS based on silicon germanium technology. *IEEE Trans. Electron Dev.* 43(6): 911–918, 1996.

35. XD Chen, QQ Ouyang, SK Jayanarayanan, F Prins, and S Banerjee. Vertical p-type high-mobility heterojunction metal-oxide-semiconductor field-effect transistors. *Appl. Phys. Lett.* 78(21): 3334–3336, 2001.

36. XD Chen, Q Ouyang, SK Jayanarayanan, and SK Banerjee. An asymmetric $Si/Si_{1-x}Ge_x$ channel vertical p-type metal-oxide-semiconductor field-effect transistor. *Solid State Electron.* 45(2): 281–285, 2001.

37. X Chen, KC Liu, SK Jayanarayanan, and S Banerjee. Electron mobility enhancement in strained SiGe vertical n-type metal-oxide semiconductor field-effect transistors. *Appl. Phys. Lett.* 78(3), 377, 2001.

38. XD Chen, QC Ouyang, G Wang, AF Tasch, and SK Banerjee. Improved hot-carrier and short-channel performance in vertical nMOSFETs with graded channel doping. *IEEE Trans. Electron Dev.* 49(11): 1962–1968, 2002.

39. XD Chen, KC Liu, QC Ouyang, and SK Banerjee. Hole and electron mobility enhancement in strained SiGe vertical MOSFETs. *IEEE Trans. Electron Dev.* 48(9): 1975–1980, 2001.

21

Industry Examples at the State-of-the-Art: Intel's 90 nm Logic Technologies

Scott E. Thompson
University of Florida

21.1 Introduction

This chapter describes uniaxial strained Si and $Si_{1-x}Ge_x$ heteroepitaxy introduced for the first time at the 90 nm technology generation into high-volume manufacturing. For more than 30 years, CMOS device technologies have improved at a dramatic rate due to dimension scaling. Scaling the vertical and horizontal MOSFET dimension reduces channel resistance through increased inversion charge and lower source to drain resistance, respectively. It is this unique property of higher performance and lower cost through dimension scaling that has established the MOSFET as the clearly dominant solid-state device. The semiconductor and microelectronic industry has made remarkable and nearly unprecedented progress during the last 30 years. During this time, the MOSFET gate length scaled from 10 μm to 45 nm and now contains many features at the nanoscale. Figure 21.1 shows the evolution pictorially with Lilenfield's MOSFET concept, the first experimental transistor in 1947, and the present day 45 nm transistor which incorporates $Si_{1-x}Ge_x$ in the source and drain to strain the Si channel [1–3].

Equally impressive are the improvements in cost and density made by the industry. For the 90 nm technology generation, greater than 200 billion transistors are fabricated on a standard 300 mm wafer. Hundreds of millions of transistors can be fabricated on a single chip with a manufacturing cost of only a few dollars. Figure 21.2 shows the wafer size history for the semiconductor industry and the size of the 300 mm wafers now state-of-the-art for 90 nm technology manufacturing. Because of these massive improvements in productivity, Gordon Moore observed that transistors are basically free [4].

The introduction of strained Si at the 90 nm technology generation with selective heteroepitaxy represents a significant departure from the historical feature size scaling. This deviation from traditional scaling is needed because conventional MOSFETs have reached some atomic-level limits. During the early years of transistor scaling, Gordon Moore (in 1965) observed that the number of transistors on a chip increased exponentially over time [5,6], which has become known as Moore's Law. However,

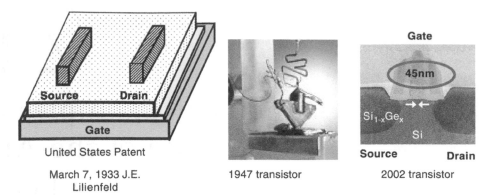

FIGURE 21.1 History of transistor development.

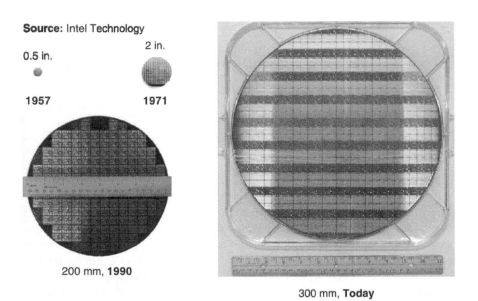

FIGURE 21.2 History of production wafer size.

according to Moore himself, "no exponential is forever" [4]. Because of high off-state leakage, the scaling limit for the planar MOSFET is approximately 20 nm. Planar MOSFETs as small as 10 nm have been fabricated; however, they do not appear practical due to high leakage [7].

The undesirable off-state leakage results from many mechanisms with source-to-drain subthreshold and gate tunneling current as the largest contributors. Aggressive gate oxide scaling during the last 30 years has resulted in a 1.2 nm physical oxide at the 90 nm generation [3], which is at the gate tunneling leakage limit for SiO_2. This is significant since some consider SiO_2 (as opposed to the Si channel) the foundation of the modern MOSFET. At present, high k gates are not ready for manufacturing due to degraded mobility [8] especially at low gate bias. Thus, the semiconductor industry needs some other material change to continue MOSFET scaling. Starting at the 90 nm technology generation, strained Si is one such material change which has been widely adopted [1–3,9]. At present, strained Si offers performance gains much larger than any other new material options. Figure 21.3 plots the feature size progress during the last 30 years and highlights the unprecedented number of new materials needed to maintain historical improvements going forward.

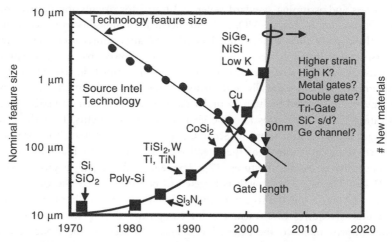

FIGURE 21.3   Technology and gate feature size and new materials needed versus time.

This chapter is organized as follows. Section 21.2 describes a strained silicon process flow in commercial production at the 90 nm logic technology generation. This section also shows the industry power and leakage trends and justifies why new materials like strained Si are needed. Section 21.3 briefly looks at the future of strained Si concepts and direction.

21.2   90 nm Strained Silicon Technology

Smith first measured the mobility enhancement through lattice strain of single-crystal silicon and germanium in 1954 [10]. Since then p-type Si has found wide application in mechanical sensors. However, until recently strain has not been incorporated into a production CMOS logic technology for several reasons. First, for the last 30 years, gate oxide and shallow junction scaling obtained adequate MOSFET improvement. Second, biaxial and uniaxal stress are difficult to integrate. The difficulty with biaxial stress is that it is introduced early in the process flow before gate formation requiring significant adjustments to lower the entire midsection thermal cycles. The difficulty with uniaxial mechanical stress is failure to improve both n-type and p-type MOSFETs simultaneously. Finally, heteroepitaxy processes are high-yield risk due to threading dislocations. In this section, the strain silicon process flow used in a 90 nm CMOS logic technology is discussed along with yield and process integration issues.

The unique advantage of this uniaxial stressed Si process flow is that (on the same wafer) compressive strain is introduced into the p-type and tensile strain in the n-type MOSFETs to improve both the electron and hole mobilities. The flow differs from past uniaxial stress work by introducing hetero-epitaxy $Si_{1-x}Ge_x$ in the p-channel source and drain. The use of heterojunctions in the p-channel MOSFET has previously been proposed [11–16] for several reasons. Ozturk [14,15] first introduced $Si_{1-x}Ge_x$ into the source–drain for the purposes of higher boron activation and abrupt profile. Banerjee [11,12,16] introduced strained $Si_{1-x}Ge_x$ into the source and drain for bandgap engineering. These past advantages are valid and provide some additional benefit in this work. However, heteroepitaxy in the source and drain in nanoscale devices results in uniaxial Si channel stress, which can significantly enhance the mobility.

Process Flow

Only slight modifications to a standard CMOS logic technology process flow are required to insert the compressive strain into the p-type and tensile strain into the n-type MOSFETs.

For uniaxial stress and assuming standard wafer and transistor orientation, lattice compression for p-channel and tension for n-channel MOSFETs are needed for mobility enhancement. To fabricate the strained Si, the process flow is nearly unchanged until after source–drain extension and spacer formation. Postspacer etch, an Si recess etch is inserted followed by selective epitaxial $Si_{1-x}Ge_x$ deposition (Figure 21.4a). The silicon etch is blocked from n-channel devices and poly-Si gates. The Si recess etch removes 100 nm vertically and 70 nm laterally from the p-channel source and drain. The etch is intentionally targeted to laterally under cut the spacer to bring the $Si_{1-x}Ge_x$ closer to the channel. Fabricating the $Si_{1-x}Ge_x$ closer to the channel has two benefits. First the channel stress is increased for larger mobility enhancement. Second, and equally important, the external resistance is reduced. However, stringent controls of the lateral etch are needed to maintain drive current and performance uniformity across all structures [17]. Next, epitaxial $Si_{1-x}Ge_x$ is grown in the source and drain (Figure 21.4b). The $Si_{1-x}Ge_x$ growth is targeted to be raised above the gate plan such that for the first time raised source and drains are introduced at the 90 nm technology generation. Raising the source and drains requires little extra cost or complexity during the $Si_{1-x}Ge_x$ deposition but offers significant improvement to the external resistance. As seen in the device cross section and mentioned previously, the $Si_{1-x}Ge_x$ is blocked from the poly-Si gates even though $Si_{1-x}Ge_x$ gates offer improved p-channel performance. The $Si_{1-x}Ge_x$ is blocked from the poly-Si gate out of concern for mushroom growth and degraded contact to gate design rule margin. The remaining process flow is conventional except for the salicide (Figure 21.4c). $Si_{1-x}Ge_x$ in the source and drain requires extensive changes to the salicide since Ge inhibits the $CoSi_2$ transition to the low-resistivity disilicide phase [18]. To solve this problem, NiSi instead of $CoSi_2$ is used. Nickel silicide requires extensive changes since all formation and postformation process steps need to be less than 500°C.

After salicide formation, longitudinal uniaxial tensile strain is introduced into the n-type MOSFET by engineering the tensile stress and thickness of the Si nitride-capping layer [19,20] present to support unlanded contacts (Figure 21.4d). Stress-induced tensile capping films are widely adopted at the 90 nm technology generation [1,2,9] and can improve n-channel device saturated drive current 10 to 15% [1,9,19,20]. However, the tensile stress from the capping layer needs to be relaxed from the p-channel device since it causes significant degradation [1,9,19,20]. There are several techniques to almost completely neutralize the capping layer strain; one is the use of a Ge implant and masking layer [20].

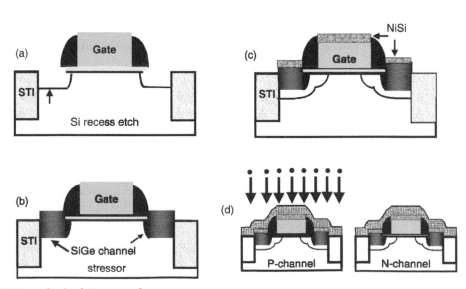

FIGURE 21.4 Strained Si process flow.

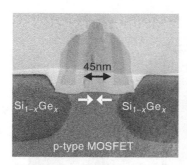 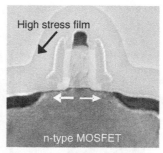

FIGURE 21.5 Transistor cross section of 45 nm gate length strained Si transistors used at 90 nm technology generation.

Raising the p-channel source and drain also reduces the negative effect of the tensile capping layer on the p-MOSFET. At the 90 nm technology generation, the thickness of the capping layer is approximately 80 nm and chosen as a balance between transistor performance and contact etch integration requirements.

The use of (i) $Si_{1-x}Ge_x$ in the p-channel source and drain, (ii) a tensile capping film, and (iii) capping film strain relaxation off the p-MOSFET, allows independent targeting of the n- and p-type MOSFET Si channel strain (by adjusting capping films stress for n-type and Ge source–drain concentration for p-type). Also, the n- and p-channel stresses are predominantly uniaxial, which are desirable since uniaxial stress offers many advantages in electrical performance over biaxial stress (see Chapter 19).

Strained Si Process Flow: Results and Discussion

TEM micrographs of 45 nm p- and n-type MOSFETs are shown in Figure 21.5, which are patterned with 193 nm lithography. At the 90 nm technology generation, 17% germanium concentration ($Si_{0.83}Ge_{0.17}$) is used, which has a lattice spacing ~1% larger than Si. The mismatch in the $Si_{1-x}Ge_x$ to Si lattice causes the smaller lattice constant Si channel to be under compressive strain. The tensile capping layers' stress on the n-channel is more complicated. The capping layer introduces longitudinal tensile and compressive out-of-plane (z) stress. Using Florida object oriented process simulator (FLOOPS) three-dimensional finite element analysis [21,22], the channel stress is calculated for the n- and p-channel MOSFETs. For the nominal 45 nm transistor used at the 90 nm technology generation, ~500 MPa of uniaxial longitudinal compressive stress is introduced into the p-channel MOSFET in the inversion layer. Stress contours for the p-channel device are shown in Figure 21.6. For the n-channel MOSFET, ~300 MPa of longitudinal tensile and out-of-plane compressive stress is present in the silicon inversion layer. Fortunately, both tensile longitudinal stress and out-of-plane compressive stress raise the energies of the x and y conduction band valleys and lowers the energy of the z valleys. Once the energy separation of the valleys is greater than kT, valley repopulation to the z valleys occurs. The z valleys have the desirable low in-plane and high out-of-plane effective mass and enhance the electron mobility.

As a point of comparison, the uniaxial stress process flow described here introduces significantly less n- and p-channel stress compared to biaxial (~1 to 2 GPa) stress. For hole mobility enhancement the uniaxial stress level is adequate since uniaxial stress offers much larger hole mobility enhancement (as compared to biaxial stress) for reasons described in Chapter 19. Hole mobility enhancements greater than 50% occur in the 90 nm generation strained Si process flow. Even higher hole mobility enhancements have been demonstrated on experimental flows but have not yet been integrated into a production flow [23]. The uniaxial stress electron mobility enhancement is significant at 20% [1–3] but less than the best reported biaxial stressed n-channel MOSFETs. For short-channel devices, 10 and 25% improvements in the saturated drive current are obtained for the n- and p-channel MOSFETs, respectively.

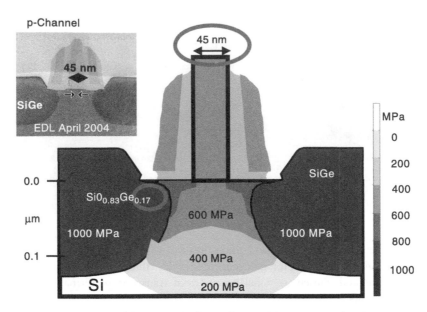

FLOOPS (FLorida Object Oriented Process Simulator)

Source: 2004 Tran. Elec. Dev., Symp. VLSI

FIGURE 21.6 Stress contours for 45 nm gate length transistor with $Si_{0.83}Ge_{0.17}$ in the source and drain.

Yield and Integration

New technology features used to introduce strained silicon (selective strained $Si_{1-x}Ge_x$ and NiSi) all have unique yield issues but can be resolved, leading to historical defect density trends. Introducing selective heteroepitaxy has many yield risks: dislocations, loss of selectivity, and blocked $Si_{1-x}Ge_x$ epitaxial growth. Since a uniaxial strained silicon structure requires a thick strained $Si_{1-x}Ge_x$ layer ($\sim$100 nm), misfit dislocations in the strained-layer are a major concern. Three yield vehicles using 90 nm design rules have been previously described: 52 Mbit CMOS SRAM [3] and two next generation microprocessors [1]. Figure 21.7 shows an SEM die photo of the SRAM and microprocessors. The SRAM contains 330 million transistors. Strained Si is fabricated on all transistors in the die including the transistors in the 6T–SRAM cell. Strained silicon and nickel silicide can yield at historical levels as shown in Figure 21.8 [1]. The defect density trend for the 90 nm technology compares favorably to past technology nodes with a two-year offset.

Perhaps the biggest yield risk for selective heteroepitaxy is dislocation in the strained layer. Thus for any strained Si structure, it is important to have a device structure that can tolerate slip dislocations without yield loss. Energy is required to form dislocations and this sets the slip system which is {1 1 1} ⟨1 1 0⟩ for heteroepitaxial grown on (1 0 0) silicon [24]. For this system, the dominant type of misfit dislocation is the so-called 60° dislocations that form 60° from the (1 0 0) plane (⟨1 1 0⟩ dislocation-line direction). An example is shown in Figure 21.9 obtained from a 90 nm microprocessor construction report [25]. Since these misfit dislocations are contained in the neutral region of the source and drain, they are expected to have minimal impact on yield, performance, or reliability.

Need for Strained Si in the 90 nm Technology Generation

The end of transistor scaling and Moore's Law has been the topic of many discussions starting in 1970 just shortly after Moore proposed the law [26–29]. This work will not be another prediction about the end of Moore's law, rather a look at key limiting factors which highlight the need for new material

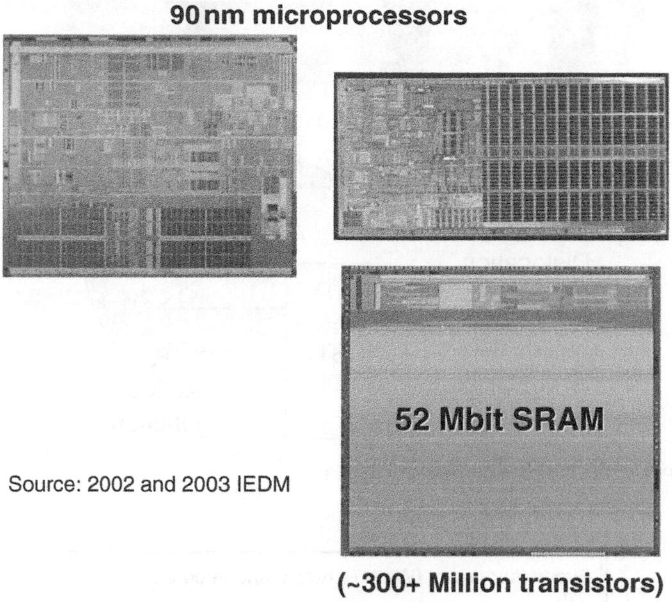

90 nm microprocessors

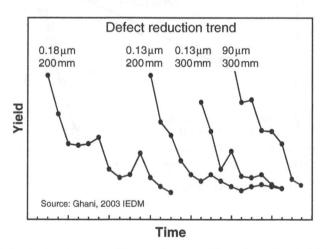

Source: 2002 and 2003 IEDM

52 Mbit SRAM

(~300+ Million transistors)

FIGURE 21.7   Die photo of 90 nm SRAM and two microprocessors.

Defect reduction trend

0.18 μm   0.13 μm   0.13 μm   90 μm
200 mm   200 mm   300 mm   300 mm

Yield

Source: Ghani, 2003 IEDM

Time

FIGURE 21.8   Defect density trend showing 90 nm technology with strained Si and nickel silicide at historical yields.

solutions for continued scaling. As pointed out earlier, planar MOSFETs as small as 10 nm have been fabricated. The MOSFET does not have a hard limit, rather practical considerations require the leakage to be less than 10 to 25% of the total power. Product data show that we are approaching this leakage power limit in the 90 nm technology generation for 45 nm MOSFETs due to a combination of gate and subthreshold leakage [3]. Figure 21.10 plots the active power and leakage power for Intel microprocessors [3]. In Figure 21.10, from 1960 to 1990, the gate and subthreshold leakage were negligible, which allowed CMOS to dissipate near zero power in the standby mode and resulted in an ideal Si technology. However, in the decade since 1990s, gate silicon dioxide and channel length scaling to the nanoscale region resulted in the off-state leakage approaching 10 to 25% of the total power. Figure 21.11 shows the

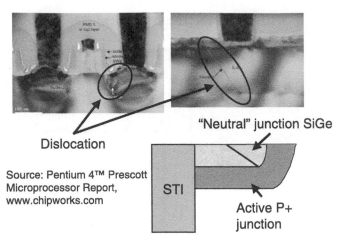

FIGURE 21.9   Dislocation in epitaxial strained $Si_{0.83}Ge_{0.17}$.

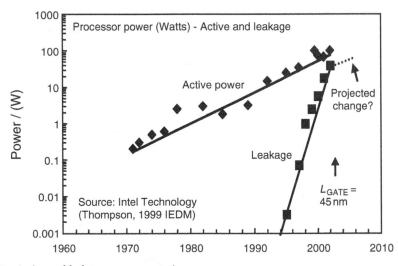

FIGURE 21.10   Active and leakage power versus time.

SiO_2 gate-scaling trend, which has reached a tunneling leakage limit at 1.2 nm. Fortunately, new materials can circumvent the gate and subthreshold leakage limits by improving performance without oxide or channel length scaling (strained Si being one technique).

21.3   Future Direction of Strained Silicon

In the current 90 nm generation, moderate levels of strain have resulted in mobility enhancements of 20 and 50% for n and p channels, respectively. Experimental and theoretical work suggests much larger mobility enhancement is achievable at higher strain. Ratios of stressed-to-unstressed mobilities of 4 and 1.7 have been reported experimentally for holes and electrons [30], respectively. Thus, an obvious evolution of this process flow going forward is integrating higher levels of strain. Higher channel strain is possible by increased strain in the (i) nitride-capping layer, (ii) epitaxial $Si_{1-x}Ge_x$ by higher Ge concentration, or (iii) fabrication of the epitaxial $Si_{1-x}Ge_x$ closer to the Si channel. More complex

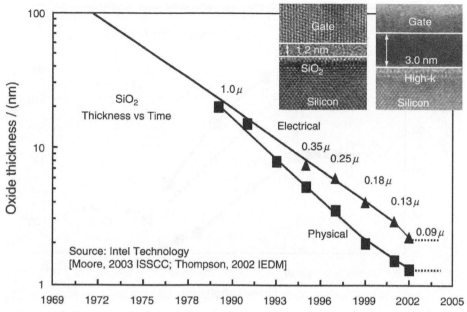

FIGURE 21.11 SiO$_2$ gate oxide thickness versus time.

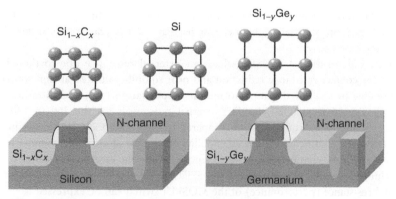

FIGURE 21.12 Two new device structures to introduce channel stress.

structures to integrate strain are also possible. Two examples are shown in Figure 21.12. Epitaxial SiC in the source and drain of n-channel MOSFETs offer higher channel strain than the capping-layer approach. Similarly, strained Si$_{1-x}$Ge$_x$ in the source and drain of a Ge channel device can be used for uniaxial tensile stress. There are also numerous other techniques and process steps to introduce strain. A few examples are high-strain capping layers introduced before poly-Si gate crystallization [31], high-stress shallow-trench isolation, and silicide [32]. In the near term for future technology generations, combinations of all these techniques will be used. This will place additional restrictions and requirements on new structures and materials. For example, for alternate device structures and materials such as FinFET, tri-gate and high *k*-gate dielectrics to be competitive with strained planar CMOS, strain or some other mobility enhancing technique [33] is needed in these structures as well.

After integrating high levels of strain for significant mobility enhancement, the external resistance of the MOSFET will next need to be addressed. This can be seen already in Figure 21.13 where the strained, improved linear drive current is plotted versus channel length [34]. As seen in Figure 21.13, for channel

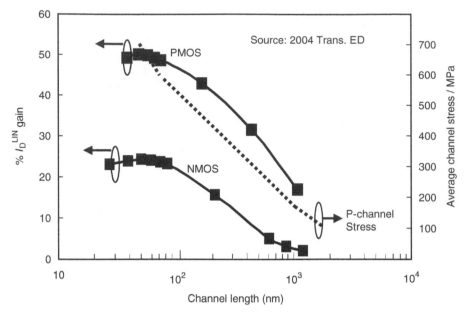

FIGURE 21.13   Strained Si improved linear drive current and average p-channel stress versus channel length.

lengths below 40 nm, the linear current improvement saturates, even though the $Si_{1-x}Ge_x$ and capping layer increases the strain at smaller channel lengths. Part of the reason this occurs is the growing importance of the external resistance.

Until recently, state-of-the-art MOSFETs had low external resistance due to the adoption of self-aligned silicide for contact resistance reduction and near solubility limited abrupt source–drain extensions made possible by shallow dopant implants, co-implantation [35], and ultrashort rapid thermal processing [36]. Typical n-channel MOSFET external resistances for the 250 to 180 nm technology generations are ~200 $\Omega\,\mu$m, which is small compared to the 1500 to 2000 $\Omega\,\mu$m channel resistance at these nodes. However, significant enhanced mobility via strain along with channel length scaling is dramatically reducing the channel resistance to a point where the external resistance can no longer be neglected. Including external resistance, the MOSFET switch can be represented as shown in Figure 21.14. The total resistance (performance) of the MOSFET switch can be expressed as

$$R_{\mathrm{TOTAL}} = \frac{V_{\mathrm{DS}}}{I_{\mathrm{D}}} = R_{\mathrm{CHANNEL}} + R_{\mathrm{SD}} = \left(\frac{L_{\mathrm{EFF}}}{W_{\mathrm{EFF}}\mu C_{\mathrm{OX}}(V_{\mathrm{GS}} - V_{\mathrm{T}})} + R_{\mathrm{SD}} \right)$$

$$= \left(\frac{L_{\mathrm{EFF}}}{W_{\mathrm{EFF}}\mu C_{\mathrm{OX}}(V_{\mathrm{GS}} - V_{\mathrm{T}})} + R_{\mathrm{SD}} \right)$$

Historically, the ratio of $R_{\mathrm{SD}}/R_{\mathrm{CHANNEL}}$ has been less than 20%. However, as pointed out in the previous sections, standard channel length scaling and strained Si all significantly lowered R_{CHANNEL}. The net effect of these trends is MOSFETs soon will be severely limited by the source–drain resistance. Extrapolation of the current trends shows that in a few year the ratio of $R_{\mathrm{SD}}/R_{\mathrm{CHANNEL}}$ will be greater than 1.

21.4   Summary

The era of simple MOSFET dimension scaling for improved performance is over. Strained Si is the next material change to extend Moore's law. At the 90 nm technology generation, selective strain epitaxial $Si_{1-x}Ge_x$ and a tensile capping layer are used to introduce strain.

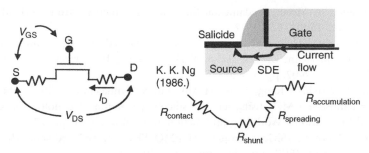

FIGURE 21.14   MOSFET switch include extrinsic and intrinsic resistance.

In the near term, advance logic technologies mobility enhancement through strain is expected to be a key performance enabler. The addition of strained Si to the planar MOSFET raises the bar for any nonclassical device structure to replace the industry workhorse.

Acknowledgments

The author would like to thank the efforts of his former colleagues in the Portland Technology Development, Technology Computer Aided Design, and in the Corporate Quality and Reliability Groups of Intel. The author also acknowledges the support and encouragement from Melanie Pecins-Thompson, Mark Bohr, Robert Chau, Tahir Ghani, Kaizad Mistry, Sunit Tyagi, and William Holt.

References

1. T Ghani et al. A 90 nm high volume manufacturing logic technology featuring novel 45 nm gate length strained silicon CMOS transistors. Technical Digest of the IEEE International Electron Devices Meeting, San Francisco, 2003, pp. 978–980.
2. SE Thompson et al. A logic nanotechnology featuring strained silicon. *IEEE Electron Dev. Lett.* 25:191–193, 2004.
3. S Thompson et al. A 90 nm logic technology featuring 50 nm strained silicon channel transistors, 7 layers of Cu interconnects, low *k* ILD, and $1 \mu m^2$ SRAM Cell. Technical Digest of the IEEE International Electron Devices Meeting, Washington, 2002, pp. 61–64.
4. GE Moore. No exponential is forever. ISSCC, San Francisco, CA, 2003.
5. GE Moore. Progress in digital integrated electronics. *IEDM Tech. Dig.* 11–13, 1975.
6. GE Moore. Cramming more components onto integrated circuits. *Electronics* 38:114–117, 1965.
7. B Doyle, R Arghavani, D Barlage, S Datta, M Doczy, J Kavalieros, A Murthy and R Chau. Transistor elements for 30nm physical gate lengths and beyond. *Intel Technol. J.*, Vol.6, 2002.
8. Z Ren, MV Fischetti, EP Gusev, EA Cartier, and M Chudzik. Inversion channel mobility in high-[kappa] high performance MOSFETs. Technical Digest of the IEEE International Electron Devices Meeting, San Francisco, 2003, pp. 33.2.1–33.2.4.
9. V Chan et al. High speed 45 nm gate length CMOSFETs integrated into a 90 nm bulk technology incorporating strain engineering. Technical Digest of the IEEE International Electron Devices Meeting, San Francisco, 2003, pp. 77–80.
10. CS Smith. Piezoresistance effect in germanium and silicon. *Phys. Rev.* 94:42–49, 1954.
11. C Xiangdong, L Kou-chen, S Ray, and S Banerjee. Bandgap engineering in vertical P-MOSFETs. *Solid-State Electron.* 45:1939–1943, 2001.
12. C Xiangdong, L Kou-Chen, QC Ouyang, SK Jayanarayanan, and SK Banerjee. Hole and electron mobility enhancement in strained SiGe vertical MOSFETs. *IEEE Trans. Electron Dev.* 48:1975–1980, 2001.

13. B Hoeneisen and CA Mead. Fundamental limitations in microelectronics—I. MOS technology. *Solid-State Electron.* 15:819–829, 1972.

14. MC Ozturk, L Jing, M Hongxiang, and N Pesovic. Advanced $Si_{1-x}Ge_x$ source/drain and contact technologies for sub-70 nm CMOS. Technical Digest of the IEEE International Electron Devices Meeting, Washington, 2002, pp. 375–378.

15. MC Ozturk, N Pesovic, I Kang, J Liu, H Mo, and S Gannavaram. Ultra-shallow source/drain junctions for nanoscale CMOS using selective silicon–germanium technology. Extended Abstracts of the Second International Workshop on Junction Technology, 2001, pp. 77–82.

16. O Qiqing et al. A novel Si/SiGe heterojunction pMOSFET with reduced short-channel effects and enhanced drive current. *IEEE Trans. Electron Dev.* 47:1943–1949, 2000.

17. K Mistry et al. Delaying forever: Uniaxail strained silicon transistors in a 90 nm CMOS technology. Symp. VLSI Tech. Dig., Hawaii, 2004.

18. K Rim et al. Characteristics and device design of sub-100 nm strained Si N- and pMOSFETs. Symp. VLSI Tech. Dig., Hawaii, 2002, pp. 98–99.

19. S Ito et al. Mechanical stress effect of etch-stop nitride and its impact on deep submicron transistor design. Technical Digest of the IEEE International Electron Devices Meeting, Washington, 2000, pp. 247–250.

20. A Shimizu et al. Local mechanical-stress control (LMC): a new technique for CMOS-performance enhancement. Technical Digest of the IEEE International Electron Devices Meeting, San Francisco, 2001, pp. 433–436.

21. ME Law and SM Cea. Continuum based modeling of silicon integrated circuit processing: an object oriented approach. *Computat. Mater. Sci.* 12:289–308, 1998.

22. S Cea and M Law. Multidimensional nonlinear viscoelastic oxidation modeling. *Simulat. Semicond. Dev. Process.* 6: 139–142, 1995.

23. PR Chidambaram et al. 35% drive current improvement from recessed-SiGe drain extensions on 37 nm gate length PMOS. Symp. VLSI Tech. Dig., Hawaii, 2004.

24. EA Fitzgerald. Dislocations in strained-layer epitaxy: theory, experiments, and applications. *Mater. Sci. Rep.* 7:88–142, 1991.

25. www.chipworks.com. Intel BX80546PG2800E Pentium™ 4 Prescott Microprocessor Structural Analysis, 2004.

26. B Hoeneisen and CA Mead. Limitations in microelectronics—II. Bipolar technology*1. *Solid-State Electron.* 15:891–897, 1972.

27. B Hoeneisen and CA Mead. Limitations in microelectronics—II. Bipolar technology. *Solid-State Electron.* 15:891–897, 1972.

28. B Hoeneisen and CA Mead. Fundamental limitations in microelectronics—I. MOS technology*1. *Solid-State Electron.* 15:819–829, 1972.

29. B Hoeneisen and CA Mead. Fundamental limitations in microelectronics—I. MOS technology. *Solid-State Electron.* 15:819–829, 1972.

30. JL Hoyt et al. Strained silicon MOSFET technology. Technical Digest of the IEEE International Electron Devices Meeting, Washington, 2002, pp. 23–26.

31. C-H Chen et al. Stress memorization technique (SMT) by selectively strained-nitride capping for sub-65 nm high-performance strained-Si device application. Symp. VLSI Tech. Dig., Hawaii, 2004.

32. A Steegen et al. Silicide induced pattern density and orientation dependent transconductance in MOS transistors. *IEDM Tech. Dig.* 497–500, 1999.

33. M Yang et al. High performance CMOS fabricated on hybrid substrate with different crystal orientations. Technical Digest of the IEEE International Electron Devices Meeting, San Francisco, 2003.

34. SE Thompson, M Armstrong, C Auth, M Alavi, M Buehler, R Chau, S Cea, T Ghani, G Glass, T Hoffman, CH Jan, C Kenyon, J Klaus, K Kuhn, Z Ma, B Mcintyre, K Mistry, A Murthy, B Obradovic, R Nagisetty, P Nguyen, S Sivakumar, R Shaheed, L Shifren, B Tufts, S Tyagi, M Bohr,

and Y El-Mansy. A 90-nm logic technology featuring strained-silicon. *IEEE Electron Dev. Lett.* 51(11): 1790–1797, 2004.

35. ME Law, GH Gilmer, and M Jaraiz. Simulation of defects and diffusion phenomena in silicon. *MRS Bulletin* 25:45–50, 2000.

36. R Singh, M Fakhruddin, and KF Poole. Rapid photothermal processing as a semiconductor manufacturing technology for the 21st century. *Appl. Surf. Sci.* 168:198–203, 2000.

22

Overview: Other Heterostructure Devices

John D. Cressler
Georgia Institute of Technology

It is perhaps not surprising that when crystal growers begin to achieve even moderate success in producing device-quality films, device engineers swoop down and rush to demonstrate a host of devices from such materials. In practice, the proof that crystal growers *have* achieved success often lies in the ability of such functional devices to be realized! While, clearly, transistors are the core building blocks of modern electronic systems, there are a number of other novel device types that become accessible with bandgap engineering techniques. Two such examples are represented here. In Chapter 23, "Resonant Tunneling Devices," S. Tsujino of the Paul Scherrer Institute discusses achieving negative differential resistance in Si–SiGe heterostructures and the types of devices that can be produced from them. In Chapter 24, "IMPATT Diodes," by E. Kasper et al. of the University of Stuttgart, gives an overview of this important class of microwave sources. Clearly, it is a healthy sign for the silicon heterostructure field that success has been achieved in a number of such novel devices. While those devices may not as yet challenge the performance of their III–V cousins, refinements are underway, and much work remains to be done.

Substrate engineering is an important subject in all of bandgap engineering, regardless of the material system. For silicon heterostructures, in particular, the underlying substrate lies at the very core of the FET performance, and bandgap engineering is unique in the sense that enormous crystals can be grown (300 mm in 2005 and counting), potentially giving rise to new and interesting applications. In Chapter 25, "Engineered Substrates for Electronic and Optoelectronic Systems," by E. Fitzgerald of MIT, a number of unique substrate techniques are described for potential electronic and optoelectronic applications. Finally, the cliched buzzword "nanotechnology" means many things to many people, but it is worthwhile noting that nanostructures can in fact be produced in Si–SiGe. In Chapter 26, "Self-Assembling Nanostructures in Ge(Si)–Si Heteroepitaxy," by R. Hull of the University of Virginia, discusses self-assembled nanostructures and their potential for applications.

23

Resonant Tunneling Devices

Soichiro Tsujino and
Detlev Grützmacher
Paul Scherrer Institute

Ulf Gennser
CNRS-LPN

23.1 Introduction

When electrons are confined within a semiconductor thin film with a thickness of the order of the de Broglie wavelength, the wave nature of the electrons becomes important, thus quantum size and tunneling effects will influence the optical and electronic properties of the material. In semiconductor heterostructures, fabricated by stacking films of several compatible semiconductor materials with different bandgaps, one can create almost arbitrary potential profiles. Suitable designs of semiconductor heterostructures permit the observation of various tunneling effects, in particular resonant tunneling. The principle of resonant tunneling has been known for a long time [1]. To realize semiconductor resonant tunneling devices such as resonant tunneling diodes (RTDs), the control of the film deposition with atomic layer thickness accuracy and the realization of semiconductor quantum well–superlattice structures with planar and sharp heterointerfaces are essential prerequisites. These are warranted by modern crystal growth technology such as molecular beam epitaxy (MBE) or chemical vapor deposition (CVD).

In Si–SiGe heterostructures, the band alignment has been shown to be of type-II, i.e., the electrons and holes are confined in different layers. The built-in strain has a large influence on the band alignment. Therefore, the potential profile can be tailored by depositing heterostructures on Si substrate or on relaxed SiGe buffer layers.

The majority of Si–SiGe tunneling devices have been fabricated in the valence band or through interband tunneling. Fewer papers have reported on electron RTDs since the conduction band offset is too small unless heterostructures are deposited on relaxed buffers and because of the large electron tunneling mass, a factor of 4 to 5 larger than hole mass, which may require very thin barriers.

This chapter is organized in the following manner. In Section 23.2, we describe the principle of resonant tunneling in a simplified model, and in Section 23.3 possible applications of the resonant tunneling devices are summarized. In Section 23.4, we review reported studies on resonant tunneling in double barrier structures, superlattices, and quantum dots. Finally, in Section 23.5, we summarize recent developments on resonant interband tunneling devices.

23.2   Principle of Resonant Tunneling

Here, we briefly describe the resonant tunneling observed in the current–voltage (I–V) characteristics of double barrier RTDs [2–5]. In a quantum well, i.e., in a thin film sandwiched by two barrier layers, quasi-confined states are formed at energies determined by the thickness W of the center layer, that is, when half integer multiples of the electron wavelength are approximately equal to W. A typical situation of a double barrier RTD at a small bias is sketched in Figure 23.1a. When a bias is applied to the structure, a small current (J_{ex}) flows, which is attributed to tunneling through the whole stack of layers comprising two barriers and the quantum well. The current increases strongly (J_{RT} in Figure 23.1b), whenever the energy of the incident electron coincides with the energy of one of the confined states, for example the E_r state in Figure 23.1a. This resonance enhancement of the electron transmission is analogous to that for photons in Fabry–Perot interferometers. At resonance, the amplitude of the incident wave builds up in the center layer by the positive interference between the incident wave and the wave reflected from the second barrier, and consequently the transmission through the structure is reinforced. Further increase of the bias detunes the resonance and the current decreases sharply, creating the negative differential resistance (NDR), which is sketched in Figure 23.1b.

The transmission function for a carrier at an energy close to the resonance $E \sim E_r$ can be written approximately by a Lorenzian,

$$T(E) \approx \frac{\Gamma_L \Gamma_R}{\Gamma_L + \Gamma_R} \frac{\Gamma}{(E - E_r)^2 + (\Gamma/2)^2} \tag{23.1}$$

The width $\Gamma = \Gamma_L + \Gamma_R$ of the level is given by the decay rate of the resonant state. $\Gamma_L/\hbar$ and $\Gamma_R/\hbar$ represent the tunneling escape rate of a carrier in the confined state via the left and the right barrier, respectively [3,4].

In layered structures, the motion of the carriers in the plane of the layer is characterized by plane waves and the tunneling process conserves the in-plane momentum $\hbar k_\parallel$ in the absence of scattering [2,4]. For $E_C^L < E_r < E_F^L$, where E_C^L is the band edge in the emitter and E_F^L is the Fermi level of the emitter, tunneling is possible only for carriers whose momenta $\hbar k_z$ lie in a disk corresponding to $k_z = q_R$ (shaded disk in Figure 23.1c), where q_R is given by $\{2m^*(E_r - E_C^L)/\hbar^2\}^{1/2}$ and m^* is the effective mass of the carrier. In this range, the current is approximately given by $(2e/\hbar)T_0 N(\Delta E)$ where $T_0 = \Gamma_L \Gamma_R/(\Gamma_L + \Gamma_R)$, $N(\Delta E) = m^* \Delta E/(\pi \hbar^2)$ the supply function and $\Delta E = (E_F^L - E_r)$. As the emitter potential rises, the

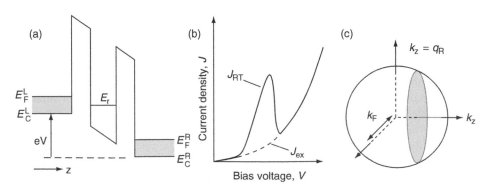

FIGURE 23.1   (a) Schematic band-diagram of a double barrier resonant tunneling diode when bias V is applied between the emitter (denoted by L) and the collector (denoted by R). (b) Expected current–voltage characteristics of the device. (c) A Fermi sphere of electrons in the emitter.

number of available electrons increases and $N(\Delta E)$ reaches the maximum value equal to $m^* (E_F^L - E_C^L)/ (\pi\hbar^2)$ when q_R is equal to 0. When E_C^L rises above E_r, there are no electrons in the emitter at zero temperature, which can tunnel while conserving $\hbar k_\parallel$. In real devices, the external voltage V needed to shift the emitter states relative to the quantum well states, ΔE, depends not only on the thicknesses of the different layers (quantum well, barriers, and doping offset), but also on space–charge effects due to charge accumulation in the emitter and in the quantum well. A linear relationship is often assumed, $\Delta E \propto \alpha V$, with the lever factor α determined empirically.

The first Si–SiGe RTDs were grown on (1 0 0) oriented Si substrates and consisted of a compressively strained $Si_{1-x}Ge_x$ quantum well, Si barriers and SiGe emitter and collector layers [6]. An example of the heavy hole valence band potential profile of such a device ($x = 0.26$) calculated self-consistently at a bias of 0.13 V is shown in Figure 23.2 [7]. The Ge compositions of the emitter and the collector layers of the device are graded toward the surface and the substrate, respectively, in order to create a smooth potential profile. The $I–V$ character of the RTD shows resonances induced by the confined heavy and light hole states in the quantum well (Figure 23.3). By varying the thickness of the quantum well, the resonant peak positions shift systematically (Figure 23.4). A good agreement between the calculated peak positions (solid curves in Figure 23.4) and the experiment is achieved by using a lever factor α of $\sim$4.2.

Although the above description of the $I–V$ characteristics of RTDs is simple and intuitive, a quantitative description of the $I–V$ curves requires the inclusion of other effects [5], such as non-resonant background current (J_{ex} in Figure 23.1b), elastic and inelastic scattering during the tunneling processes [3,4], charge build-up and multiband effects. The inclusion of band mixing in the presence of strain and scattering is important, especially for hole tunneling, since the quantized states of holes in quantum wells originate from three different bands: heavy hole (HH), light hole (LH), and split-off hole (SO). The in-plane dispersions of these states are strongly influenced by the interaction between the

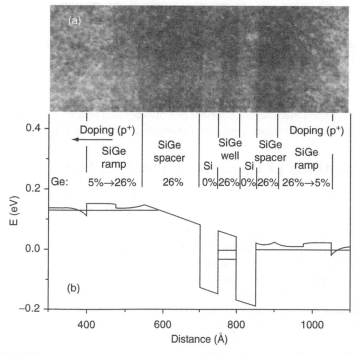

FIGURE 23.2 (a) A TEM picture of a cross section of an RTD with asymmetric spacer layers. (b) The layer sequence and the valence band edge calculated self-consistently for a bias of 0.13 V, including heavy hole states only. (From G. Dehlinger, U. Gennser, D. Grützmacher, T. Ihn, E. Müller, and K. Ensslin. *Thin Solid Films* 369: 390–393, 2000. With permission.)

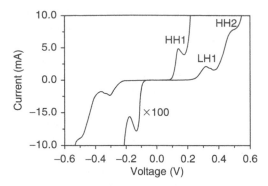

FIGURE 23.3 Current–voltage characteristics of a symmetric RTD measured at 4 K. The low voltage region is enlarged by a factor of 100 to show the HH$_1$ peak. (From G. Dehlinger, U. Gennser, D. Grützmacher, T. Ihn, E. Müller, and K. Ensslin. *Thin Solid Films* 369: 390–393, 2000. With permission.)

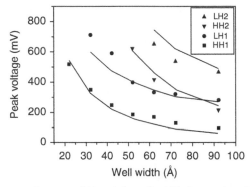

FIGURE 23.4 Relation between confinement shift and the well width for a series of RTDs. Dots show the measured resonance-positions and the solid lines are calculated values. (From G. Dehlinger, U. Gennser, D. Grützmacher, T. Ihn, E. Müller, and K. Ensslin. *Thin Solid Films* 369: 390–393, 2000. With permission.)

three valence bands [8]. Furthermore, a triangular potential is often formed in the emitter, leading to the formation of quantized emitter states. Therefore, the supply function has to be modified accordingly [5].

23.3 Applications of Resonant Tunneling Devices

RTDs have been an important instrument for exploring the physics of semiconductor nanostructures. At the same time, because of the extreme nonlinearity of the $I-V$ characteristics and high operation frequencies, up to THz [9], RTDs have been intensively studied for several high-speed logic circuit applications [4,5,10] and in high-frequency oscillator-switching applications [11]. The employment of RTDs in integrated circuits has several advantages, including reduced number of components and circuit complexity for a given function, reduced power consumption, and high speed. Therefore, many applications have been studied: a low power consumption SRAM cell with a single transistor [12], multivalued logic [13], monostable–bistable transition logic gates [14], a compact RTD/HBT circuit with high-frequency and low power consumption operation for wireless communication [15], and analog-to-digital converters [16]. Two-terminal logic circuits can be constructed using solely RTDs [17], but lack of current gain restricts the fan-out of the circuits and limits the applications. Most promising is the integration of RTDs with high-frequency transistors such as heterobipolar transistors (HBT) and heterostructure field-effect transistors (HFET), which would offer flexibility in circuit design, and enable a more extensive basic function library [10].

One of the figures-of-merit of RTDs for device application is the peak-to-valley current ratio (PVR), which measures the ratio of a current maximum (at resonance) to the following current minimum at higher bias. High PVRs (>100) have been achieved in III–V semiconductor RTDs and using those, working prototypes of RTD-based circuits have been demonstrated. For commercial applications, however, technical challenges such as the uniformity of the layer deposition across the wafers have to be surmounted [5,10]. The PVR of Si–SiGe double barrier RTDs has not reached a competitive level, mainly due to rather low band offsets and high effective carrier masses compared to III–V semiconductors, even though high-speed SiGe transistors, especially HBTs, have been successfully developed (see the chapter on HBTs). As an alternative approach to RTDs, several groups have developed a SiGe resonant interband tunneling diode (RITD) (see section "Resonant Interband Tunneling Diodes").

Another important application of resonant tunneling is in quantum cascade lasers based on intersubband optical transitions [18]. In this unipolar device, the active layer consists of a series connection of several cells. Each cell is essentially a four-level system, where population inversion is realized by resonant tunneling to the upper transition state, and the upper state itself is spatially confined by Fabry–Perot reflection between superlattice barriers to achieve a high injection efficiency and long lifetime. Si–SiGe quantum cascade structures have been studied to realize Si-based lasers [19–21].

23.4 Resonant Tunneling in Si–SiGe Heterostructures

Resonant Tunneling in Double Barrier Structures

Resonant tunneling in Si–SiGe double barrier structures has been reported for both electrons and holes. Among these, most have focused on hole transport in the valence band. In fact, the first SiGe RTDs were demonstrated in p-type double barrier structures by Liu et al. [6] and by Rhee et al. [22] by taking advantage of the larger valence band discontinuity compared to the conduction band offset. Devices have been prepared either by MBE or by various CVD methods. For example, Zaslavsky et al. [23] reported a double barrier diode with a 2.3-nm thick compressibly strained $Si_{0.75}Ge_{0.25}$ quantum well and 5-nm thick unstrained Si barriers deposited by atmospheric CVD on Si substrates. This RTD showed a PVR of 4 at 4.2 K, comparable to the best results observed in MBE-grown devices. The observation of NDR at room temperature has not been possible in p-type Si–SiGe structures, due to the increased valley current at high temperature by thermally assisted tunneling through higher resonant states [24,25].

Compared to the hole transport, the resonant tunneling of electrons in the conduction band has attracted less attention [26–31]. Although observation of NDR at room temperature has been reported in these devices, the physics of electron resonant tunneling is not well understood and requires further study. In particular, a demonstration of confinement shifts of the resonances is still lacking. Electron RTDs using alternative barriers such as SiO_2 [32,33] or CaF_2 lattice matched to Si(1 1 1) substrates [34] have been studied because of the large conduction band offsets available in those systems.

Momentum conservation of the tunneling process has been studied by magnetotunneling experiments. Magnetotunneling is especially useful in valence band RTDs because the mixing between HH, LH, and SO bands leads to highly nonparabolic and anisotropic subbands [8,35]. Resonant tunneling with a magnetic field B applied parallel to the current, along the growth direction, is supposed to occur between Landau levels having the same Landau-level index. A modulation of the $I–V$ characteristics of n-type RTDs has been explained by phonon-assisted tunneling which breaks the Landau-level index conservation. In p-type Si–SiGe RTDs Landau-level tunneling has been shown to occur even without phonons due to the mixing of Landau levels in the valence bands [23,36,37]. With the B-field perpendicular to the current, the resonant tunneling peak is shifted because the conservation of the canonical momentum in the tunneling process results in a displacement of the energy dispersion in the emitter states with respect to the quantum well states (Figure 23.5, left). This has allowed the mapping of the dispersion relations of the hole-subbands. The anisotropy of the hole-subbands has been studied by following the relation between the resonant peak position and the angle of the B-field within the

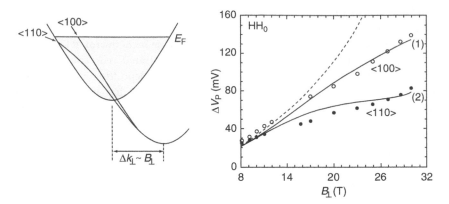

FIGURE 23.5 Schematic diagram (left) of the energy alignment of the occupied heavy-hole states in the emitter (shaded) and a HH_0 subband in the quantum well under magnetic field $B_\perp$ applied parallel to the quantum well plane, and HH_0 peak voltage shift (right) ΔV_p versus in-plane $B_\perp$-field oriented along the $\langle 1\,1\,0 \rangle$ and $\langle 1\,0\,0 \rangle$ directions in a pseudomorphic Si–SiGe RTD. (From A. Zaslavsky, T.P. Smith III, D.A. Grützmacher, S.Y. Lin, and T.O. Sedgwick. *Phys. Rev. B* 48: 15112–15115, 1993.)

quantum well plane [38,39]. Using this technique, a large anisotropy of the dispersion between $\langle 1\,0\,0 \rangle$ and $\langle 1\,1\,0 \rangle$ direction was detected as shown in Figure 23.5 (right).

To describe the hole tunneling properly, a multiband model must be employed. In such models, the momentum $\hbar k_\parallel$ parallel to the layers needs to be taken into account since the mixing between HH, LH, and SO states is very sensitive to it, and mixing changes the transmission probability dramatically [8,35]. Band mixing occurs between HH and LH subbands only at $k_\parallel \neq 0$ and depends strongly on the separation of the states and their nonparabolicity, thus the strain, the quantum well width, and the band offset have an impact on the strength of the band mixing. The dominant peak in the $I–V$ characteristics of pseudomorphic RTDs (see Figure 23.3) that is assigned to tunneling via light hole states is susceptible to severe band-mixing effects. It has been shown that the peak current in p-type GaAs–AlGaAs RTDs is dominated by tunneling via off-zone center ($k_\parallel \neq 0$) states [40], though a similar demonstration for SiGe RTDs has not yet been done. In contrast to the pseudomorphic RTDs, recent experiments in high-Ge concentration strain-compensated RTDs suggest that the resonant tunneling from HH emitter states through LH quantum well states is extremely weak. This might be attributed to the difference in strain and to the splitting of the HH and LH/SO band by more than 80 meV [41].

Resonant Tunneling in Strain-Compensated and High-Ge Concentration Si–SiGe Superlattices and Quantum Cascade Structures

Resonant tunneling and miniband transport are key ingredients for quantum cascade structures [18,42]. Recently, vertical hole transport in high Ge content strain-compensated SiGe–Si quantum wells and superlattices on relaxed buffer substrates have been studied for application in Si-based quantum cascade lasers for the midinfrared wavelength range [21,43]. The samples are composed of alternating compressively strained QWs and tensile strained Si barriers with an average Ge concentration equal to the value in the relaxed buffer substrate [44]. In this way, it is possible to grow a thick active layer without suffering from the critical thickness limitation [45–47]. Strain-compensated superlattices and quantum cascade structures exceeding 1 to 2 μm in thickness have been demonstrated by depositing Si–$Si_{0.2}Ge_{0.8}$ layers on $Si_{0.5}Ge_{0.5}$ relaxed buffer substrates using solid source MBE. By employing low growth temperatures of ~300°C, islanding within the high germanium content SiGe layers was kinematically suppressed. Consequently, highly planar and atomically sharp Si–$Si_{0.2}Ge_{0.8}$ interfaces were realized (Figure 23.6). The interface roughness of the samples was found to be less than 0.3 to 0.4 nm [48,49].

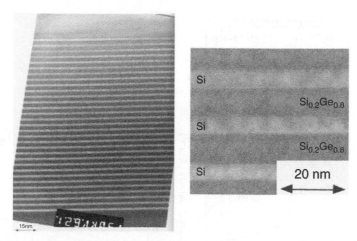

FIGURE 23.6 High-resolution cross-sectional TEM picture of a strain-compensated superlattice with 30 periods of 8.3-nm thick $Si_{0.2}Ge_{0.8}$ quantum wells separated by 5-nm thick Si barriers deposited on (1 0 0) $Si_{0.5}Ge_{0.5}$ relaxed buffer substrate by molecular beam epitaxy at low temperature. Overview (left) and magnified view (right). (Courtesy of E. Müller.)

Vertical transport in symmetrically strained superlattices was first studied by Park et al. [50]. They observed NDR in their p-type $Ge_{0.4}Si_{0.6}$—Si superlattice on $Ge_{0.2}Si_{0.8}$ relaxed buffer, which was ascribed to the tunneling via LH minibands. Tsujino et al. [43] studied strain-compensated Si—$Si_{0.2}Ge_{0.8}$ superlattices on $Si_{0.5}Ge_{0.5}$ relaxed buffers having different tunneling coupling strength (Figure 23.7). They observed current peaks originating from sequential resonant tunneling between neighboring quantum wells in weakly coupled superlattices and transport via ~50 meV wide HH miniband in strongly coupled superlattices [51].

Resonant Tunneling in Quantum Dot Structures

Resonant tunneling through quantized states at lower dimensions has been explored by reducing the lateral dimension of RTDs. When the diameter of a p-type pseudomorphic Si—$Si_{0.75}Ge_{0.25}$ RTD was reduced below ~0.1 μm, the differential conductance showed additional structure, ascribed to the quantization of the in-plane motion [52,53]. Further study revealed that these conductance peaks are due to the resonant tunneling through quantum ring states along the rim of the mesas created by inhomogeneous strain relaxation on the lateral surface of the device [54]. In a smaller device with a diameter of 45 nm, single-hole tunneling and Coulomb blockade via quantum ring states were observed [55].

Resonant tunneling through self-assembled Si quantum dots buried in SiO_2 [56] and self-assembled Ge quantum dots [57] has been reported (see the chapter on self-assembled quantum structures in SiGe–Si). Tunneling and Coulomb blockade through laterally defined SiGe quantum dots have also been detected by in-plane transport [58,59].

23.5 Resonant Interband Tunneling Diodes

The main idea of resonant interband tunneling diodes (RITD) is to combine interband tunneling in degenerately doped p–n diodes, i.e., Esaki tunnel diodes [60], with the confined states in quantum wells [61]. At a bias close to zero, resonant current flows from a quantized electron state in the n-side to a quantized hole state in the p-side by interband tunneling. When the bias is set at the out-of-resonance condition, the large interband energy gap blocks the current. Therefore, a large PVR is expected in RITDs compared to RTDs where the available barrier heights are only a fraction of the bandgap.

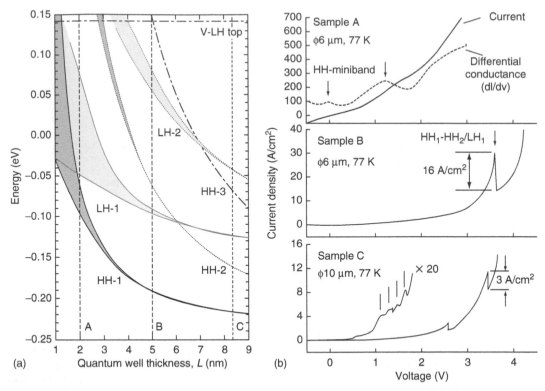

FIGURE 23.7 (a) The dependence of energetic position of hole states on the quantum well thickness W in strain-compensated Si—$Si_{0.2}Ge_{0.8}$ superlattices on $Si_{0.5}Ge_{0.5}$ relax buffer substrates with constant ratio between W and the thickness L_b of the Si barriers ($L_b = 0.6W$). The energies are with respect to the valence band edge of unstrained $Si_{0.5}Ge_{0.5}$. Shaded area shows the miniband formation. (b) Current–voltage (*I–V*) characteristics of strain-compensated Si–$Si_{0.2}Ge_{0.8}$ superlattice samples A, B, and C with 30 periods of quantum wells ($W = 2$, 5, and 8.3 nm respectively) at 77 K. Their expected hole states are marked by vertical lines in (a). The observed *I–V*s show peaks originating from sequential resonant tunneling between neighboring wells (samples B, C) and periodical peaks indicating the formation of the electrical field domains (sample C in the range of 1–2 V). When W and L_b are further reduced, HH_1 states form a ~50 meV wide miniband, giving rise to a conductance peak around 0 V (sample A).

Recently SiGe RITDs have been intensively studied because of their potential to fulfill the requirements of room temperature operation and high-peak current densities [62–68]. A basic structure consists of an undoped SiGe tunneling layer, sandwiched by δ-doped p and n layers with or without spacer layers (Figure 23.8). Two-dimensional states are formed at the δ-doped layers due to the Coulomb potential. Very high sheet doping concentrations of the order of $10^{14}\,cm^{-2}$ with sharp doping profiles are realized by MBE at low growth temperatures. Maximum dopant incorporation with sufficiently sharp profiles is achieved at growth temperatures of 460°C for Si:B and 370°C for Si:P [67]. Because of the low growth temperature, postgrowth annealing at moderate temperatures (~600°C) has proven to be crucial to lower the valley current and obtain diodes exhibiting pronounced NDR at room temperature (Figure 23.9). Duschl et al. [67] reported SiGe RITDs with a PVR of 4.8 and a peak current density J_p of 30 kA cm^{-2}. They also fabricated RITDs with PVR of 6 but at reduced J_p of ~1 kA cm^{-2} [68]. The trade-off between PVR and J_p was systematically studied by Jin et al. (Figure 23.10) [69]. Narrowing down the film thickness between the two δ-doped layers leads to an increase of J_p but to a reduction of the PVR due to enhanced tunneling via defect states. By optimizing the sample structure, they fabricated SiGe RITD having J_p up to 151 kA cm^{-2} with a PVR of 2 at room temperature.

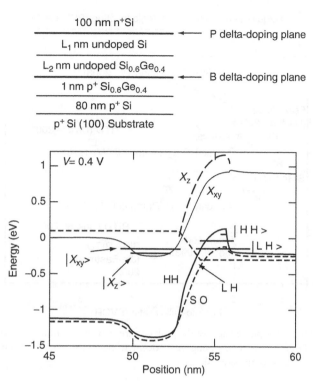

FIGURE 23.8 Sample structure (top) and (b) band diagram of a Si–SiGe resonant interband tunneling diode at an applied bias of 0.4 V showing conduction band longitudinal (X_z) and transverse (X_{xy}) valleys and HH, LH, and SO bands. (From N. Jin, S.-Y. Chung, A.T. Rice, P.R. Berger, R. Yu, P.E. Thomson, and R. Lake. *Appl. Phys. Lett.* 83: 3308–3310, 2003. With permission.)

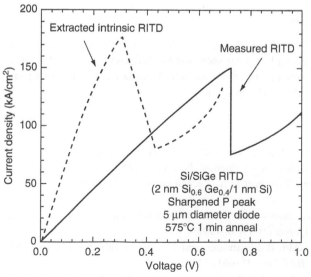

FIGURE 23.9 Measured and extracted intrinsic *I–V* characteristics of a Si–SiGe RITD with elevated and sharpened P δ-doping. (From N. Jin, S.-Y. Chung, A.T. Rice, P.R. Berger, R. Yu, P.E. Thomson, and R. Lake. *Appl. Phys. Lett.* 83: 3308–3310, 2003. With permission.)

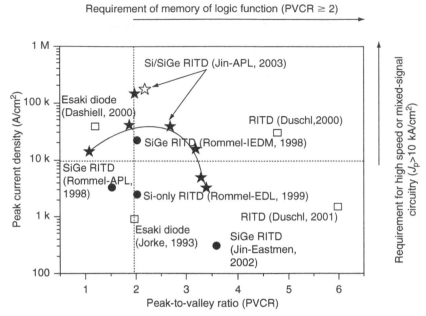

FIGURE 23.10 A graph of PVR plotted against the peak current density of a number of Si-based interband tunnel diodes. (Adapted from From N. Jin, S.-Y. Chung, A.T. Rice, P.R. Berger, R. Yu, P.E. Thomson, and R. Lake. *Appl. Phys. Lett.* 83: 3308–3310, 2003. With permission.)

Due to the required narrow spacing of the n- and p-type δ-doped layers, the capacitance of RITD devices is large. Therefore, the operation of these interband-tunneling devices will be limited at the high frequency and most likely be slower than double barrier RTDs. However, as summarized in Figure 23.10, the performance of reported RITDs has achieved the requirements for digital device applications, a PVR larger than ~2 and J_p larger than ~10 kA cm^{-2}, in a wide range [69].

Acknowledgments

The authors would like to thank H. Sigg for helpful discussions and careful reading of the manuscript, and A. Borak, J. Gobrecht, J.F. van der Veen, and M. Rüfenacht for careful reading of the manuscript. This work is partially supported by the Swiss National Foundation and the European community within the SiGeNET project and the SHINE project.

References

1. D. Bohm. *Quantum Theory.* New York, NY: Prentice-Hall, 1966.
2. L. Esaki. Long journey into tunneling. *Rev. Mod. Phys.* 46: 237–244, 1974.
3. S. Datta. *Electronic Transport in Mesoscopic Systems.* Cambridge, UK: Cambridge University Press, 1995.
4. H. Mizuta and T. Tanoue. *The Physics and Applications of Resonant Tunneling Diodes.* Cambridge, UK: Cambridge University Press, 1995.
5. J.P. Sun, G.I. Haddad, P. Mazumder, and J.N. Schulman. Resonant tunneling diodes: models and properties. *Proc. IEEE* 86: 641–661, 1998.
6. H.C. Liu, D. Landheer, M. Buchanan, and D.C. Houghton. Resonant tunneling in Si/Si$_{1-x}$Ge$_x$ double-barrier structures. *Appl. Phys. Lett.* 52: 1809–1811, 1988.
7. G. Dehlinger, U. Gennser, D. Grützmacher, T. Ihn, E. Müller, and K. Ensslin. Investigation of the emitter structure in SiGe/Si resonant tunneling structures. *Thin Solid Films* 369: 390–393, 2000.

8. R. Wessel and M. Altarelli. Analytic solution of the effective-mass equation in strained Si–Si$_{1-x}$Ge$_x$ heterostructures applied to resonant tunneling. *Phys. Rev. B* 40: 12457–12462, 1989.
9. T.C.L.G. Sollner, W.D. Goodhue, P.E. Tannenwald, C.D. Parker, and D.D. Peck. Resonant tunneling through quantum wells at frequencies up to 2.5 THz. *Appl. Phys. Lett.* 43: 588–590, 1983.
10. P. Mazumder, S. Kulkarni, M. Bhattacharya, J.P. Sun, and G.I. Haddad. Digital circuit applications of resonant tunneling devices. *Proc. IEEE* 86: 664–686, 1998.
11. H.C. Liu and T.C.L.G. Sollner. High-frequency resonant-tunneling devices. In *Semiconductor and Semimetals High Speed Heterostructure Devices*, Vol. 41, R.A. Kiehl and T.C.L. Gerhard Sollner, eds. San Diego, CA: Academic Press, 1994, pp. 359–419.
12. P. van der Wagt, A. Seabaugh, and E. Beam III. RTD/HFET low standby power SRAM gain cell. *IEDM Tech. Dig.* 1996: 425–428, 1996.
13. F. Cappaso. Quantum electron devices: physics and applications. In *Semiconductor and Semimetals, High Speed Heterostructure Devices*, Vol. 41, R.A. Kiehl and T.C.L. Gerhard Sollner, eds. San Diego, CA: Academic Press, 1994, pp. 1–77.
14. K.J. Chen, K. Maezawa, and M. Yamamoto. InP-based high-performance monostable–bistable transition logic elements (MOBILEs) using integrated multiple-input resonant-tunneling devices. *IEEE Electron Dev. Lett.* 17: 127–129, 1996.
15. H.J. De Los Santos, K.K. Chui, D.H. Chow, and H.L. Dunlap. An efficient HBT/RTD oscillator for wireless applications. *IEEE Microwave. Wireless. Comp. Lett.* 11: 193–195, 2001.
16. T.P.E. Broekaert, B. Brar, J.P.A. van der Wagt, A.C. Seabaugh, F.J. Morris, T.S. Moise, and E.A. Beam III. A monolithic 4-bit 2-Gsps resonant tunneling analog-to-digital converter. *IEEE Solid-State Circuits* 33: 1342–1349, 1998.
17. RCA tunnel diode manual. Sommerville, NJ: RCA Semiconductor and Materials Division, 1963.
18. J. Faist, F. Capasso, D.L. Sivco, C. Sirtori, A.L. Hutchinson, and A.Y. Cho. Quantum cascade laser. *Science* 264: 553–556, 1994.
19. G. Dehlinger, L. Diehl, U. Gennser, H. Sigg, J. Faist, K. Ensslin, D. Grutzmacher, and E. Muller. Intersubband electroluminescence from silicon-based quantum cascade structures. *Science* 290: 2277–2280, 2000.
20. S.A. Lynch, R. Bates, D.J. Paul, D.J. Norris, A.G. Cullis, Z. Ikonic, R.W. Kelsall, P. Harrison, D.D. Arnone, and C.R. Pidgeon. Intersubband electroluminescence from Si/SiGe cascade emitters at terahertz frequencies. *Appl. Phys. Lett.* 81: 1543–1545, 2002.
21. L. Diehl, S. Mentese, E. Muller, D. Grutzmacher, H. Sigg, U. Gennser, I. Sagnes, Y. Campidelli, O. Kermarrec, D. Bensahel, and J. Faist. Electroluminescence from strain-compensated Si$_{0.2}$Ge$_{0.8}$/Si quantum-cascade structures based on a bound-to-continuum transition. *Appl. Phys. Lett.* 81: 4700–4702, 2002.
22. S.S. Rhee, J.S. Park, R.P.G. Karunasiri, Q. Ye, and K.L. Wang. Resonant tunneling through a Si/Ge$_x$Si$_{1-x}$/Si heterostructure on a GeSi buffer layer. *Appl. Phys. Lett.* 53: 204–206, 1988.
23. A. Zaslavsky, D.A. Grützmacher, S.Y. Lin, T.P. Smith III, R.A. Kiehl, and T.O. Sedgwick. Observation of valence-band Landau-level mixing by resonant magnetotunneling. *Phys. Rev. B* 47: 16036–16039, 1993.
24. U. Gennser, V.P. Kesan, S.S. Iyer, T.J. Bucelot, and E.S. Yang. Temperature dependent transport measurements on strained Si/Si$_{1-x}$Ge$_x$ resonant tunneling devices. *J. Vac. Sci. Technol. B* 9:2059–2063, 1991.
25. G.D. Schen, D.X. Xu, M. Willander, G.V. Hansson, and Y.M. Wang. Temperature effects for current transport in resonant tunneling structures. *Appl. Phys. Lett.* 58: 738–740, 1991.
26. K. Ismail, B.S. Meyerson, and P.J. Wang. Electron resonant tunneling in Si/SiGe double barrier diodes. *Appl. Phys. Lett.* 59: 973–975, 1991.
27. Z. Matutinovic-Krstelj, C.W. Liu, X. Xiao, and J.C. Strum. Symmetric Si/Si$_{1-x}$Ge$_x$ electron resonant tunneling diodes with an anomalous temperature behavior. *Appl. Phys. Lett.* 62: 603–605, 1993.
28. D.J. Paul, P. See, I.V. Zozoulenko, K.-F. Berggren, B. Kabius, B. Holländer, and S. Mantl. Si/SiGe electron resonant tunneling diodes. *Appl. Phys. Lett.* 77: 1653–1655, 2000.

29. P. See and D. J. Paul, The scaled performance of Si/Si$_{1-x}$Ge$_x$ resonant tunneling diodes. *IEEE Electron Device Lett.* 22: 582–584, 2001.

30. D.J. Paul, P. See, R. Bates, N. Griffin, B.P. Coonan, G. Redmond, G.M. Crean, I.V. Zozoulenko, K.-F. Berggren, B. Holländer, and S. Mantl. Si/SiGe electron resonant tunneling diodes with graded spacer wells. *Appl. Phys. Lett.* 78: 4184–4186, 2000.

31. Y. Suda and H. Koyama. Electron resonant tunneling with a high peak-to-valley ratio at room temperature in Si$_{1-x}$Ge$_x$/Si triple barrier diodes. *Appl. Phys. Lett.* 79:2273–2275, 2001.

32. R. Tsu. Silicon-based quantum wells. *Nature* 364: 19, 1993.

33. Y. Wei, R.M. Wallace, and A.C. Seabaugh. Controlled growth of SiO$_2$ tunnel barrier and crystalline Si quantum wells for Si resonant tunneling diodes. *J. Appl. Phys.* 81: 6415–6424, 1997.

34. M. Watanabe, Y. Iketani, and M. Asada. Epitaxial growth and electrical characteristics of CaF$_2$/Si/ CaF$_2$ resonant tunneling diode structures grown on Si(1 1 1) 1°-off Substrate. *Jpn. J. Appl. Phys.* 39: L964–L967, 2000.

35. C.Y.-P. Chao and S.L. Chuang. Resonant tunneling of holes in the multiband effective-mass approximation. *Phys. Rev. B* 43: 7027–7039, 1991.

36. H.C. Liu, D. Landheer, M. Buchanan, D.C. Houghton, M. D'Iorio, and S. Kechang. Hole resonant tunneling in Si/SiGe heterosturctures. *Superlattices Microstruc.* 5: 213–217, 1989.

37. G. Schuberth, G. Abstreiter, E. Gornik, F. Schäffler, and J.F. Luy. Resonant tunneling of holes in Si/ Si$_x$Ge$_{1-x}$ quantum-well structures. *Phys. Rev. B* 43: 2280–2284, 1991.

38. U. Gennser, V.P. Kesan, D.A. Syphers, T.P. Smith III, S.S. Iyer, and E.S. Yang. Probing band structure anisotropy in quantum wells via magnetotunneling. *Phys. Rev. Lett.* 67: 3828–3831, 1991.

39. A. Zaslavsky, T.P. Smith III, D.A. Grützmacher, S.Y. Lin, and T.O. Sedgwick. In-plane valence-band nonparabolicity and anisotropy in strained Si–Ge quantum wells. *Phys. Rev. B* 48: 15112–15115, 1993.

40. Y.C. Chung, T. Reker, A.R. Glanfield, and P.C. Klipstein. Dominance of Fermi-surface holes in p-type tunneling. *Phys. Rev. Lett.* 88: 126802-1–126802-4, 2002.

41. U. Gennser, M. Scheinert, L. Diehl, S. Tsujino, A. Borak, C.V. Falub, D. Grützmacher, A. Weber, D.K. Maude, Y. Campidelli, O. Kermarrec, and D. Bensahel. Total angular momentum conservation during tunnelling through semiconductor barriers. cond-mat/0501212.

42. C. Sirtori, F. Capasso, J. Faist, A.L. Hutchinson, D.L. Sivco, and A.Y. Cho. Resonant tunneling in quantum cascade lasers. *IEEE J-QE* 34: 1722–1729, 1998.

43. S. Tsujino, S. Mentese, L. Diehl, E. Müller, B. Haas, D. Bächle, S. Stutz, D. Grützmacher, Y. Campidelli, O. Kermarrec, and D. Bensahel. Resonant tunneling in Si–SiGe superlattices on relaxed buffer substrates. *Appl. Surf. Sci.* 224: 377–381, 2004.

44. D. Grützmacher, S. Mentese, E. Müller, L. Diehl, H. Sigg, Y. Campidelli, O. Kermarrec, D. Bensahel, T. Roch, J. Stangl, and G. Bauer. Strain compensated Si/Si$_{0.2}$Ge$_{0.8}$ quantum cascade structures grown by low temperature molecular beam epitaxy. *J. Cryst. Growth* 251: 707–717, 2003.

45. E. Kasper, H.-J. Herzog, H. Jorke, and G. Abstreiter. Strained layer Si/SiGe superlattices. *Superlattice Microstruct.* 3: 141–146, 1987.

46. E.A. Fitzgerald, Y.H. Xie, M.L. Green, D. Brasen, A.R. Kortan, J. Michel, Y.J. Mii, and B.E. Weir. Totally relaxed Ge$_x$Si$_{1-x}$ layers with low threading dislocation densities grown on Si substrates. *Appl. Phys. Lett.* 59: 811–813, 1991.

47. E. Koppensteiner, P. Hamberger, G. Bauer, A. Pesek, H. Kibbel, H. Presting, and E. Kasper. X-ray-diffraction investigation of single step and step-graded SiGe alloy buffers for the growth of short-period Si(m)Ge(n) superlattices using reciprocal space mapping. *Appl. Phys. Lett.* 62: 1783–1785, 1993.

48. M. Meduna, J. Novak, G. Bauer, V. Holy, C.V. Falub, S. Tsujino, E. Müller, D. Grützmacher, Y. Campidelli, O. Kermarrec, and D. Bensahel. Annealing studies of high Ge composition Si/SiGe multilayers. *Z. Kristallogr.* 219: 195–200, 2004.

49. S. Tsujino, C.V. Falub, E. Müller, M. Scheinert, L. Diehl, U. Gennser, T. Fromherz, A. Borak, H. Sigg, D. Grützmacher, Y. Campidelli, O. Kermarrec, and D. Bensahel. Hall mobility of narrow Si$_{0.2}$Ge$_{0.8}$-Si quantum wells on Si$_{0.5}$Ge$_{0.5}$ relaxed buffer substrates. *Appl. Phys. Lett.* 84: 2829–2831, 2004.

50. J.S. Park, R.P.G. Karunasiri, K.L. Wang, S.S. Rhee, and C.H. Chern. Hole transport through minibands of a symmetrically strained Ge_xSi_{1-x}/Si superlattice. *Appl. Phys. Lett.* 54: 1564–1566, 1989.

51. S. Tsujino, C.V. Falub, E. Müller, and D. Grützmacher, unpublished.

52. A. Zaslavsky, K.R. Milkove, Y.H. Lee. B. Ferland, and T.O. Sedgwick. Strain relaxation in silicon–germanium microstructures observed by resonant tunneling spectroscopy. *Appl. Phys. Lett.* 67: 3921–3923, 1995.

53. P.W. Lukey, J. Caro, T. Zijlstra, E. van der Drift, and S. Radelaar. Observation of strain-relaxation-induced size effects in p-type Si/SiGe resonant-tunneling diodes. *Phys. Rev. B* 57: 7132–7140, 1998.

54. J. Liu, A. Zaslavsky, and L.B. Freund. Strain-induced quantum ring hole states in a gated vertical quantum dot. *Phys. Rev. Lett.* 89: 096804-1–096804-4, 2002.

55. J. Liu, A. Zaslavsky, B.R. Perkins, C. Aydin, and L.B. Freund. Single-hole tunneling into a strain-induced SiGe quantum ring. *Phys. Rev. B* 66: 161304-1–161304-4, 2002.

56. M. Fukuda, K. Nakagawa, S. Miyazaki, and M. Hirose. Resonant tunneling through a self-assembled Si quantum dot. *Appl. Phys. Lett.* 70: 2291–2293, 1997.

57. O.G. Schmidt, U. Denker, K. Eberl, O. Kienzle, F. Ernst, and R.J. Haug. Resonant tunneling diodes made up of stacked self-assembled Ge/Si islands. *Appl. Phys. Lett.* 77: 4341–4343, 2000.

58. U. Dötsch, U. Gennser, C. David, G. Dehlinger, D. Grützmacher, T. Heinzel, S. Lüscher, and K. Ensslin. Single-hole transistor in a p-Si/SiGe quantum well. *Appl. Phys. Lett.* 78: 341–343, 2001.

59. X.-Z. Bo, L.P. Rokhinson, H. Yin, D.C. Tsui, and J.C. Sturm. Nanopatterning of Si/SiGe electrical devices by atomic force microscopy oxidation. *Appl. Phys. Lett.* 81: 3263–3265, 2002.

60. L. Esaki. New phenomenon in narrow germanium p–n junctions. *Phys. Rev.* 109: 603–604, 1958.

61. M. Sweeny and J. Xu. Resonant interband tunnel diodes. *Appl. Phys. Lett.* 54: 546–548, 1989.

62. H. Jorke, H. Kibbel, K. Strohm, and E. Kasper. Forward-bias characteristics of Si bipolar junctions grown by molecular beam epitaxy at low temperatures. *Appl. Phys. Lett.* 63: 2408–2410,1993.

63. M.R. Sardela Jr., H.H. Radamson, and G.V. Hansson. Negative differential resistance at room temperature in δ-doped diodes grown by Si-molecular beam epitaxy. *Appl. Phys. Lett.* 64: 1711–1713, 1994.

64. S.L. Rommel, T.E. Dillon, M.W. Dashiell, H. Feng, J. Kolodzey, P.B. Berger, P.E. Thompson, K.D. Hobart, R. Lake, A.C. Seabaugh, G. Klimeck, and D.K. Blanks. Room temperature operation of epitaxially grown $Si/Si_{0.5}Ge_{0.5}$/Si resonant interband tunneling diodes. *Appl. Phys. Lett.* 73: 2191–2193, 1998.

65. S.L. Rommel, T.E. Dillon, P.R. Berger, P.E. Thompson, K.D. Hobart, R. Lake, and A.C. Seabaugh. Epitaxially grown Si resonant interband tunnel diodes exhibiting high current densities. *IEEE Electron Dev. Lett.* 20: 329–331, 1999.

66. R. Duschl, O.G. Schmidt, G. Reitemann, E. Kasper, and K. Eberl. High room temperature peak-to-valley current ratio in Si based Esaki diodes. *Electron. Lett.* 35: 1111–1112, 1999.

67. R. Duschl, O.G. Schmidt, and K. Eberl. Room temperature I–V characteristics of $Si/Si_{1-x}Ge_x$/Si interband tunneling diodes. *Physica E* 7: 836–839, 2000.

68. K. Eberl, R. Duschl, O.G. Schmidt, U. Denker, and R. Haug. Si-based resonant inter- and intraband tunneling diodes. *J. Cryst. Growth* 227–228: 770–776, 2001.

69. N. Jin, S.-Y. Chung, A.T. Rice, P.R. Berger, R. Yu, P.E. Thomson, and R. Lake. $151\,kA/cm^2$ peak current densities in Si/SiGe resonant interband tunneling diodes for high-power mixed signal applications. *Appl. Phys. Lett.* 83: 3308–3310, 2003.

24

IMPATT Diodes

Erich Kasper and
Michael Oehme
University of Stuttgart

24.1 Introduction

*Imp*act *a*valanche *t*ransit *t*ime (IMPATT) devices diodes are known from silicon and III–V material. As discrete devices mounted on special heatsinks, they are very powerful sources of microwave radiation. At 100 GHz CW-power of 1 W and pulsed power of 50 W can be obtained. The technology is complicated but well established after three decades of development. The negative resistance level of these discrete devices is rather low (at the order of a few ohms), which causes high efforts in designing appropriate resonators.

With increasing demand on mm-wave electronics for contactless sensors, security systems, and automobile applications, the pull for monolithic integration rises due to cost, weight, accuracy, and reliability issues. SiGe-SIMMWICs (silicon monolithic mm-wave integrated circuits) offer competitive responses onto these requirements. Planar IMPATTs can be monolithically integrated into SIMMWICs. The integrated IMPATTs aim to a much lower power level (mW regime) and offer much higher negative resistances (typically 10 to 100 Ω), which allows easier resonator and oscillator design. Small and simple designs are demonstrated up to 100 GHz operation frequency. Research activities focus on the implementation of heterostructures, on the replacement of the noisy avalanche multiplication by other injection mechanism, on the transfer of the delay concept to transistors, and on extension into the terahertz frequency regime.

In this chapter, we report the activities of integrated devices. Discrete classical IMPATTs are not treated.

24.2 Structure and Principle Function

In common diodes and transistors the output phase delay between voltage and current is kept small because with increasing phase delay the output power decreases. The IMPATT diode stands for a separate class of devices where a large phase delay is intentionally aimed. The preferred delay is around 180° (π), which characterizes a negative resistance used to overcome the positive load resistance.

Generally, the phase delay in this class of devices consists of two contributions: one from an injection phase Φ and one from a transit angle Θ. The total phase delay φ is given by

$$\varphi = \Phi + \frac{\Theta}{2} \tag{24.1}$$

The transit angle Θ counts only half because a displacement current occurs during the whole transit of a current pulse. For a space–charge region with electrical fields high enough for saturated carrier velocity the transit angle Θ is simply given by [1]

$$\Theta = \omega \frac{l_d}{v_S} \tag{24.2}$$

with $\omega = 2\pi f$ representing frequency, l_d the length of drift region, and v_S the saturation velocity of carriers. All transit angles may be realized by a proper choice of the length l_d, but the amplitude decreases because of the pulse broadening by the displacement current. Therefore, the injection phase Φ is essential for a large-phase delay φ and a large negative amplitude. In the IMPATT diode the injection phase $\Phi = \pi/2$ is produced by the avalanche multiplication process in a rather small avalanche region with width l_a (Figure 24.1).

The frequency-dependent impedance $Z = R + jX$ of such a simple diode structure with $l_a \ll l_d$ is given by

$$R = \mathrm{Re}(Z) = R_S - \frac{v_S(1 - \cos\Theta)}{A\varepsilon\omega^2}\left(\frac{1}{\omega^2/\omega_a^2 - 1}\right) \tag{24.3}$$

$$X\left(\frac{\omega A\varepsilon}{l_d}\right) = \left(\frac{\sin\Theta}{\Theta} - 1\right) - \frac{\dfrac{l_a}{l_d} + \dfrac{\sin\Theta}{\Theta}}{(\omega_a/\omega)^2 - 1} \tag{24.4}$$

The avalanche frequency ω_a depends on material constants as velocity v_S, permittivity ε, ionization coefficient α (E), and on the current density J_0. The avalanche frequency is slightly temperature dependent because of the temperature dependence of $v_S(T)$ and $\alpha(T)$

$$\omega_a^2 = 2\frac{d\alpha}{dE}v_S J_0/\varepsilon \tag{24.5}$$

The principle structure of a single drift diode and the resulting impedance levels are given in the following figures. Single drift is the term when only one carrier type—as in Figure 24.2, the electrons—contributes to the drift current. The doping sequence given in Figure 24.2 is known as Read structure where avalanche multiplication and drift are rather clearly separated (Read).

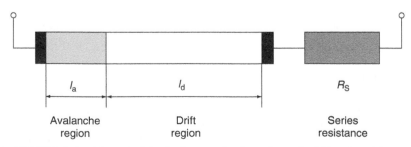

FIGURE 24.1　IMPATT diode with carrier injection by an avalanche region and a drift region with saturation velocity.

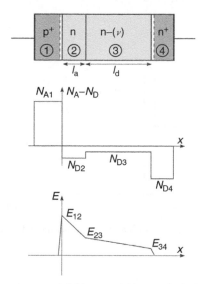

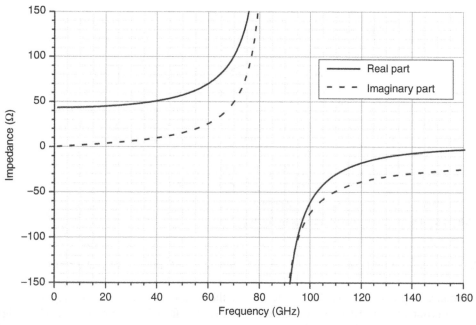

FIGURE 24.2 Structure, doping sequence, and field strength in a single drift Read structure.

FIGURE 24.3 Impedance (real part R solid line, imaginary part X broken line) as a function of frequency ($f_a = 85$ GHz).

At the avalanche frequency, the sign of the impedance switches from positive to negative values. Usually at frequencies slightly above the avalanche frequency the IMPATT is utilized as oscillator or amplifier. The idealized (loss less) impedance curve is characterized by a rapid decay of the negative impedance at frequencies above f_a (see Figure 24.3).

The injection mechanism may be changed to tunneling (TUNETT), thermoionic emission across a barrier (BARITT), or coherent transport [2] in a resonance-phase transistor (RPT) [3]. Figure 24.4

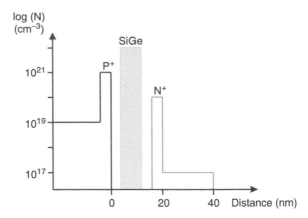

FIGURE 24.4 Injection region of a MITATT with a mixed tunneling and avalanche multiplication injection.

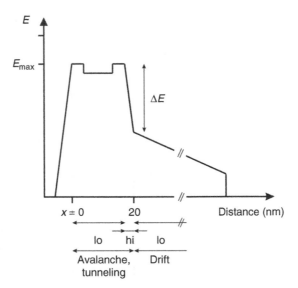

FIGURE 24.5 Electric field strength in a lo–hi–lo IMPATT.

exhibits the typical profile of a MITATT (mixed tunneling and impact avalanche transit time device) diode [4].

The tunneling probability may be adjusted [5] by the distance of the p^+ and n^+ regions and by the germanium content of the SiGe layer in between. The electrical field distribution in such a low–high–low (lo–hi–lo) active region is shown in Figure 24.5. The maximum field strength E_{max} (around 5×10^7 to 10^8 V/m) is obtained at the intrinsic region between the doping spikes. At the N^+-doping spike the field strength is reduced by ΔE and the electric field enters the drift region with a field strength $E_{max} - \Delta E$. The field step ΔE is correlated with the sheet concentration N_S in the spike by

$$N_S = N_D d_D = \frac{\Delta E \varepsilon}{l} \tag{24.6}$$

Sheet concentration in the order of 2 to 3×10^{16} m^{-2} is required to obtain the necessary field steps of 3.2 to 4.8×10^7 V/m.

24.3 SiGe—SIMMWIC

SIMMWIC were already proposed in 1984 [6] but the availability of silicon-based high-speed devices added recently to the attractiveness of integration concepts. Due to the reduced wavelength in the mm-wave regime (30 to 300 GHz) the nature of waveguide propagation has to be considered for the design of SIMMWICs. For example, at 100 GHz the wavelength in silicon is roughly 1 mm and so much smaller than the typical chip dimensions.

In the full version the SIMMWIC contains antenna, different planar waveguides, passive devices, active semiconductor devices, and sometimes also microelectromechanical (MEMS) devices integrated on a low-loss silicon substrate. Usually the low-loss substrate is obtained by a high-purity float zone (FZ) growth technique, resulting in specific resistivities of 1000 Ω cm and more. Often only subsystems are monolithically integrated and described as SIMMWIC. In the following, some examples of typical layouts are given for the illustration of the reader and only the breakdown behavior and negative differential resistance (NDR) of monolithic-integrated IMPATTs is treated in depths.

Layout Examples of Waveguides, Antenna, and Passive Circuits

The most appropriate planar waveguide for integration is the coplanar waveguide (CPW), which consists of a central signal line separated on both sides by slots from the surrounding ground plate (Figure 24.6).

The signal line has to be considered as a MOS varactor to understand the propagation losses of the transmission line [7], which are only low if inversion or accumulation layers are suppressed below the lines. Resistors, different types of capacitors (Figure 24.7), and spiral inductors belong to the passive devices used in SIMMWIC circuits for filtering, frequency adjustment, and impedance adjustment. Packages and hybrid connections require sophisticated efforts with higher frequencies.

A possible integration of the antenna is therefore an attractive option of SIMMWIC designs for both receiver and transmitter circuits. A simple example (Figure 24.8) is a rectenna [9] where a Schottky-detector is integrated with a planar antenna (rectifying antenna) and the detector signal is amplified by an operation amplifier. Only low- or medium-speed signals are treated at the package pins, and therefore, a commercial package could be used.

Integrated Schottky diodes may be designed for RC-frequency limits in excess of 1 THz [10]. Schottky diodes especially when driven in the so-called Mott operation are excellent candidates for detecting, mixing, frequency multiplication in the upper mm-wave frequency regime. Mott operation means that even under forward conditions the epitaxial layer is depleted which requires good control of dopant background and abrupt transitions. Silicide Schottky-barriers well developed in Si-technology [11] need an improved understanding of mixed silicide and germanide formation on SiGe layers.

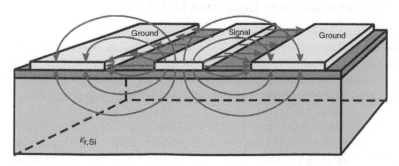

FIGURE 24.6 Coplanar waveguide with an insulating layer beneath the transmission lines.

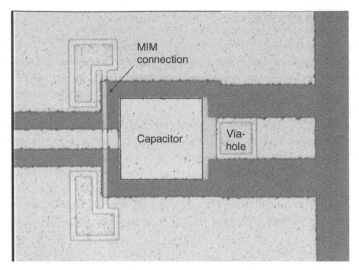

FIGURE 24.7 Metal–insulator–metal (MIM) capacitor in the intermediate frequency (IF) port of a harmonic 38 GHz mixer [7].

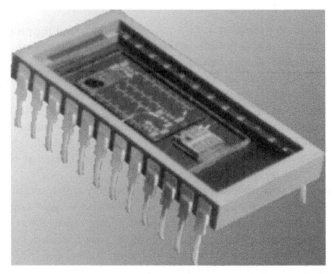

FIGURE 24.8 90 GHz receiver module consisting of a rectenna and an amplifier mounted on a multichip module.

Electrical Characterization of Integrated IMPATTs

Integrated IMPATT diodes are operated in the avalanche breakdown. They deliver a negative differential resistance above an avalanche frequency. The width of the frequency band with negative resistance is mainly determined by the series resistance because the amount of negative resistance provided by the diode decreases rapidly above the avalanche frequency. In the following, the report concentrates on breakdown behavior, S-parameter measurements of 75 to 110 GHz impedance and analysis of series resistance.

In a single drift silicon IMPATT the onset of breakdown (typically around -10 V for a 100 GHz IMPATT) is very sharp (Figure 24.10) and agrees roughly with predictions from ionization coefficients and breakdown field strengths. The predictions are explained on the example of a uniformly doped (N_D) single drift structure of thickness d_n. The electric field distribution is given in Figure 24.9. Assumed is

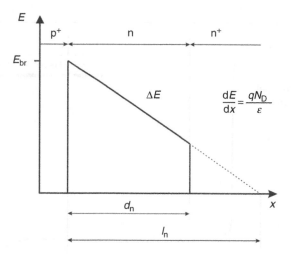

FIGURE 24.9 Trapezoidal field distribution in a single drift structure at breakdown.

a single drift structure with a reach through factor $F > 1$. The reach through factor F is defined by the ratio between projected depletion length l_n and technologically realized active layer thickness d_n

$$F = \frac{l_n}{d_n} \tag{24.7}$$

Onset of breakdown is obtained when the maximum field at the p^+n junction reaches the doping-dependent breakdown field strength $E_{br}(N_D)$. The breakdown voltage V_{br} is essentially equal to the area under the field function.

$$V_{br} + V_{bi} = \int_0^{d_n} E dx = \frac{\varepsilon}{2q} \frac{E_{br}^2(N_D)}{N_D} \frac{2F-1}{F^2} = E_{br}d_n - \frac{qN_D}{2\varepsilon} d_n^2 \tag{24.8}$$

(V_{bi}—built in voltage, q—electron charge, ε—permittivity).

The breakdown field E_{br} is a function slowly varying with doping and approximately given by

$$E_{br} = \frac{4 \times 10^7 \, \text{V/m}}{1 - \frac{1}{3} \log (N_D/10^{16} \, \text{cm}^{-3})} \tag{24.9}$$

At the high current densities where the IMPATT is operated (Figure 24.10) one sees a bend of the breakdown curve to higher voltages. This bending is mainly caused by two effects: heating and injection of carriers. The avalanche breakdown has a positive temperature coefficient because at higher temperature phonon scattering retards the carrier speed necessary for impact ionization. The injection of carriers into the space–charge region reduces the space–charge density in the depletion layer, which—like a lower doping—increases the breakdown voltage. The increase is proportional to the current and can be expressed as space–charge resistance R_{SC}:

$$R_{SC} = \frac{d_n^2}{2A\varepsilon v_S} \tag{24.10}$$

The avalanche breakdown characteristics of pseudomorphic SiGe are similar to that of silicon. But for thicker SiGe layers or higher Ge contents the SiGe layers have to be grown on virtual substrates

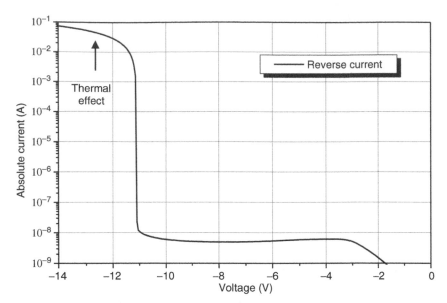

FIGURE 24.10   Reverse characteristics of the IMPATT diode.

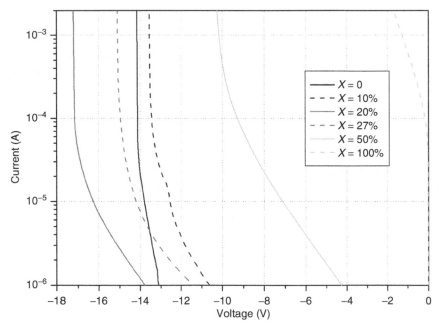

FIGURE 24.11   Breakdown of SiGe p–n junctions on virtual substrates. The Ge content X is varied between 0% and 100%. The shift in the absolute value of the breakdown is partly also caused by slight variations of the doping and thickness.

consisting of a relaxed SiGe buffer layer onto the silicon substrate. For monolithic integration with silicon circuits this buffer layer has to be thin, ideally below 100 nm [12]. The threading dislocation density of these thin buffers is in the order between 10^5 and 10^7 cm^{-2}, which degrades the breakdown behavior (Figure 24.11).

The breakdown characteristic (look at the slope, the absolute value shift is partly caused by different layer parameters) of low Ge content layers ($X = 0.1$ to 0.27) is rather similar to silicon at current levels above 0.1 mA. Even the 50% Ge layer demonstrates acceptable breakdown behavior above 1 mA current. In the 100% Ge layer on a thin virtual substrate, the breakdown is masked by high reverse current levels and these layers cannot be used for IMPATT operation. Progress in virtual substrate technology for high Ge content is necessary.

Proper dc-characteristics of the breakdown are a rapid test but the ultimate confirmation of the phase delay is given by *S*-parameter measurement in the selected microwave region. For a comparison and judgment of the results the *S*-parameter values are recalculated to impedance values (Figure 24.12). At the avalanche frequency f_a the imaginary part of the impedance switches sign and the negative real part obtains its maximum value. The negative impedance of the integrated IMPATT is quite high, for example $-7000\,\Omega$ for the 18 mA current curve at 77.5 GHz. Another important property of the frequency curves of the impedance values is the strong current dependency, which can be used to adjust the impedance level of the device. The operation in an oscillator circuit is above the avalanche frequency where the NDR is typically in the order of several tenths of ohms (Figure 24.13). The oscillating conditions require that the NDR of the device surpasses the load resistance to allow for undamped oscillations

$$Z_D + Z_L = 0 \qquad (24.11)$$

Stable oscillation is obtained when the sum of diode impedance Z_D and load impedance Z_L equals zero.

A frequency increase above the avalanche frequency reduces strongly the NDR. Up to which frequency an NDR is offered by the IMPATT depends also on the series resistance R_S. A low series resistance R_S is essential for a wide NDR regime. The series resistance in an integrated IMPATT consists of three different contributions (Figure 24.14): contact resistance R_C, epitaxy resistance R_{EPI}, and buried layer resistance R_{BL}.

Low contact resistance R_C is obtained with a highly doped semiconductor on the surface ($\geq 10^{20}\,\text{cm}^{-3}$) and a metal system with a fairly low Schottky barrier. The examples given in this chapter are with NiSi contacts a silicide metal, which also shows promises for sub-100 nm device dimensions. The buried layer resistance R_{BL} is low for a high-doped uniform layer, which is made in the given

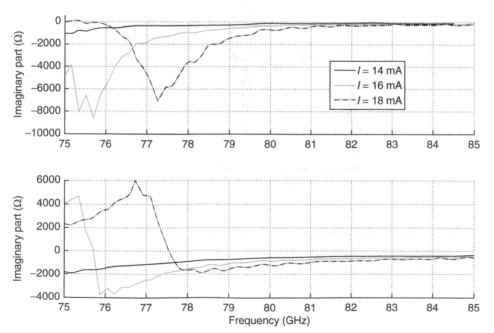

FIGURE 24.12 IMPATT impedance in the frequency band 75 to 85 GHz as a function of the current.

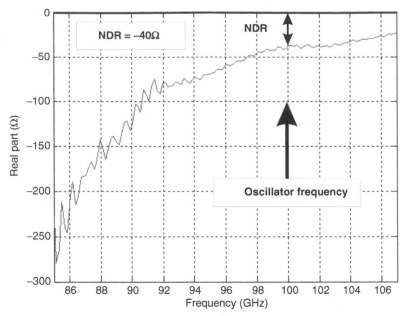

FIGURE 24.13 Impedance (real part) of the IMPATT around the oscillator frequency. Negative differential resistance (NDR) is $-40\,\Omega$ at an oscillation frequency of $100\,\mathrm{GHz}$.

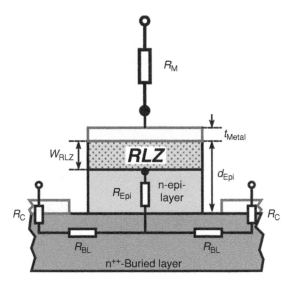

FIGURE 24.14 n^{+} buried layer for integration techniques.

examples by molecular beam epitaxy (MBE). The epitaxy resistance R_{EPI} is zero for breakdown operation because the epitaxy layer is depleted. But the extraction of the series resistance is done under forward voltage where the epitaxy layer contributes. The extracted value of the series resistance has to be reduced by

$$R_{\mathrm{EPI}} = \rho_{\mathrm{EPI}} d_{\mathrm{n}}/A \tag{24.12}$$

to account for the R_{S} value seen at breakdown.

The circuit test is performed by placing the integrated IMPATT diode into a planar resonator (Figure 24.15) on a silicon substrate. A rather straightforward coplanar resonator design matched the impedance level of the IMPATT. The spectrum of the 93 GHz oscillator is shown in Figure 24.16.

24.4 Extensions of the Concept

For circuit design the decoupling of the output terminal from the input by a three terminal device is considered as advantageous. Extensions of the phase delay concept to transistors could enhance the acceptance of these devices in more complex circuits. Recently, the successful test of a SiGe RPT was reported [13] and the basics of this transistor will be explained below. The separation of carriers in the high field of a reverse-biased junction is utilized for carrier injection in the drift region of an IMPATT. It is one of the mechanisms of the device, which together with a properly designed resonator allows for oscillations in the 100 GHz regime. A much more direct conversion in high-frequency radiation would

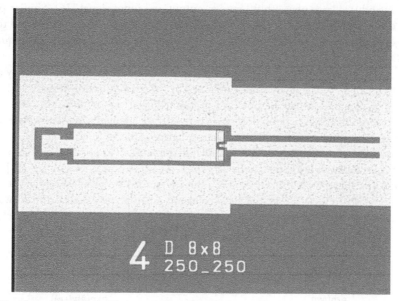

FIGURE 24.15 Planar oscillator with integrated IMPATT diode. The resonator is designed for an oscillating frequency near 93 GHz.

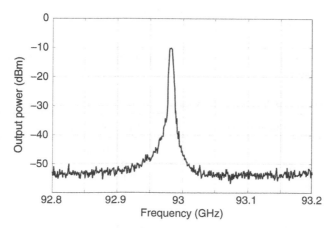

FIGURE 24.16 Spectrum of the oscillator depicted in Figure 24.15, with an oscillating frequency of 93 GHz.

be obtained by the separation of photogenerated electron–hole couples into a Hertzian dipole. Very recent and preliminary results give hints for a terahetz source [14]. The basics will be explained in the second part of this section.

Resonance-Phase Operation of HBT

The operation frequency of integrated circuits (IC) may be extended—depending on complexity and requirements—up to a certain fraction (typically 1/20 to 1/2) of the frequency limits of the transistor type used in the IC. Commonly considered frequency limits are the transit frequency f_T and the maximum oscillation frequency f_{max}. Within the common transistor paradigm the frequency limits have to be increased to allow higher circuit operation frequencies. Remarkable research results well beyond 100 GHz have been obtained with silicon-based transistors by lateral shrinking of dimensions and by use of SiGe–Si heterostructures [15]. Approaching the frequency limit, the output signal of these transistors is reduced and a phase shift between input and output takes place. In the newly proposed transistor type, the operation frequency should be increased far above the transit frequency f_T by an intentionally introduced large phase shift between output and input signal [16]. Similar principles are known from diode-type devices (e.g., IMPATT diodes) but never successfully transferred to transistors. In order to get the resonance effect within the 40 GHz setup we reduced the transit frequencies of the experimental versions to below 15 GHz.

The RPT concept is based on the achievement of NDR in a defined frequency band by a large phase delay of at least an angle π. The phase shift is obtained by a delayed injection (Figure 24.17) into a drift region [17]. Delayed injection may be obtained by tunneling or carrier diffusion [18]. Using for the technological realization a SiGe heterobipolar structure [19] we adopted a bipolar nomenclature for the electrodes emitter, base, collector for the more general terms cathode, injector, anode in Figure 24.17.

The layer structure of the processed RPT is shown in Table 24.1. To obtain the necessary phase shift both base and collector layers are chosen to be very thick. After the simulations in Ref. [18] the base layer thickness is 120 nm. Incorporated in the base is a linearly graded Ge profile from about 5% Ge at the emitter–base junction to 30% Ge at the base–collector junction to enhance the forward diffusion transport. The whole bandgap difference is about 170 meV. We use the term ultrametastable when

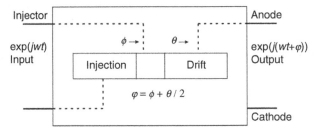

FIGURE 24.17 Two port representation of the resonance-phase transistor with the terminals cathode (emitter), injector (base), anode (collector).

TABLE 24.1 Vertical Layer Structure of the Processed RPT (Wafer A)

Structure	Thickness (nm)	Doping (cm^{-3})	Ge content (%)
Buried layer		n^{++}; 12 Ω/sq.	Si
Collector	$w_C = 1200$	N_C, n $= 3 \times 10^{16}$	Si
Spacer	$D = 10$	Intrinsic	Si$_{1-X}$Ge$_X$, X $= 30\%$
SiGe-base	$w_B = 120$	N_B, p^{++}; 2×10^{19}	Gradient
Spacer	$d = 2$	Intrinsic	Si$_{1-X}$Ge$_X$, X $= 5\%$
Emitter	$d = 70$	n; 1×10^{18}	Si
Emitter contact	$d = 230$	n^{++}; 2×10^{20}	Si

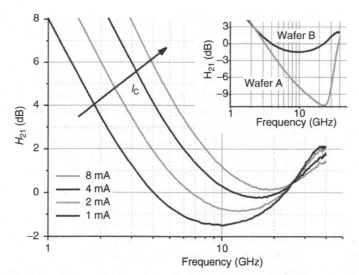

FIGURE 24.18 Common emitter current gain H_{21} over frequency for a 225 μm² emitter size RPT of wafer B, $U_{CE} = 4\,V$. The inset shows a comparison between two wafers ($I_C = 1\,mA$, $U_{CE} = 4\,V$) with different collector dopings.

thickness and Ge content are not only above the critical values of pseudomorphic growth but also beyond that of metastable growth at 550°C as measured by People and Bean [20]. Epitaxy below 550°C and low-temperature processing are necessary to get devices from ultrametastable structures.

In Figure 24.18, the current gain H_{21} in dB is shown in common emitter configuration. For frequencies below the transit frequency f_T the behavior is common. For low frequencies, the current gain approaches a constant value. Increasing the frequency results in the typical roll off of H_{21}. Up to a collector current $I_C = 10\,mA$ f_T is increasing, for higher currents the modified Kirk effect limits f_T because of the low collector doping concentration. For frequencies higher than f_T H_{21} first is decreasing partly even below 0 dB. But for still higher frequencies the resonant-phase effect turns H_{21} to increase again, reaching $H_{21} > 0$ dB at ~23 GHz and so active transistor operation at frequencies beyond f_T seems possible. At low currents ($I_C = 0.5\,mA$) the maximum gain of the resonance peak is seen with $H_{21} = 2.3$ dB at 36 GHz. The used measurement setup allowed measurements up to 40 GHz and demonstrated clearly the onset of resonance-phase effect. As expected by the model the resonance-phase effect is less current dependent than the transit frequency f_T.

Terahertz Oscillations from Optical Hot Carrier Injection

Under high electric fields photogenerated electron–hole pairs are separated with increasing velocity up to the saturation velocity or even above when velocity overshoot occurs. The ultrafast separation of the photogenerated electron–hole pairs creates a sheet of accelerating charges. The emitted radiation depends on the transport properties of the hot carriers. Basic experiments with femtosecond laser pulse excitation demonstrated oscillations around 4.5 THz. The preliminary study [14] was mainly aimed at investigating the hot carrier properties and to understand the mechanism. But obviously, carrier separation in a high electric field is a new candidate for terahertz radiation.

24.5 Growth and Process Requirements

For the integration of the IMPATT diodes a buried layer with a low sheet resistance and a good contact is required. Buried layers can be realized with different methods for example with ion implantation. We use a uniform buried layer with very high doping in the range of $1 \times 10^{20}\,cm^{-3}$. The layers are grown with the physical deposition method MBE (see Chapter 6, *SiGe and Si Strained-Layer Epitaxy for Silicon Heterostructure Devices*). This method allowed p-type and n-type doping over a wide range of concentration and with abrupt doping transitions. In the MBE system boron is used for p-type doping. This element has a large equilibrium solid solubility and a low surface segregation. For n-type doping antimony is used in MBE because this material can be directly coevaporated from a low-temperature-controlled effusion source during the growth. Because Sb has an extreme temperature sensitivity of surface segregation [21] we use for sharp doping transitions special doping strategies such as prebuild-up, flash-off techniques [22], or the doping by secondary ions (DSI) [23].

Figure 24.19 shows the layer stack for the monolithic-integrated IMPATT diode. The challenges for the growth are the abrupt doping transitions from 1×10^{20} to 1×10^{17} cm^{-3} in the n-region and from 1×10^{17} to 1×10^{20} cm^{-3} on the pn junction. Very sharp profiles over three orders of magnitudes are essential for the high-frequency performance. For the application in integrated high-frequency circuits a silicon (1 0 0) substrate with a high specific resistance greater than 1000 Ω cm is of advantage.

The complete doping structure is shown in Figure 24.20. After the thermal cleaning of the substrate [24] the growth starts with an intrinsic silicon buffer. For the doping of the 500-nm thick buried layer a prebuild-up growth strategy with constant Sb-flux at low growth temperatures (430°C) is applied. This procedure starts with the supply of an antimony adlayer of a fraction of a monolayer. At the end of the buried layer (see point 1 in Figure 24.20) one monolayer of antimony sticks on the surface. By higher growth temperatures, the segregation length Δ_S of antimony increases dramatically [21], for example, by increasing the temperature from 500°C to 600°C Δ_S increases by a factor of 1000. The bulk concentration, n_B, is determined from the ratio between surface density n_S and Δ_S. For constant n_S, n_B decreases as Δ_S increases. This is used for an abrupt dopant profile n$^+$/n. The temperature is increased during a growth interruption. Then the antimony level decreases abruptly over four to five decades and it is adjusted in the following n-silicon layer to 10^{17} cm^{-3} with the DSI method. A defined negative voltage of 100 V is applied on the substrate. By this, silicon ions are extracted from the electron gun and are accelerated to the substrate. The incoming silicon ions collide with the surface antimony atoms, so that they are incorporated in the growing layer. For a sharp doping transition on the pn junction the high-boron-doped uniform cap layer is also grown with the prebuild-up method (see point 3 in Figure 24.20). For providing the exact segregating adlayer density, we developed a method to measure this density at a fixed doping level [25].

The growth is monitored with different *in situ* analyzing methods. One powerful method is the reflection-supported pyrometric interferometry (RSPI) measurement [26]. RSPI has been proven to be capable of providing in situ real-time information concerning temperature and film thickness for

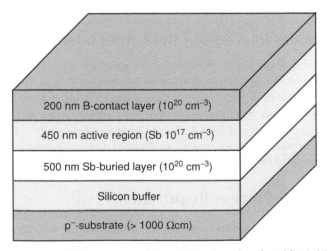

FIGURE 24.19   Layer stack for a 100 GHz integrated IMPATT diode with n-doped buried layer.

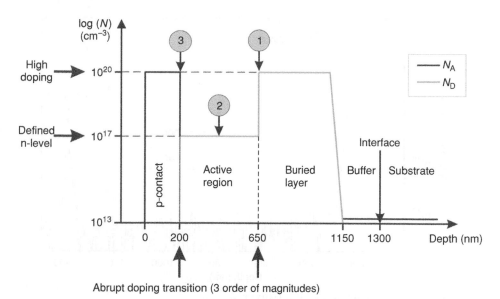

FIGURE 24.20 Vertical doping structure of the integrated IMPATT diode in Figure 24.19.

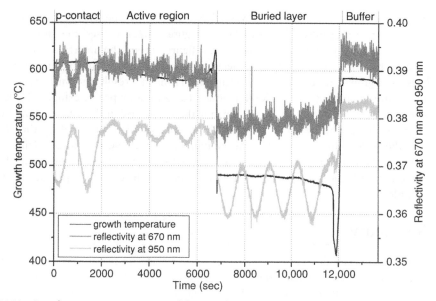

FIGURE 24.21 Interferometry measurement of the complete growth process sequence.

numerous applications in semiconductor manufacturing. Figure 24.21 shows the RSPI measurement over the complete growth process of an integrated IMPATT diode. The growth direction is from the right to the left side. After the intrinsic silicon buffer the growth temperature is decreased during a growth interruption. The reflectivity of both wavelengths decreases because the index of refraction decreases with lower temperature. During the growth of the buried layer oscillations appear, which result from the change in the index of refraction between intrinsic and very-high-doped silicon. From the distance between two maxima respectively minima the layer thickness is calculated. From the example shown in Figure 24.21, the buried layer thickness is 535 nm. With the same calculations the drift region and the p-contact are analyzed.

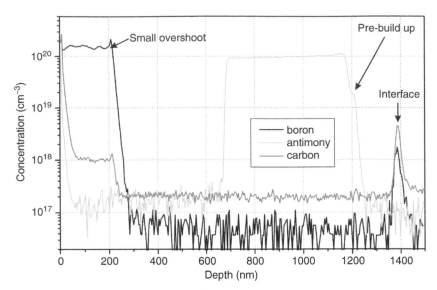

FIGURE 24.22 SIMS depth profile of a realized IMPATT structure.

A secondary ion mass spectroscopy (SIMS) depth profile of the elements boron, antimony, and carbon from the integrated IMPATT diode (Figure 24.22) proves the intended structure. The growth direction is from right to left in this figure. From the interface position (carbon- and boron-peak) the total layer thickness is determined as 1388 nm. The silicon rate is calculated from thickness and growth time as $r_{Si} = 0.107$ nm/sec. From this result, the thickness of the buried layer is with 535 nm the same as on the RSPI analysis. The antimony profile exhibits clearly the very-high-doped buried layer with 1×10^{20} cm^{-3} but not the active region, because the antimony SIMS background is too high. The transition between the n^+/n layers is dominated by the SIMS depth resolution. At the start of the buried layer, the submonolayer prebuild-up causes a faint step in the profile upraise. In the boron profile a small overshoot at the start of the very-high-doped p-contact proves the prebuild-up is too high. But the transition between drift region and p-contact is very sharp over three decades.

24.6 Summary

Monolithic integration [27–30] of semiconductor devices with passive circuits [31–38] will allow radar systems with chip dimensions. Monolithic-integrated IMPATTs offer NDR within a wide mm-wave frequency regime (30 to 150 GHz). These integrated devices strengthen the competitiveness of SiGe-SIMMWIC solutions by robust, simple and small oscillator, resonator and amplifier designs. SiGe heterostructures are especially used for alternative injection mechanisms (tunneling) and for three terminal device versions (resonance-phase transistor). Recent basic experiments on emission from field separated electron–hole pairs gave hints for extension of the frequencies into the terahertz regime.

Acknowledgments

The device processing group and the microwave group provided unpublished material, we thank especially C. Schoellhorn, M. Morschbach, and M. Jutzi. J. Hasch from Robert Bosch GmbH advised in high-frequency characterization and we acknowledge his help. We are grateful to H. Pfizenmaier for his early application driven interest in SIMMWIC realizations beyond 100 GHz.

References

1. S.M. Sze. *Physics of Semiconductor Devices.* John Wiley, New York (1981).
2. A.A. Grinberg and S. Luryi. Coherent transistor. *IEEE Trans. Electron Dev.* ED-40, 1537–1539 (2003).
3. E. Kasper and J.F. Luy. State of the Art and Future Trends in Silicon IMPATT Diodes for mm Wave Seeker Requirements, Military Microwaves Conference Proceedings. Microwave Exhibitions and Publishers, London, pp. 293–298 (1990).
4. J.F. Luy, H. Jorke, H. Kibbel, A. Casel, and E. Kasper. Si/SiGe heterostructure MITATT diode. *Electron. Lett.* 24, 1386–1387 (1988).
5. R. Duschel, O.G. Schmidt, G. Reitemann, E. Kasper, and K. Eberl. High room temperature peak to valley current ratio in Si based Esaki diodes. *Electron. Lett.* 35, 1–2 (1999).
6. P.J. Stabile, A. Rosen, W.M. Janton, A. Gombar, and M. Kolan. Millimeter wave silicon device and integrated circuit technology. IEEE MTT-S, Int. Microwave Symp. Digest, pp. 448–450 (1984).
7. W. Zhao, C. Schöllhorn, and E. Kasper. Interface loss mechanism of millimeter-wave coplanar waveguides on silicon. *IEEE Trans. Microwave Theory Tech.* 50, 407–410 (2002).
8. W. Zhao, C. Schöllhorn, E. Kasper, and C. Rheinfelder. 38 GHz coplanar harmonic mixer on silicon. Proceedings of the Third Topical Meeting on Silicon Monolithic Integrated Circuits in RF Systems, Ann Arbor, Michigan, pp. 138–141 (2001).
9. M. Herrmann, D. Beck, E. Kasper, J.-F. Luy, K.M. Strohm, and J. Buechler. Hybrid 90 GHz rectenna chip with CMOS preamplifier. Proceedings of the 26th European Solid State Research Conference, ESSDERC, Bologna, Italy, pp. 527–530 (1996).
10. M. Morschbach, C. Schöllhorn, M. Oehme, E. Kasper, A. Müller, and T. Buck. Integrated Schottky mixer diodes with cut off frequencies above 1 THz^2. Proceedings of the 34th European Microwave Conf., Amsterdam, Netherlands. October 2004, pp. 1133–1136.
11. C. Schoellhorn and E. Kasper. *Recent Res. Devel. Microwave Theory Tech.* 2, 155–182 (2004).
12. K. Lyutovich, E. Kasper, F. Ernst, M. Bauer, and M. Oehme. Relaxed SiGe buffer layer growth with point defect injection. *Mater. Sci. Eng. B* 71, 14–19 (2000).
13. E. Kasper, J. Eberhardt, H. Jorke, J.-F. Luy, H. Kibbel, M.W. Dashiell, O.G. Schmidt, and M. Stoffel. SiGe resonance phase transistor: Active transistor operation beyond the transit frequency f_T. *Solid-State Electron.* 48, 837–840 (2004).
14. A. Brodschelm, C. Schoellhorn, E. Kasper, and A. Leitensdorfer. Ultrafast high-field transport after 10 fs hot carrier injection in Si and SiGe. *Semicond. Sci. Technol.* 19, 267–269 (2004).
15. J. Eberhardt and E. Kasper. Modelling of SiGe hetero-bipolar transistor: 200 GHz frequencies with symmetrical delaytimes. *Solid State Electron.* 45, 2097–2100 (2001).
16. J. Weller, H. Jorke, K.M. Strohm, J.-F. Luy, and R. Sauer. Transistor action by transit time shift— a study for SiGe based devices. 28th European Microwave Conference Proceedings, pp. 745–750 (1998).
17. E. Kasper and G. Reitemann. Physics of future ultra high speed transistors: new concepts. 29th European Microwave Conference Proceedings, Vol. 1, pp. 155–157 (1999).
18. H. Jorke, M. Schäfer, and J.-F. Luy. Resonance phase transistor—concepts and perspectives. Topical Meeting on Si Monolithic Integrated Circuits in RF Systems, IEEE, pp. 149–156 (2001).
19. E. Kasper. Silicon germanium trends on process integration. ULSI Process Integration II, ECS Proc., Vol. 2, pp. 143–154 (2001).
20. R. People and J.C. Bean. Calculation of critical thickness versus lattice mismatch for SiGe/Si strained layer heterostructures. *Appl. Phys. Lett.* 47, 322–324 (1985) and 49, 229 (1986).
21. H. Jorke. Surface segregation of Sb on Si(100) during molecular beam epitaxy growth. *Surf. Sci.* 193, 569–578 (1988).
22. S.S. Iyer, R.A. Metzger, and F.G. Allen. Sharp profiles with high and low doping levels in silicon grown by molecular beam epitaxy. *J. Appl. Phys.* 52, 5608–5613 (1981).
23. H. Jorke, H.-J. Herzog, and H. Kibbel. Secondary implantation of Sb into Si molecular beam epitaxy layers. *Appl. Phys. Lett.* 47, 511–513 (1985).

24. E. Kasper, M. Bauer, and M. Oehme. Quantitative secondary ion mass spectrometry analysis of SiO$_2$ desorption during in situ cleaning. *Thin Solid Films* 321, 148–152 (1998).

25. M. Oehme, M. Bauer, T. Grasby, and E. Kasper. A novel measurement method of segregating adlayers in MBE. *Thin Solid Films* 369, 138–142 (2000).

26. M. Bauer, M. Oehme, M. Sauter, G. Eifler, and E. Kasper. Time resolved reflectivity measurements of silicon solid phase epitaxial regrowth. *Thin Solid Films* 369, 228–232 (2000).

27. A. Rosen, M. Caulton, P. Stabile, A.M. Gombar, W.M. Janton, C.P. Wu, J.F. Corboy, and C.W. Magee. Silicon as a millimeter-wave monolithically integrated substrate. *RCA Rev.* 42, 633–660 (1981).

28. P. Russer. Si and SiGe millimeter-wave integrated circuits. *IEEE Trans. Microwave Theory Tech.* 46(5) (1998).

29. Y. Konishi, ed. *Microwave Integrated Circuits.* Marcel Dekker, New York (1991).

30. N.K. Das and H.L. Bertoni, eds. *Direction for the Next Generation of MMIC Devices and Systems.* Plenum Press, New York (1997).

31. W. Heinrich, J. Gerdes, F.J. Schmückle, C. Rheinfelder, and K.M. Strohm. Coplanar passive elements on Si for frequencies up to 110 GHz. *IEEE Trans. Microwave Theory Tech.* 46(5), 709–712 (1998).

32. K.C. Gupta, R. Garg, I. Bahl, and P. Bhartia. *Microstrip Lines and Slotlines.* Artech House, Boston (1996).

33. H. Hasegawa, M. Furukawa, and H. Yanai. Properties of microstrip line on Si–SiO$_2$ system. *IEEE Trans. Microwave Theory Tech.* MTT-19(11), 869–881 (1971).

34. Interdigital capacitors and their application to lumped-element microwave integrated circuits. *IEEE Trans. Microwave Theory Tech.* MTT-18, 1028–1033 (1970).

35. A. Gopinath. Losses in coplanar waveguides. *IEEE Trans. Microwave Theory Tech.* MTT-30, 1101–1104 (1982).

36. R.K. Ulrich, W.D. Brown, S.S. Ang, F.D. Barlow, A. Elshabini, T.G. Lenihan, H.A. Naseem, D.M. Nelms, J. Parkerson, L.W. Shaper, and G. Morcan. Getting aggressive with passive devices. *IEEE Circ. Dev.* 16, 17–25 (2000).

37. J.N. Burghartz, D.C. Edelstein, K.A. Jenkins, and Y.H. Kwark. Spiral inductors and transmission lines in silicon technology using copper–damascene interconnects and low-loss substrates. *IEEE Trans. Microwave Theory Tech.* 45, 1961–1968 (1997).

38. H.A. Wheeler. Simple inductance formulas for radio coils. *Proc. IRE* 16, 1398–1400 (1928).

25

Engineered Substrates for Electronic and Optoelectronic Systems

Eugene A. Fitzgerald
Massachusetts Institute of Technology

25.1 Lattice-Matched Substrate World

Contemporarily, we analyze the semiconductor systems by a substrate materials class, i.e., "Si" or "III–Vs." The III–V classification is ultimately split into "GaAs" or "InP," for example. This nomenclature contains two subtle but important connotations. First, we recognize the importance of the material on the performance of the system. Second, it is assumed inherently that materials choices are confined to a particular bulk substrate material, e.g., a Si substrate. Bulk is defined here as a substrate grown by a bulk crystal growth technique, which in general employs pulling a crystal from a melt. Semiconductor compounds that form elemental or binary compounds, referred to below as "bulk semiconductors," are amenable to bulk crystal growth, but miscible alloys of the bulk compounds are not. Thus, all current semiconductor-based systems are built on particular lattice constants allowed by nature. Early in epitaxial growth research, many lattice-mismatched films (i.e., films composed of materials that are alloys of bulk semiconductors) were deposited on bulk semiconductors, but the lattice-mismatch between the film and substrate led to poor material quality in the thin film. A consequence of this epitaxial research was that lattice-mismatched epitaxy, i.e., achieving lattice constants in-between bulk semiconductors, was considered impractical. Thus, nearly all electronic and optoelectronic systems are built on lattice constants of the bulk semiconductors. These materials combinations can be seen in Figure 25.1 by following the vertical lines of constant lattice-constant up and down the diagram. In Si technology, today a "silicon wafer" is often a silicon epitaxial layer on a bulk substrate. In CD lasers and other applications involving optoelectronics near the 870 nm wavelength, the AlGaAs alloy system on GaAs was employed, as heterostructure devices could be designed without creating lattice mismatch. And for telecommunications applications, InP substrates became the bulk semiconductor of choice, as InGaAsP alloys could be grown lattice-matched to the InP lattice constant and also achieve the desired 1.3 and 1.55 μm wavelength emission required for low-loss transmission in optical fiber. Thus, all commercial and defense semiconductor systems have been developed on bulk semiconductors using lattice constants of the substrate.

25.2 Limitations of Lattice-Matched Systems

There are limitations of lattice-matched semiconductor systems. The first and most obvious is the rich nature of semiconductor bandgaps and lattice constants in Figure 25.1 that have not been accessed due to the restriction of building on the lattice constant of the substrate. In general, electronic performance improves as the lattice constant increases within a class of semiconductor materials. For long wavelength applications, optical properties improve as well, in that the bandgap shrinks as lattice constants increase, allowing ultra-long-wavelength detection. Yet electronic complexity (i.e., device integration density) of the semiconductor system decreases as lattice constant is increased, leading to a forced trade-off between integration density and performance.

A related but somewhat different limitation is that the electronic and optoelectronic systems built on these platforms create a separation of platforms throughout the system. For example, Si is the basis for all digital computation, whereas III–V and II–VI materials are the basis for most high-frequency RF and optical interfaces. Thus, electronic systems are now limited by board-level consequences induced by the separation of semiconductor platforms. For example, even if a system is created with the capability of gathering a large amount of optical or RF data with a III–V-based platform, getting that information into Si-based computing platforms is a board-level performance and cost issue.

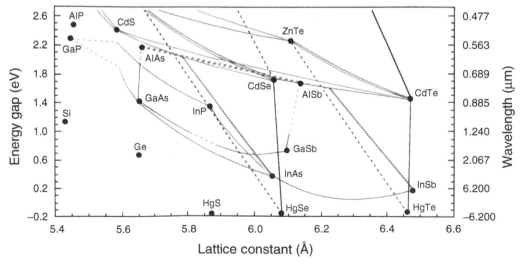

FIGURE 25.1 Energy gap versus lattice constant of most II–VI, III–V, and IV semiconductors. (From Bell Laboratories.)

The performance limitations described above can be seen in the forced evolution of lattice-matched systems. Staying within the constraints of lattice-matching, researchers in the mid-1980s began to explore incorporating slightly mismatched semiconductor films by keeping the level of mismatch and the film thickness below the critical thickness for dislocation introduction. This effort to squeeze even the slightest enhancement from materials with slightly different lattice constant is evidence of the immense desire to move away from mature, common-lattice-constant systems. Examples of such materials and devices that have migrated into applications are the strained SiGe base heterojunction bipolar transistor (covered extensively elsewhere in this volume), and the pseudomorphic strained InGaAs channel HEMT (pHEMT). In optoelectronic applications, the 980 nm strained InGaAs pump laser is an example of pushing the limit on strained layer critical thickness. It is important to realize that the added performance in these devices is severely limited by the critical layer thickness constraint, and highlights the need for the increased performance that can be released by new materials with new lattice constants.

25.3   Age of Lattice-Mismatched Substrate Engineered Materials

Parallel research efforts over the past 20 years in epitaxy, strain-relaxed semiconductor epitaxy, and wafer-bonding technology are ushering in a new age of lattice-mismatched substrate engineering. Advances in fundamental materials science as well as an increase in demand for electronic and optoelectronic systems uninhibited by lattice-constant constraints show that new engineered substrates can have a large impact in the near and far term. We are entering an age of "anything on anything," i.e., relaxed buffer technology and wafer bonding together allow a limitless ability to move laterally in Figure 25.1. The age of lattice-mismatched materials will allow the integration of any material on any bulk semiconductor substrate. In particular, the most visible area of engineered substrates is currently in relaxed SiGe–Si. The lattice constants in between Si and Ge offer a host of materials and devices that can be constructed in the Si CMOS manufacturing infrastructure, such as low-power or high-frequency CMOS and the integration of III–V photonics with Si. We first describe the three cornerstones that have created this opportunity in engineered substrates, and then describe potential future devices and systems, which may be constructed on these nanoengineered substrates.

Strain-Relaxed Layers on Substrates through Composition Grading

About 15 to 30 years ago, the field of lattice-mismatched semiconductors was dominated by experiments and theory elucidating the critical layer thickness. When a slightly lattice-mismatched semiconductor film is grown on top of a substrate, a certain thickness of strained material is deposited before it is energetically favorable to introduce misfit dislocations to relieve the strain, and this thickness is termed the critical thickness. Much of the experimental data collected to elucidate the concept of critical thickness were collected in the InGaAs/GaAs [1] and SiGe/Si [2] materials systems, such experiments focusing on exploring when small levels of mismatch resulted in dislocation formation. This focus arose from the interest in the science of lattice-mismatch, as well as realization that high levels of strain in a film, without relaxation, may be beneficial to devices [3]. This early research led to some eventual commercial successes like the InGaAs pHEMT and the SiGe HBT [4]. Early progress in lattice-mismatched semiconductors has been previously reviewed [5].

As strained layer devices were approaching serious commercial interest, research continued into the critical thickness issue. How misfit dislocations are introduced, i.e., the kinetics of dislocation introduction, became an active area of research, especially experimentally measuring the velocities of dislocations in strained layers [6]. In addition to the velocity of dislocations, it was also shown that nucleation plays an important role, specifically the presence or absence of heterogeneous nucleation sources [7]. It was shown that the critical thickness could be extensively exceeded if a substrate area lacked a nucleation site for dislocation introduction. At the time, this result was still of interest only to extend the degree of strain or thickness that one could contain in a completely strained film. However, the experiments showed a path to creating completely relaxed layers with low threading dislocation density, a long-sought goal that was evasive due to the previous lack of understanding in dislocation kinetics.

Using this new dislocation information, it was shown in 1991 that it was possible to create low threading dislocation density relaxed SiGe layers on Si [8]. This result opened the door to high mobility strained Si [9] as well as the potential of creating relaxed Ge and GaAs layers on Si for the integration of optoelectronic devices on Si [10]. The key to creating the relaxed alloys like SiGe on Si with low threading dislocation density at the top surface was the use of layers of graded composition grown at relatively high temperatures. Such a structure minimizes dislocation nucleation and encourages maximum threading dislocation propagation, leading to high levels of strain relaxation with relatively low threading dislocation density.

In early work, it was recognized that maximum threading dislocation flow in the material during graded buffer deposition was critical to obtaining the combined desire of high relaxation and low threading density. A model was created, predicting a counter-intuitive feature, which was that higher growth temperatures would lead to lower threading dislocation density [10]. Under conditions of gradual grading and relatively thick layers, the model can produce a practical result that can be used to predict threading dislocation density in relaxed buffers [11]:

$$\rho_t = \frac{2R_g R_{gr} e^{U/kT}}{bBY^m \varepsilon_{eff}^m} \tag{25.1}$$

where ρ_t is the threading density at the surface of the relaxed buffer layer, R_g is the growth rate, R_{gr} is the grading rate (i.e., strain relieved per unit thickness), U is the activation energy for dislocation glide, b is the Burgers vector, B is a constant, Y is the biaxial modulus, m is a number typically between 1 and 2, and ε_{eff} is the effective strain the threading dislocation experiences during glide. Note that the strongest factor in reducing threading dislocation density is temperature, since temperature is in the exponent of Equation 25.1. The model was first confirmed in the InGaP/GaP [12] system and later in the SiGe/Si system [13].

There are important consequences of Equation 25.1 with regard to typical threading dislocation densities as well as the cost of producing relaxed SiGe substrates (also referred to as "virtual substrates," as the wafer is Si, but the graded composition layer converts the surface to a relaxed SiGe lattice constant, thus creating a surface which would be reminiscent of a bulk SiGe substrate). First, once the highest temperatures are achieved from a practical perspective, the threading dislocation density cannot be influenced drastically by any other variable. For example, at temperatures of 850°C and higher, it is typical that relaxed SiGe layers on Si have threading dislocation densities on the order of 10^4 to $10^5 \, cm^{-2}$, and further significant reduction just by manipulating growth variables is not possible. Fortunately, this threading dislocation density is low enough for both majority carrier devices in SiGe and also low enough for minority carrier devices in GaAs–SiGe–Si [14]. Second, the lowest cost SiGe substrates will have the greatest perfection, i.e., lowest threading dislocation density. This relationship is embedded in Equation 25.1 since the growth rate can increase drastically with increased temperature as long as the activation energy for CVD decomposition is less than the activation energy for dislocation glide. Thus, increased temperature lowers threading dislocation density and increases growth rate, leading to less costly wafers. Recently, relaxed SiGe–Si has been produced commercially using Cl-based chemistry at higher growth temperatures, leading to low-cost substrates with threading disloca-tion densities less than $10^5 \, cm^{-2}$.

Transfer of Relaxed Lattice Constants via Wafer Bonding of Virtual Substrates

In addition, moving laterally in between bulk semiconductor substrate lattice constants, the relaxed epitaxial layers on a conventional substrate offer another advantage: the potential usefulness of wafer bonding is released. Traditionally, there were hopes that wafer bonding could be used to at least create engineered substrates with one lattice constant of a bulk substrate on another. For example, a bulk GaAs wafer might be bonded to a Si wafer, and a thin layer of GaAs could be transferred by using processing techniques to remove much of the original GaAs substrate, for example. There are application limitations in traditional wafer bonding, such as not being able to create any lattice constant in between bulk substrate lattice constants. In addition, there are processing limitations that limit even the possible applications like GaAs on Si. First, the end markets that drive Si and GaAs are different, and therefore, the scaling of the industry infrastructure is different, resulting in Si wafers being substantially larger than GaAs wafers at any point in time. Thus, economies of scale are not captured as the area of usable Si substrate would be less since the bonded GaAs wafer would be of smaller width. Of course, one could always use smaller Si wafers, but then the CMOS fabrication facility used for the CMOS electronics would be trailing edge and have larger transistors, thus limiting the combined integrated platform from using the best computation platform. Second, the bonding of two bulk substrates with different lattice constants leads to problems due to the materials also having a different coefficient of thermal expansion. For example, InP–Si wafer bonding at high bonding temperatures leads to a shattering of the material due to the fusion of thick materials with differing coefficients of thermal expansion; upon cooling to room temperature, tremendous stress is created, shattering the material or creating a high degree of curvature or fragility in the composite.

The use of high-quality relaxed epitaxial layers circumvents the issues that originally faced wafer-bonding technology. The bulk of both wafers can be of the same material, and second, the substrates are the same size. We have been able to demonstrate that this relaxed buffer bonding is possible even for virtual Ge on Si [15]. We have transferred Ge from the surface of a virtual Ge wafer on Si to an SiO_2–Si substrate, creating Ge–oxide–Si, called germanium-on-insulator (GOI). Other examples are silicon–germanium-on-insulator (SGOI) and strained-silicon-on-silicon (SSOS). Both are structures that can only be created with relaxed buffer bonding. With further research in this area, one could potentially create any semiconductor material on another, at any wafer diameter [16]. Furthermore, judicious selection of the virtual substrate platform could allow the transferred layer to be in either relaxed or strained form, thereby adding another degree of freedom to the process.

Low-Temperature CVD Device Layers

The third area that has allowed the creation of nanoengineered substrates is low-temperature CVD epitaxy [17]. We have demonstrated that at low enough growth temperatures, thin, flat 2% compressive and 2% tensile films can be created in the SiGe materials system. An example of why low-temperature growth is needed to suppress adatom surface migration is shown in Figure 25.2. As previously shown [18] and reconfirmed here, compressive layers are difficult to deposit in a very planar way for significant strains. Compressed layers of $Si_{1-y}Ge_y$ on relaxed buffers of $Si_{1-x}Ge_x$ ($y > x$) are required for very high hole mobility as shown below. Flat layers are critical for obtaining high hole mobility, as coherent strain relaxation will produce a roughened interface that degrades hole mobility drastically. Figure 25.2 shows that for a compressively strained Ge layer on relaxed $Si_{0.3}Ge_{0.7}$, a growth temperature of 400°C or less is needed to prevent relaxation through surface roughening, which can then also lead to dislocation nucleation as seen in Figure 25.2. Figure 25.2c shows that flat, thin, highly compressed Ge layers are possible to deposit at low enough temperatures.

25.4 Nanoengineered Substrates for MOSFETS

Epitaxial relaxed $Si_{1-x}Ge_x$ buffer layers [19] create a larger lattice constant on a Si substrate, allowing subsequently grown $Si_{1-y}Ge_y$ layers to be strained in tension ($y < x$) or compression ($y > x$). Early work in application of strain via relaxed SiGe concentrated on investigating elevated carrier mobility in pure tensile Si layers deposited on relaxed $Si_{1-x}Ge_x$ [20,21]. Relatively short-channel MOSFETs containing strained Si have shown that higher mobility and drain current measured in long-channel devices are retained at shorter channel lengths [22,23]. Also, recently a strained Si ring oscillator composed of 35-nm gate transistors operated 20% to 40% faster than the control Si ring oscillator [24]. A quantitative

(a) 550°C

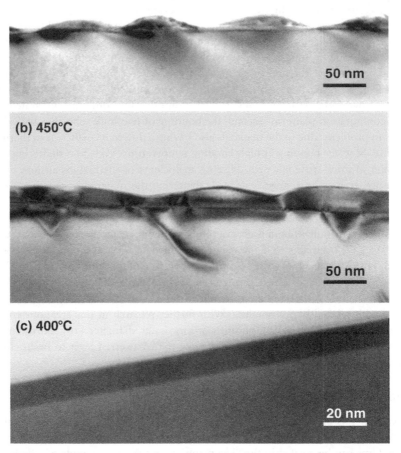

(b) 450°C

(c) 400°C

FIGURE 25.2 XTEM of ε-Ge grown on $Si_{0.3}Ge_{0.7}$ at (a) 550°C, (b) 450°C, and (c) 400°C showing evolution of morphology with T (ε-Ge is the dark layer). The heterostructure in (c) is appropriate for use in surface-channel devices.

method to correlate the effect of mobility enhancement in long and short channels shows that approximately 50% of the long-channel drain current enhancement is obtained in shorter channels [25]. Thus, large MOSFET devices can be used to rapidly probe heterostructures for channel enhancement, as well as limits to processing [26–29]. In this summary, we report on probing advanced SiGe MOSFETs on nanoengineered substrates imparting tensile and compressive strains.

The MOSFET fabrication process was selected to speed processing and allow extraction of real channel mobility through measurement of MOSFET drain current [30]. The long-channel MOSFETs were formed in a single mask step and utilized thick (300 nm) deposited gate oxide. No other aspects of the MOSFET were optimized. As demonstrated in many benchmarks to date, these large MOSFETs produce channel mobility data versus effective vertical field or inversion charge identical to more fully processed devices [26].

Single Strained Si Channels on Relaxed $Si_{1-x}Ge_x$

The most common strained Si structure for surface-channel MOSFETs is a 10 to 20 nm strained Si layer deposited on a relaxed $Si_{1-x}Ge_x$ buffer with $x \sim 0.20$. nMOSFETs fabricated from such a structure show an enhancement in electron mobility of about 1.8 times as compared to the Si control nMOSFETs [22–26]. With very low-temperature processing, pMOSFETs can show 10% to 20% enhancement [26], although short-channel MOSFETs processed in more commercial processes show the same performance as control Si [25]. Thus, the "first generation" of strained Si substrate will give nMOSFETs an 80% increase in electron mobility and a 20% to 40% increase in nMOS drive current, but the pMOSFET will not see much of an enhancement.

Further increases in strain do not increase the electron mobility enhancement in strained Si; however, hole mobility enhancement continues to increase with increasing strain. Early recognition of this potential enhancement for holes had led to investigations using higher Ge compositions in the relaxed buffer to enhance the hole mobility in pMOSFETs. Figure 25.3 is a plot of hole mobility enhancement (as compared to control Si MOS devices) for relaxed buffer Ge concentrations greater than $x = 0.3$ [27]. For structures with $x \geq 0.40$, the strained Si layer thickness exceeds the critical thickness for misfit dislocation introduction at the strained Si–SiGe interface.

In Figure 25.3, first note that in the strained Si–$Si_{0.65}Ge_{0.35}$ structure, the mobility enhancement of holes in the channel decreases as the vertical field (i.e., the inversion charge) is increased (such a decrease is not seen in the electron mobility enhancement). This decrease is the typical problem associated with hole mobility enhancement in strained Si. At high fields in strongly scaled devices (>1 MV/cm), one can see that there will only be a small enhancement remaining, if at all. For $x = 0.4$ to 0.5, hole mobility enhancement as large as two times can be seen, and it is likely some enhancement will be retained at higher fields despite the presence of misfit dislocations at the strained Si–SiGe interface. Since the enhancements are equal for all buffer concentrations at $E_{eff} \sim 0.6$ MV/cm, there appears to be no incentive for further increases in buffer Ge concentrations. However, note for the first time that the rate of enhancement decreases with vertical field has somehow been affected in the strained Si–$Si_{0.5}Ge_{0.5}$ structure.

Dual-Channel Heterostructures on Relaxed $Si_{1-x}Ge_x$

In investigating the potential source of the hole mobility decrease with vertical field, we have noticed that the out-of-plane hole effective mass ($m_\perp$) is as light, and can be even lighter, than the in-plane effective mass ($m_\parallel$) in strained Si [31,32]. For the electron in strained Si, $m_\perp > m_\parallel$, the preferred situation for an inversion charge at the SiO_2–Si interface. Thus, as the hole mass is lightened by the strain, so is the vertical mass, and, in fact, it may be very light and difficult to contain the DeBroglie wavelength in the

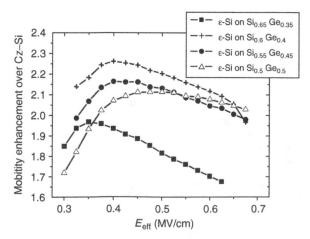

FIGURE 25.3   The mobility enhancement of holes versus vertical field under the gate in pMOSFETs in strained Si for different Ge concentrations in the relaxed $Si_{1-x}Ge_x$ buffer.

strained Si layer. Therefore, it seems reasonable to try and contain the hole wave function in the vertical direction.

Inserting a compressed $Si_{1-y}Ge_y$ layer ($y > x$) below the strained Si layer accomplishes this goal, and further increases the in-plane mobility additionally. The compressive strain breaks the degeneracy of the hole valence band, increasing the scattering time. We term the strained Si-compressed SiGe structure a "dual-channel" heterostructure. Figure 25.4 is a graph showing the mobility enhancement for a set of dual-channel MOSFETs as well as single-channel MOSFETs for comparison (data from Ref. [28]).

Figure 25.4 shows that indeed, the dual-channel heterostructures can support very high hole mobility at relatively high vertical fields. Also, the slope of the hole enhancement decrease with field can be less as well, further showing the scalability of these structures. A significant observation of enhancement versus field plots for dual heterostructures is that the rate of hole degradation with field is decreased when the dual-channel layers are kept thin. For example, the maximum hole enhancement for the dual-channel structure on $Si_{0.7}Ge_{0.3}$ shown in Figure 25.4 was obtained in a structure in which both the strained Si layer and buried compressed SiGe layer were ~4 nm. At this thickness, we estimate the hole wave function must be spread across both the strained Si layer and compressed SiGe layer, even at the largest vertical fields we can create in these structures. Thus, it appears that the mixed character of the hole spread across both layers is beneficial. When the structure has thicker layers (~8.5 nm), and therefore, a structure closer to true buried-channel structure, the mobility enhancement is less at high field and the slope of the mobility decrease with field is similar to conventional strained Si.

Note also in Figure 25.4 that the dual-channel structures were created with the same strain level incorporated into the compressed SiGe layer (the difference between the buffer composition and compressed SiGe composition is always held at $\sim x = 0.30$). The comparison of the data shows that when there is enough strain present to split the valence band degeneracy, the Ge concentration in the compressed layer is the most important factor for hole enhancement. The increased curvature of the valence bands from the increased Ge concentration leads to much lighter holes and increased hole mobility.

The two observations noted above led to more advanced single-channel structures: hole wave function hybridizing across two layers and high valence band curvature from Ge. The dual-channel data suggest

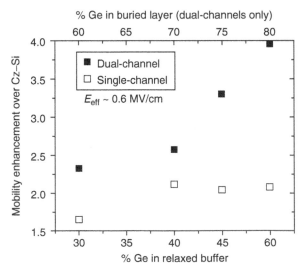

FIGURE 25.4 The mobility enhancement of holes versus concentration of Ge in the relaxed buffer. The data for single channels are shown in the open squares, whereas the dual-channel structures are the filled squares. The data are taken with approximately $E_{eff} = 0.6$ MV/cm.

that single-channel heterostructures with very high concentrations of Ge and thin strained Si surface layers may support high enhancements at high field.

Finally, we note here that dual-channel structures with a compressed pure Ge layer and a dislocated Si cap layer have shown PMOS mobility enhancement factors greater than eight times [29].

Nanostructured SiGe Channels

We speculate that a hole forced to exist across a thin strained Si layer and a relaxed high-Ge content film will have a character similar to a hole that exists in a structure with a compressed SiGe layer (dual channel). This behavior will occur at a relatively high vertical field, in which the hole can be forced to occupy both the surface strained Si channel and the relaxed SiGe alloy below. The wave function should be hybridized between the two layers. Thus, a split valence band from the strained Si is hybridized with a degenerate high curvature band from the relaxed SiGe alloy. The result should be a split valence band with high curvature, which resembles a compressed SiGe layer band structure.

To test this hypothesis, strained Si layers were deposited on relaxed buffers with $x = 0.60$ and 0.70. The strained Si layers were kept to approximately 4 nm in thickness. The layers were greater than the critical thickness for misfit dislocation introduction. Figure 25.5 is a plot of the hole mobility enhancement versus inversion charge. We use inversion charge for the x-axis due to the fact that the exact vertical field experienced by the carriers is not easily determined due to the large band offsets close to the SiO_2-Si interface. High inversion charge occurs when there is a high vertical field, as in Figure 25.3 and Figure 25.4, but the exact quantitative relationship requires detailed Poisson–Schrodinger solutions, which incorporate full band structures, band alignments, and three-dimensional effective masses.

For both the $x = 0.60$ and 0.70 single-channel structures, we observe a unique phenomenon in which the hole mobility enhancement factor increases with inversion charge or effective vertical field. Thus, it is possible to have a structure that can host both nMOS and pMOS channels with high carrier mobility enhancements at high vertical fields. We interpret this increase in carrier mobility enhancement as a result of the wave function averaged between the two valence structures. At low fields, the hole resembles the relaxed $Si_{0.4}Ge_{0.6}$ hole, since the band offset at the strained $Si-Si_{0.4}Ge_{0.6}$ interface forces most of the hole in the buried relaxed $Si_{0.4}Ge_{0.6}$ material. The enhancement is not large due to alloy scattering; it is well known that the alloy scattering in relaxed SiGe alloys suppresses mobility for most of the alloy compositions, and mobility rises sharply very near the pure Ge and pure Si concentrations [32]. As voltage is applied to the gate and vertical field and inversion charge are increased, more of the hole is

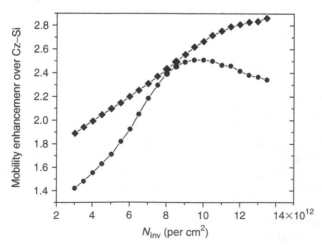

FIGURE 25.5 The mobility enhancement versus inversion charge for single-channel strained Si on relaxed $Si_{1-x}Ge_x$ alloys with $x = 0.60$ (circles) and $x = 0.70$ (squares).

forced to sample the surface strained Si, thus hybridizing the wave function as described above. The valence band splitting inherited from the strained Si now decreases hole scattering, and mobility is enhanced. At some larger field, the hole should once again occupy mostly the top strained Si layer and exhibit, once again, a decrease in enhancement with vertical field (Figure 25.5).

Figure 25.5 clearly shows that the $x = 0.60$ structure behaves accordingly. At first, the mobility enhancement rises as the vertical field and inversion charge are increased. At about $10^{13}\,\mathrm{cm}^{-2}$, the enhancement starts to level off and decrease slightly with further increases in field. Thus, we suspect that in the leveling-off phase, much of the hole is becoming "strained-Si-like," and the decrease in mobility enhancement with vertical field is once again observed. A stunning result is that for $x = 0.70$, the point of leveling-off has been pushed out to very high inversion charge, and therefore, the mobility enhancement nears three times at $1.4 \times 10^{13}\,\mathrm{cm}^{-2}$. It is interesting to note that this enhancement is larger than the best dual-channel structure results at these high inversion charge densities. The larger band offset between the strained Si and the $\mathrm{Si}_{0.30}\mathrm{Ge}_{0.70}$ dictates that a higher field will be required to pull the majority of the hole into the strained Si, thus creating a higher inflection point for the enhancement versus inversion charge curve.

Finally, we must speculate that wave function penetration into the oxide may play an important role in hole scattering in these structures. One can interpret much of the data also with an oxide-penetration perspective. For example, as more of the hole begins to reside in the strained Si layer, recall that the vertical mass is also quite light in the vertical direction. As the hole moves into the strained Si, it becomes larger and more "unwieldy" in the vertical direction, penetrating the gate oxide to a greater extent and decreasing mobility drastically. From this perspective, larger band offsets below the strained Si will keep the hole from penetrating as far into the oxide for a given field or inversion charge.

Although the exact reason for the enhancement versus inversion charge curves is unknown, and much future analysis and modeling will be necessary to ascertain the exact mechanisms of enhancement, it is clear that the empirical view regarding a single carrier and its interaction with the layer structure is sufficient to converge on a plethora of very high-mobility pMOS structures, and has shown the ability to engineer the enhancement versus inversion slope in strained Si pMOS.

In continuing with the empirical design guidelines discussed so far, a further test of our hypothesis would be to create an environment for the hole such that the band structure "seen" by the hole would be invariant with respect to vertical field. We have constructed, therefore, a MOS structure with a "digital alloy" channel, i.e., one in which the wavelength of the hole is greater than the periodicity of the digital alloy, or superlattice [33]. To apply the appropriate comparison to the other data, the structure consisted of the $x = 0.70$ substrate, and an approximately 10 nm superlattice was constructed with approximately 1 nm periodicity. The superlattice layers were composed of pure strained Si and $x = 0.70$ relaxed layers. Thus, the superlattice is nothing more than an ordered intermixing of the $x = 0.70$ single-channel structure discussed in the last section. However, by distributing the strained Si through the $x = 0.70$ in the 10 nm superlattice, we have created a thicker layer of material that will provide the hole the same environment, independent of vertical field.

The pMOSFETs fabricated from the digital alloy channel described above indeed have a hole enhancement, approximately two times. This is a surprising result, since if we consider the average potential of the digital alloy, it is essentially an ordered version of a tensile, $y = 0.35$ alloy on relaxed $x = 0.70$ (the digital alloy is composed of equal thickness $x = 0.70$ and 0, thus averaging to $x = 0.35$). Other experiments in our laboratory have shown that such a random alloy structure of $y = 0.35$ on relaxed $x = 0.70$ will result in no enhancement as compared to control Si MOS (i.e., tensile alloys on relaxed SiGe for p-channels are generally not useful). Therefore, we conclude that the ordering in the digital alloy is responsible for increasing the scattering time of holes by removing the random alloy scattering of SiGe, in one direction.

Figure 25.6 is a summary plot of the enhancement factor versus inversion charge for three prototypical structures. The strained Si single-channel structure on relaxed $\mathrm{Si}_{0.60}\mathrm{Ge}_{0.40}$ represents the conventional strained Si PMOS structure, in which the hole mobility enhancement is lost with vertical field. The advanced strained Si single channel on $\mathrm{Si}_{0.30}\mathrm{Ge}_{0.70}$ shows that the enhancement factor can be engineered

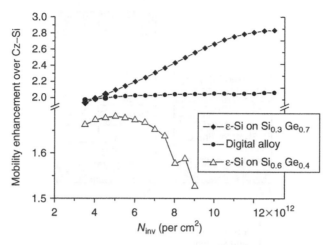

FIGURE 25.6 The mobility in enhancement in pMOSFETs versus inversion charge for the advanced strained Si single-channel structure, the digital alloy structure, and the conventional strained Si channel with a relatively high Ge concentration of $x = 0.4$.

to increase with increasing inversion charge. Finally, the pMOSFET hole mobility enhancement as a function of inversion charge is also shown. Note that the channel has indeed been engineered such that the mobility enhancement factor is independent of inversion charge, or vertical field. Figure 25.6 shows the versatility of the empirical method followed in this work for understanding the band structure effects on a single hole extended over multiple-channel layers.

SiGe Nanostructured Channels on Insulator

In the section "Transfer of Relaxed Lattice Constants via Wafer Bonding of Virtual Substrates," it was mentioned that relaxed buffer bonding opened doors to new engineered substrates, as unattainable bulk lattice constants can be produced on a virtual substrate on Si and then the full-diameter layer of that material can be transferred to a host Si wafer. The first demonstration of this was the creation of SGOI [34–37]. The process is shown in Figure 25.7. As described previously, the virtual buffer, in this case SiGe, is bonded to a layer with SiO_2 on Si, and the original substrate and graded layer is removed with a standard etch back or exfoliation technique. The result is the ability to transfer SiGe or any strained or unstrained SiGe heterostructure to another wafer. Figure 25.8 is a picture showing the transfer of a strained Si–SiGe layer and the resulting heterostructure on insulator. Any of the high-mobility SiGe heterostructures described previously in this chapter can be transferred to "OI," thus combining the benefits of high mobility with the benefits of SOI.

As the relaxed SiGe buffer can be tuned to any lattice constant, Ge on Si can be created and the surface of the virtual Ge wafer can be transferred, creating GOI [15]. There are complications as the planarization of thin virtual Ge layers is more difficult than it is for lower Ge concentrations, but nonetheless it is possible as shown in Figure 25.9. Ge-on-insulator can be used for Ge-based electronics or, as is the case for this thicker GOI, for optical devices.

Future Potential of Nanoengineered SiGe MOSFETs

Figure 25.10 is a summary of all nMOS and pMOS data accumulated to date on the large variety of heterostructures fabricated into MOSFETs in our laboratory. Note that although this paper has concentrated on progress in the hole mobility issue (and therefore, on pMOSFETs), Figure 25.10 includes electron mobility enhancements extracted from nMOSFETs *fabricated from the same material structure as the pMOSFETs*. Thus, at any relaxed buffer Ge concentration, the total enhancements in both nMOS

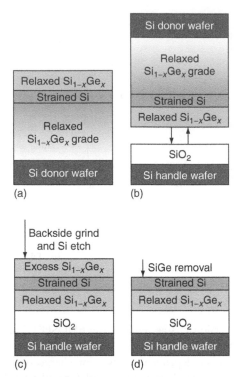

FIGURE 25.7   Schematic of one method of using relaxed buffer bonding to create SGOI or SSOI.

FIGURE 25.8   Cross-sectional TEM of a strained Si–Si$_{0.75}$Ge$_{0.25}$ on OI.

FIGURE 25.9 TEM cross section of germanium on insulator, produced from the bonding of a virtual Ge wafer to a SiO$_2$–Si wafer. There is a thin bonding Si layer between the Ge and insulator, and a thin etch stop layer remaining on the top of the Ge layer.

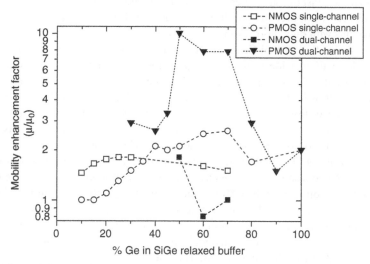

FIGURE 25.10 The mobility in enhancements in strained SiGe nMOSFETs and pMOSFETs versus relaxed buffer Ge concentration. Data for both single-channel and dual-channel structures are shown.

and pMOS for the structure can be estimated by looking at the acquired points vertically for the same structure. All of the data were extracted at a vertical field equivalent in the Si controls of about 0.6 MV/cm.

The plot reveals some interesting conclusions for the SiGe material system. First, note that the strained Si commercialized today ($x \sim 0.20$) obtains the highest electron enhancement on the chart, and therefore further improvements via new strained Si substrates require PMOS mobility enhancements, which in turn require relaxed buffers with higher Ge concentrations. Also note that the single channel advanced structures with $x = 0.50$ to 0.70 also support high electron enhancements, nearly the full enhancement observed at lower Ge concentrations, despite the dislocations introduced from the large lattice mismatch between the strained Si and relaxed SiGe buffer.

Another observation in Figure 25.10 is that the highest PMOS enhancements occur on relaxed buffers with intermediate Ge concentrations ($x = 0.40$ to 0.70), and not at the end-points (i.e., Si and Ge lattice constants). We believe this indicates the importance of strain in enhancing hole mobility in inversion layers in this materials system. Other issues with lattice constants near the Ge-end of the chart are that nMOS performance is typically very poor when $x > 0.70$. This degradation of nMOS channels may be related to the use of the strained Si–SiO$_2$ gate stack used in all MOSFETs in Figure 25.10. Since electrons, like the holes discussed in this chapter, would invariably be mixed over both Si and Ge layers, the poor nMOS performance may be related to the very different band structures of Si and Ge. Having an electron averaging over Si and Ge conduction bands would seemingly create much scattering, as Si has the minimum energy conduction valleys in the $\langle 1\,0\,0 \rangle$ directions, whereas Ge has conduction band minima in the $\langle 1\,1\,1 \rangle$ direction.

Finally, note that the highest enhancement observed in our laboratory to date is a dual channel, pure strained Si–pure strained Ge–relaxed Si$_{0.5}$Ge$_{0.5}$ structure, demonstrating an electron mobility enhancement of 1.7× for the nMOS, and about a 10× hole mobility enhancement for the PMOS. These data suggest that mobility can be enhanced by approximately 1000% over conventional Si MOS mobility, assuming that short-channel device optimization and low-temperature processing are possible.

25.5   Engineered Substrates for III–V–Si Integration

Although we have concentrated on describing the impact of relaxed SiGe on channel mobility in MOSFETs, SiGe lattice constants also offer a pathway to integrating III–V materials with Si technology. An engineered substrate composed of both III–V materials and Si could be used to host high-performance optoelectronic circuits with digital processing capability.

Early work has shown that virtual Ge on Si can be high enough quality for minority carrier devices, and Ge photodiodes with near-ideal reverse leakage currents were obtained [38]. Since GaAs is nearly the same lattice constant as Ge (see Figure 25.1), high-quality Ge virtual substrates can be converted to a GaAs–Si substrate by growth of a lattice-matched layer on Ge (Figure 25.11). This heterovalent interface can be deleterious if deposition is not initiated properly, as exposure to As at certain temperatures can create antiphase boundaries [39–41]. GaAs grown on Ge–SiGe–Si has high minority carrier lifetime [14], allowing the fabrication of high-efficiency GaAs solar cells on Si [42].

The GaAs–SiGe–Si material is high enough quality that it supports room temperature, continuous wave lasing in GaAs-based lasers on Si. GaAs lasers on Si are considered a test vehicle for the eventual integration of optoelectronics with Si CMOS, since room temperature continuous lasing cannot be achieved with poor quality materials. We have achieved room temperature, continuous wave lasing of AlGaAs/GaAs and AlGaAs/InGaAs quantum well lasers on Si substrates and fabricated primitive optical links [43–45] (Figure 25.12). Many materials and processing challenges were overcome to achieve lasing as described in detail in the publications. The laser structures were very primitive, gain-guided broad stripe lasers. Such high threshold lasers were constructed for ease and as a demonstration of the material quality. For example, conventional GaAs deposited directly on Si would not have sufficient quality for lasing action.

Figure 25.13 shows also the improved lifetime of the laser. The first lasers created on Ge–SiGe–Si that lased already lasted more than 20 min (as shown in the figure). This laser lifetime was remarkable for a room temperature, continuous wave laser on Si. And the technology is robust, i.e., unlike earlier reports on GaAs lasers on Si, such lasers could be now created at will. With this advance, we quickly improved the lifetime, as shown in the figure.

The improved laser device achieved a lifetime of 4 h. Again, this result is very reproducible, and further improvement is expected as improved laser designs are implemented. Mesa ridge lasers with low threshold currents should improve laser lifetime drastically.

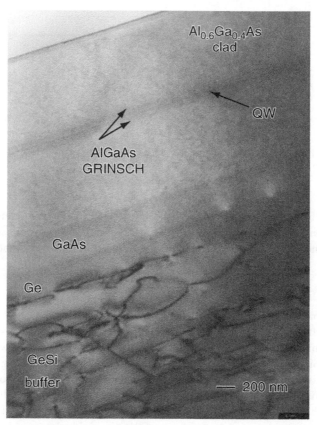

FIGURE 25.11 Cross-sectional TEM image of the first structure to lase, and Figure 25.12 shows the improvement of our laser process once lasing was achieved. Shown are the light-voltage curves for the laser diodes on Ge–SiGe–Si and the control GaAs substrate. The y-axis is current in the photodetector, which is detecting the laser light, and that current is linearly related to light from the laser. The L–V curves from the improved GaAs lasers on Si are indistinguishable from the control lasers on GaAs substrates, and the thresholds and quantum efficiency were also nearly identical.

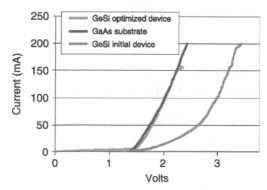

FIGURE 25.12 Light intensity (represented on graph by photodetector current) as a function of voltage applied to laser diodes on GaAs and Ge–SiGe–Si substrates ("SiGe substrates").

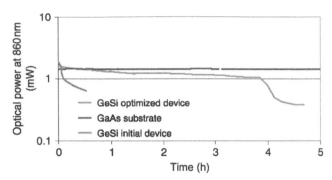

FIGURE 25.13 Laser lifetime at constant optical power out. Initial devices lasted as long as 20 min, whereas new improved devices lasted 4 h.

The window for true monolithic optoelectronic integration on Si has opened. With a properly designed III–V–Si engineered substrate, it will be possible to process the substrate in a Si CMOS fabrication facility and produce Si CMOS digital ICs with optical input and output.

25.6 Future Engineered Substrates

With the great flexibility to combine thin highly strained materials and materials composed of previously unattainable lattice constants, we can imagine new materials that have not existed before that may have application in microelectronics. As an example, we have recently created a new engineered substrate platform, strained Si on Si (SSOS) [46]. The community has been focusing on combining SOI technology with strained Si, and that motivated the work to demonstrate such combinations early on. However, the potential goes far beyond such linear combinations of existing advanced materials. We have demonstrated that a strained Si layer can be transferred to another silicon wafer, creating a strained Si bulk wafer with no SiGe present. A cross section of the material is shown in Figure 25.14. Note that the interface is still bonded with a strength near that of a semiconductor bond, and therefore, the strained Si can be held in place without the need for the original relaxed SiGe and host substrate. There should be an edge-dislocation array at the strained Si–Si interface, since there is a difference in in-plane lattice spacing between strained Si and Si. Figure 25.15 is a plan-view TEM image of the strained Si–Si interface, and $g*b$ analysis of the interface shows that the dislocations are edge-type. Also, the spacing of the dislocations is correct for the difference in lattice constant for strained Si–Si. To our knowledge, this is the first material created at this scale that is compositionally the same yet abruptly variant in strain state; i.e., it is a heterojunction that is defined by strain difference only and not by composition difference; it is the first homochemical heterojunction. Although a basic study at this point, it shows that relaxed buffer bonding has great potential for creating new, previously unattainable materials and heterostructures.

25.7 Market Adoption of Engineered Substrates

For the past 30 years, the material of choice for the semiconductor industry has been Si, and Moore's Law has been pursued by exchanging the manufacturing infrastructure in the supply chain with newer versions of equipment. Each new factory possessed equipment and processes to create higher density circuits on Si, and enough market growth occurred to justify reinvestment into the next factory.

Today we are at crossroads. Increased transistor density no longer brings sufficient market growth to reinvest in larger factories. New engineered substrate materials, like strained Si, can be interpreted as aiding the extension of Moore's Law. Alternatively, strained Si can be considered the beginning of a new

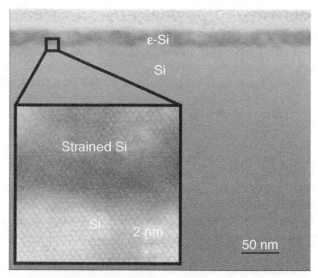

FIGURE 25.14 Cross-sectional TEM image of the SSOS structure. Shown at inset is a high-resolution image of the interface.

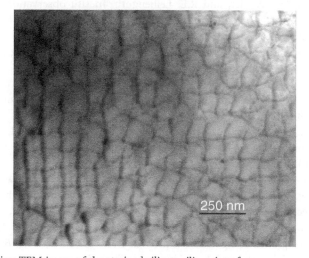

FIGURE 25.15 Plan-view TEM image of the strained silicon–silicon interface.

roadmap, in which semiconductor value is delivered through novel circuits on new engineered substrates.

In the first case, engineered substrates must be absorbed into the existing large Si CMOS supply chain. This naturally takes time, as the scale of the industry and the lack of knowledge about other parts of the supply chain result in scaling only occurring when critical mass in education of the technology is reached. In the later case, high-performance engineered substrates can be implemented earlier in an integrated supply chain environment for smaller markets. In this case, the market must grow and result in the increasing size of the integrated manufacturer and the supply chain. In either case, bottom–up innovation in semiconductors will take some time, but will allow the semiconductor industry to continue to grow into previously unattainable markets.

Acknowledgments

The author would like to thank Matt Currie, Mike Groenert, Saurabh Gupta, David Isaacson, Andy Kim, Larry Lee, Chris Leitz, Arthur Pitera, Gianni Taraschi, and Vicky Yang. The author is very grateful to the sponsors of the research reviewed in this chapter: The Army Research Office, DARPA Heterogeneous Integration, Singapore-MIT Alliance, MARCO MSD, MARCO IFC, and NSF CMSE central facilities at MIT.

References

1. P.L. Gourley, I.J. Fritz, and L.R. Davidson. Controversy of critical layer thickness for InGaAs/GaAs strained-layer epitaxy. *Appl. Phys. Lett.* 52(5): 377–379, 1988.
2. A.T. Fiory, J.C. Bean, R. Hull, and S. Nakahara. Thermal relaxation of metastable strained-layer Ge_xS_{1-x}/Si epitaxy. *Phys. Rev. B* 31(6): 4063, 1985.
3. J.M. Woodall, P.D. Kirchner, D.L. Rogers, M. Chisolm, and J.J. Rosenberg. Proceedings of IEEE/Cornell Conference on Advanced Concepts in High Speed Semiconductor Devices and Circuits, 1987, p. 3.
4. D.L. Harame, J.M.C. Stork, G.L. Patton, S.S. Iyer, B.S. Meyerson, G.J. Scilla, E.F. Crabbe, and E. Ganin. IEDM Tech. Digest, 1988, p. 889.
5. E.A. Fitzgerald. Dislocations in strained-layer epitaxy: theory, experiment, and applications. *Mater. Sci. Rep.* 7: 87, 1991.
6. R. Hull, J.C. Bean, D.J. Werder, and R.E. Leibenguth. In situ observations of misfit dislocation propagation in Ge_xSi_{1-x}/Si (1 0 0). *Appl. Phys. Lett.* 52(19): 1605–1607, 1988; C.G. Tuppen and C.J. Gibbings. *Thin Solid Films* 183: 133, 1989.
7. E.A. Fitzgerald. The effect of substrate growth area on misfit and threading dislocation densities in mismatched heterostructures. *J. Vac. Sci. Technol. B* 7(4): 782–788, 1989.
8. E.A. Fitzgerald , Y.-H. Xie, M.L. Green, D. Brasen, A.R. Kortan, J. Michel, Y.-J. Mii, and B.E. Weir. Totally relaxed Ge_xSi_{1-x} layers with low threading dislocation densities grown on Si substrates. *Appl. Phys. Lett.* 59(7): 811–813, 1991.
9. Y.-J. Mii, Y.-H. Xie, E. A. Fitzgerald , D. Monroe, F.A. Thiel, B.E. Weir, and L.C. Feldman. Extremely high electron-mobility in Si/Ge_xSi_{1-x} structures grown by molecular-beam epitaxy. *Appl. Phys Lett.* 59(13): 1611–1613, 1991.
10. E.A. Fitzgerald , Y.-H. Xie, D. Monroe, P.J. Silverman, J.-M. Kuo, A.R. Kortan, F.A. Thiel, B.E. Weir, and L.C. Feldman. Relaxed Ge_xSi_{1-x} structures for III–V integration with Si and high mobility 2-dimensional electron gases in Si. *J. Vac. Sci. Technol. B* 10(4): 1807–1819, 1992.
11. E.A. Fitzgerald, A.Y. Kim, M.T. Currie, T.A. Langdo, G. Taraschi, and M.T. Bulsara. Dislocation dynamics in relaxed graded composition semiconductors. *Mater. Sci. Eng. B* 67: 53, 1999.
12. A.Y. Kim, W.S. McCullough, and E.A. Fitzgerald. Evolution of microstructure and dislocation dynamics in In_xGa_{1-x}P graded buffers Grown on GaP by MOVPE: engineering device-quality substrate materials. *J. Vac. Sci. Technol. B* 17: 1485, 1999.
13. C.W. Leitz, M.T. Currie, A.Y. Kim, J. Lai, E. Robbins, and E.A. Fitzgerald. Dislocation glide and blocking kinetics in compositionally graded SiGe/Si. *J. Appl. Phys.* 90: 2730–2736, 2001.
14. R.M. Sieg, J.A. Carlin, S.A. Ringel, M.T. Currie, S.M. Ting, T.A. Langdo, G. Taraschi, E.A. Fitzgerald, and B.M. Keyes. High minority-carrier lifetimes in GaAs grown on low-defect-density Ge/GeSi/Si substrates. *Appl. Phys. Lett.* 73(21): 3111, 1998.
15. A.J. Pitera, G. Taraschi, M.L. Lee, C.W. Leitz, Z.-Y. Cheng, and E.A. Fitzgerald. Coplanar integration of lattice-mismatched semiconductors with silicon by wafer bonding Ge/$Si_{1-x}Ge_x$/Si virtual substrates. *J. Electrochem. Soc.* 151: G443, 2004.
16. U.S. Patents: #6,602,613; #6,703,144; #6,680,495; #6,677,655.
17. B.S. Meyerson. UHV/CVD growth of Si and Si:Ge alloys: chemistry, physics, and device applications. *Proc. IEEE* 80: 1592, 1992.

18. Y.-H. Xie, G.H. Gilmer, C. Roland, P.J. Silverman, S.K. Buratto, J.Y. Cheng, E.A. Fitzgerald, A.R. Kortan, S. Schuppler, M.A. Marcus, and P.H. Citrin. Semiconductor surface roughness: dependence on sign and magnitude of bulk strain. *Phys. Rev. Lett.* 72: 3006, 1994.

19. E.A. Fitzgerald, Y.H. Xie, M.L. Green, D. Brasen, A.R. Kortan, J. Michel, Y.J. Mii, and B.E. Weir. Totally relaxed Ge_xSi_{1-x} layers with low threading dislocation densities grown on Si substrates. *Appl. Phys. Lett.* 59(7): 811–813, 1991.

20. Y.J. Mii, Y.H. Xie, E.A. Fitzgerald, D. Monroe, F.A. Thiel, B.E. Weir, and L.C. Feldman. Extremely high electron-mobility in Si/Ge_xSi_{1-x} structures grown by molecular-beam epitaxy. *Appl. Phys. Lett.* 59(13): 1611–1613, 1991.

21. F. Schaffler, D. Tobben, H.J. Herzog, G. Abstreiter, and B. Hollander. High-electron-mobility Si/SiGe heterostructures—influence of the relaxed SiGe buffer layer. *Semicond. Sci. Technol.* 7(2): 260–266, 1992.

22. J. Welser, J.L. Hoyt, and J.F. Gibbons. IIA-4 Temperature and scaling behavior of strained-Si N-Mosfets. *IEEE Trans. Electron Dev.* 40(11): 2101, 1993.

23. K. Rim, J.L. Hoyt, and J.F. Gibbons. *IEEE Trans. Electron Dev.* 47: 1406, 2000.

24. Q. Xiang, J.-S. Goo, H. Wang, Y. Takamura, B. Yu, J. Pan, A. Nayfeh, A. Holbrook, F. Arasnia, E. Paton, P. Besser, M. Sidorov, E. Adem, A. Lochtefeld, G. Braithwaite, M.Currie, R. Hammond, M. Bulsara, and M.-R. Lin. Proceedings of the First International SiGe Technology and Device Meeting, 2003, p. 13.

25. A. Lochtefeld and D.A. Antoniadis. On experimental determination of carrier velocity in deeply scaled NMOS: how close to the thermal limit? *IEEE Electron Dev. Lett.* 22(2): 95–97, 2001.

26. M.T. Currie, C.W. Leitz, T.A. Langdo, G. Taraschi, E.A. Fitzgerald, and D.A. Antoniadis. Carrier mobilities and process stability of strained Si n- and p-MOSFETs on SiGe virtual substrates. *J. Vac. Sci. Technol. B* 19: 2268, 2001.

27. C.W. Leitz, M.T. Currie, M.L. Lee, Z.Y. Cheng, D.A. Antoniadis, and E.A. Fitzgerald. Hole mobility enhancements and alloy scattering-limited mobility in tensile strained Si/SiGe surface channel metal-oxide-semiconductor field-effect transistors. *J. Appl. Phys.* 92(7): 3745–3751, 2002.

28. C.W. Leitz, M.T. Currie, M.L. Lee, Z.-Y. Cheng, D.A. Antoniadis, and E.A. Fitzgerald. Hole mobility enhancements in strained $Si/Si_{1-y}Ge_y$ p-type metal-oxide-semiconductor field-effect transistors grown on relaxed $Si_{1-x}Ge_x$ $(x < y)$ virtual substrates. *Appl. Phys. Lett.* 79(25): 4246–4248, 2001.

29. M.L. Lee, C.W. Leitz, Z.-Y. Cheng, A. Pitera, T.A. Langdo, M.T. Currie, G. Taraschi, E.A. Fitzgerald, and D.A. Antoniadis. Strained Ge channel p-type metal-oxide-semiconductor field-effect transistors grown on $Si_{1-x}Ge_x/Si$ virtual substrates. *Appl. Phys. Lett.* 79(20): 3344–3346, 2001.

30. M.A. Armstrong. PhD Thesis. Technology for SiGe heterostructure-based CMOS devices MIT, 1999, p. 171.

31. D.K. Nayak and S.K. Chun. Low-field hole mobility of strained Si on (1 0 0) $Si_{1-x}Ge_x$ substrate. *Appl. Phys. Lett.* 64(19): 2514–2516, 1994.

32. M.V. Fischetti and S.E. Laux. Band structure, deformation potentials, and carrier mobility in strained Si, Ge, and SiGe alloys. *J. Appl. Phys.* 80(4): 2234–2252, 1996.

33. M.L. Lee and E.A. Fitzgerald. Hole mobility enhancements in nanometer-scale strained silicon heterostructures grown on Ge-rich relaxed $Si_{1-x}Ge_x$. *J. Appl. Phys.* 94(4): 2590–2596, 2003.

34. G. Taraschi, T.A. Langdo, M.T. Currie, E.A. Fitzgerald, and D.A. Antoniadis. Relaxed SiGe-on-insulator fabricated via wafer bonding and etch back. *J. Vac. Sci. Technol. B* 20(2): 725–727, 2002.

35. L.J. Huang, J.O. Chu, D.F. Canaperi, C.P. D'Emic, R.M. Anderson, S.J. Koester, and H.-S.-P. Wong. SiGe-on-insulator prepared by wafer bonding and layer transfer for high-performance field-effect transistors. *Appl. Phys. Lett.* 78(9): 1267–1269, 2001.

36. Z. Cheng, G. Taraschi, M.T. Currie, C.W. Leitz, M.L. Lee, A. Pitera, T.A. Langdo, J.L. Hoyt, D.A. Antoniadis, and E.A. Fitzgerald. Relaxed silicon–germanium on insulator substrate by layer transfer. *J. Electron. Mater.* 30(12): L37, 2001.

37. T. Tezuka, N. Sugiyama, and S. Takagi. Fabrication of strained Si on an ultrathin SiGe-on-insulator virtual substrate with a high-Ge fraction. *Appl. Phys. Lett.* 79(12): 1798–1800, 2001.

38. S. Samavedam, M. Currie, T. Langdo, and E.A. Fitzgerald. High-quality germanium photodiodes integrated on silicon substrates using optimized relaxed graded buffers. *Appl. Phys. Lett.* 73: 2125, 1998.

39. E.A. Fitzgerald , J.-M. Kuo, Y.-H. Xie, and P.J. Silvennan. Necessity of Ga pre-layers in GaAs/Ge growth using gas-source molecular beam epitaxy. *Appl. Phys. Lett.* 64: 733, 1994.

40. S. Ting, R. Sieg, S. Ringel, and E. A. Fitzgerald. Range of defect morphologies on GaAs grown on Offcut(0 0 1) Ge substrates. *J. Electron. Mater.* 27: 451, 1998.

41. R. Sieg, S. Ringel, S. Ting, S. Samavedam, M. Currie, T. Langdo, and E.A. Fitzgerald. Toward device-quality GaAs growth by molecular beam epitaxy on offcut Ge/Si$_{1-x}$Ge$_x$/Si substrates. *J. Vac. Sci. Technol. B* 16(3):1471, 1998.

42. S. Ringel, J.A. Carlin, C.L. Andre, M.K. Hudait, M. Gonzalez, D.M. Wilt, E.B. Clark, P. Jenkins, D. Scheiman, A. Allerman, E.A. Fitzgerald, and C.W. Leitz. Single junction InGaP/GaAs solar cells grown on Si substrates with SiGe buffer layers. *Prog Photovolt.: Res. Appl.* 10: 417, 2002.

43. M.E. Groenert, C.W. Leitz, A.J. Pitera, V.K. Yang, H. Lee, R.J. Ram, and E.A. Fitzgerald. Monolithic integration of room-temperature cw GaAs/AlGaAs lasers on Si substrates via relaxed graded GeSi buffer layers. *J. Appl. Phys.* 93(1): 362–367, 2003.

44. M. Groenert, A. Pitera, R.J. Ram, and E.A. Fitzgerald. Improved room-temperature continuous wave GaAs/AlGaAs and InGaAs/GaAs/AlGaAs lasers fabricated on Si substrates via relaxed graded Ge$_x$Si$_{1-x}$ buffer layers. *J. Vac. Sci. Technol. B* 21(3): 1064–1069, 2003.

45. V.K. Yang, M.E. Groenert, G. Taraschi, C.W. Leitz, A.J. Pitera, M.T. Currie, Z. Cheng, and E.A. Fitzgerald. Monolithic integration of III–V optical interconnects on Si using SiGe virtual substrates. *J. Mater. Sci.: Mater. Electron.* 13(7): 377–380, 2002.

46. D.M. Isaacson and E.A. Fitzgerald, patents pending.

26

Self-Assembling Nanostructures in Ge(Si)–Si Heteroepitaxy

Robert Hull
University of Virginia

26.1 Scope of This Chapter

In this chapter, strain-driven morphological instabilities and transitions in the Ge(Si)–Si($1\,0\,0$) system are described, and current understanding of the fundamental mechanisms governing these phenomena summarized. The degree to which these processes may be controlled and organized to produce arrays of semiconductor "quantum dots" (QDs) with potential nanoelectronic or nanophotonic applications is also discussed. Due to length limitations these discussions are necessarily limited to key concepts and phenomena; for more detailed discussions of fundamental phenomena, several excellent, longer reviews exist [For example, 1–3]. *A note on nomenclature:* in the subsequent discussion, the format Ge_xSi_{1-x} means that the structures under discussion are comprised explicitly of an alloy of Ge and Si, the format Ge(Si) means they are comprised of either pure Ge or Ge_xSi_{1-x} alloy material.

26.2 Introduction

As described elsewhere in this volume, the lattice parameter mismatch between Ge and Si—about 4.1% at room temperature—produces enormous stored strain energies in epitaxial Ge–Si heterostructures. For the case of Ge_xSi_{1-x}–Si heterostructures, the lattice parameter difference scales approximately as the Ge fraction, x. Linear, isotropic elasticity theory shows that the biaxial strain, ε; the stress, σ; and the stored elastic strain energy per unit interface area, E; in a coherent (i.e., dislocation-free), epitaxial film of Ge_xSi_{1-x} on a rigid Si substrate is given by

$$\varepsilon = 0.041x \tag{26.1}$$

$$\sigma = 2G\varepsilon(1+v)/(1-v) \tag{26.2}$$

$$E = 2Gh\varepsilon^2(1+v)/(1-v) \tag{26.3}$$

Here G, v, and h are the shear modulus, Poisson's ratio, and thickness of the epitaxial film, respectively. Note that the quoted elastic strain value is at room temperature—differential thermal expansion coefficients between Ge(Si) and Si can amount to a few percent of the lattice mismatch strain at the growth temperature [4], and are additive to the elastic strain for Ge_xSi_{1-x}–Si (i.e., the thermal expansion coefficient for Ge, $5.9 \times 10^{-6}\,K^{-1}$ is greater than that for Si, $2.2 \times 10^{-6}\,K^{-1}$). The elastic strain is compressive, i.e., the lattice parameter of Ge(Si) is greater than that of Si. Using tabulated values for G and v, the biaxial stress in the coherent Ge_xSi_{1-x} film is of order $10x$ GPa, and the volumetric stored elastic strain energy is several hundred megajoules per m^3 for pure Ge/Si. These enormous strain energies and stresses will seek routes to relax, and the main relaxation mechanisms are summarized in Figure 26.1. In the absence of any relaxation mechanisms (Figure 26.1a) the epitaxial layer grows coherently and in planar fashion. The lattice strain causes a tetragonal distortion of the unit cell, and the epitaxial layer strain, stress and elastic stored energy are given in Equation 26.1 to Equation 26.3. A general mechanism for epitaxial strain relaxation is roughening or islanding of the surface, and is the main mechanism of interest in this chapter. As shown in Figure 26.1b, this allows the interatomic bonds at the surface to relax toward their equilibrium lengths. Other strain-relaxation mechanisms are interfacial misfit dislocation injection, as described elsewhere in this volume, and interfacial interdiffusion (Figure 26.1d and c) respectively. Of course, all three relaxation mechanisms can and do operate to differing degrees in parallel. The primary goal of this chapter is to describe in some detail the mechanism (Figure 26.1b), and how it can be used to controllably generate heteroepitaxial clusters of Ge(Si) on Si surfaces that can be viewed as *quantum dots*, i.e., as individual electronic or optoelectronic device elements that can store, transfer, absorb or emit electrons, holes, or photons.

26.3 Roughening and Islanding as a Strain-Relief Mechanism

The introduction of surface topography into a compressively strained film allows relaxation of bond length and angles in the surface region, and thereby allows the relaxation of strain energy in the system.

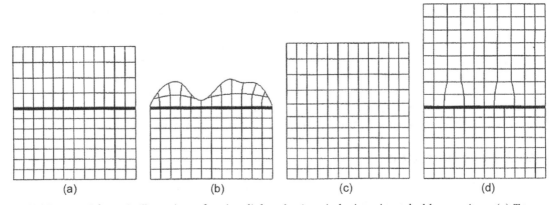

(a)   (b)   (c)   (d)

FIGURE 26.1 Schematic illustrations of strain-relief mechanisms in lattice-mismatched heteroepitaxy. (a) Tetragonal coherent straining of the epitaxial layer. (b) Strain relaxation through surface topography. (c) Strain relaxation through interdiffusion. (d) Strain relaxation through misfit dislocation injection.

Of the mechanisms described in Figure 26.1, this particular mechanism tends to be the most dominant at lower epitaxial film thicknesses, where the perturbed surface region is a greater fraction of the total epitaxial film thickness. Surface topography does also persist to greater epitaxial film thicknesses, and generally couples with misfit dislocation injection in this limit. Islanding in the Ge(Si)–Si system is also frequently coupled with interdiffusion with the Si substrate, as will be discussed later in this chapter. As first demonstrated by Eaglesham and Cerullo [5], deformation of the local substrate region also occurs.

The formation of coherent islands in strained layer epitaxy—at least in systems with relatively high lattice mismatch strain (greater than a few percent or so) generally occurs in the Stranski–Krastanov mode, as is the case for Ge(Si)–Si [5]. In this growth mode, the heteroepitaxial deposit initially wets the substrate as a planar thin film. In epitaxy of pure Ge on Si(100), this "wetting layer" is generally a few atomic monolayers thick. Subsequent growth of the heteroepitaxial film then occurs by growth of strained islands on the wetting layer. It is these islands that can be viewed as potential "quantum dots" for nanoelectronic or nanophotonic applications.

In epitaxy of pure Ge–Si(100), the subsequent morphological evolution follows a well-documented series of transitions of island geometries that minimize the combined contributions of surface and strain energies, as will be described in the next section of this chapter. In principle, the same series of transitions occur for Ge_xSi_{1-x}–Si(100) epitaxy, but it is observed that the associated length scales increase substantially with decreasing strain [6–8]. Thus, a key issue becomes whether there is sufficient adatom mobility at the operative growth rates and growth temperatures to realize surface transport over the necessary length scales to attain the equilibrium structures. In the limit of the equilibrium structures attained, the observed surface morphological states relax of order 30 to 50% of the elastic strain [6]. In regimes of more limited adatom mobility, more complex morphological states are observed, as will be discussed in a later section.

For lower misfit (ca. $\leq 2\%$ lattice mismatch) strain, wetting layer thicknesses are sufficiently large that morphological strain relaxation can be viewed as a roughening, rather than as an islanding transition. In either event, models of strain relief are generally formulated in continuum or atomistic frameworks. Continuum approaches generally balance the relief of elastic strain energy with the increase in surface energy caused by surface roughening [9–14]. Atomistic formulations generally minimize the total system energy, including contributions from step and step interaction energies [15–17].

Strain-driven surface roughening may be understood in terms of diffusive mass transport from regions of high to low strain energy density. Consider a coherent lattice-mismatched film with an undulated surface. The lattice mismatch strain will produce a nonuniform stress distribution, with relaxation (and hence lower strain energy density) at the peaks of the perturbation and stress concentrations (higher strain energy density) in the troughs (Figure 26.2). The resulting lateral variation in strain energy density drives atomic transport from troughs to peaks. Opposing this transport is the resulting increase in surface energy. Continuum models show that above a critical wavelength of the surface morphology (i.e., such that the amount of additional surface area created is less than for a topology of the same amplitude but lower wavelength), the relief of strain energy is the dominant process in the system [11,12]. Thus, both the amplitude of the perturbation and the magnitude of the variation of stress distribution between peak and trough are predicted to grow. In the case of Ge(Si)–Si(100) epitaxy, this, in turn, leads to preferential bonding of Ge atoms at the peaks, where the local lattice parameter is larger [18]. These continuum approaches predict that any planar strained film is unstable with respect to roughening as a strain-relief mechanism.

Atomistic frameworks inherently incorporate energy barriers to roughening or islanding transitions, because of the energy associated with the atomic steps necessary to create surface morphology. They, thus, predict that planar-strained films are metastable with respect to roughening. Of course, atomic steps are always present on a surface, particularly in the case of standard Si(100) substrates, which are typically only specified to azimuthal directions within a few tenths of a degree of (100)—thus necessarily containing relatively high-step densities (interspersed by regions of (100) terrace) to accommodate the substrate misorientation. In fact, interaction energies between these steps produce a likely mechanism for surface roughening or islanding. In the case of compressively strained Si (and, by

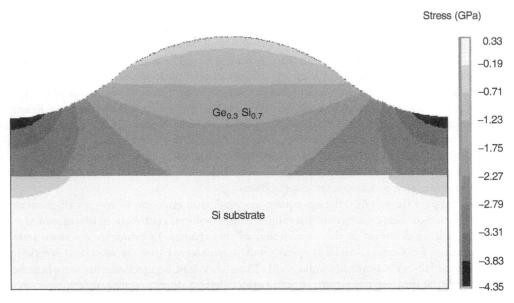

FIGURE 26.2 Finite element calculation using anisotropic elastic constants of the stress distribution (σ_{xx}, perpendicular to the sinusoidal modulations) in a $Ge_{0.3}Si_{0.7}$ film with a mean thickness of 50 nm and a one-dimensional sinusoidal surface modulation with an amplitude of 40 nm and a period of 250 nm. Stress contours are in GPa, where negative values correspond to compressive stress in the film. Calculations are performed with ANSYS 7.1. (Courtesy of C.C. Wu.)

extension, Ge and Ge_xSi_{1-x}), step interaction energies encourages bunching of surface steps [19,20]. This provides a potential mechanism for the initial formation of islands and roughened surfaces in strained epitaxial films [7,8]. Such atomistic models provide more realistic treatments of the initial formation of rough surfaces, and help explain the specific morphological transitions that occur in the Ge(Si)–Si(1 0 0) system. Continuum approaches, however, have provided key insight into the balance between strain energy and surface energy in roughening transitions. In summary, both sets of models have provided a strong foundation for both qualitative and quantitative understanding of morphological transitions in Ge(Si)–Si(1 0 0), and many other strained layer systems.

26.4   Equilibrium Strain-Driven Morphological Transitions in the Ge(Si)–Si System

With the background of the previous section, observed morphological transitions in the Ge(Si)–Si(1 0 0) system may be qualitatively (and, to a degree, quantitatively) understood. We first concentrate on the series of equilibrium transitions in Ge–Si(1 0 0), and then extrapolate to Ge_xSi_{1-x}–Si(1 0 0).

As described previously, the fundamental growth mode in Ge(Si)–Si is the Stranski–Krastanov mode, whereby a thin wetting layer is first formed, followed by island formation. The thickness of the wetting layer for Ge–Si(1 0 0) is of order a few monolayers, depending somewhat upon the growth conditions and the exact surface chemistry. Atomistic details of the evolution of this Ge wetting layer on Si(1 0 0) are described in Ref. [2]. The initial islands that form on the wetting layer surface are of relatively low aspect (height–diameter) ratio, and evolve relatively rapidly into a geometry that is bounded primarily by {5 1 0} facets, the so-called "hut cluster" geometry, as described by Mo et al. [21]. (Note that intermediate stages between the planar wetting layer and the hut cluster geometry also exist, e.g., Ref. [22].) The hut clusters are coherent to the substrate (i.e., free of misfit dislocations). In this first stage of the morphological evolution, minimization of surface energy (that scales with surface area) is more significant than minimization of strain energy (an approximately volumetric term). In strained Ge(Si), the {5 1 0} surface has a particularly low energy [23], discouraging growth to higher aspect ratio facets which, while they would be more efficient in reducing strain energy, have higher surface energy. As the individual clusters grow, volumetric energy terms (i.e., strain energy) become increasingly important with respect to areal energy terms (i.e., surface energy), and the clusters undergo a transition to a higher aspect ratio geometry, termed "dome clusters" bounded predominantly by {3 1 1} facet [24]. This facet represents the next major cusp in the energy-orientation diagram, and thus produces a dome cluster configuration whose decrease in surface energy with respect to a hut cluster of the same volume more than compensates for the increase in strain energy. For pure Ge–Si(1 0 0) the initial dome clusters are still coherent to the substrate. With increasing growth of the dome clusters, they eventually dislocate. Subsequent growth beyond this coherent–incoherent transition has been shown to be cyclical, according to the need to digitally introduce additional dislocations, each of which has an energy barrier associated with its introduction [25]. These morphological transitions are summarized in Figure 26.3.

The length scales associated with these transitions have been studied by multiple authors. In particular, the transition between the cluster and dome states has received much attention, and generally occurs at lateral island dimensions of order a few tens of nanometers [26,27]. The transition between coherent and dislocated domes generally occurs at dimensions of order 60 to 70 nm [28,29]. It should be stressed that these dimensions pertain to growth on clean Si(1 0 0) surfaces, and for Ge growth temperatures and deposition rates where the time at temperature is such that significant diffusion does not occur between Ge clusters or wetting layer and the Si substrate. Relatively rapid stress-enhanced diffusion occurs between these entities, as been documented by multiple authors [30–33]. With increasing Si incorporation into the clusters, the cluster strain is reduced, and the corresponding length scales of transitions (i.e., from hut to dome, and from coherent dome to incoherent dome) increase. The chemical state of the substrate is also key in determining the relevant length scales. The effects of surfactant species, such as P, Ga, and B, have been studied by several authors [34–37], and have shown that cluster facets, aspect ratios, and transition dimensions are affected by the presence of the surfactant

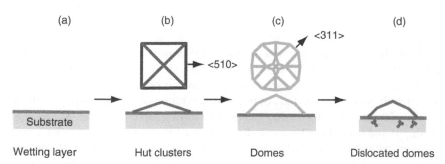

FIGURE 26.3 Equilibrium series of morphological transitions for Ge(Si)–Si(1 0 0). (a) Ge wetting layer, (b) {5 1 0} facetted hut cluster, (c) coherent dome cluster with dominant {3 1 1} facet, and (d) dislocated dome cluster.

species. For example, the transition from coherent to dislocated clusters was found to be reduced to 20–30 nm as a function of the presence of Ga from near-surface focused ion beam (FIB) implantation [35], corresponding to a significant increase in the cluster aspect ratios with respect to classic hut and dome geometries on the clean Si(1 0 0) surface. In addition, chemical surfactants can also cause the growth mode to change from the Stranski–Krastanow to Frank van der Merwe (i.e., three-dimensional clusters without a wetting layer).

Essentially similar sets of transitions (wetting layer–hut clusters–dome clusters) have been observed for Ge_xSi_{1-x}–Si(1 0 0) epitaxy for Ge compositions as low as $x = 0.2$ ($\varepsilon = 0.01$) in the limit where the growth temperature is sufficiently high that adatom mobility is high enough to allow the equilibrium lengths scales to be attained [6]. Evolution of the wetting layer into {5 1 0} facetted hut clusters via growth and increasing contact angle of stepped mounds in Ge_xSi_{1-x}–Si(1 0 0) has been observed by real-time LEEM imaging [7,8]. A number of studies have shown that length scales (wetting layer thickness, cluster transition dimensions, etc.) scale inversely with strain [6–8], so reducing the strain in the system means that adatoms have to migrate further to achieve equilibrium structures—for example achieving the wetting layer–hut cluster–coherent dome cluster sequence at $x = 0.2$ in Ge_xSi_{1-x}–Si(1 0 0) requires growth temperatures 700°C or higher [6]. At lower growth temperatures, with more limited adatom migration lengths, different configurations are observed, as summarized in the next section.

26.5 Non-Equilibrium Strain-Driven Morphological Transitions in the Ge(Si)–Si System

We have established in the previous section that a well-defined series of morphological transitions occurs under conditions of sufficient adatom mobility to allow attainment of equilibrium microstructures in the Ge(Si)–Si(1 0 0) system. However, under conditions of reduced adatom mobility, different microstructures form. At sufficiently reduced growth temperatures (whose magnitudes depend strongly upon lattice mismatch strain, but are typically below about 350 to 400°C), surface migration lengths and islanding or roughening are largely suppressed. Under these conditions the film necessarily grows in a quasi-planar fashion, but with high densities of point defects (and for sufficiently high epitaxial layer thicknesses, interfacial misfit dislocations). Such structures contain extremely high strain energy densities.

A more complex regime is the region of intermediate adatom migration lengths, i.e., where adatoms can diffuse over significant distances, but not over sufficient lengths to attain the equilibrium microstructures described in the previous section. In this intermediate regime, periodic surface roughening is frequently observed, as described in Section 26.3 [18,38–40]. However, under specific ranges of kinetically limited growth conditions, well-defined but more complex microstructures may exist. We have recently described a new metastable morphological transition state that occurs during growth of Ge_xSi_{1-x}–Si(1 0 0) with $0.2 < x < 0.4$ within the specific ranges of epilayer growth temperature and growth rate [41–44], the "quantum dot molecule." The essential microstructural evolution is shown in Figure 26.4. First, shallow pits are observed to form in the growing epitaxial layer. These pits are square (bounded by $\langle 0\,0\,1 \rangle$ directions on the growth surface), are strain relieving, do not penetrate nearly as deep as the original Si–Ge_xSi_{1-x} interface, and so far as can be ascertained are not associated with extrinsic effects such as crystalline defects or impurities. With subsequent epilayer growth, small islands of Ge_xSi_{1-x} material form at each pit edge, and grow to form a continuous wall. At this point, the facet angles of both pits and bounding island walls have stabilized at {5 1 0}. In subsequent growth, both the "molecule" geometry and size (which depends upon the magnitude of the strain in the epilayer, but is ca. 220 nm for $Ge_{0.3}Si_{0.7}$) are fixed, i.e., during subsequent epilayer growth, i.e., the structure essentially conformally "floats" on the growth surface [44]. Such microstructures may have significant application to exploratory nanoelectronic architectures such as quantum cellular automata (QCAs), as described in later sections of this chapter. A similar geometry microstructure has been previously reported in the

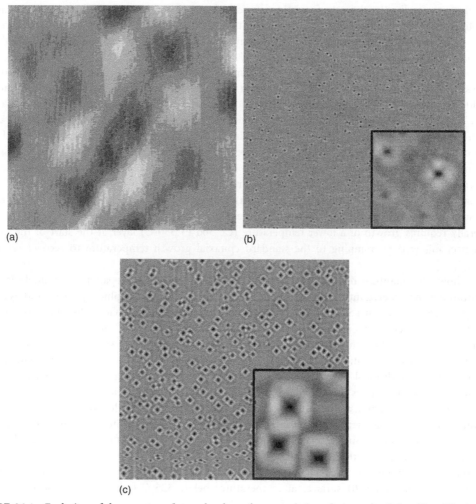

FIGURE 26.4 Evolution of the quantum dot molecule surface morphology for growth of $Ge_{0.3}Si_{0.7}$–$Si(1\ 0\ 0)$ at a growth temperature of 550°C and a growth rate of 0.9 A/sec. AFM scans are 5 μm × 5 μm in area. $Ge_{0.3}Si_{0.7}$ film thicknesses are (a) 0, (b) 15, and (c) 30 nm. Image (a) shows that immediately following growth of the Si buffer layer, the surface is close to atomically flat morphology with no pre-existing pits. Enlarged images of quantum dot molecule structures are shown in the insets to (b) and (c). The RMS surface morphology amplitudes are 0.14, 0.21, and 1.1 nm in (a), (b), and (c), respectively. (Images by J. Gray, J. Floro, and R. Hull. With permission.)

literature, but in that case was linked directly to carbon contamination at the Si substrate–Si buffer layer interface [45]. Neither the geometry, nor the evolution, of the present quantum dot molecules are consistent with that microstructural history.

26.6 Controlled Growth of Quantum Dot Arrays

For the great majority of electronic or nanoelectronic device architectures that might employ quantum dots, ordering of the quantum dots, either into single spatial frequencies or into more complex patterns, is desirable or necessary. Thus, there has been a great deal of work in the literature focusing upon techniques to control the nucleation sites for quantum dots, or seeking mechanisms by which the quantum dots self-organize into ordered arrays.

Perhaps the earliest work on the formation of ordered quantum dot arrays was based upon harnessing the strain field interactions between quantum dots. In "quantum dot superlattices," where successive layers of quantum dots are separated by intervening spacer layers of the same material as the substrate (i.e., in the present case, layers of coherent Ge(Si) quantum dots separated by Si spacer layers), strain field interactions between quantum dots increasingly organize the two-dimensional arrays of dots in successive layers [46,47]. This is because the strain field associated with a Ge quantum dot in one layer produces a surface strain (dilation) field that encourages localized nucleation of a QD in the successive layer. Thus the QDs form in columns. Further, the three-dimensional minimization of the total crystal strain energy encourages the QDs to organize within the two dimensions of each QD layer [46] (this can be accomplished by tilting, or even by the disappearance of individual columns). These processes are illustrated by three-dimensional tomographic FIB images in Figure 26.5. It should be emphasized that one challenge in creating such QD superlattice structures is that on Si capping, individual dots often dramatically decrease their aspect ratio (i.e., flatten), particularly for dots with higher Ge concentrations and at higher growth temperatures [48,49]. Maintenance of Ge(Si) quantum dot aspect ratios therefore generally requires growth of a lower temperature (ca. 300 to 350°C) layer at the start of the capping sequence, followed by ramping to the standard epitaxial growth temperature to recover crystalline quality [50].

But how can quantum dots be organized into either more complex patterns, or single layers of quantum dots be ordered into a periodic array? One method is to lithographically open windows in, for example, an SiO$_2$ layer on a Si substrate, and to achieve local nucleation of Ge in the windows [51,52]. A broader set of mechanisms is available through appropriate "templating" of the substrate surface. As illustrated in Figure 26.6, there are multiple mechanisms by which local surface modification can be expected to control nucleation sites for quantum dots. One method is to introduce topography into the surface. The additional degree of freedom associated with a free edge creates a logical site for a strained nucleus to form, as the underlying substrate can more readily deform. The use of mesa edges and facets has been demonstrated to be highly effective in localizing nucleation [53–56]. Similarly, at the more atomistic scale, steps or step bunches are preferred attachment sites [57]. Additional methods are local modification of surface crystallinity or chemistry. Both may affect local adatom diffusivity and attachment energies, thereby localizing the probability of forming a critical nucleus. Another method of modifying the local chemistry is desorption using a scanning probe microscope tip of hydrogen from Si(1 0 0) surfaces at temperatures below 535°C (while the hydrogen-terminated

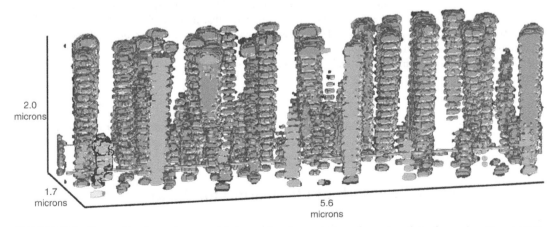

FIGURE 26.5 Focused ion beam tomographic image showing evolution of quantum dot columns in a Ge quantum dot superlattice sample. Ge quantum dot layers are grown with an equivalent planar layer thickness of 1.4 nm at 750°C. Intervening Si spacer layers are grown with 20 nm at 300°C, and 80 nm at 750°C. The spatial resolution in the vertical direction was intentionally decreased so that the evolution of the columns could more easily be observed. The column in blue terminates during the growth of the superlattice structure. (Image courtesy of A. Kubis. With permission.)

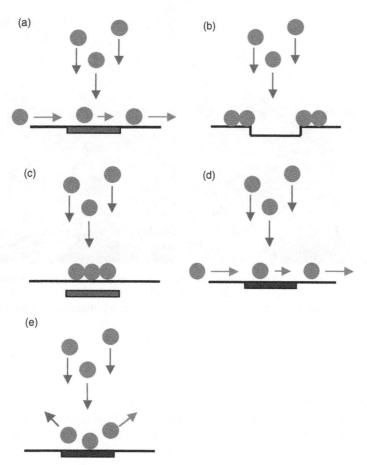

FIGURE 26.6 Schematic illustration of basic mechanisms for local modification of substrate surface structure for templated heteroepitaxial assembly. Local modification of (a) crystallinity, (b) topography, (c) surface lattice parameter from buried stressors, (d) surface chemistry, and (e) surface reactivity. "Atoms" are shown as solid circles. The lengths of lateral arrows in (a) and (d) indicate variations in local surface adatom mobility. The solid rectangular regions in (a) and (c) indicate regions of disturbed crystallinity (or second phase inclusions in the case of (c)). The solid rectangular regions in (d) and (e) represent regions of locally modified chemistry.

surface remains stable elsewhere), allowing local epitaxy of Ge in the desorbed regions [58]. Buried strain centers can also localize nucleation because of the creation of associated dilation or compression fields at the crystal surface. For example, both buried SiO_2 precipitates [59] and buried misfit dislocations [60–62] have been shown to localize Ge quantum dot nucleation on Si(1 0 0) surfaces. Finally, for the case of chemical vapor deposition (or other related thermally activated deposition mechanisms) variations in local surface chemistry can affect the surface reactivity and hence local deposition rates [34,63].

One method by which all these mechanisms can be explored is through local Ga^+ FIB modification of the substrate surface. The FIB can create controlled local surface topography through sputtering, modifies the surface crystallinity (creating surface amorphization for implantation into Si at room temperature, which can be fully recovered by moderate time–temperature annealing cycles), modifies the local surface chemistry (and hence reactivity) through the implanted species, and creates local strain centers due to residual defects. In Figure 26.7, we show how such FIB implantation (performed in conjunction with ultrahigh vacuum chemical vapor deposition, and transmission electron microscope

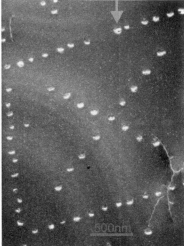

FIGURE 26.7   Local control of Ge–Si(1 0 0) cluster nucleation by focused ion beam prepatterning. Images are from transmission electron microscope imaging *in situ* to the FIB and deposition capabilities. Left-hand image shows *in situ* as-patterned Si(1 0 0) surface (25 keV Ga$^+$, 10 pA, 100 μs per feature). Right-hand image, following annealing and Ge deposition by chemical vapor deposition (digermane source) at 600°C (Ge clusters are the bright dots as indicated by the red arrow). (Images by M. Kammler, F. Ross, and R. Hull. With permission.)

imaging, all within the same integrated instrument), can successfully localize nucleation in Ge–Si(1 0 0) [35,42,63]. In fact this method convolutes possible effects from local modification of surface crystallinity, surface reactivity and strain, but extensive experiments suggest that localized nano-topography and strain fields are the primary mechanism for controlling nucleation [35,63].

26.7   Potential Nanoelectronic and Nanophotonic Device Applications

Nanoelectronic and nanophotonic device applications of Ge$_x$Si$_{1-x}$–Si heterostructures are dealt with extensively elsewhere in this volume, but specific opportunities exist for harnessing the properties of Ge(Si) quantum dots. One exploratory application is the use of QD quadruplets as cells for QCAs [64–66]. In this architecture, extra charges (electrons or holes) will occupy opposite corners of the cell, creating bistable states that can form the basis of digital logic. The quantum dot molecule geometry discussed in Section 26.5 has the appropriate geometry to form individual quantum dot cells, providing groups of cells can be organized into the correct patterns for QCA circuits. We have demonstrated the ability to organize the quantum dot molecules using topographic forcing functions on the substrate surface, such that the dimensions of interstices between holes in a two-dimensional array match to the dimensions of the quantum dot molecules (Figure 26.8). Other potential uses of quantum dot arrays include single electron transistors [67,68] and memory elements, either in epitaxial structures or embedded in oxides [69–71]. In all cases, the ability to accurately control the spatial distribution of quantum dots is key to successful realization of these nanoelectronic applications. While Ge, Si, and Ge$_x$Si$_{1-x}$ are of course indirect semiconductors, the optical properties of embedded Ge(Si) quantum dots are also a topic of active research, including the ability to tune optical emission wavelength through the dot dimensions, for example in photovoltaic applications [72–75]. Applications to potential quantum computing applications [58], and thermoelectric properties of Ge quantum dot superlattices are also being explored [76].

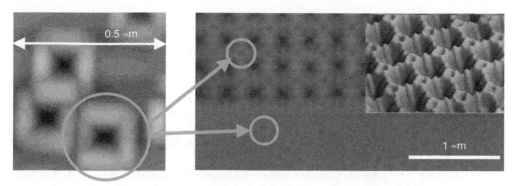

FIGURE 26.8 Controlled formation of quantum dot molecules by topographic forcing functions (surface topography is created *ex situ* to the deposition chamber with a focused ion beam). Images are recorded in the atomic force microscope. Left-hand image shows the quantum dot molecule structure. Right-hand image shows ordered arrays formed on patterned substrates with original (i.e., prior to deposition) hole spacings of 250 nm. Inset shows the perspective image of patterned region. In the unpatterned region, quantum dot molecules form at apparently random locations. (Images by J. Gray, J. Floro, and R. Hull. With permission.)

26.8 Summary

The high mismatch strain in the Ge_xSi_{1-x}–Si system provides a natural driving force for cluster (quantum dot) assembly during heteroepitaxy. The equilibrium set of morphological transitions that occur during growth in this system are well documented and reasonably well understood, at least for the (1 0 0) surface. Transitions under nonequilibrium (i.e., kinetically limited) growth conditions are less well explored and understood, although it is clear that fascinating metastable structures, such as the quantum dot molecule, can occur. Substantial progress has been made by several groups in localizing cluster nucleation into desired arrays or patterns, although no technique to date has fully demonstrated the necessary combinations of spatial accuracy, field of view, and acceptably low error rates necessary for practical application of most potential nanoelectronic device concepts. Further advances in this area are necessary to accelerate the development of applications.

Acknowledgments

The author acknowledges invaluable discussions and collaborations with many individuals, relevant to the material in this chapter: J. Floro (Sandia); F. Ross, M. Reuter, J. Tersoff, and R. Tromp (IBM); I. Berbezier (CNRS, Marseilles); E. Stach (Lawrence Berkeley); and S. Atha, J. Bean, J. Gray, A. Kubis, M. Kammler, A. Portavoce, T. Pernell, T. Vandervelde, and C.-C. Wu (U. Virginia). Work at the University of Virginia described in this chapter was performed under a Materials Research Science and Engineering Center (DMR-0080016), and a Focused Research Group (DMR-0075116), funded by the National Science Foundation.

References

1. A.R. Woll, P. Rugheimer, and M.G. Lagally. Self organized quantum dots. *Int. J. High Speed Electron. Syst.* 12, 45–78 (2002).
2. D.E. Savage, F. Liu, V. Zielask, and M.G. Lagally. Fundamental mechanisms of thin film growth. In *Germanium Silicon: Physics and Materials*, R. Hull and J.C. Bean, eds. Academic Press, San Diego, CA, 1999, pp. 49–101.
3. R.M. Tromp and J.B. Hannon. Thermodynamics of nucleation and growth. *Surf. Rev. Lett.* 9, 1565–1593 (2002).
4. K.L. Wang and X. Zheng. In *Properties of Strained and Relaxed GeSi*, E. Kasper, ed. IEEE EMIS Datareview Series No. 12, London, England, 1995, pp. 78–80, and references therein.

5. D.J. Eaglesham and M. Cerullo. Dislocation-free Stranski–Krastanow growth of Ge on Si(1 0 0). *Phys. Rev. Lett.* 64, 1943–1946 (1990).

6. J.A. Floro, E. Chason, L.B. Freund, R.D. Twesten, R.Q. Hwang, and G.A. Lucadamo. Evolution of coherent islands in $Si_{1-x}Ge_x/Si(0\ 0\ 1)$. *Phys. Rev. B* 59, 1990–1998 (1999).

7. P. Sutter and M.G. Lagally. Nucleationless three-dimensional island formation in low-misfit heteroepitaxy. *Phys. Rev. Lett.* 84, 4637–4640 (2000).

8. R.M. Tromp, F.M. Ross, and M.C. Reuter. Instability-driven SiGe island growth. *Phys. Rev. Lett.* 84, 4641–4644 (2000).

9. M. Grinfeld. Two-dimensional islanding atop stressed solid helium and epitaxial films. *Phys. Rev B* 49, 8310–8319 (1994).

10. W.H. Yang and D.J. Srolovitz. Cracklike surface instabilities in stressed solids. *Phys. Rev. Lett.* 71, 1593–1596 (1993).

11. B.J. Spencer, P.W. Voorhees, and S.H. Davis. Morphological instability in epitaxially strained dislocation-free solid films: linear stability theory. *J. Appl. Phys.* 73, 4955–4970 (1993).

12. B.J. Spencer, P.W. Voorhees, and S.H. Davis. Morphological instability in epitaxially strained dislocation-free solid films: nonlinear evolution. *Phys. Rev. B* 47, 9760–9777 (1993).

13. H. Gao. Stress concentration at slightly undulating surfaces. *J. Mech. Phys. Solids* 39, 443–458 (1991).

14. L.B. Freund and F. Jonsdottir. Instability of a biaxially stressed thin film on a substrate due to material diffusion over its free surface. *J. Mech. Phys. Solids* 41, 1245–1264 (1993).

15. J. Tersoff and R.M. Tromp. Shape transition in growth of strained islands: spontaneous formation of quantum wires. *Phys. Rev. Lett.* 70, 2782–2785 (1994).

16. K.M Chen, D.E. Jesson, S.J. Pennycook, T. Thundat, and R.J. Warmack. Critical nuclei shapes in the stress-driven 2D-to-3D transition. *Phys. Rev. B Rapid. Commun.* 56, R1700–R1703 (1997).

17. J. Tersoff, Y.H. Phang, Z. Zhang, and M.G. Lagally. Step-bunching instability of vicinal surfaces under stress. *Phys. Rev. Lett.* 75, 2730–2733 (1995).

18. A.G. Cullis, D.J. Robbins, A.J. Pidduck, and P.W. Smith. The characteristics of strain-modulated surface undulations formed upon epitaxial $Si_{1-x}Ge_x$ alloy layers on Si. *J. Cryst. Growth* 123, 333–343 (1992).

19. N.C. Bartelt, R.M. Tromp, and E.D. Williams. Step capillary waves and equilibrium island shapes on Si(0 0 1). *Phys. Rev. Lett.* 73, 1656–1659 (1994).

20. C. Duport, P. Politi, and J. Villain. Growth instabilities induced by elasticity in a vicinal surface. *J. Phys.* 5, 1317–1350 (1995).

21. Y.-W. Mo, D.E. Savage, B.S. Swartzentruber, and M.G. Lagally. Kinetic pathway in Stranski–Krastanov growth of Ge on Si(0 0 1). *Phys. Rev. Lett.* 65, 1020–1023 (1990).

22. A. Vailionis, B. Cho, C. Glass, P. Desjardins, D.G. Cahill, and J.E. Greene. Pathway for the strain-driven two-dimensional to three-dimensional transition during growth of Ge on Si(1 0 0). *Phys. Rev. Lett.* 85, 3672–3675 (2000).

23. V.B. Shenoy, C.V. Ciobanu, and L.B. Freund. Strain induced stabilization of stepped Si and Ge surfaces near (0 0 1). *Appl. Phys. Lett.* 81, 364–366 (2002), and references therein.

24. M. Tomitori, K. Watanabe, M. Kobayashi, and O. Nishikawa. STM study of the Ge growth mode on Si(0 0 1) substrates. *Appl. Surf. Sci.* 76/77, 322–328 (1994).

25. F.K. LeGoues, M.C. Reuter, J. Tersoff, M. Hammar, and R.M. Tromp. Cyclic growth of strain-relaxed islands. *Phys. Rev. Lett.* 73, 300–303 (1994).

26. G. Medeiros-Ribeiro, A.M. Bratkovski, T.I. Kamins, D.A.A. Ohlberg, and R.S. Williams. Shape transition of germanium nanocrystals on a silicon (0 0 1) surface from pyramids to domes. *Science* 279, 353–355 (1998).

27. F.M. Ross, R.M. Tromp, and M.C. Reuter. Transition states between pyramids and domes during Ge/Si island growth. *Science* 286, 1931–1934 (1999).

28. F.M. Ross, J. Tersoff, and R.M. Tromp. Coarsening of self-assembled Ge quantum dots on Si(0 0 1). *Phys. Rev. Lett.* 80, 984–987 (1998).

29. T.I. Kamins, E.C. Carr, R.S. Williams, and S.J. Rosner. Deposition of three-dimensional Ge islands on Si(0 0 1) by chemical vapor deposition at atmospheric and reduced pressures. *J. Appl. Phys.* 81, 211–219 (1997).

30. J. Tersoff. Enhanced nucleation and enrichment of strained-alloy quantum dots. *Phys. Rev. Lett.* 81, 3183–3186 (1998).

31. S.A. Chapparro, J. Drucker, Y. Zhang, D. Chandrasekhar, M.R. McCartney, and D.J. Smith. Strain-driven alloying in Ge/Si(1 0 0) coherent islands. *Phys. Rev. Lett.* 83, 1199–1202 (1999).

32. T.I. Kamins, G. Medeiros-Ribeiro, D.A.A. Ohlberg, and R.S. Williams. Dome-to-pyramid transition induced by alloying of Ge islands on Si(0 0 1). *Appl. Phys A* 67, 727–730 (1998).

33. Y. Zhang, M. Floyd, K.P. Driver, J. Drucker, P.A. Crozier, and D.J. Smith. Evolution of Ge/Si(1 0 0) island morphology at high temperature. *Appl. Phys. Lett.* 80, 3623–3625 (2002).

34. T.I. Kamins, G. Medeiros-Ribeiro, D.A.A. Ohlberg, and R.S. Williams. Influence of phosphine on Ge/Si(0 0 1) island growth by chemical vapor deposition. *J. Appl. Phys.* 94, 4215–4224 (2003).

35. M. Kammler, R. Hull, M.C. Reuter, and F.M. Ross. Lateral control of self-assembled island nucleation by focused-ion-beam-micropatterning. *Appl. Phys. Lett.* 82, 1093–1095 (2003).

36. H. Takamiya, N. Miura, N. Usami, T. Hattori, and Y. Shiraki. Drastic modification of the growth mode of Ge quantum dots on Si by using boron adlayer. *Thin Solid Films* 369, 84–87 (2000).

37. X. Zhou, B. Shi, Z. Jiang, W. Jiang, D. Hu, D. Gong, Y. Fan, X. Zhang, X. Wang, and Y. Li. Boron-mediated growth of Ge quantum dots on Si(1 0 0) substrate. *Thin Solid Films* 369, 92–95 (2000).

38. N.-E. Lee, D.G. Cahill, and J.E. Greene. Evolution of surface roughness in epitaxial $Si_{0.7}Ge_{0.3}/Si(0\ 0\ 1)$ as a function of growth temperature (200–600 degrees C) and Si(0 0 1) substrate miscut. *J. Appl. Phys.* 80, 2199–2210 (1996).

39. A.J. Pidduck, D.J. Robbins, A.G. Cullis, W.Y. Leong, and A.M. Pitt. Evolution of surface morphology and strain during SiGe epitaxy. *Thin Solid Films* 222, 78–84 (1992).

40. D.E. Jesson, S.J. Pennycook, J.-M. Baribeau, and D.C. Houghton. Direct imaging of surface cusp evolution during strained-layer epitaxy and implications for strain relaxation. *Phys. Rev. Lett.* 71, 1744–1747 (1993).

41. J.L. Gray, R. Hull and J.A. Floro. Control of surface morphology through variation of growth rate in SiGe/Si(1 0 0) epitaxial films: nucleation of quantum fortresses. *Appl. Phys. Lett.* 81, 2445–2447 (2002).

42. R. Hull, J.L. Gray, M. Kammler, S. Atha, P. Kumar, T. Vandervelde, J.C. Bean, J.A. Floro, and F.M. Ross. Precision placement of heteroepitaxial semiconductor quantum dots. *Mater. Sci. Eng. B* 101, 1–8 (2003).

43. T.E. Vandervelde, P. Kumar, T. Kobayashi, J.L. Gray, T. Pernell, J.A. Floro, R. Hull, and J.C. Bean. Growth of quantum fortress structures in $Si_{1-x}Ge_x/Si$ via combinatorial deposition. *Appl. Phys. Lett.* 83, 5205–5207 (2003).

44. J.L. Gray, N. Singh, D.M. Elzey, R. Hull, and J.A. Floro. Kinetic size selection mechanisms in intrinsic quantum dot molecules. *Phys. Rev. Lett.* 92, 135504:1–4 (2004).

45. X. Deng and M. Krishnamurthy. Self-assembly of quantum dot molecules: heterogeneous nucleation of SiGe islands on Si(1 0 0). *Phys. Rev. Lett.* 81, 1473–1476 ((1998).

46. J. Tersoff, C. Teichert, and M.G. Lagally. Self-organization in growth of quantum dot superlattices. *Phys. Rev. Lett.* 76, 1675–1678 (1996).

47. C. Teichert, M.G. Lagally, L.J. Peticolas, J.C. Bean, and J. Tersoff. Stress-induced self-organization of nanoscale structures in SiGe/Si multilayer films. *Phys. Rev. B* 53, 16334–16337 (1996).

48. J.L. Bischoff, C. Pirri, D. Dentel, L. Simon, D. Bolmont, and L. Kubler. AFM and RHEED study of Ge/Si(0 0 1) quantum dot modification by Si capping. *Mater. Sci. Eng. B* 69/70, 374–379 (2000).

49. P. Sutter and M.G. Lagally. Embedding of nanoscale 3D SiGe islands in a Si matrix. *Phys. Rev. Lett.* 81, 3471–3474 (1998).

50. A. Rastelli, E. Muller, and H. von Kanel. Shape preservation of Ge/Si(0 0 1) islands during Si capping. *Appl. Phys. Lett.* 80, 1438–1440 (2002).

51. A.A. Shklyaev, M. Shibata, and M. Ichikawa. High-density ultrasmall epitaxial Ge islands on Si(1 1 1) surfaces with a SiO$_2$ coverage. *Phys. Rev. B* 62, 1540–1543 (2000).

52. E.S. Kim, N. Usami, and Y. Shiraki. Selective epitaxial growth of dot structures on patterned Si substrates by gas source molecular beam epitaxy. *Semicond. Sci. Technol.* 14, 257–265 (1999).

53. L. Vescan. Lateral ordering of Ge islands along facets. *J. Cryst. Growth* 194, 173–177 (1998).

54. T.E. Kamins and S. Williams. Lithographic positioning of self-assembled Ge islands on Si(0 0 1). *Appl. Phys. Lett.* 71, 1201–1203 (1997).

55. G. Jin, J. Wan, Y.H. Luo, J.L. Liu, and K.L. Wang. Uniform and ordered self-assembled Ge dots on patterned Si substrates with selectively epitaxial growth technique. *J. Cryst. Growth* 227–228, 1100–1105 (2001).

56. L. Vescan, T. Stoica, and B. Hollander. Lateral ordering of Ge islands on Si mesas made by selective epitaxial growth. *Mater. Sci. Eng. B* 89, 49–53 (2002).

57. I. Berbezier, A. Ronda, A. Portavoce, and N. Motta. Ge dots self-assembling: surfactant mediated growth of Ge on SiGe (1 1 8) stress-induced kinetic instabilities. *Appl. Phys. Lett.* 83, 4833–4835 (2003).

58. J.R. Tucker and T.-C Shen. Can single electronic integrated circuits and quantum computers be fabricated in silicon? *Int. J. Circuit Theory Appl.* 28, 553–562 (2000).

59. H. Omi, D.J. Bottomley, and T. Ogino. Strain distribution control on the silicon wafer scale for advanced nanostructure fabrication. *Appl. Phys. Lett.* 80, 1073–1075 (2003).

60. F.M. Ross. Growth processes and phase transformations studied by *in situ* transmission electron microscopy. *IBM J. Res. Develop.* 44, 489–501 (2000).

61. C. Teichert, C. Hofer, K. Lyutovich, M. Bauer, and E. Kasper. Interplay of dislocation network and island arrangement in SiGe films grown on Si(1 0 0). *Thin Solid Films* 380, 25–28 (2000).

62. H.J. Kim, Z.M. Zhao, and Y.H. Xie. Three-stage nucleation and growth of Ge self-assembled quantum dots grown on partially relaxed SiGe buffer layers. *Phys. Rev. B* 68, 205312:1–7 (2003).

63. A. Portavoce, M. Kammler, R. Hull, M.C. Reuter, M. Copel, and F.M. Ross. Growth kinetics of Ge islands during Ga surfactant-mediated UHV-CVD on Si(0 0 1). *Phys. Rev. B* 70, 195306: 1–9 (2004).

64. C.S. Lent, P.D. Tougaw, W. Porod, and G.H. Bernstein. Quantum cellular automata. *Nanotechnology* 4, 49–57 (1993).

65. C.S. Lent and P.D. Tougaw. A device architecture for computing with quantum dots. *Proc. IEEE* 85, 541–557 (1997).

66. I. Amlani, A.O. Orlov, G. Toth, G.H. Bernstein, C.S. Lent, and G.L. Snider. Digital logic gate using quantum-dot cellular automata. *Science* 284, 289–291 (1999).

67. S. Kanjanachuchai, J.M. Bonar, and H. Ahmed. Single-charge tunnelling in n- and p-type strained silicon germanium on silicon-on-insulator. *Semicond. Sci. Technol.* 14, 1065–1068 (1999).

68. L.J. Klein, K.A. Slinker, J.L. Truitt, S. Goswami, K.L.M. Lewis, S.N. Coppersmith, D.W. van der Weide, M. Friesen, R.H. Blick, D.E. Savage, M.G. Lagally, C. Tahan, R. Joynt, M.A. Eriksson, J.O. Chu, J.A. Ott, and P.M. Mooney. Coulomb blockade in a silicon/silicon–germanium two-dimensional electron gas quantum dot. *Appl. Phys. Lett.* 84, 4047–4049 (2004).

69. C.L. Heng, Y.J. Liu, A.T.S Wee, and T.G. Finstad. The formation of Ge nanocrystals in a metal–insulator–semiconductor structure and its memory effect. *J. Cryst. Growth* 262, 95–1004 (2004).

70. D.-W. Kim, T. Kim, Y. Liu, L. Weltzer, and S. Banerjee. SiGe quantum dots memory devices with HfO$_2$ tunneling oxide. Proceedings of the 61st Device Research Conference. Conference Digest. IEEE, Piscataway, NJ, USA, Cat. No. 03TH8663, 2003, pp. 131–132.

71. N. Deng, L. Pan, L. Zhang, and P. Chen. Ge quantum dot memory realized with vertical Si/SiGe resonant tunneling structure. Proceedings of the Fourth International Workshop on Junction Technology. IEEE, Piscataway, NJ, USA, Cat. No. 04EX762, 2004, pp. 256–258.

72. L. Chu, A. Zrenner, M. Bichler, and G. Abstreiter. Quantum-dot infrared photodetector with lateral carrier transport. *Appl. Phys. Lett.* 79, 2249–2251 (2001).

73. S. Tong, J. Liu, J. Wan, R. Faez, V. Pouyet, and K.L. Wang. Normal-incidence near-1.55 μm Ge quantum dot photodetectors on Si substrate. *SPIE Proc.* 4580, 193–201 (2001).

74. B-C Hsu, S.T. Chang, T-C Chen, P-S Kuo, PS Chen, Z Pei, and C.W. Liu. A high efficient 820 nm MOS Ge quantum dot photodetector. *IEEE Electron Dev. Lett.* 24, 318–320 (2003).

75. N. Usami, A. Alguno, T. Ujihara, K. Fujiwara, G. Sazaki, K. Nakajima, K. Sawano, and Y. Shiraki. Stacked Ge islands for photovoltaic applications. *Sci. Technol. Adv. Mater.* 4, 367–370 (2003).

76. B. Yang, J.L. Liu, K.L. Wang, and G. Chen. Simultaneous measurements of Seebeck coefficient and thermal conductivity across superlattice. *Appl. Phys. Lett.* 80, 1758–1160 (2002).

27

Overview: Optoelectronic Components

John D. Cressler

Georgia Institute of Technology

Given the natural ease with which monolithic integration can be realized in silicon, the merger of the photonics world and the electronics world would seem to offer compelling advantages, at least for the long-haul. Clearly, with respect to light emission and detection, Si (or SiGe) with its indirect bandgap is at a significant disadvantage with respect to their III–V counterparts. Nevertheless, significant progress has been made in Si-based light emitting diodes, as disucssed by K.L. Wang of UCLA in Chapter 28, "Si–Si LEDs." For shorter wavelength applications, much has been accomplished in realm of light detectors, as discussed in Chapter 29, "Near-Infrared Detectors," by L. Colace of the University of Rome, and Chapter 30, "Si-Based Photonic Transistors Devices for Integrated Optoelectronics," by W. Li of Linköping University. The quest for useful levels of light emission in Si is an old one, and while at first glance it might seem a laughable prospect, bandgap engineering facilitates a number of interesting possibilities that is fueling serious interest. In Chapter 31, for instance, "Si–SiGe Quantum Cascade Emitters," by D. Paul of the University of Cambridge, the potential for light emission is the Si–SiGe system is addressed using quantum cascade techniques. In addition to this substantial collection of material, and the numerous references contained in each chapter, a number of review articles and books detailing the operation and modeling of various optoelectronic components exist, including Refs. [1–5].

References

1. R. People. Physics and applications of Ge_xSi_{1-x}/Si strained layer heterostructures. *IEEE Journal of Quantum Electronics* 22:1696–1710, 1986.
2. JC Bean. Silicon-based semiconductor heterostructures: column IV bandgap engineering. *Proceedings of the IEEE* 80:571–587, 1992.
3. RA Soref. Silicon-based optoelectronics. *Proceedings of the IEEE* 81:1687–1706, 1993.
4. CK Maiti and GA Armstrong. *Applications of Silicon–Germanium Heterostructure Devices*. London: Institute of Physics Publishing, 2001.
5. CK Maiti, NB Chakrabarti, and SK Ray. *Strained Silicon Heterostructures: Materials and Devices*. London: The Institution of Electrical Engineers, 2001.

28

Si–SiGe LEDs

Kang L. Wang, S. Tong,
and H.J. Kim
University of California at Los Angeles

28.1 Introduction

Light emitting diodes (LEDs) are one of the major components in realizing optoelectronic functions. However, it is well known that Si is an indirect bandgap material, and therefore, its luminescence efficiency is quite poor compared to III–V direct bandgap materials. Due to the nature of its indirect bandgap, the radiative recombination in Si at the band edge needs phonon assistance to maintain the momentum conservation. The transverse optical (TO) phonon-assisted peak dominates the emission spectrum. Due to the multiparticle nature of the recombination process, the radiative lifetime of the carriers is much longer compared to direct transition recombinations. Germanium bandgap is also indirect, but the direct valley at Γ point is only 0.15 eV higher than the indirect valley. Thus, excited electrons can enter both valleys, and recombination occurs via direct and indirect emission channels. Thus, it is possible to observe direct transitions from thin samples [1]. However, the output is weak due to reabsorption by the material, phonon scattering and low carrier density in direct valley. Nevertheless, people were persistently exploring possible solutions for efficient radiation in Si- and Ge-based materials on silicon substrate, such as porous silicon [2], Si or Ge quantum dots embedded in larger bandgap matrix, e.g., SiO_x [3,4] and a-Si:H [5]. The incorporation of rare-earth atoms into silicon has also been studied. Erbium-based emitters take advantage of the intra-4f shell transitions at 1.54 mm [6]. Room temperature electroluminescence (EL) of erbium-doped silicon LEDs has been reported [7].

Ge and SiGe can be epitaxially grown on a Si wafer with higher quality and fewer problems than III–V materials, since Ge and Si all belong to group IV. SiGe structures are fabricated by MBE or LPCVD techniques to maintain their lattice structure integrity, though defects such as misfit dislocations and point defects always appear in a certain degree. These materials include Si_nGe_m strained layer superlattices (SLSs), $Si_{1-x}Ge_x$ quantum wells, and $Si_{1-x}Ge_x$ quantum dots [8–12].

28.2 LED Design

The idea of Brillouin-zone folding was first proposed by Gnutzmann and Clausecker [13]. A two-material superlattice with a period of several monolayers can fold the X valley back to the Γ point. For Si, the conduction band minima is at $0.85(\pi/a)$. By growing Si_nGe_m SLS, the conduction band minima can be folded to the Γ point due to the artificially increased lattice constant in the growth direction, so that an enhancement of luminance is expected due to the direct transition in the SLS. Zone-folding ideas have been studied theoretically [13,14] and experimentally [15–19]. Low temperature and room temperature [19] EL and photoluminescence (PL) have been observed from Si–Ge superlattice structures. With the study of absorption threshold behavior [20], transition mechanisms of the EL have been explored. It was shown that bandgap-related transitions as well as defect-related transition might contribute to the EL spectra. However, the transition matrix element calculation shows that the oscillator strength of the direct transition is still at least one order of magnitude lower than in direct III–V semiconductors [21].

Other approaches include realizing quantum structures inspired by the quantum confinement effect. Since the bandgap of Si (1.15 eV) is larger than that of Ge (0.67 eV), it is possible to form wells for carriers in SiGe heterostructures. The ideal situation is to have a type-I (as shown in Figure 28.1a) band alignment between the two materials, i.e., both electrons and holes are confined in Ge wells. However, in reality, the band alignment between Si and Ge is typically type-II (as shown in Figure 28.1b), that is, holes are confined in Ge and electrons in Si. Nevertheless, researchers were able to achieve light radiation in such structures up to room temperature. In quantum dots, due to strong carrier localization within a space of less than 10 nm, the momentum is a poorly defined physical quantity and does not have to be conserved, resulting in enhanced no-phonon luminescence.

In this section, we present the experimental results of PL and EL in this field. The results from Si_nGe_m SLSs $Si_{1-x}Ge_x$ quantum wells, and Ge QDs will be presented separately.

Studies show that if SLS is grown on a Si substrate, due to the strain effect on the band structure, the in-plane four-fold conduction band minima, instead of the two zone-folding conduction band minima, are lowest in energy. So the superlattice needs to be grown on a SiGe substrate. Figure 28.2 shows the calculation results of Si_5Ge_5 grown on the Si structure and $Si_{0.5}Ge_{0.5}$ alloy buffer layer, respectively, using the effective mass approximation. The upper graph shows that the lowest interband transition for the SLS on Si substrate is an indirect transition between the nonfolded conduction band minima to the heavy hole states. And the lower graph shows direct transitions with both the conduction band minimum and valance band maximum in the SLS region. One way to obtain the SiGe substrate is to grow a SiGe-completed relaxed buffer layer with the strain of Si and Ge in the superlattice canceling each other [15,22]. The major problem of the fully relaxed buffer layers is the large number of misfit dislocations generated during strain relief [17]. One way to minimize the defect density in the buffer layer so as to improve the superlattice film quality is to grow a thick buffer layer aswell as a partially strained buffer layer [23]. It is also known that higher growth temperature is good for defect-free films. One monolayer (ML) of Sb is used to reduce the intermixing of Si and Ge at interface under high growth temperature [24].

Figure 28.3 shows the EL and PL spectra for 250 nm Si_5Ge_5 grown on 1 μm thick $Si_{0.5}Ge_{0.5}$ buffer layer. The samples were grown as p–i–n structure and processed with standard silicon technology for waveguide LED structures. The EL was measured at an injected electrical power of 25 mW at 5 K. The

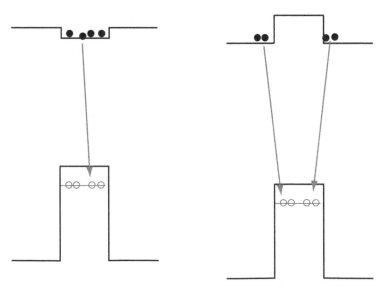

FIGURE 28.1 (a) Type-I band alignment, where holes and electrons are confined spatially in the same location. (b) Type-II band alignment, where holes and electrons are spatially separated.

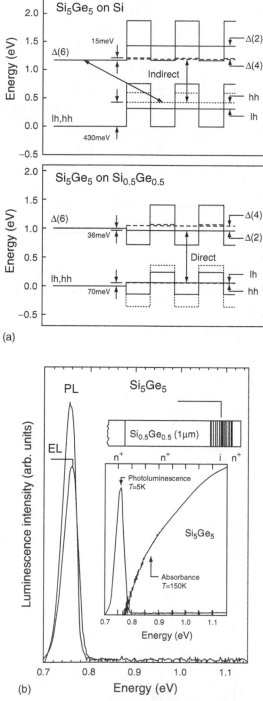

FIGURE 28.2 (a) Bandgap and band aligments for Si_5Ge_5 superlattices grown pseudomorphically on a silicon substrate and on a fully relaxed $Si_{0.5}Ge_{0.5}$ alloy buffer layer. (b) Comparison between EL and PL for a Si_5Ge_5 superlattice grown on a 1-μm thick fully relaxed $Si_{0.5}Ge_{0.5}$ alloy buffer layer. The inset shows the comparison between absorption and photoluminescence spectra. The absorption spectrum is plotted on a logarithmic scale. (After U Menczigar, G Abstreiter, J Olajos, H Grimmeiss, H Kibbel, H Presting, and E Kasper. *Phys Rev B* 47:4099–4102, 1993. With permission.)

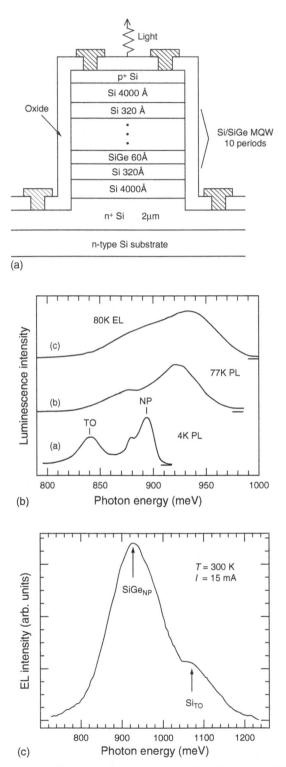

FIGURE 28.3 (a) Device schematic. The active regions consists of ten layers of Si (320 A)/Si$_{0.65}$Ge$_{0.35}$(60 A). (b) Photoluminescence spectra at (a) 4.2 K and (b) 77 K, and (c) electroluminescence spectrum with 10 mA drive current and heat sink temperature at 80 K. (c) Electroluminescence spectra with room temperature heat sink with 15 mA drive current. (After Q Mi, X Xiao, JC Sturm, LC Lenchyshyn, and MLW Thewalt. *Appl Phys Lett* 60:3177–3129, 1992. With permission.)

light emission is attributed to the intrinsic superlattice region because of the same energy peak for the PL and EL. The absorption spectrum measurements show that the resonance energy is in agreement with PL result of 0.76 eV. Thus the PL and EL spectra of the strained Si_5Ge_5 superlattice are likely due to bandgap-related transitions [21].

Another structure for light emission is SiGe quantum wells. Characteristic features of such luminescence are the phonon replicas found in indirect bandgap semiconductors plus a signal near the bandgap due to direct exciton recombination without any phonons involved [25]. At $T < 20\,K$ the PL is dominated by shallow bound excitons (BE), while at $T > 20\,K$ it is dominated by free excitons (FE). No-phonon (NP) emission from FE is the result of alloy effects and was first observed in bulk unstrained alloys [26]. PL and EL of pseudomorphic $Si/Si_{1-x}Ge_x$ multiple quantum wells (MQW) and single quantum wells (SQW) grown by molecular beam epitaxy (MBE) at 400°C were first reported by Noel et al. PL signal with a line width of about 100 meV was observed after thermal annealing at temperatures above 550°C. The PL peaks, however, were observed about 100 meV below the expected bandgap [27].

Band edge PL due to exciton recombination was first obtained from strained $Si_{0.958}Ge_{0.042}$ by Terashima et al. [28]. EL at 1.2 μm from a pseudomorphic $Si_{0.8}Ge_{0.2}$ alloy grown by CVD at 610°C was demonstrated by Robbins et al. at 77 K [29]. EL (1.3 μm) from strained $Si_{0.65}Ge_{0.35}$ was obtained at room temperature with the internal quantum efficiency having a lower limit of 2×10^{-4} [25]. Then EL was observed at temperature up to 60°C in p-type strained $Si_{0.65}Ge_{0.35}$–Si MQW grown on Si(1 1 1) substrates by Fukatsu et al. The no-phonon (NP) line and its TO phonon replica were well resolved in the room temperature EL spectrum at growth temperature of 800°C. These experimental results also show that for a structure grown at a temperature lower than 500°C, the defect-related line dominates the luminescence spectra, while for high-temperature grown sample, bandgap-related PL signals can be observed [21].

As shown in Figure 28.3a, the sample consisted of ten layers of $Si_{0.65}Ge_{0.35}$ quantum wells placed inside the intrinsic region of a p–i–n diode. Figure 28.3b shows the PL results at 4 K and 77 K. The peaks near 896 and 842 meV in the 4.2 K spectrum were identified as the NP and the TO replica, respectively. The 77 K spectrum has a clear NP peak at 924 meV. The EL spectrum ($I = 10$ mA, 400 Hz modulation, 50% duty cycle) at a heat sink temperature of 80 K is also shown. Based on the similarity of the EL and PL spectra, we conclude that the physical mechanisms responsible for recombination are transitions of electrons directly from conduction band to valence band, not involving deep levels. It is also important to note that the peak emitted from the NP line is at 1.34 μm. Figure 28.3c shows room temperature EL, which is insensitive to temperature from 77 K to room temperature. Although some emissions due to the TO replica from recombination in the Si layer are observed, well over 70% of the emitted spectrum is from the $Si_{0.65}Ge_{0.35}$. Internal quantum efficiency was estimated to have a lower limit of 2×10^{-4} at room temperature [25].

We now discuss the fabrication and measured results of LEDs based on Ge quantum dots. Ge QDs can be grown on Si substrates using both solid source and gas source MBE as well as LPCVD. The dots are formed via the Stranski–Krastanov mode, beginning with layer-by-layer growth followed by island formation. The key driving force of the dot formation is the 4.16% lattice mismatch between Si and Ge. At the beginning of the growth of Ge on Si, misfit strain is built up but is fully accommodated. Once the Ge film thickness exceeds its critical thickness of a few (three to four) monolayers, the strain starts to relax by forming three-dimensional pyramidal islands, and the surface of the layer becomes rougher. Those small islands may evolve into large domes upon subsequent Ge deposition. When the Ge thickness exceeds 20 Å or so, misfit dislocations form to relax the additional strain arising from the accumulation of the film thickness [30]. The areal dot density is typically on the order of 10^8 to 10^{10} cm^{-2}. Some groups reported dots density as high as 10^{11} cm^{-1} by carbon deposition prior to Ge growth [31,32]. Usually dots have variable shapes, i.e., pyramid, dome, and superdome. The dots' height ranges from 3 to 10 nm, and the dot base ranges from 25 to 100 nm. With increasing growth temperature, there is significant interdiffusion between the grown Si and Ge. Thus, the dots are not pure Ge and the content inside the dot is not uniform. Investigations indicate that the dots have a Ge-rich core with less Ge content in the outer side [33]. The island formation results in partial relaxation, thus there is some residual strain inside the dot. The bottom of the dot is compressively strained while the top is more relaxed. These dot properties can be controlled by modifying the growth conditions, including substrate temperature, growth rate, and doping. After the formation of Ge dots, there are three monolayers of Ge remaining on the Si surface in addition to the dot locations, which is called the wetting layer. Multilayers of Ge dots are often grown to enhance the optical properties, which can be achieved by repeatedly growing a layer of Ge dot covered by a Si spacer layer. It is well known that the subsequent Ge quantum dot is preferentially nucleated on top of the underlying one when the Si space layer is thin [34,35].

Electroluminescence studies were carried on p–i–n or p–n LED structures; the substrates can be either p-type or n-type. The Ge dots are grown on the intrinsic or lightly doped region. Holes and electrons are injected from the p- and n-type contact Si layers, and recombination occurs in the center region. Since holes can be confined in Ge dots very easily due to their large effective mass and the large valence band offset, at low temperatures, the luminescence from Ge dots is significant. The transition is similar to the PL transition. That is, holes from the dots and electrons from the surrounding Si layers are within the electron de Broglie wavelength. This is also due to the type-II band alignment between the two materials obtained by these growth methods. Electrons cannot stay in Ge dots since within them the energy is higher than in Si.

Photoluminescence can be used to determine the transition energy within such Ge dot materials. With a high-energy excitation, the photogenerated carriers relax to their ground states and recombine. Figure 28.4 shows the low-temperature PL spectrum at 4.5 K for a typical dot sample, which consists of ten-period Ge dot (1.6 nm)/Si (50 nm) superlattices [36]. The broad PL peak located around 0.83 eV was attributed to the PL from the island and can be decomposed into two Gaussian line-shaped peaks at 824 and 866 meV, respectively. They were attributed to the NP transition and TO replica of the Ge islands. The two peaks located at 1007 and 949 meV are attributed to the NP_{WL} transition and its TO_{WL} phonon-assisted transitions of upper pseudomorphic wetting layers. The inset shows the possible mechanism for the NP peak arising from the Ge quantum dots. The Ge–Si quantum dot system has been shown to have

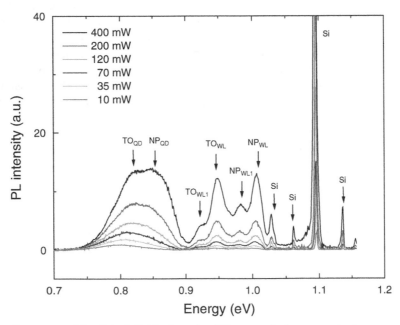

FIGURE 28.4 PL spectra of Ge–Si (0 0 1) islands under different excitation power levels measured at 4.5 K. In addition to the PL peaks from Si, there are two separate components, which come from the islands and the wetting layer, respectively. The TO and NP PL lines originating from the first wetting layer and the upper wetting layer are indicated by TO_{WL1}, NP_{WL1}, TO_{WL}, and NP_{WL}, respectively. The PL band from the islands can be deconvoluted into two Gaussian line-shaped peaks that are indicated by TO_{ID} and NP_{ID}, respectively. (After XZ Liao, J Zou, DJH Cockayne, J Wan, ZM Jiang, G Jin, and KL Wang. *Phys Rev B* 65:153306–153309, 2002. With permission.)

a type-II band alignment. The radiative recombination occurs between electrons in the Si layers and holes confined in the Ge dots, giving rise to the Ge quantum dot peak in the PL spectrum. The TO phonon-assisted peak is about 50 meV lower in energy. The enhanced no-phonon transition is mainly a consequence of relaxed momentum conservation due to alloy-induced disorder. In addition to the Ge peaks, peaks at 1.153, 1.095, 1.061, and 1.027 eV originate from Si and correspond to nonphonon (NP) replica, transverse acoustic (TA), TO, 2TA+TO, and TO+OΓ peaks, respectively.

For the Ge dots grown in high density with carbon predeposition, the dot size can be reduced to 10 nm in diameter and approximately 2 nm in height [31]. Due to the much smaller dot height, there is a blueshift of PL spectrum compared to the large Ge islands. It is assumed that for this kind of dot, electrons are confined in the underlying carbon-rich layer, while the heavy holes are in the Ge-rich upper part of the Ge islands. This band alignment is obtained from the well-known situation in neighboring $Si_{1-x}Ge_x/Si_{1-y}C_y$ quantum wells, so the recombination is spatially indirect between the two regions. However, a temperature-dependence study showed that the thermal activation energy is only ~30 meV, due to much stronger confinement in the small size dot. The luminescence was quenched at 50 K. This will hinder the application of LEDs with this kind of dot at room temperature.

For fully strained Ge on Si, there is a maximum band offset of 700 meV in the valence band. From PL measurement, however, we can only obtain Ge dots with energy ~400 meV below the Si bandgap. This is believed to be mainly due to Si alloying into the islands and quantum confinement effect. In addition, the Ge island is not fully strained, so part of the strain is also transferred into the surrounding Si matrix, further reducing the valence band offset.

The EL peak corresponds to the energy difference between the Si conduction band edge and the ground state in the Ge dots. At low injection, recombination from those high excited states of holes in the Ge dots is unlikely since the relaxation of holes is very fast. Since heavy holes have a large effective mass, and the dot lateral dimension is usually big, in the order of 50 to 100 nm, the quantized energy-level separations in Ge dots are very small, only several meV. This is much smaller than the optical phonon in Ge, which is 34 meV. Thus, holes can efficiently relax to ground or low excited states via optical and acoustic phonons. EL spectra are usually broader than kT (k is the Boltzmann constant), which is ascribed to the nonuniformity of the dot size as well as their Ge content. Chretien et al. [37] studied the spectrum dependence upon injection current density. It was discovered that with increasing the current density from 60 to 6000 A/cm², there is significant blueshift of the peak. This shift is due to a band-filling effect. They also grew Ge dots on different sizes Si mesas. At low injection current density (60 A/cm²), the Ge dot peak experiences a blueshift with increasing the mesa size (lateral size from 5 to 500 μm).

Generally speaking, in the recombination process, energy and momentum conservation should both be observed. But for Ge dots, the height is usually in the range of 6 to 10 nm, so the holes are strongly confined in the growth direction. This strong confinement will relax the requirement of the momentum conservation. It is also suggested that the alloy effect of Si and Ge will loosen this conservation requirement. These effects improve the luminescence efficiency of Ge dots. The integrated intensity from dots is much higher than that from Si at low temperatures. However, the efficiency is still orders of magnitude lower than that of III–V materials. Chang et al. [12] measured the external quantum efficiency at room temperature, 4.6×10^{-6} was obtained at injection current density of 65 A/cm², and the estimated internal efficiency was 1.5×10^{-4}. This low efficiency is mainly due to the long radiative lifetime and poor carrier confinement as well as the presence of nonradiative centers such as defects in the active region.

EL spectra from Ge dot samples change with measurement temperature. Shown in Figure 28.5 are the EL spectra at different temperatures [11]. Typically, at low temperature, the luminescence from Ge is weak while that from Si is strong. With increasing temperature, the Ge peak increases and the Si peak quenches. The EL is relatively constant up to 225 K and then starts to decrease after this temperature. This is because that the confined holes can be thermally excited out of the Ge dot wells at high temperature. Through measured EL quenching characteristics, an activation energy of 230 meV was estimated, which corresponds to the effective barrier height for holes in Ge dots. The decrease of

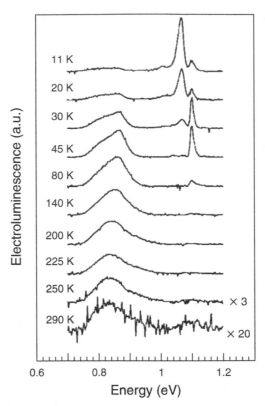

FIGURE 28.5 Temperature dependence of the electroluminescence of five layers of Ge QDs separated by 35-nm thick Si barrier layers. The sample was prepared by a lamp-heated single wafer chemical vapor deposition reactor. The dot density is around 1.7×10^{10} cm^{-2}. The quantum dots of a single Ge layer grown in the same conditions have a typical base size of 50 nm and a height of 4 nm. (After T Brunhes, P Boucaud, S Sauvage, F Aniel, JM Lourtioz, C Hernandez, Y Campidelli, O Kermarrec, D Bensahel, G Faini, and I Sagnes. *Appl Phys Lett* 77:1822–1824, 2000. With permission.)

luminescence efficiency at room temperature limits the application of such LEDs. However, there is a report claiming that the integrated intensity of Ge peak remains the same at room temperature [12]. They attributed the enhanced efficiency to the low nonradiative recombination center density that is caused by the surface passivation and thermal treatment in the processing.

The current-dependent EL intensity ($L \sim J^m$) reveals the competition of radiative and nonradiative processes in the active region. Figure 28.6 shows the dependence of integrated EL intensity (I) on current density (j) [8]. A linear dependence was shown at low temperatures and low injection, indicating that most of the injected carriers are radiatively recombined. At higher injection current density ($>$20 A/ cm^2), $m = 0.67$, which implies Auger processes ($m = 2/3$). At room temperature, $m = 1.3$ and 1 for low ($<$20 A/cm^2) and high ($>$20 A/cm^2) injection, respectively. The superlinear dependence indicates that a Shockley–Read–Hall (SHR) process is important due to the increased capture probability of nonradiative centers. The linear dependence at high injection suggests that the nonradiative traps are saturated. Talalaev et al. [38] studied this $L \sim J^m$ dependence and found that m to be 4.8 and 3.1 for $\sim$1 and $\sim$2 A/cm^2 current densities, respectively. The variation of m was interpreted as a band-bending change, the redistribution of the electrons within the near-contact region and progressive filling of the electron miniband with an increase of the forward bias. A comparison of the PL and EL intensities of the QD-related peak demonstrates a higher efficiency for an electrical excitation.

One group also tried to confine electrons near the Ge dots by decreasing the thickness of Si spacer layer to 13 nm [10]. This is to produce a combination of self-assembled Ge dots with tensile-strained Si close to the center of dots, which should have provided electron localization and the overlap of electron

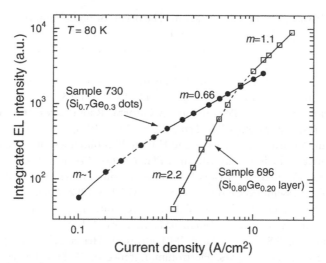

FIGURE 28.6 Dependence of integrated EL intensity *I* on current density *j* for Ge QDs (sample 730) at 80 K. Result of Si$_{0.8}$Ge$_{0.2}$ (sample 696) is also shown. (After R Apetz, L Vescan, A Hartmann, C Dieker, and H Luth. *Appl Phys Lett* 66:445–447, 1995. With permission.)

and hole wave functions. With this modification, a significantly enhanced Ge dot-related PL signal up to room temperature at 1.55 mm wavelength was obtained.

28.3 Summary

Despite great efforts contributing to the development of SiGe-based LEDs, there are still two major problems remaining. The first one is the low efficiency. As we have seen, the external quantum efficiency is in the order of 10^{-4} to 10^{-6}. This is mainly due to the long lifetime of radiative recombination process in the material. Although it is improved with the aforementioned methods, the low efficiency still remains the biggest obstacle for such LED devices. The other problem is the luminescence quenching at room temperature. For most of the studies to date, the EL intensity drops at higher temperatures (e.g., >250 K). Relatively shallow carrier confinement is insufficient for devices operating at room temperature.

In the future, for achieving high-performance LED based on SiGe materials, carrier confinement has to be improved. This may be achieved by a different design of the structure. The formation of very small (1 to 2 nm in size) dots is also a possible approach because the confinement can be drastically enhanced, thus increasing the recombination rate [39]. In device processing, roughing the surface to increase the coupling-out of the emission from inside of the active region to the outer surface may also be important, since Si and Ge have large refractive indexes, so the angle for the light to emit from the device is small ($\sim$17° for Si). By roughing the surface, photons have more chances to escape from the body. Since the emission spectra are very broad from Ge dots, it may also help to apply a Febry–Perrot resonant cavity to select the wavelength of interest, thus prohibiting those unwanted emissions. This should increase the efficiency at selected wavelengths.

References

1. JI Pankove. *Optical Processes in Semiconductors*, Chapter 6. Englewood Cliffs, NJ, Prentice-Hall, 1971.
2. LT Canham. Silicon quantum wire array fabrication by electrochemical and chemical dissolution of wafers. *Appl Phys Lett* 57:1046–1050, 1990.
3. DJ DiMaria, JR Kirtley, EJ Pakulis, DW Dong, TS Kuan, FL Pesavento, TN Theis, JA Cutro, and SD Brorson. Electroluminescence studies in silicon dioxide films containing tiny silicon islands. *J Appl Phys* 56:401–416, 1984.

4. AJ Kenyon, PF Trwoga, CW Pitt, and G Rehm. The origin of photoluminescence from thin films of silicon-rich silica. *J Appl Phys* 79:9291–9300, 1996.

5. S Tong, XN Liu, and XM Bao. Study of photoluminescence in nanocrystalline silicon/amorphous silicon multilayers. *Appl Phys Lett* 66:469–471, 1995.

6. H Ennen, J Schneider, G Pomrenke, and A Axmann. 1.54-μm luminescence of erbium-implanted III–V semiconductors and silicon. *Appl Phys Lett* 43:943–945, 1983.

7. G Franzo, F Priolo, S Coffa, A Polman, and A Carnera. Room-temperature electroluminescence from Er-doped crystalline Si. *Appl Phys Lett* 64:2235–2237, 1994.

8. R Apetz, L Vescan, A Hartmann, C Dieker, and H Luth. Photoluminescence and electroluminescence of SiGe dots fabricated by island growth. *Appl Phys Lett* 66:445–447, 1995.

9. L Vescan, T Stoica, O Chretien, M Goryll, E Mateeva, and A Muck. Size distribution and electroluminescence of self-assembled Ge dots. *J Phys Lett* 87:7275–7282, 2000.

10. K Eberl, OG Schmidt, O Kienzle, and F Ernst. Preparation and optical properties of Ge can C-induced Ge quantum dots on Si. *Thin Solid Films* 373:164–169, 2000.

11. T Brunhes, P Boucaud, S Sauvage, F Aniel, JM Lourtioz, C Hernandez, Y Campidelli, O Kermarrec, D Bensahel, G Faini, and I Sagnes. Electroluminescence of Ge/Si self-assembled quantum dots grown by chemical vapor deposition. *Appl Phys Lett* 77:1822–1824, 2000.

12. WH Chang, AT Chou, WY Chen, HS Chang, TM Hsu, Z Pei, PS Chen, SW Lee, LS Lai, SC Lu, and MJ Tsai. Room-temperature electroluminescence at 1.3 and 1.5 μm from Ge/Si self-assembled quantum dots. *Appl Phys Lett* 83:2958–2960, 2003.

13. U Gnutzmann and K Clausecker. Theory of direct optical transitions in an optical indirect semiconductor with a superlattice structure. *Appl Phys* 3:9–14, 1974.

14. RJ Turton, M. Jaros, and I Morrison. Electronic band structure and nonparabolicity in strained-layer $Si–Si_{1-x}Ge_x$ superlattices. *Phys Rev B* 38:8397–8405, 1988.

15. SJ Chang, CF Huang, MA Kallel, KL Wang, RC Bowman Jr, and PM Adama. Growth and characterization of Ge/Si strained-layer superlattices. *Appl Phys Lett* 53:1835–1837, 1988.

16. T Pearsall, JM Vandenberg, R. Hull, and JM Bonar. Structure and optical properties of strained Ge–Si superlattices grown on (0 0 1) Ge. *Phys Rev Lett* 63:2104–2107, 1987.

17. R Zachai, K Eberl, G Abstreiter, E Kasper, and H Kibbel. Photoluminescence in short-period Si/Ge strained-layer superlattices. *Phys Rev Lett* 64:1055–1058, 1990.

18. MA Kallel, VA Engels, KL Wang, and RPG Karunasiri. MBE Si_mGe_n strained monelayer superlattices. *J Cryst Growth* 111:897–901, 1991.

19. J Engvall, J Olajos, HG Grimmeiss, H Presting, H Kibbel, and E Kasper. Electroluminescence at room temperature of a Si_nGe_m strained-layer superlattice. *Appl Phys Lett* 63:491–493, 1993.

20. J Olajos, J Engvall, HG Grimmeiss, H Presting, H Kibbel, and E Kasper. Band-to-band transitions in strain-symmetrized, short-period Si/Ge superlattices. *Thin Solid Films* 222:243–245, 1992.

21. U Menczigar, J Brunner, E Friess, M Gail, G Abstreiter, H Kibbel, H Presting, and E Kasper. Photoluminescence studies of $Si/Si_{1-x}Ge_x$ quantum wells and Si_mGe_n superlattices. *Thin Solid Films* 222:227–233, 1992.

22. E Kasper, H Kibbel, H Jorke, H Brugger, E Friess, and F Abstreiter. Symmetrically strained Si/Ge superlattices on Si substrates. *Phys Rev B* 38:3599–3601, 1988.

23. U Menczigar, G Abstreiter, J Olajos, H Grimmeiss, H Kibbel, H Presting, and E Kasper. Enhanced bandgap luminescence in strain-symmetrized $(Si)m/(Ge)n$ superlattices. *Phys Rev B* 47:4099–4102, 1993.

24. J Olajos, J Engvalt, HG Grimmeiss, U Menczigar, M Gail, G Abstreiter, H Kibbel, E Kasper, and H Presting. Photo- and electroluminescence in short-period Si/Ge superlattice structures. *Semicond Sci Technol* 9:2011–2016, 1994.

25. Q Mi, X Xiao, JC Sturm, LC Lenchyshyn, and MLW Thewalt. Room-temperature 1.3 μm electroluminescence from strained $Si_{1-x}Ge_x/Si$ quantum wells. *Appl Phys Lett* 60: 3177–3129, 1992.

26. J Weber and MI Alonso. Near-band-gap photoluminescence of Si–Ge alloys. *Phys Rev B* 40:5683–5693, 1989.

27. JP Noel, NL Rowell, DC Houghton, and DD Perovic. Intense photoluminescence between 1.3 and 1.8 μm from strained $Si_{1-x}Ge_x$ alloys. *Appl Phys Lett* 57:1037–1039, 1990.

28. K Terashima, M Tajima, and T Tatsumi. Near-band-gap photoluminescence of $Si_{1-x}Ge_x$ alloys grown on Si (1 0 0) by molecular beam epitaxy. *Appl Phys Lett* 57:1925–1927, 1990.

29. DJ Robbins, P Calcott, and WY Leong. Electroluminescence from a pseudomorphic $Si_{0.8}Ge_{0.2}$ alloy. *Appl Phys Lett* 59:1350–1352, 1991.

30. DJ Eaglesham and M Cerullo. Dislocation-free Stranski–Krastanow growth of Ge on Si (1 0 0). *Phys Rev Lett* 64:1943–1946, 1990.

31. OG Schmidt, C Lange, K Eberl, O Kienzle, and F Ernst. Formation of carbon-induced germanium dots. *Appl Phys Lett* 71:2340–2342, 1997.

32. AI Yakimov, AV Dvurechenskii, YY Proskuryakov, AI Nikiforov, OP Pchelyakov, SA Teys, and AK Gutakovskii. Normal-incidence infrared photoconductivity in Si p–i–n diode with embedded Ge self-assembled quantum dots. *Appl Phys Lett* 75:1413–1415, 1999.

33. XZ Liao, J Zou, DJH Cockayne, J Wan, ZM Jiang, G Jin, and KL Wang. Alloying, elemental enrichment, and interdiffusion during the growth of Ge(Si)/Si (0 0 1) quantum dots. *Phys Rev B* 65:153306–153309, 2002.

34. GS Solomon, JA Trezza, AF Marshall, and JS Harris Jr. Vertically aligned and electronically coupled growth induced InAs islands in GaAs. *Phys Rev Lett* 76:952–955, 1996.

35. Y Sugiyama, Y Nakata, K Imamura, S Muto, and N Yokoyama. InAs self-assembled quantum dots on (0 0 1) GaAs grown by molecular beam epitaxy. *Jpn J Appl Phys* 35:1320–1324, 1996.

36. J Wan, GL Jin, ZM Jiang, YH Luo, JL Liu, and KL Wang. Band alignments and photon-induced carrier transfer from wetting layers to Ge islands grown on Si (0 0 1). *Appl Phys Lett* 78:1763–1765, 2001.

37. O Chretien, T Stoica, D Dentel, E Mateeva, and L Vescan. Influence of the mesa size on Ge island electroluminescence properties. *Semicond Sci Technol* 15:920–925, 2000.

38. VG Talalaev, GE Cirlin1, AA Tonkikh, ND Zakharov, and P Werner. Room temperature electroluminescence from Ge/Si quantum dots superlattice close to 1.6 μm. *Phys Stat Sol (a)* 198:R4–R6, 2003.

39. MS Hybertsen. Absorption and emission of light in nanoscale silicon structures. *Phys Rev Lett* 72:1514–1517, 1994.

29

Near-Infrared Detectors

Lorenzo Colace,
Gianlorenzo Masini, and
Gaetano Assanto
University "Roma Tre"

29.1 Introduction

The term "near-infrared" or NIR is most commonly used with reference to a wavelength spectral range between 0.7 and 2.0 μm. This portion of infrared is of paramount importance in several applications, first among them are the optical communications with transmission windows located at 0.85, 1.3, and 1.55 μm, corresponding to GaAs-based laser emission and to two minima in attenuation for standard silica fibers, respectively. Moreover, wavelength division multiplexing for high-capacity links encourages to using the whole interval between 1.3 and 1.6 μm (S, C, and L bands). Due to the growing demand for wideband internet and massive data transmission, applications have shifted from long-haul point-to-point connections to local networks down to subscribers, softening the specifications and opening entirely new markets [1]. In addition to communications, NIR spectroscopy is employed in remote sensing of the environment, monitoring of industrial processes, biology, and medicine. For example, water has absorption lines that allow to detect its content in the flora for fire prevention [2], various gas species exhibit NIR absorption bands useful for emission or toxicity analysis [3]. In addition, NIR spectroscopy has been exploited for DNA sequencing [4], brain activity mapping [5], and cancer detection [6]. However, since optical fiber communications remain the main field driving research in NIR detectors, this chapter will focus on NIR detectors on silicon from the receiver standpoint.

Among semiconductor materials, SiGe alloys have allowed to fabricate novel, high performance electronic devices such as heterojunction bipolar and field-effect transistors [7], opening new perspectives also in NIR optoelectronics by exploiting the bandgap of germanium (0.66 eV), much lower than that of silicon (1.12 eV), which is otherwise useless for detection at these wavelengths. Numerous NIR devices have been proposed, from emitters to waveguides, couplers, modulators, and detectors [8]. The centrosymmetry and the indirect nature of Si, Ge, and SiGe pose important limitations to SiGe devices as compared to III–Vs; nevertheless, exploiting a few structurally related modifications such as quantum confinement, acceptable performances have been foreseen, with the significant advantage of the compatibility with the unsurpassed silicon VLSI technology. The key to the success of SiGe-based optoelectronics is the ability to conveniently compromise between material performance and monolithic integration. The fabrication of embedded optoelectronics and electronics in the same process does considerably reduce alignment and interconnection problems, offering improved reliability, yield, compactness, and reduced parasitics. Conversely, monolithic integration with SiGe requires heteroepitaxy, which is known to be critical for lattice-mismatched materials such as Si and Ge, with mismatch of about 4%. In the specific case of photodetectors, material quality requirements are less stringent than in other devices, thus encouraging toward the realization of innovative and competitive devices.

This chapter is organized as follows. After a short introduction on photodetector operation and figures of merit in Section 29.2, Section 29.3 deals with the SiGe technology for photodiodes, focusing on the quality of both strained and relaxed epilayers, their optical and electrical characteristics and their bearing in device performances. Design strategies and fabrication are discussed in Section 29.4 along with an overview of the most relevant approaches that were proven successful to date.

29.2 NIR Photodetectors

Near-infrared light detection in semiconductors is obtained by the creation of electron–hole pairs through the absorption of photons with energy greater than the bandgap. Germanium, with its gap of 0.66 eV, allows to reveal wavelengths as large as 1.88 μm. A photocurrent is generated when a built-in or applied electric field sweeps the carriers toward a couple of contacts (anode and cathode). Photodetectors can be subdivided into three categories: photomultipliers, photoconductors, and photodiodes, the latter being p–n or p–i–n junctions, avalanche and metal–semiconductor–metal (MSM) diodes. The pin photodiode is by far the most commonly used in optical receivers and we will focus on it, keeping in mind that a large portion of the following can be extrapolated to other devices.

The pin photodiode consists of an intrinsic semiconductor (active layer) sandwiched between two heavily doped p$^+$ and n$^+$ regions (contacts). One of the most important figures related to the light–current conversion is the responsivity R, defined as the photocurrent flowing per incident watt of radiation and expressed by

$$R = \frac{\lambda}{1.24}(1 - \Theta_R)\eta_{\text{int}}(1 - e^{-\alpha W}) \tag{29.1}$$

where λ is the vacuum wavelength in μm, Θ_R the reflectivity, η_{int} (internal quantum efficiency) the number of collected electron–hole pairs per incident photon, α the optical absorption, and W the active layer thickness. The maximum theoretical responsivity of $\lambda/1.24$ can be approached with suitable antireflection coatings to minimize Θ_R, good electronic quality to grant a drift length large enough to prevent recombination and a large product αW to maximize absorption efficiency.

The speed of the photoresponse depends on three main factors: the finite time needed by carriers to reach the contacts, the RC time-constant, and the diffusion time of the photogenerated electron–hole pairs outside the intrinsic layer. The transit time can be approximately written as

$$t_{\text{tr}} = W/\mu E \tag{29.1a}$$

where μ is the carrier mobility and E the electric field. At large enough fields (10^4 V/cm), velocity saturation occurs and (29.1a) becomes

$$t_{\text{tr}} = W/v_{\text{sat}} \tag{29.1b}$$

i.e., transit time limitations can be reduced by decreasing W.

In the RC time constant above, R is the combination of diode series-resistance and load, while C is the junction capacitance with the addition of parasitics. The junction capacitance can be expressed as

$$C_j = \frac{\varepsilon A}{W} \tag{29.2}$$

where ε is the dielectric constant of the absorbing layer and A the area of the detector. Balancing the opposite effects of W on transit time and capacitance, an optimization can be pursued. However, a conventional high-speed photodiode is commonly designed to be small enough (area) to be transit-time limited, while W is set by a compromise between responsivity and speed, usually leading to W comprised between $1/\alpha$ and $2/\alpha$. More device-oriented speed parameters are the -3 dB bandwidth $f_{3\,dB}$ and the full width at half maximum (FWHM) of the pulse photoresponse, with $2f_{3\,dB} = 1/\text{FWHM}$.

The two main sources of noise are dark current and quantum noise, both regarded as shot noise. Neglecting the background radiation, the total rms shot noise current is

$$\overline{i_s^2} = 2qB(I_p + I_d) \tag{29.3}$$

where q is the electron charge, B the bandwidth, and I_p and I_d photocurrent and dark current, respectively. The pertinent figure of merit is the noise equivalent power (NEP), defined as the minimum detectable rms optical power corresponding to a unity signal-to-noise ratio. The NEP is inversely proportional to the responsivity and scales with the rms shot noise in expression (29.3) above. In order to reduce both the transit time and the diffusion capacitance while, sometimes, improving their linearity, photodiodes are often operated at high reverse bias, thereby increasing the dark current and affecting negatively their noise performance.

29.3 SiGe: A Material for NIR Detection

In this section, we discuss the SiGe material issues drawing specific attention to near-infrared light detection. Although Si and Ge have the same crystal structure, the growth of SiGe epitaxial layers on Si is affected by a large lattice mismatch (up to 4.2% for pure Ge on Si). At the deposition start-off, the initial SiGe layers adjust their lattice through compression (in the growth plane) and tensile strain (along the normal), while the substrate remains substantially undistorted. As the growth proceeds, however, the large strain is abruptly released and Ge recovers its own lattice spacing. This process, referred to as *relaxation*, takes place once the strain reaches a threshold corresponding to a critical thickness h_c, and it is usually associated with the generation of a large amount of defects both in the growth plane (misfit dislocations) and perpendicularly to it (threading dislocations). The critical layer thickness h_c as a function of Ge-concentration is graphed in Figure 29.1 (solid line) according to Bean's model [9]. Results obtained with more conservative models and accounting for strain stability over temperature cycles typical of silicon processes are represented by a dashed line. The figure emphasizes the main problem in strained-layer epitaxy: high-quality epilayers of suitable thickness for electronic devices must have a low Ge content; on the other hand, if Ge-rich alloys are needed, one has to deal with a relaxed material. For a long time, relaxed epilayers were not considered suitable for devices, but they were recently reconsidered taking into account the tolerances to defects in different applications [10]. Defects, such as dislocations, affect the electrical properties of semiconductors in two main ways: first, they act as scattering centers thus impairing free carriers mobility; second, they introduce levels in the forbidden gap, which can serve as recombination centers or carrier traps. Majority-carrier devices (such as FET), when working in the active region, are less sensitive to carrier lifetime than minority-carrier devices (e.g., HBT, BJT) and, therefore, tolerate higher defect densities. Photogenerated carriers, in properly designed and biased p–i–n devices, travel at the saturation velocity and are collected by drift: for this reason, short lifetimes and impaired mobilities are still acceptable. The only concern with defects-induced levels in the gap relates to dark current, which becomes governed by Shockley–Read–Hall generation. In Equation 29.1 the factor $(1 - e^{-\alpha W})$ is the fraction of light absorbed in the intrinsic region of the p–i–n photodiode and approaches unity when W exceeds the absorption length. In Figure 29.2 we display the absorption coefficients at 1.3 and 1.55 μm, respectively, versus Ge-content for both unstrained (circles) and strained (lines) $Si_{1-x}Ge_x$ alloys. Absorption coefficients of unstrained alloys are after Potter [11] and Braunstein et al. [12] while strained-alloy data were calculated by Naval et al. [13]. The considerably

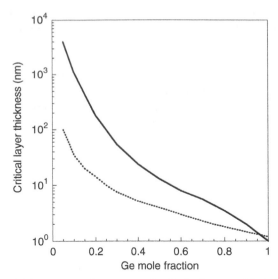

FIGURE 29.1 Critical layer thickness versus Ge-concentration for both metastable (solid line) and unconditionally stable (dashed line) cases.

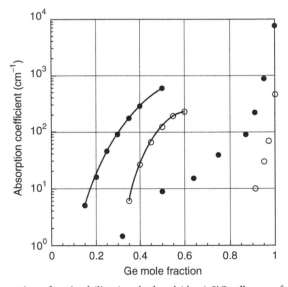

FIGURE 29.2 Optical absorption of strained (lines) and relaxed (dots) SiGe alloy as a function of Ge-content. The former are calculated, the latter are measured. Both 1.3 and 1.55 μm data are shown with filled and empty circles, respectively.

larger values of the latter are expected as a consequence of the strain-related bandgap shrinkage, as predicted by People [14] and experimentally demonstrated by Lang et al. [15].

From Figure 29.1 and 29.2 it is apparent that, for each composition of either strained or unstrained SiGe, the thickness required to obtain adequate absorption efficiencies is much larger than the corresponding critical thickness. For example, for $x = 0.2$ at 1.3 μm, the absorption is 20 cm^1 (strained), requiring more than 1 mm to absorb 90% of the light while the critical thickness is only 200 nm.

The limitation described above suggested alternative strategies for fabricating sensitive and fast photodetectors: (i) the use of strained, high-quality SiGe alloys with low Ge-content to compensate the low responsivity with excellent noise performances, thanks to good transport properties and low dark currents; (ii) the use of relaxed pure-Ge layers to exploit the high absorption of the pure material at the cost of larger noise. In the case (i), materials can be obtained with a crystal quality comparable with the best achievable in conventional III–V heterostructure devices [9]. In strained SiGe layers the formation of dislocations is suppressed by keeping thicknesses below the critical value. The relevant drawback remains the previously mentioned low absorption. The adoption of multiple heterostructures or different geometries (guided-wave detectors) to overcome such limitation will be addressed in Section 29.4. Since a significant portion of this handbook is devoted to strained-layer epitaxy, we will not further discuss it, but turn to the less known relaxed material. In this case (ii), large αW products can be obtained through high Ge concentration and thick layers, but at the expense of a poorer crystal quality. The competition between relaxed (ii) and strained (i) approaches rests on the development of either a defect engineering strategy, in order to lower the defect density to an acceptable level, or an advanced design strategy.

In general, due to lattice mismatch, the growth of Ge on Si is expected to be two-dimensional for the first few monolayers, after which islands are formed [16]. Islanding during epitaxy (or three-dimensional growth), unless induced to exploit the quantum properties of nanostructures, is detrimental because it results in a nonflat, inhomogeneous film. Therefore, large efforts have been devoted to identify growth methods and parameters aiming at the release of strain (rather than island formation) through a controlled insertion of dislocations.

The growth of Ge-rich epilayers above critical thickness has been attempted by several groups. Early heteroepitaxial growths by both chemical vapor deposition (CVD) [17] and physical vapor deposition (PVD) were carried out at a pressure of 10^{-10} Torr on heated substrates: they exhibited similar characteristics with dislocation densities around $10^9\,cm^{-2}$. Pure-Ge single crystals were also obtained by evaporation at 10^{-6} Torr from a boron-nitride crucible at low temperatures (375°C to 425°C) [18]. Using an analogous technique Ohmachi et al. reported a remarkably high carrier mobility of 1000 cm^2/V sec [19].

One of the first MBE-grown pure-Ge epilayers was reported with a threading dislocation density between 10^7 and $10^8\,cm^{-2}$ [20], and a comparable density was obtained by Fujinaga via low-pressure CVD and thermal annealing [21]. In 1991 at IBM, for the first time, Cunningham et al. reported the UHV–CVD epitaxial growth of pure Ge on Si [22].

Despite these initial and promising results, the reported threading dislocation (TD) densities were considered too large for practical purposes, and several techniques were proposed for improving the growth: graded buffer layers, surfactants, carbon incorporation, growth on processed substrates, and low-temperature buffer layers.

Strain relaxation can be accomplished by the insertion of dislocations through a completely or partially relaxed buffer layer [23], either stepwise or linearly graded. The effectiveness of linearly graded buffers was first demonstrated by Fitzgerald et al. at AT&T Bell Laboratories [24] investigating both MBE and CVD growth techniques. They adopted a high growth temperature to completely relax the first layer and chose the grading rate in order to maintain a low strain throughout it. Their buffers with a grading rate of 10% Ge/μm up to final Ge-compositions of 23% and 50% exhibited TD densities of 4.4×10^5 and $3 \times 10^6\,cm^{-2}$, respectively. They also obtained CVD epitaxial $Si_{0.7}Ge_{0.3}$ films with TD densities between 10^5 and $10^6\,cm^{-2}$ with grading of 10% to 40% Ge/μm [25]; nevertheless, the extension to pure-Ge turned out much more difficult. Only in the mid-1990s, combining a slowly varying graded buffer (10- to 20-μm thick) and a chemi-mechanical polishing, a pure-Ge epilayer was reported with a threading dislocation density as low as $2 \times 10^6\,cm^{-2}$ using CVD [26].

The use of surfactants has been proposed to reduce the surface-free energies of both substrate and epilayer and thereby ease two-dimensional growth in the presence of a large mismatch. A remarkable result was obtained by Liu et al. with MBE by introducing a single Sb monolayer before epitaxy [27]. They fabricated a graded buffer with up to 50% Ge, followed by a 0.3-μm thick $Si_{0.5}Ge_{0.5}$ cap-layer, with a TD density of $1.5 \times 10^4\,cm^{-2}$, two orders of magnitude lower than without Sb. H_2 has also been proposed as a surfactant [28], and its effectiveness demonstrated in the growth of thick Ge films [29]. Despite their beneficial effects, however, surfactants are often prone to the drawback of residual doping [30].

Another attempt to grow SiGe epilayers above critical thickness consists in the incorporation of carbon atoms to form SiGeC compounds. The strain can be compensated due to the smaller lattice parameter of C with respect to Si. The viability of this technique was first demonstrated by Eberl et al. in the growth of SiGeC epilayers with 25% Ge [31]. A TD density of $10^5 \, \text{cm}^{-2}$ was obtained in a 1-μm-thick relaxed step-graded buffer (up to 30% Ge) in a combination of $Si_{1-x}Ge_x$ and $Si_{1-x-y}Ge_x C_y$ [32]. Unfortunately, the C dose is limited to about 4% by the low solubility of C in Si and by SiC precipitation, preventing its use in Ge-rich epilayers. Moreover, as C is added to a SiGe alloy keeping constant the Ge concentration, the bandgap is expected to increase, partially thwarting the effect of Ge [33].

Improved crystals can be obtained if misfit dislocations, rather than pinned and swept across the epilayer, are terminated on a free surface without affecting the film. Since this depends on the sample size, a decrease in TD density can be expected when the area is reduced, or the substrate is processed in order to obtain selective epitaxy. The effectiveness of this approach was proved by growing defect-free $Si_{0.9}Ge_{0.1}$ layers up to 3 μm on silicon oxide patterned Si [34] and high-quality pure-Ge on Si patterned through interferometric lithography [35].

The remarkable results obtained by graded buffers can be ascribed to strain release by means of the insertion of dislocations, which are expected not to propagate in the epilayer. A simpler way to achieve the same is the use of low-temperature Ge-buffer layers. The substrate temperature is crucial in Ge-rich epitaxy on Si: at low temperature (LT) the growth mode can be layer-by-layer, but above a certain temperature it turns into three dimensional (e.g., islands). In addition, if the surface is hydrogenated (as in CVD at low temperatures), the anisotropy of the surface energy is reduced, thus promoting a layer-by-layer growth (anisotropy is the main agent in island formation). This can be accomplished by keeping the substrate at a temperature T_{sub} below the desorption value T_{des} of H from Ge [36].

Initially, only silicon LT-buffers were considered. Chen et al. reported a TD density of about $10^6 \, \text{cm}^{-2}$ in a 500 nm thick $Si_{0.7}Ge_{0.3}$ MBE grown on a 200 nm LT Si buffer [37]. A slightly lower TD density ($10^5 \, \text{cm}^{-2}$) was reported by Li et al. using a thinner (50 nm) LT Si buffer before the MBE deposition of $Si_{0.7}Ge_{0.3}$ [38]. Linder et al. obtained TD densities of the order of $10^4 \, \text{cm}^{-2}$ in MBE $Si_{0.85}Ge_{0.15}$ on 100 nm LT Si buffers [39], but Si buffers were effective only for low-concentration SiGe epilayers. A combination of Si LT buffer and step-graded buffer was employed in $Si_{0.1}Ge_{0.9}$ on Si with a TD density of $3 \times 10^6 \, \text{cm}^{-2}$ [40].

A new approach, based on an LT Ge buffer, has been recently proposed for the epitaxy of pure Ge on silicon via UHV-CVD [41]. At the beginning, the substrate is kept at low temperature in order to grow a flat and relaxed epitaxial layer of thickness suitable for total strain relaxation. The temperature is chosen in order to prevent H desorption from the Ge surface, thus avoiding the nucleation of three-dimensional Ge islands due to the H surfactant. Following the LT GE buffer, the deposition can proceed as in the homoepitaxial case, thus allowing a temperature increase for a faster growth rate. The good crystal quality was assessed by RHEED spectra, and the defect distribution in Ge evaluated by TEM. As expected, most defects were confined in the buffer, although a few dislocations penetrated the Ge epilayer up to the free surface.

Luan et al. proposed a method to considerably reduce residual threading dislocations based on postgrowth thermal annealing [42]. They used an LT Ge-buffer and cyclic annealing between high and low temperatures, obtaining TD densities of $10^7 \, \text{cm}^{-2}$. The TD reduction is due to enhanced dislocation glide and annihilation, as promoted by thermal stress. Following these results, the same group demonstrated a further TD reduction by performing a similar growth on small mesas of dimensions from 10 to 100 μm. The average TD density was about $2 \times 10^6 \, \text{cm}^{-2}$ [42].

An entirely novel approach employed the granular structure of polycrystalline Ge as an alternative to solve the problem of the large strain [43]. Thin films of poly-Ge were deposited on silicon by thermal evaporation in a vacuum of 10^{-6} Torr at different substrate temperatures. Raman spectroscopy and absorption measurements showed a clear transition between amorphous and polycrystalline around $T_{sub} = 300°C$ and poly-Ge optical absorption comparable to crystalline germanium [44]. The electronic properties of poly-Ge on Si are affected by a large acceptor-like defect density, leading to short carrier

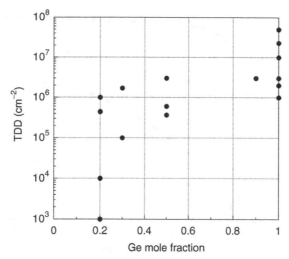

FIGURE 29.3 Threading dislocation density versus Ge content.

lifetime and severely limiting the doping. Nevertheless, the large absorption in the NIR (about $10^4 \, \mathrm{cm}^{-1}$ at 1300 nm) makes this material quite interesting for NIR photodetectors [45]. Moreover, the required $T_{sub} = 300°C$ ensures good compatibility with silicon technology with respect to both CVD and MBE heteroepitaxy, where substrate cleaning, deposition, and annealing require temperatures in the range 600°C to 1000°C.

Figure 29.3 summarizes the reported TD densities versus Ge concentration in films above critical thickness [46]. The graph confirms that, even in the relaxed regime, the larger the mismatch (Ge concentration) the more critical is the epitaxial growth. The visible spread in values for a fixed Ge-concentration depends on the method employed for defect reduction and on the typical sensitivity of heteroepitaxy to apparatus and parameters.

While TD density is a commonly accepted figure-of-merit for epitaxial films, from a device viewpoint its impact in terms of transport parameters must be investigated. Early studies on plastically deformed germanium demonstrated a linear relationship between TD and carrier lifetime, suggesting that lifetime is limited by recombination at dislocations [47]. More recently, deep-level transient spectroscopy has pointed out to a linear relationship between TD and deep-trap densities, confirming the role of dislocations in the enhancement of minority carrier recombination and generation [48]. Quantitative correlations between leakage current and TD density have been demonstrated in SiGe p–n and p–i–n diodes [49,50]. For the latter, dark currents of 1 mA/cm² were measured for TD of $10^7 \, \mathrm{cm}^{-2}$ in 0.5-μm thick $Si_{0.75}Ge_{0.25}$, with a linear increase with TD density. This behavior is expected in photodiodes with thick intrinsic layers, because the dark current is dominated by thermal generation, inversely proportional to carrier lifetime and directly proportional to the intrinsic layer width. Figure 29.3 shows that pure-Ge epilayers can be obtained with TD spanning from 10^6 to $10^8 \, \mathrm{cm}^{-2}$. Using the relationship in Ref. [50], dark currents of 0.2 to 20 mA/cm² can be expected in this density range for a 1-μm-thick intrinsic layer. Although an extrapolation of the scaling law from $Si_{0.75}Ge_{0.25}$ to pure-Ge is questionable, a dark current of 20 mA/cm² was measured on 4 μm p–i–n diodes with TD densities of $2 \times 10^7 \, \mathrm{cm}^2$, in close agreement with the predicted value of 16 mA/cm² [51].

At zero or low reverse bias, the TD-related carrier-lifetime reduction can lead to enhanced recombination in the depleted active layer of a p–i–n photodiode, thus affecting the internal quantum efficiency (η_{int}). A suitable approach to investigate this is to measure photocurrent versus bias on a metal–semiconductor–metal structure, where the semiconductor is the SiGe epilayer under test. Figure 29.4 shows the photocurrent versus bias for closely spaced (10 μm) interdigited metal–Ge–metal photodiodes with different TD densities [52]. The TD effect is clear in both the magnitude of

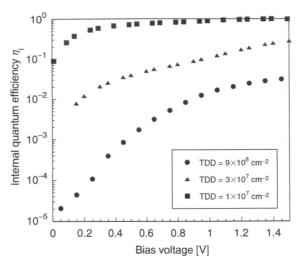

FIGURE 29.4   Internal quantum efficiency for various TD as a function of applied voltage.

η_{int} and the higher applied voltage required in lower quality epilayers. Quantitatively, the experimental data can be fit to extract the $\mu\tau$ product by

$$\eta_{\text{int}} = \frac{L(E)}{d} \left[1 - e^{-\frac{d}{L(E)}} \right] \qquad (29.4)$$

with d the interelectrode spacing and $L(E) = \mu\tau E$ the drift length, and μ, τ, and E the carrier mobility, lifetime, and electric field amplitude, respectively [53]. The possibility (demonstrated in Figure 29.4) of reaching total internal quantum efficiency at very low voltage bias for TDD of 10^7 cm^2 is quite remarkable and could help as a reference for the evaluation of the acceptable TDD for photodetector application. As will become clear in the next paragraph, the dark current and the internal quantum efficiency can be regarded as important factors in the evaluation of a certain SiGe technology for NIR applications.

In this section, after addressing the material issues of SiGe as a semiconductor for the realization of the active layer of photodetectors, we have presented a wide and updated review of the several growth techniques that demonstrate the large variety of SiGe and the different level of compromise between crystal quality and NIR absorption.

29.4   SiGe Detectors: Design and Fabrication

According to the epilayer used, SiGe light detectors can be divided in the two main categories: strained and relaxed devices. As already pointed out, small αW values achievable in the former force the adoption of waveguide geometries where the light, confined in the growth plane, is absorbed in propagation rather than across the thickness (as for normal incidence). This approach is usually pursued in conjunction with multiple heterostructures: the growth of a SiGe layer is followed, before critical thickness, by a Si-layer to partially release the strain, repeating the steps up to the desired thickness. Figure 29.5 represents the most common structures: (a) mesa photodiode at normal incidence, (b) photodiode embedded in, or (c) above a waveguide.

Waveguide Photodetectors (WPD)

Guided-wave geometries can compensate for the low αW products in strained epilayers. While these devices are only suitable as fiber-optic receivers, the latter constitute a vast portion of the detector market.

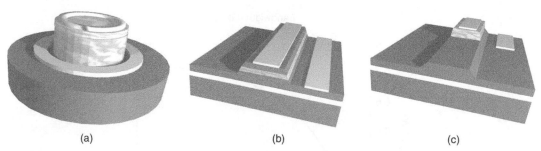

FIGURE 29.5 Schematic of the most commonly used photodetector geometries: (a) normal incidence, (b) waveguide photodiode, and (c) passive waveguide with photodiode.

In WPD the responsivity, at variance with (29.1), can be expressed as

$$R = \frac{\lambda}{1.24} C (1 - \Theta_R) \eta_{\text{int}} (1 - e^{-\alpha \Gamma f L}) \tag{29.5}$$

where C is the coupling efficiency between the optical fiber and the guide, and depends on the two-dimensional overlap integral of the two corresponding modes. C can be optimized and eventually approach unity if tapered waveguide or fibers are employed. Antireflection coating of the input facet is possible but not common, because the process is not planar as the device fabrication. The exponent $\alpha \Gamma f L$ can be regarded as an absorption–length product, with L the device length and $\alpha \Gamma f$ an effective absorption given by the product of the optical absorption α in the alloy, the guide confinement Γ and the ratio f between the SiGe and the total semiconductor volume of the waveguide. The bandwidth is limited by both junction capacitance and transit time, as mentioned in Section 29.2. To avoid RC speed-limitations due to large capacitances, the WPD length must be kept as short as possible.

WPDs are commonly integrated on silicon-on-insulator (SOI) wafers, with dielectric confinement perpendicular to the growth plane. Lateral confinement is obtained by etching ridges of size compatible with the fiber core. The need for n^+ and p^+ top and bottom contacts, respectively, complicates the guide design due to free-carrier absorption losses, which imposes intrinsic Si-spacers around the SiGe heterostructure. Several design parameters are involved in the optimization of the WPD using (29.5) as a guideline while keeping low the capacitance. We will focus mostly on material-related parameters, e.g., the thickness h of the SiGe alloy, the thickness W of the silicon epilayer between two adjacent SiGe layers, the number N of SiGe–Si repetitions, Ge-concentration x, and device length L. Even if the alloy absorption increases with x, quantum size effects eventually arise due to the limits on critical thickness, widening the bandgap and, consequently, lowering α.

Once typical alloy concentrations for highest absorption are chosen in the 0.5 to 0.6 μm range with thickness set to the maximum allowable h_c, then silicon thickness and number of periods have to be selected. The overall structure thickness, however, is limited by the critical value corresponding to the average Ge composition or by fabrication issues. Figure 29.6 is an example of calculated SiGe WPD following such design guidelines. The responsivity is derived for both 1.3 and 1.55 μm for SiGe alloy at 50% using Equation 29.5 assuming unity internal quantum efficiency and employing a vectorial modal solver for evaluating the factor $\alpha \Gamma f$. As for the absorption coefficients, strained values are used from Figure 29.1, while the critical thicknesses are those metastable (solid line of Figure 29.2). The responsivity clearly increases with detector length at the expense of the bandwidth, indicated in the upper side of the plot, but a considerably lower absorption poses severe limitations to the WPD at 1.55 μm. A systematic SiGe design technique for the optimization of 1.3 μm waveguide photodiode has been developed by Naval et al. and the results are reported in Ref. [13]. The following is a review of the most relevant SiGe WPD presented to date. A first SiGe waveguide device was proposed by Temkin et al. [54]: the active layer (which also is the guide core) was a strained Ge_xSi_{1-x}/Si quantum well repeated 20

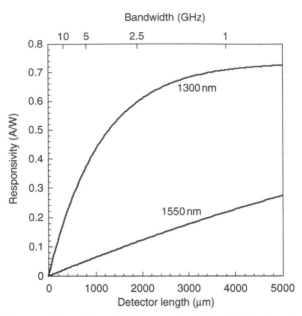

FIGURE 29.6 Calculated responsivity and bandwidth versus device length. The investigated structure is made of three QW (10 nm $Si_{0.5}Ge_{0.5}$ + 500 nm Si) between two Si spacers (1 μm). Width is taken as 10 μm.

times, and various Ge contents in the 0.4 to 0.6 range were realized adjusting the thickness to the critical value. The maximum external quantum efficiency at 1.3 μm was 10.2% for a 10 V reverse bias. At this bias the dark current was 7.1 mA/cm^2, with a bandwidth close to 1 GHz in a 300-μm-long detector. A similar heterostructure (3.3 nm $Ge_{0.6}Si_{0.4}$ and 29 nm Si) was used in an avalanche diode embedded in waveguides 50-μm wide and 50 to 500 μm long [55] with a maximum responsivity of 1.1 A/W at 1.3 μm. A successive device with inverted doping (n^+pp^+ on an n^+ type substrate) exhibited more stable electrical characteristics and an external responsivity as high as 4 A/W at 1.3 μm for 30 V reverse bias, with a remarkably fast (100 ps) response time [56]. As new optical transition of SiGe superlattices was predicted in the NIR [57], short-period (few monolayers) GeSi heterostructures were also attempted in waveguides [58]. The lack of responsivity improvements, probably due to the large quantum size effect, did not encourage further efforts in this direction. A waveguide p–i–n photodiode was fabricated in UCLA/AT&T with a 1-μm thick Ge-rich ($x = 0.7$) multi-quantum-well absorbing layer, equipped with two 1-μm thick Si spacers. The external quantum efficiency at 1.32 μm reached 7% for 14 V, while the dark current density (at the same bias) was 2.7 mA/cm^2 [59]. Open eye-diagrams for 0.5 and 1.5 Gb/s pseudorandom NRZ (nonreturn to zero) optical signals were also reported. Splett et al. [60] introduced a novel WPD SiGe detector: SiGe alloys of different compositions were used for the waveguide ($x = 0.02$) and for the detector ($x = 0.45$), the latter grown above the guide as sketched in Figure 29.5c. For a 7 V reverse bias the maximum external efficiency at 1.28 μm was 11%, with a dark current of about 1 mA/cm^2 and a GHz bandwidth. The first SiGe detector on a SOI waveguide was reported by Kesan et al. [61]. The absorbing layer was a typical SiGe MQW structure: the device exhibited a good responsivity of 0.4 A/W at 1.1 μm, but poor above 1.2 μm.

A SiGeC waveguide detector was proposed by Huang et al. [62]. It consisted of an 80-nm-thick SiGeC absorbing layer with a Ge-content of about 50%. The measured maximum external quantum efficiency at 0.3 V was 0.2 and 8% at 1.55 and 1.3 μm, respectively, with dark current density of 4 mA/cm^2. Li et al. reported a large (one of the largest) responsivity of 80 mA/W at 1.55 μm for a SiGe device operated at 10 V. The device (2-mm long) consisted of a 50%SiGe MQW above a low Ge-content SiGe waveguide [63]. Undulating MQWs of $Si_{0.5}Ge_{0.5}$ 5 nm layers, sandwiched in 12.5-nm thick Si barriers, formed the absorption layer of an MSM WPD fabricated on SOI with a 2 μm thickness, a 65 μm width, and 240 μm length. This device exhibited responsivities as high as 1.6 and 0.1 A/W at 1.3 and 1.55 μm, respectively [64]. In this case the large quantum efficiency at 1.3 μm, the observed sublinear dependence of responsivity on light intensity and the large dark current density pointed to the role of photoconductive gain, expected to affect noise and bandwidth. Comparable responsivities (0.16 A/W at 1.55 μm) were reported in similar guided-wave structures. The increased absorption was explained in terms of high local Ge-concentration, typical of a three-dimensional growth mode [65].

In 1997, the NEC Corporation realized a waveguide detector on SOI employing a 30-period MQW with 3 nm $Si_{0.9}Ge_{0.1}$ and 32 nm Si. The low Ge-content, imposed for compatibility with the high temperature (950°C) used in Si technology for bipolar and MOS transistors, resulted in an NIR photoresponse only slightly larger than in silicon. For a 5 V reverse bias at 980 nm the detector featured an external quantum efficiency of 25% to 29%, a dark current below 50 μA/cm² and a frequency response extending up to 10.5 GHz and indicating good material quality [66]. Two specific features of this device were the selective growth of SiGe layers in a previously opened trench in SOI overlayer and the groove receptacle for fiber alignment.

In conclusion, while most SiGe WPDs in literature are consistent with the predictions in Figure 29.6, lower than expected responsivities can be often associated to coupling and waveguide losses. Figure 29.7 shows responsivity versus reverse bias as reported by various authors. The need for a large voltage is associated to the trapping of photogenerated carriers by the SiGe wells, a common feature of most MQW detectors.

Normal Incidence Photodetectors

Normal incidence photodetectors (NIP) are the most common, because light from any source can be easily coupled and the fabrication is entirely planar. In this case, light propagation and carrier transport proceed in the same direction (as schematically shown in Figure 29.5a).

The design of NIPs is based on the optimization of the responsivity, as evaluated from (29.1), and resembles the design of conventional photodetectors except for the need of dealing with relaxed materials. If the device can be operated at a fixed wavelength, a SiO_2 layer of suitable thickness can effectively reduce reflection losses, which would otherwise amount to about 36%. The collection

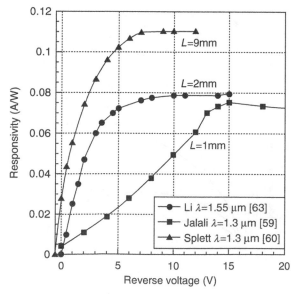

FIGURE 29.7 Selection of some relevant results in terms of responsivity versus reverse bias voltage. The device length *L* is also indicated.

efficiency, which depends on transport properties (through the mobility–lifetime product $\mu\tau$), can be made close to unity by applying a large enough reverse bias. It has been shown that 100% carrier collection can be achieved at <1 V [52] and that the built-in voltage of a p–i–n diode is high enough to efficiently operate at short circuit if the TD density is around $10^7\,cm^{-2}$ [51].

When using relaxed layers the material of choice is pure-Ge, therefore the maximum α is available and the absorption efficiency is maximized by employing the largest thickness W. Upper limits to W relate to the longest acceptable deposition time (two-dimensional heteropitaxy in the presence of large strain imposes small growth rates) or are dictated by trade-off with transit-time limitations. The charts in Figure 29.8 can be used to design a Ge photodetector based on bandwidth and responsivity (upper horizontal axis). In the figure, the iso-bandwidth curves are traced versus active layer thickness and photodiode diameter, for velocity saturation and absorption as in Figure 29.1. In the second plot, the large required thickness is associated with the rather low absorption of Ge at 1.55 μm (460 cm^{-1}). It should be noted that the spectral range between 1.45 and 1.54 μm (S and part of C bands) is exploited for WDM optical communications: in this window Ge-absorption spans between 5800 and 1470 cm^{-1}. Recently, bandgap narrowing of Ge-layers grown on Si and induced by strain accumulated during growth and due to different thermal expansion of Si and Ge was reported [67]. This opens new perspectives for exploiting Ge-detectors in the L band, as well.

Following the design guidelines above, below we review the most relevant results in NIP obtained with relaxed material.

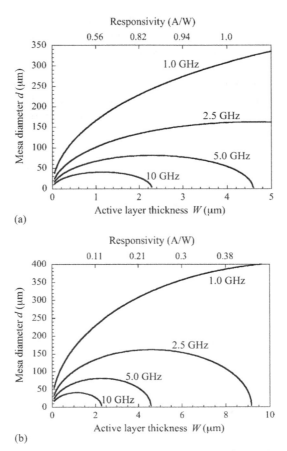

FIGURE 29.8 Calculated responsivity and bandwidth versus device area and active layer thickness for illumination at (a) 1.3 μm and (b) 1.55 μm.

The first pure-Ge p–i–n detector was MBE grown on a thick step-graded buffer to separate the intrinsic active layer from the highly dislocated Si–Ge interface. The diode exhibited good quantum efficiency (40% at 1.3 μm) in the photovoltaic mode, indicating satisfactory transport properties of the Ge-intrinsic layer. The large (50 mA/cm^2) reverse dark current was attributed to the residual dislocation density. A significant improvement was obtained with a SiGe superlattice embedded in the buffer layer, reducing TD down to 5×10^7 cm^{-2}. Unfortunately, even the quantum efficiency dropped to 3% at 1.3 μm [68].

After a long inactivity due to a largest interest to the second window (1.3 μm) where strained SiGe were competitive, pure-Ge has been revived for use in an extended portion of the NIR. Adopting a two-phase growth Sutter et al. tried to bypass the time-consuming growth of a thick buffer graded up to pure-Ge [69]. The growth rate was low during the evaporation of a thin layer, then the temperature was raised for the remaining deposition. The Ge layer was 4-μm-thick, and the first 2 μm highly p-doped the p–i–n junction formed by subsequent diffusion. Material characterization confirmed the good crystalline quality with TD of 5×10^6 cm^{-2}. Performances were very similar to those by Luryi et al. (43% peak external quantum efficiency at 1.55 μm in the photovoltaic mode with dark current of 51 mA/cm^2). Among NIP based on strained SiGe, Huang et al. reported p–i–n photodiodes employing a MQW absorbing layer with 10 nm $Si_{0.5}Ge_{0.5}$ wells and 40 nm Si spacers up to an overall thickness of 500 nm [70]. The device was a p$^+$–i–n$^+$ structure, antireflection coated with SiO_2 to optimize light coupling at 1.3 μm. A 1% external quantum efficiency was reported at 1.3 μm and dark current densities of about 3 mA/cm^2 for a reverse bias of 4 V. The same group tested the incorporation of carbon to compensate for SiGe–Si lattice mismatch [71]. This normal incidence p–i–n detector with a SiGeC active layer was grown on p$^+$ Si using 80 nm $Si_{0.4-x}Ge_{0.6}C_x$ alloy with $x = 1.5\%$. The normal incidence quantum efficiency at 1.3 μm was about 1%, remarkable when compared to the similar efficiency obtained in the previous work with a 500 nm SiGe [70]. Only slight improvements (1.3% to 2.2% at 1.3 μm) were subsequently achieved with thicker active layers [72]. The early experiments in SiGeC for NIR NIPs confirmed the expected limitations: due to the counteracting effects of energy gap increase in the alloy and strain balance associated with C-concentration, only modest absorptions or small enlargements of critical thickness could be obtained. An alternative approach to NIPs consists of strained layer photodetectors embedding SiGe in a resonant cavity. A fourfold responsivity enhancement at 1.3 μm (6.5 mA/W) was demonstrated employing the silicon–oxide interface of a SOI substrate as the bottom mirror and a SiO_2–Si Bragg reflector on the top. The SiGe-layer was a MQW of 20 periods of 8 nm $Si_{0.65}Ge_{0.35}$ and 19 nm Si [73]. A similar approach was used to thin the Ge-layer and achieve good responsivities at 1.55 μm. Around 740-nm-thick Ge was deposited on a double-SOI, which served as the bottom mirror, while the Ge–air interface defined the top reflector, yielding $R = 0.19$ A/W [74].

To exploit the benefits of a low-temperature buffer along with the surfactant action of hydrogen to reduce TD, pure-Ge was CVD deposited on Si for MSM [75]. The NIP exhibited a 1.3 μm responsivity of 240 mA/W at 1 V bias.

More recently, Ge-on-Si MSM detectors with small finger-spacing (1 to 3 μm) [76] based on a 300 nm pure-Ge layer (MBE grown using Sb as surfactant) exhibited responsivities as high as 140 and 90 mA/W at 1.3 and 1.55 μm, respectively, with dark current densities exceeding a few A/cm^2. This major drawback of MSM geometries is associated with low Schottky barriers, although MSM are the fastest Ge-on-Si NIPs demonstrated to date, with responses as short as 12 psec. Interdigited Ge p–i–n photodetectors with small spacing (1 μm) and 1 μm pure-Ge on a thick SiGe graded-buffer provided responsivities of about 0.51 A/W at 1.3 μm, dark currents of about 0.7 A/cm^2 and a 3 dB-bandwidth of 3.8 GHz [77]. Later on, low-dislocation Ge-on-Si NIPs was realized by a combination of a low-temperature buffer and postgrowth annealing [78]. The devices consisted of 1 μm unintentionally doped Ge on p-type (1 0 0) Si. Due to the improved material quality, the photodetectors exhibited a highly saturated responsivity of 550 and 250 mA/W at 1.3 and 1.55 μm, respectively, at reverse biases of a few hundred mV. The measured speed of 850 psec was RC limited even in the smallest 200×200 μm^2 device, and the dark saturated current was 30 mA/cm^2. Remarkable improvements were demonstrated by fabricating p–i–n structures with intrinsic layer from 1 to 4 μm in thickness. This, along with antireflection coatings, allowed the highest reported Ge-on-Si responsivities of 0.89 and 0.75 A/W at 1.3 and 1.55 μm, respectively, at <1 V bias, with RC limited pulse response <200 psec at 1.3 μm and dark currents as low as 15 mA/cm^2 [79,80]. The remarkable performances for this devices are reported in Figure 29.9, where the spectral responsivity and the 3 dB bandwidth are shown. The lowest dark current in Ge-on-Si pn-junctions was reached with low TD density by an optimized graded-buffer complemented by chemi-mechanical polishing at an intermediate composition of $Si_{0.5}Ge_{0.5}$. Dark current densities were lower than 0.2 mA/cm^2, comparable with what expected for a bulk-Ge diode with the same doping profile. A maximum $R = 133$ mA/W was obtained at 1.3 μm in short circuit, denoting excellent transport properties in the junction; its absolute value was limited by the small (0.24 μm) thickness of the absorbing layer [81].

A comparison between the two approaches, LT and graded buffers, respectively, was conducted by Jiang et al. [82]. They evaluated the responsivity of two p–i–n devices with active layers of (a) 400-nm-thick $Si_{0.15}Ge_{0.85}$ on LT Si buffer, (b) 300-nmthick $Si_{0.65}Ge_{0.35}$ on 2 μm SiGe graded buffer. Saturated values of 70 and 150 mA/W were measured at 1.3 μm in samples (a) and (b), respectively. The short-circuit responsivity of the LT sample was much larger than on the graded buffer, suggesting a better quality or a more efficient collection due to the shorter depletion region.

NIR photodetectors have been demonstrated in poly-Ge–Si with $R = 16$ and 6 mA/W at 1.3 and 1.55 μm, respectively [44]. The limitation was attributed to the short diffusion length of poly-Ge and to the unintentional high p-doping. These NIPs have been operated at >2.5 Gbit/sec with a dark current of 2 mA/cm^2 at 1 V [83].

Toward Monolithic Integration

The pioneering work by People [9] and Pearsall [84] on electronic and optical characterization of SiGe heterostructures opened new perspectives in the fabrication of silicon-based photonic devices, the key

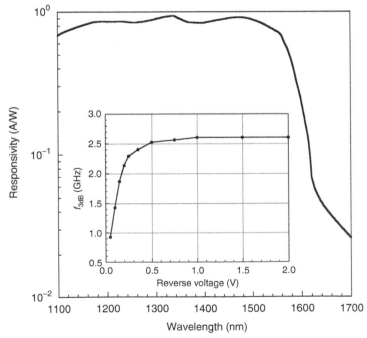

FIGURE 29.9 Experimental data from Famà et al. [80] showing the spectral responsivity of a 4-μm thick pure-Ge mesa on Si. Bandwidth as a function of reverse bias is displayed in the inset.

issue is the compatibility with the unsurpassed VLSI technology. Since then, large effort has been devoted to the exploitation of SiGe heterostructures in optoelectronics, as reported in Refs. [8,85].

Nevertheless, to date only two SiGe photonic devices have been monolithically integrated with Si electronics. A monolithic SiGe–Si p–i–n and front-end transimpedance amplifier circuit was demonstrated in 1998, with SiGe HBTs exhibiting $f_{max} = 34\,GHz$ and DC-gain $= 25$ [86]. The SiGe p–i–n shared the HBT base and collector and provided $R = 0.3\,A/W$ at 850 nm, with a 450 MHz bandwidth. Although the SiGe contribution in the p–i–n was negligible (the absorption layer is the Si collector), this was the first demonstration of monolithic feasibility using a standard process. The first SiGe optoelectronic integrated circuit (IC) operating in the NIR was a linear array of eight NIPs, connected to a transimpedance amplifier through an analog multiplexer. The IC was fabricated in the ALCATEL 2.0 μm CMOS technology and, after the CMOS process, a polycrystalline Ge-layer was evaporated at low temperature in properly windowed n-wells to form the p–n photodiodes. Due to the low thermal budget required for poly-Ge deposition, the silicon electronics preserved both functionality and performance, and the NIP responsivity and dark current density were 16 mA/W and 1 mA/cm², respectively [87]. Figure 29.10 is a photograph of the chip. The poly-Ge film was deposited on a silicon tub (cathode) and extended over a metal pad (anode), as shown in the enlargement on the right. The same authors are currently working at a more advanced chip containing a linear array of poly-Ge photodiodes provided with A/D conversion circuitry and serial digital output.

29.5 Summary

In this chapter, we have attempted to provide the reader with a comprehensive introduction to NIR detection with SiGe. Since this relatively new material has been investigated at levels typical of actual device engineering, we reviewed its most relevant properties with specific attention to photodiode implementations. Both guided-wave (WPD) and normal incidence (NIP) detectors have been discussed, pointing to basic design guidelines and focusing on limitations and trade-offs typical of the SiGe–Si heterostructures. We have presented a critical overview of the best devices fabricated in the past 20 years.

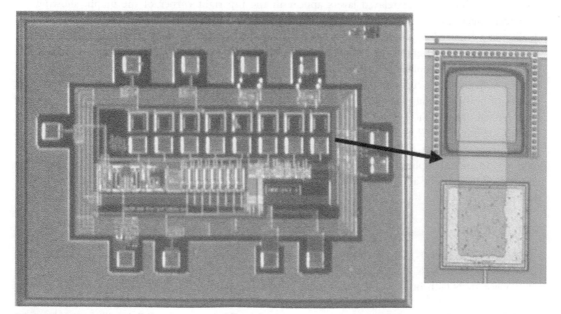

FIGURE 29.10 Photograph of the integrated eight-pixel linear array with amplifier and addressing electronics. The poly-Ge film on a silicon tub is visible in the enlargement on the right.

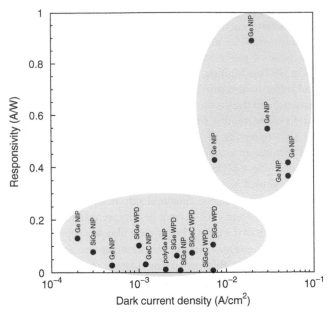

FIGURE 29.11 Responsivities and corresponding dark current densities for most reported photodetectors. NIP stands for normal incidence, WPD for waveguide detectors. The top-right (bottom-left) oval collects the relaxed-Ge (strained-alloy) photodiodes.

At the end of this chapter, finally, in Figure 29.11, we collect responsivities and dark current densities of a number of SiGe devices reported to date and cited in the bibliography. The picture clearly shows the trade-off between efficiency and material quality. Data for SiGe alloys are grouped in the bottom-left, showing that high-quality strained-epilayers with low Ge-alloys are obtained at the expense of a trade-off with absorption efficiency. To overcome the thickness limitation, they can be employed in waveguide geometries. Thick Ge-rich relaxed layers appear in the top-right corner of the graph, showing the beneficial effect of the Ge-absorption associated to large thicknesses. However, they exhibit lower crystal quality, as witnessed by the large current density, and are commonly used at normal incidence. It is worth noting that attempts in reducing the dark current of pure-Ge devices turned into severe quenching of the responsivity (as seen in Figure 29.11). Despite the fact that the ideal region with $R > 0.5$ A/W and dark currents below 1 mA/cm^2 have not yet been accessed, we trust that the current quality of both materials and devices is close to meet the requirements of industry developments and commercialization of SiGe-based NIR detectors for a large number of applications. Most pure Ge-based devices exhibit large R at both 1.3 and 1.55 μm, while the corresponding dark currents are often negligible with respect to both thermal and excess noise of the transimpedance amplifiers in most receivers. In critical applications where the shot noise associated to the dark current is unacceptable, the strained-alloy needs be adopted, eventually aiming at higher sensitivities in suitably tailored waveguide geometries.

References

1. AE Willner and Y Xie. Wavelength domain multiplexed (WDM) fiber optic communication network. In M Bass, ed. *Handbook of Optics*. NewYork: McGraw Hill, 2001, pp. 1–27.
2. P Ceccato, N Gobron, S Flasse, B Pinty, and S Tarantola. *Remote Sensing of Environment* 82:188–192, 2002.
3. AC Stanton, JA Silver, DS Bomse, DB Oh, DC Hovde, DJ Kane, and ME Paige. Trace gas detection using near-infrared diode lasers. *Proc. IEEE-LEOS Ann. Meet.* 2:324–325, 1995.

4. SA Soper, JH Flanagan Jr, BL Legendre, DC Williams, and RP Hammer. Near-infrared, laser-induced fluorescence detection for DNA sequencing applications. *IEEE J. Sel. Top. Quantum Electron.* 2: 1129–1139, 1996.

5. PJ Kirkpatrick, P Smielewski, JMK Lam, and P Al-Rawi. Use of near infrared spectroscopy for the clinical monitoring of adult brain. *J. Biomed. Opt.* 1:363–374, 1996.

6. TD Tosteson, BW Pogue, E Demidenko, TO McBride, and KD Paulsen. Confidence maps and confidence intervals for near infrared images in breast cancer. *IEEE Trans. Medical Imaging* 18:1188–1193, 1999.

7. JD Cressler. SiGe technology. In WK Chen, ed. *The VLSI Handbook.* Boca Raton: CRC Press, 2000, pp. 1–10.

8. RA Soref. Silicon-based Optoelectronics. *Proc. IEEE* 81:1687–1706, 1993.

9. R People. Physics and applications of GeSi/Si strained-layer heterostructures. *IEEE J. Quantum Electron.* 22:1696–1710, 1986.

10. FK LeGoues. The effect of strain on the formation of dislocations at the SiGe/Si interface. *MRS Bull.* 21:38–44, 1996.

11. RF Potter. Germanium. In ED Palik, ed. *Handbook of Optical Constants of Solids.* New York: Academic Press, 1985, pp. 465–469.

12. R Braunstein, AR Moore, and F Herman. Intrinsic optical absorption in germanium–silicon alloys. *Phys. Rev.* 109:695–710, 1958.

13. L Naval, B Jalali, L Gomelsky, and JM Liu. Optimization of SiGe/Si waveguide photodetectors operating at $\lambda = 1.3\,\mu m$. *J. Lightwave Technol.* 14:787–797, 1996.

14. R People. Indirect band gap of coherently strained GeSi bulk alloys on $\langle 0\,0\,1 \rangle$ silicon substrates. *Phys. Rev. B* 32:1405–1408, 1985.

15. DV Lang, R People, JC Bean, and AM Sergent. Measurement of the band gap GeSi/Si strained-layer heterostructures. *Appl. Phys. Lett.* 47:1333–1335, 1985.

16. I Daruka, J Tersoff, and AL Barabasi. Shape transition in growth of strained islands. *Phys. Rev. Lett.* 82:2753–2756, 1999.

17. TF Kuech and M Maenpaa. Epitaxial growth of Ge on $\langle 1\,0\,0 \rangle$ Si by a simple chemical vapor deposition technique. *Appl. Phys. Lett.* 39:245–247, 1981.

18. M Garozzo, G Conte, F Evangelisti, and G Vitali. Heteroepitaxial growth of Ge on $\langle 1\,1\,1 \rangle$ Si by vacuum evaporation. *Appl. Phys. Lett.* 41:1070–1072, 1982.

19. Y Ohmachi, T Nishioka, and Y Shinoda. The heteroepitaxy of Ge on Si(1 0 0) by vacuum evaporation. *J. Appl. Phys.* 54:5466–5469, 1983.

20. JM Baribeau, TE Jackman, DC Houghton, P Maigne, and MW Denhoff. Growth and characterization of SiGe and Ge epilayers on (1 0 0) Si. *J. Appl. Phys.* 63:5738–5746, 1988.

21. K Fujinaga. Low-temperature heteroepitaxy of Ge on Si by GeH_4 gas low-pressure chemical vapor deposition. *J. Vac. Sci. Technol. B* 9:1511–1516, 1991.

22. B Cunningham, JO Chu, and S Akbar. Heteroepitaxial growth of Ge on (1 0 0) Si by ultrahigh vacuum, chemical vapor deposition. *Appl. Phys. Lett.* 59:3574–3376, 1991.

23. E Kasper. Growth and properties of Si/SiGe superlattices. *Surf. Sci.* 174:630–639, 1986.

24. EA Fitzgerald, YH Xie, ML Green, D Brasen, AR Kortan, J Michel, YJ Mii, and BE Weir. Totally relaxed GeSi layers with low threading dislocation densities on Si substrates. *Appl. Phys. Lett.* 59:811–813, 1991.

25. GP Watson, EA Fitzgerald, YH Xie, and D Monroe. Relaxed, low defect density $Si_{0.7}Ge_{0.3}$ epitaxial layers grown on Si by rapid thermal chemical vapor deposition. *J. Appl. Phys.* 75:263–269, 1994.

26. MT Currie, SB Samavedam, TA Langdo, CW Leitz, and EA Fitzgerald. Controlling threading dislocation densities in Ge on Si using graded SiGe layers and chemical–mechanical polishing. *Appl. Phys. Lett.* 72:1718–1720, 1998.

27. JL Liu, CD Moore, GD U'Ren, YH Luo, Y Lu, G Jin, SG Thomas, MS Goorsky, and KL Wang. A surfactant-mediated relaxed $Si_{0.5}Ge_{0.5}$ graded layer with a very low threading dislocation density and smooth surface. *Appl. Phys. Lett.* 75:1586–1588, 1999.

28. M Copel and RM Tromp. Are bare surfaces detrimental in epitaxial growth? *Appl. Phys. Lett.* 58:2648–2650, 1991.

29. A Sakai and T Tatsumi. Ge growth on Si using atomic hydrogen as a surfactant. *Appl. Phys. Lett.* 64:52–54, 1994.

30. D Reinking, M Kammler, M Horn-von-Hoegen, and KR Hofmann. Enhanced Sb segregation in surfactant-mediated-heteroepitaxy: high-mobility, low-doped Ge on Si. *Appl. Phys. Lett.* 71:924–926, 1997.

31. K Eberl, SS Iyer, S Zollner, JC Tsang, and FK LeGoues. Growth and strain compensation effects in the ternary $Si_{1-x-y}Ge_xC_y$ alloy system. *Appl. Phys. Lett.* 60:3033–3035, 1992.

32. HJ Osten and E Bugiel. Relaxed $Si_{1-x}Ge_x/Si_{1-x-y}Ge_xC_y$ buffer structures with low threading dislocation density. *Appl. Phys. Lett.* 70:2813–2815, 1997.

33. RA Soref. Silicon-based group IV heterostructures for optoelectronic applications. *J. Vac. Sci. Technol. A* 14:913–918, 1996.

34. F Banhart and A Gutjahr. Stress relaxation in SiGe layers grown on oxide-patterned Si substrates. *J. Appl. Phys.* 80:6223–6228, 1996.

35. TA Langdo, CW Leitz, MT Currie, EA Fitzgerald, A Lochtfeld, and DA Antoniadis. High quality Ge on Si by epitaxial necking. *Appl. Phys. Lett.* 76:3700–3702, 2000.

36. L Di Gaspare, G Capellini, E Palange, F Evangelisti, L Colace, G Masini, F Galluzzi, and G Assanto. Ge on Si(1 0 0) near infrared detectors: material issues and results. Proc. of the 24th Int. Conf. Phys. Semicond. (CD), Jerusalem, 1998.

37. H Chen, LW Guo, Q Cui, Q Hu, Q Huang, and JM Zhou. Low-temperature buffer layer for growth of a low dislocation-density SiGe layer on Si by molecular-beam epitaxy. *J. Appl. Phys.* 79:1167–1169, 1996.

38. JH Li, CS Peng, Y Wu, DY Dai, JM Zhou, and ZH Mai. Relaxed $Si_{0.7}Ge_{0.3}$ layers grown on low-temperature Si buffers with low threading dislocation density. *Appl. Phys. Lett.* 71:3132–3134, 1997.

39. KK Linder, FC Zhang, JS Rieh, P Bhattacharya, and D Houghton. Reduction of dislocation density in mismatched SiGe/Si using a low-temperature Si buffer layer. *Appl. Phys. Lett.* 70:3224–3226, 1997.

40. CS Peng, ZY Zhao, H Chen, JH Li, YK Li, DY Dai, Q Huang, JM Zhou, YH Zhang, TT Sheng, and CH Tung. Relaxed $Ge_{0.9}Si_{0.1}$ alloy layers with low threading dislocation densities grown on low-temperature Si buffers. *Appl. Phys. Lett.* 72:3160–3162, 1998.

41. L Colace, G Masini, F Galluzzi, G Assanto, G Capellini, L Di Gaspare, and F Evangelisti. Ge/Si(1 0 0) photodetector for near infrared light. *Solid State Phenom.* 54:55–57, 1997.

42. HC Luan, DR Lim, KK Lee, KM Chen, JG Sandland, K Wada, and LC Kimerling. High-quality Ge epilayers on Si with low threading-dislocation densities. *Appl. Phys. Lett.* 75:2909–2911, 1999.

43. L Colace, G Masini, F Galluzzi, and G Assanto. Evaporated polycrystalline germanium for near infrared photodetection. Proc MRS Fall Meeting, Boston, 1998, pp. 469–475.

44. G Masini, L Colace, and G Assanto. Advances in the field of poly-Ge on Si near infrared photo-detectors. *Mater. Sci. Eng. B* 69:257–260, 2000.

45. L Colace, G Masini, and G Assanto. Ge-on-Si approaches to the detection of near infrared light. *IEEE J. Quantum Electron.* 35:1843–1852, 1999.

46. G Masini, L Colace, and G Assanto. Germanium thin film on silicon for detection of near-infrared light. In HS Nalwa, ed. *Handbook of Thin Film Materials.* New York: Academic Press, 2002, pp. 327–367.

47. GK Wertheim and GL Pearson. Recombination in plastically deformed germanium. *Phys. Rev.* 107:694–698, 1957.

48. PN Grillot, SA Ringel, EA Fitzgerald, GP Watson, and YH Xie. Electron trapping kinetics at dislocations in relaxed GeSi/Si heterostructures. *J. Appl. Phys.* 77:3248–3256, 1995.

49. FM Ross, R Hull, D Bahnck, JC Bean, LJ Perticolas, and CA King. Changes in electrical devices characteristics during the in situ formation of dislocations. *Appl. Phys. Lett.* 62:1426–1428, 1993.

50. LM Giovane, HC Luan, AM Agarwal, and LC Kimerling. Correlation between leakage current density and threading dislocation density in SiGe pin diodes grown on relaxed graded buffer layers. *Appl. Phys. Lett.* 78:541–543, 2001.

51. G Masini, L Colace, G Assanto, HC Luan, K Wada, and LC Kimerling. High performance p–i–n Ge on Si photodetectors for the near infrared: from model to demonstration. *IEEE Trans. Electron. Dev.* 48:1092–1096, 2001.

52. L Colace, G Masini, G Assanto, HC Luan, K Wada, and LC Kimerling. Efficient high-speed near-infrared Ge photodetectors integrated on Si substrates. *Appl. Phys. Lett.* 76:1231–1233, 2000.

53. V Chu, JP Conde, S Shen, and S Wagner. Photocurrent collection in a Schottky barrier on an amorphous silicon–germanium alloy structure with 1.23 eV optical gap. *Appl. Phys. Lett.* 55:262–264, 1989.

54. H Temkin, TP Pearsall, JC Bean, RA Logan, and S Luryi. Ge_xSi_{1-x} strained-layer superlattice waveguide photodetectors operating near 1.3 μm. *Appl. Phys. Lett.* 48:963–965, 1986.

55. S Luryi, TP Pearsall, H Temkin, and JC Bean. Waveguide infrared photodetectors on a silicon chip. *IEEE Electron Dev. Lett.* 7:104–107, 1986.

56. H Temkin, A Antreasyan, NA Olsson, TP Pearsall, and JC Bean. $Ge_{0.6}Si_{0.4}$ rib waveguide avalanche photodetector for 1.3 μm operation. *Appl. Phys. Lett.* 49:809–811, 1986.

57. TP Pearsall, J Bevk, LC Feldman, JM Bonar, and JP Mannaerts. Structurally induced optical transitions in Ge–Si superlattices. *Phys. Rev. Lett.* 58:729–732, 1987.

58. TP Pearsall, EA Beam, H Temkin, and JC Bean. Ge–Si/Si infrared zone folded superlattice detectors. *Electron. Lett.* 24:685–686, 1988.

59. B Jalali, L Naval, and AFJ Levi. Si-based receivers for optical data links. *J. Lightwave Technol.* 12:930–935, 1994.

60. A Splett, T Zinke, K Petermann, E Kasper, H Kibbel, HJ Herzog, and H Presting. Integration of waveguides and photodetectors in SiGe for 1.3 μm operation. *IEEE Photon. Technol. Lett.* 6:59–61, 1994.

61. VP Kesan, PG May, E Bassous, and SS Iyer. Integrated waveguide photodetector using Si/SiGe multiple quantum wells for long wavelength applications. *IEDM Techn. Dig.* 637–639, 1990.

62. FY Huang, K Sakamoto, KL Wang, P Trinh, and B Jalali. Epitaxial SiGeC waveguide photodetector grown on Si substrate with response in the 1.3–1.55 μm wavelength range. *IEEE Photon. Technol. Lett.* 9:229–231, 1997.

63. B Li, G Li, E Liu, Z Jiang, J Qin, and X Wang. Monolithic integration of a SiGe/Si modulator and multiple quantum well photodetector for 1.55 mm operation. *Appl. Phys. Lett.* 73:3504–3506, 1998.

64. H Lafontaine, NL Rowell, and S Janz. Growth of undulating $Si_{0.5}Ge_{0.5}$ layers for photodetectors at lambda = 1.55 μm. *Appl. Phys. Lett.* 72:2430–2432, 1998.

65. S Janz, JM Baribeau, DJ Lockwood, JP McCaffrey, S Moisa, NL Rowell, DX Xu, H Lafontaine, and MRT Pearson. Si/SiGe photodetectors using three-dimensional growth modes to enhance photo-response at λ = 1550 nm. *J. Vac. Sci. Technol.* 18:588–592, 2000.

66. T Tashiro, T Tatsumi, M Sujiyama, T Hashimoto, and T Morikawa. *IEEE Trans. Electron Dev.* 44:545–549, 1997.

67. DD Cannon, J Liu, Y Ishikawa, K Wada, DT Danielson, S Jongthammanurak, J Michel, and LC Kimerling. Tensile strained epitaxial Ge films on Si (1 0 0) substrates with potential application in L-band telecommunications. *Appl. Phys. Lett.* 84:906–908, 2004.

68. A Kastalsky, S Luryi, JC Bean, and TT Sheng. Single-crystal Ge/Si infraredphotodetector for fiber optics communications. *Proc. Electrochem. Soc.* PV85-7:406–411, 1985.

69. P Sutter, U Kafader, and H von Känel. Thin film photodetectors grown epitaxially on silicon. *Solar Energy Mater. Solar Cells* 31:541–547, 1994.

70. FY Huang, X Zhu, MO Tanner, and KL Wang. Normal-incidence strained-layer superlattice $Ge_{0.5}Si_{0.5}$/Si photodiodes near 1.3 μm. *Appl. Phys. Lett.* 67:566–568, 1995.

71. FY Huang and KL Wang. Normal-incidence epitaxial SiGeC photodetector near 1.3 μm wavelength grown on Si substrate. *Appl. Phys. Lett.* 69:2330–2232, 1996.

72. X Shao, SL Rommel, BA Orner, H Feng, MW Dashiell, RT Troeger, J Kolodzey, PR Berger, and T Laursen. 1.3 μm photoresponsivity in Si-based GeC photodiodes. *Appl. Phys. Lett.* 72:1860–1862, 1998.

73. C Li, Q Yang, H Wang, J Zhu, L Luo, J Yu, and Q Wang. SiGe/Si resonant cavity enhanced photodetectors with a silicon on oxide reflector operating near 1.3 μm. *Appl. Phys. Lett.* 77:157–159, 2000.

74. O Dosunmu, MK Ernsley, DD Cannon, B Ghyselen, LC Kimerling, and S Unlu. Germanium on double-SOI photodetectors for 1550 nm operation. *Proc. of the IEEE-LEOS 16th Ann. Meet.*, Vol. 2, 2003, pp. 853–854.

75. L Colace, G Masini, F Galluzzi, G Assanto, G Capellini, L Di Gaspare, E Palange, and F Evangelisti. Metal–semiconductor–metal near-infrared light detector based on epitaxial Ge/Si. *Appl. Phys. Lett.* 72:3175–3177, 1998.

76. D Buca, S Winnerl, S Lenk, S Mantl, and Ch Buchal. Metal–germanium–metal ultrafast infrared detectors. *J. Appl. Phys.* 92:7599–7605, 2002.

77. J Oh, JC Campbell, SG Thomas, S Bharatan, R Thoma, C Jasper, RE Jones, and TE Zirkle. Interdigitated Ge p–i–n photodetectors fabricated on a Si substrate using graded SiGe buffer layers. *IEEE J. Quantum Electron.* 38:1238–1241, 2002.

78. G Masini, L Colace, G Assanto, HC Luan, K Wada, and LC Kimerling. High responsivity near infrared Ge photodetectors integrated on Si. *Electron. Lett.* 35:1467–1468, 1999.

79. G Masini, L Colace, G Assanto, HC Luan, and LC Kimerling. *Electron. Lett.* 36:2095–2096, 2000.

80. S Famà, L Colace, G Masini, G Assanto, and HC Luan. *Appl. Phys. Lett.* 81:586–588, 2002.

81. SB Samavedam, MT Currie, TA Langdo, and EA Fitzgerald. High quality germanium photodiodes integrated on silicon substrates using optimized relaxed graded buffers. *Appl. Phys. Lett.* 73:2125–2127, 1998.

82. RL Jiang, ZY Lo, WM Chen, L Zang, SM Zhu, XB Liu, and XM Cheng. Normal-incidence SiGe/Si photodetectors with different buffer layers. *J. Vac. Sci. Technol. B* 18:1251–1253, 2000.

83. G Masini, L Colace, and G Assanto. 2.5 Gbit/s polycrystalline germanium-on-silicon photodetector operating from 1.3 to 1.55 μm. *Appl. Phys. Lett.* 82:2524–2526, 2003.

84. TP Pearsall. Silicon–germanium alloys and heterostructures: Optical and electronic properties. *CRC Crit. Rev. Solid State Mater. Sci.* 15:551–599, 1989.

85. G Masini, L Colace, and G Assanto. Si based optoelectronics for communications. *Mater. Sci. Eng. B* 89:2–9, 2002.

86. JS Rieh, D Klotzkin, O Qasaimeh, LH Lu, K Yang, LPB Katehi, P Bhattacharya, and ET Croke. Monolithically integrated SiGe/Si PIN-HBT front-end photoreceivers. *IEEE Photon. Technol. Lett.* 10:415–417, 1998.

87. G Masini, V Cencelli, L Colace, F DeNotaristefani, and G Assanto. Monolithic integration of near-infrared Ge photodetectors with Si complementary metal–oxide–semiconductor readout electronics. *Appl. Phys. Lett.* 80:3268–3270, 2002.

30

Si-Based Photonic Transistor Devices for Integrated Optoelectronics

Wei-Xin Ni and
Anders Elfving

Linköping University

30.1 Introduction

It is known that, due to its indirect bandgap, Si is an optically inefficient material, although many efforts have been made in manipulating the materials in order to improve the efficiency. For example, by incorporating SiGe quantum well layers or self-assembled Ge dots in the Si structure, one can push the absorption edge into the wavelength range of 1.3 to 1.55 μm due to the reduced bandgap [1]. The basic optical properties cannot be improved, however, since the nature of the indirect bandgap remains for these materials. In this chapter, we give a summary of the efforts made using another approach, i.e., although the materials are not very efficient due to physical limitations, one can instead find new types of device solutions. Therefore, one can in a more optimized way use the material potential for fabrication of practically useful Si-based optoelectronics. In this context, three-terminal photonic devices, namely photonic transistors, are considered. In the following sections, we select several examples to demonstrate how three-terminal transistors can be implemented for such a purpose of integrated optoelectronics.

30.2 Light Emitters

It is of high interest for achieving efficient Si-based light emitters, which are the key components for realizing all Si-based optoelectronics, but presently are still unavailable. Er^{3+} ions can emit near infrared light at 1.54 μm at room temperature when incorporated together with oxygen in Si. Transistor solutions implemented for the Si:Er emitters are mainly motivated for an efficient pumping mechanism of Er excitation. Hot electron injectors have been studied using both bipolar and MOS transistors.

Si–SiGe–Si:Er Heterojunction Light-Emitting Transistors

As established, Er emission at 1.54 μm is due to an intra-4f transition of Er^{3+} ions. Different from the optical or electrical pumping mode in Si for emitting light directly via interband recombination, excitation of the Er-doped Si system is a process of energy transfer from carriers to Er ions. There exist two main excitation processes of Er^{3+} ions: (i) excitation by electron–hole recombination-mediated energy transfer at an Er-related defect level, in cases of a forward biased p–n junction [2] or carrier generation due to laser irradiation; and (ii) hot carrier direct impact excitation in case of a reverse biased p–n junction [3–6].

However, since the spontaneous radiative decay time of Er ions is very long (~1 msec) [7], non-radiative de-excitation processes strongly compete with the radiative decay, causing a significant reduction of the luminescence intensity. For the reverse biased devices, although the thermally activated energy back transfer process at elevated temperatures can be suppressed [3,4], the Auger de-excitation induced by the excess free carriers (because of Er and O dopants) as well as injected carriers may set a limit on high luminescence intensity. The high density of electrons and holes could be generated by hot electron impact ionization of the Si matrix [5]. Therefore, Er excitation would never be efficient if the device is operating in the avalanche breakdown regime.

Consequently, there is a necessary tradeoff when controlling the electron kinetic energy for impact excitation while avoiding avalanche breakdown, which is however a difficult task to realize in a conventional diode structure. It has been impossible to de-couple the effects of two correlated variables, i.e., the applied voltage and the injection current. SiGe–Si:Er heterojunction light transistors containing a thin SiGe base layer and an Er-doped active layer in the collector were thus studied [5,7], aiming at achieving high electrical pumping efficiency for Er excitation. In these devices, one can in a controlled way introduce hot electrons from the transistor emitter with a collector bias voltage below avalanche breakdown for improved impact excitation efficiency.

The SiGe–Si:Er light-emitting transistors were fabricated using an emitter-down structure (as schematically shown in Figure 30.1). The layer structures were grown through the pre-patterned oxide windows by differential molecular beam epitaxy (MBE) [5] for achieving a freestanding external base contact. Er ions together with oxygen, supplied by sublimation of Er and SiO during the MBE growth, were incorporated in the B–C junction with the area aligned to the emitter, which permits all incorporated Er ions to be electronically pumped by injecting hot electrons from the emitter.

SiGe–Si:Er HBTs are typically operated with the common-emitter configuration. Intense electroluminescence (EL) was observed from Er-doped HBTs measured at low injection current (~0.17 A/cm²) for two base layer thicknesses (Figure 30.2). The determined impact excitation cross section was ~5 × 10^{-15} cm², which was a 50-fold increase compared to the values reported from conventional diode structures. The external quantum efficiency was ~1.5 × 10^{-4}, which was increased due to an efficient excitation process with a controlled acceleration condition avoiding impact ionization [7].

The HBT-type Si–SiGe–Si:Er:O light-emitting devices are also a useful tool for studying the excitation and de-excitation mechanisms, since the device is able to separately control the applied bias across the B–C junction (the hot electron acceleration field), and the injection current density (the electron flux) during an impact excitation process.

The influence of the Auger effect due to carriers from ionized dopants on the EL intensity was clearly revealed by the EL decay measurements on the SiGe–Si:Er:O-HBT [7]. Under common-emitter configuration, a long 1/e decay time (e.g., 190 μsec for I_c = 1.2 mA) was measured, when applying a DC bias (marked as 5-5 V in Figure 30.3) across C–E for the electron acceleration. However, the measured decay time decreased to 4 μsec, when V_{ce} was only applied in pulses synchronized with the V_{be} pulses. A longer decay time constant observed in the former case is due to suppression of the Auger effect because of carrier depletion in the space–charge region under the constant bias. When V_{ce} and V_{be} were switched-off simultaneously, the excited Er ions are quickly embedded in a region where the carriers due to dopant ionization act as de-excitation centers via an Auger transfer process, thus causing a fast decay. The understanding of the Auger carrier effect is crucial for an efficient Si:Er emitter.

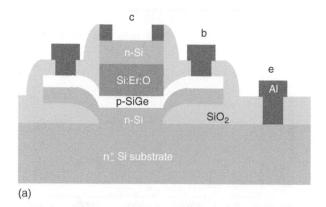

(a)

(b)

FIGURE 30.1 (a) A schematic cross section and (b) an SEM micrograph of the SiGe/Si:Er:O HBT-type light-emitting device prepared by differential MBE.

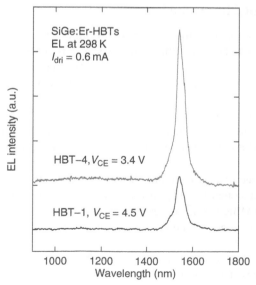

FIGURE 30.2 EL spectra measured at 300 K from SiGe–Si:Er:O-HBTs with the base thickness of 50 nm (HBT-1) and 30 nm (HBT-4), respectively, with a very low driving current.

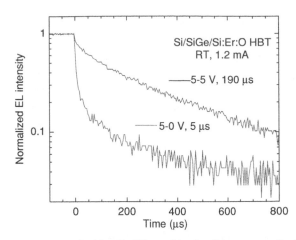

FIGURE 30.3   EL decay curves measured with two different bias conditions across the B–C junction.

Si–SiO$_2$:Er MOS Emitters

Er^{3+} ions can be very efficiently excited when incorporating them in a SiO$_2$ layer using hot-electrons injected from the poly-Si gate contact. Since the material systems used for these types of devices are not single crystalline, it will thus not be discussed in this chapter. Some detailed descriptions can be found in Ref. [8] for those who are interested.

30.3   Photodetectors

The interesting wavelengths for the Si-based detector devices are mainly in the near-infrared range (e.g., 1.3 to 1.55 μm, for possible applications in optical links and chip optical interconnects, etc.), and the mid/far infrared range (3 to 20 μm, for environment monitoring, thermal imaging and night vision, etc.). SiGe-based heterojunction material systems are widely studied for these purposes.

1.3 to 1.5 μm Ge-Dot Phototransistors

The idea of the phototransistor was proposed already by Shockley et al. in 1951 [9], and the working principle was discussed in detail in Ref. [10]. For this type of detector, electron–hole pairs, generated by photoionization due to the illuminating light beam with the energy larger than the semiconductor bandgap, are separated by an applied reverse bias across the B–C junction. The holes move and contribute to the base current, which then facilitate the injection of electrons from the emitter to collector resulting in current gain, i.e., the primary photocurrent is amplified through the transistor.

High-performance phototransistors had been made using heterojunction materials. A successful example was the InP-based double heterojunction phototransistor with optical-gain cutoff frequencies of up to 135 GHz [11].

In 2002, Elfving et al. reported growth and characterization of the first phototransistor fabricated using the Si–SiGe material system [12–14], in which multiple Ge dot layers were incorporated in the B–C junction region using MBE. In this case, electron–hole pairs only generated in the base and collector with the SiGe layer and the islands by the infrared radiation below the Si bandgap, but not in the E–B junction.

The optically controlled I–V characteristics of such a Ge-dot phototransistor are depicted in Figure 30.4. As shown in the figures, the device revealed a very low dark current, ∼0.01 mA/cm^2 at −2 V [14]. A strong light modulation effect was observed, i.e., the collector current I_c was drastically increased when increasing the light power, which is similar to the case when changing the base current I_b. For the

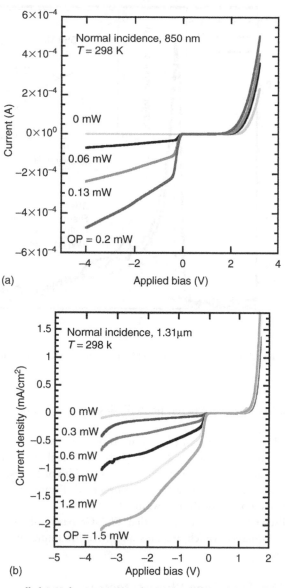

FIGURE 30.4 Optically controlled *I–V* characteristics of a Ge-dot HBT with laser illumination at different optical power at (a) 850 nm and (b) 1.31 μm. The light modulation effect was observed only in the reverse bias direction at 1.31 μm.

experiments using a 850 nm wavelength radiation source, such an effect was observed at both bias directions, but at 1.31 μm the light modulation only occurred when the Ge dot containing B–C junction was reverse biased (Figure 30.4b) while the E–B junction became photoinsensitive. This is a natural effect, because in this case Si is completely transparent for the incident infrared light, such that there is no generation of photocarriers in Si, which may bring an advantage for the low noise performance of light detection.

These Ge-dot phototransistors were measured with very high photoresponse [13], which were ~2.5 A/W at 850 nm (normal incidence, an apparent external quantum efficiency value of ~350 % at this wavelength), ~0.5 A/W (waveguide) at 1.31 μm, and 25 mA/W (waveguide) at 1.55 μm, respectively, at a bias condition of −4 V. These values are significantly higher compared to the measured

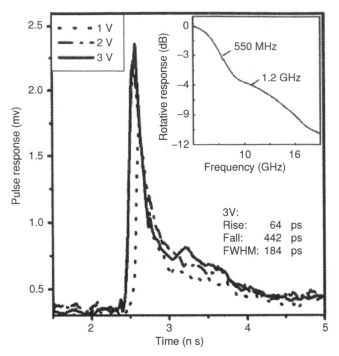

FIGURE 30.5 Pulse response of SiGe HPT at 850-nm and 50-psec pulse laser illumination with the applied voltages of 1, 2, and 3 V, respectively.

photoresponse obtained from reference p–i–n photodiodes with an identical Ge dot layer structure in the intrinsic region.

Pei et al. showed that the cutoff frequency (f_T) and maximum oscillation frequency (f_{max}) of the SiGe–Si-MQWs phototransistor were found to be 25 GHz [15], which is thus suitable for gigabit integrated circuits. In particular, the transient responsivity with the pulsewidth of 184 psec (the rise time of 64 psec and fall time of 442 psec) at a wavelength of 850 nm was observed (Figure 30.5), in spite of the fast falling of the ac response at the level of the 6-dB bandwidth at 1.2 GHz.

All of the above observed features indicate that Si–SiGe-based phototransistors have excellent electrical and optical performance, which are thus attractive for future Si-based optoelectronic integrated circuit applications.

1.3 to 1.5 μm Ge-Dot FET Type Photodetectors

The optically controlled field-effect transistors (FET) can also generate high photoresponse, which has attracted a great deal of attention to be used as sensitive detectors. Furthermore, the FET phototransistors can be easily used, for example, mixing of a microwave signal with an optically coupled local oscillator signal for oscillator tuning, etc.

The principle of using the FET as a photodetector is very straightforward. The measured photocurrent depends on the lateral carrier transport, according to the following equation:

$$I_{DS} \propto \left(\frac{W}{L}\right)\mu_{n_s}$$

where L is the conducting channel length (the source-to-drain distance) and W is the channel width, μ is the carrier mobility, and n_s is the number of charges in the conduction channel. For an undoped structure, n_s is determined by the photoionization cross section and the efficiency of carrier transfer into the conduction channel, which can be controlled by the top gate. The equation thus tells us that the primary photogenerated carries can yield a larger effect of the eventually measured I_{DS}, due to the gain factor determined by the device design.

Several types of FET photodetectors (MESFET or HEMT) were studied using III–V materials, and showed excellence performance. At the 0.6 µm wavelength range, Khalid and Rezazadeh reported that the DC photoresponsivity of GaAs MESFETs was about 4.5 A/W [16], and pulse responses with a FWHM value of 22 psec. By using an InP–InGaAs HEMT structure with semitransparent meander shaped gate (ITO), Marso et al. reported a very high DC photoresponse of 15.4/W [17], and pulse response with a FWHM of 90 nsec at 1.3 µm wavelength. RF measurements were also carried out with a frequency limit up to 20 GHz.

Studies on SiGe-based FET type detectors have just been initiated, however, but the results are still very promising and may lead to some interesting applications. Elfving et al. recently reported a SiGe-QW–Ge-dot HEMT photodetector operating at 1.3 to 1.55 µm [18]. For this detector, Ge dots were used as the absorption medium to push the cutoff wavelength into the interesting 1.55 µm regime. However, in-plane current transport, namely I_{DS}, is limited by the discrete distribution of the Ge islands and the very thin wetting layer. To solve the problem, SiGe quantum wells (QWs) were placed next to the dot layers to serve as the high mobility channel when the photogenerated carried can be transferred from the dots to the wells.

Ten periods of SiGe(6 nm)/Si(10 nm)/Ge(8 monolayers)/Si(60 nm) multiple stacks were grown at 600°C using MBE, and the detectors were processed using a multi-finger mesa design with Al source and drain contacts connected to the side edge of the mesas, which is shown in Figure 30.6. Pt was used as gate material to create a Schottky-contact.

Some preliminary experimental results based on normal incidence measurements (200 µm in diameter shinning area) showed that the responsivity was about four times higher than that observed from the reference sample without SiGe QWs. The increase of the photocurrent is proportional to $P_{op}^{0.85}$ at $V_{DS} = 5$ V (Figure 30.7), indicating an efficient photocarrier transfer process from the Ge QDs into the SiGe QWs.

The effect of the gate bias has been studied using the broadband light source with a long pass filter ranging >1.1 µm. Figure 30.8 shows the dark current and photocurrent measured at $V_G = 0$ and 2 V. With no incident photons, the dark current was small and almost independent of the gate voltage. The photoresponsivity was >200 mA/W at $V_{DS} = -2.5$ V and $V_G = 2$ V. Even though one can observe some gate leakage, the photocurrent can be modulated with V_G when near infrared photons are incident on to the transistors. By switching the direction of V_G, the detectors can quickly be switched between on-and off-state, with a decay or rise time of ~300 nsec (not shown), which was actually limited by the bandwidth of the experimental instruments.

Ge-Dot FET-Type Mid–Far Infrared Photodetectors

Similar ideas to that described in Section 3.2 can be used to fabricate photodetectors operating in the mid/far-infrared range. The main difference is that the dots must be doped, so that intersubband photoexcitation is responsible for generating photo-carriers for detection.

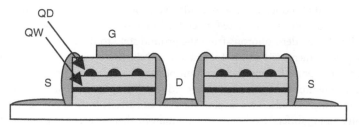

FIGURE 30.6 Schematic drawing of the device cross section.

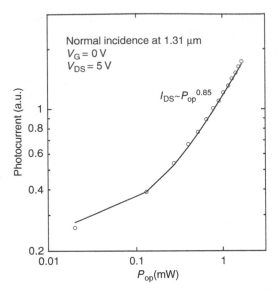

FIGURE 30.7 The photocurrent as a function of the incident optical power for the SiGe-QW–Ge-dot photo-MESFET.

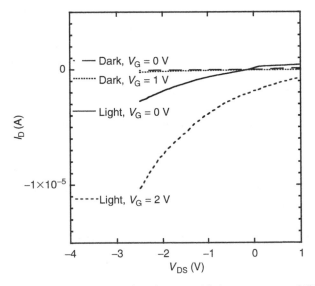

FIGURE 30.8 *I–V* characteristics of a MEFET photodetector with Pt gate contact. With light incident to the detector, current modulation was observed while the dark current was weakly dependent on V_G.

Most mid/far-infrared photodetectors were made in a conventional way that is based on intersubband transitions within the quantum well or self-assembled quantum dots, i.e., carriers within the wells or dots are first excited by incident photons from the ground state to the excited states or the continuous band, and subsequently measured as photoenhanced conductivity when these carriers are moved out from the wells and transported along the direction perpendicular to the potential wells [19].

Several physical limitations are, however, imposed for this type of vertical transport photodetector. First, the detection, which relied on removal of the excited carriers from the potential well for transport in the continuous band, suffers from thermal excitation, therefore conducting a large dark current at elevated temperatures. The smaller the transition energy, which is required for detection at longer wavelengths, the more severe the effect. Therefore, there is a tradeoff between the operation wavelength and temperature. Furthermore, during carrier transport there is a large probability that excited carriers can fall into the successive potential well, i.e., the so-called re-trapping mechanism, which limits the detection quantum efficiency.

In 2002, Bougeard et al. [20] demonstrated a novel approach on measuring the photoresponse of structures based on in-plane transport of holes photoexcited from self-assembled Ge dots in Si. As seen in Figure 30.9, the devices showed broad spectral response ranging from about 2 to 4:75 μm with a maximum around 3 μm. The addition of SiGe-QWs as conductive channels increased the photoresponse up to about 90 mA/W at 20 K, which was about 50 times better compared to the reference without using SiGe QWs.

Adnane et al. at Linköping have further developed this idea to manufacture Ge-dot-based detectors using an FET structure [21]. In these structures, one uses the larger dot size to allow detection at a longer wavelength (small transition energy), while deep trapping of photoexcited holes in the dot well with high thermal barrier may ensure a low dark current. The eventually measured source–drain current is determined by charge transfer via either the tunnelling or thermal excitation process, triggered by an emission field provided from the gate, from the discrete dots to a SiGe two-dimensional QW placed close to each dot layer as conducting channel, and would be amplified due to the geometrical design factor (*W/L* ratio) and the mobility in the channel.

The photoconductivity measurements were performed at 20 K using a glow-bar infrared light source in combination with different beam splitters forming the bandwidth in two ranges, i.e., 1.5 to 6 μm (CaF_2) and 3 to 15 μm (KBr). Some experimental results are summarized in Figure 30.10.

As revealed in Figure 30.10a, the FET detectors showed a very low dark current. The photoresponse was evidently observed when shining the device with the various bandwidths of infrared radiation. Much more pronounced photoconductivity was observed from the multifinger sample (Figure 30.10b), and a photoresponsivity value of ~100 mA/W was observed with a broadband source at 3 to 15 μm.

In summary, although the structures of transistor devices are often more complex than conventional two terminal diodes, with the implementation of the natural transistor function, one can fabricate devices with much improved photonic performance compared to the simple solutions. This has been seen in terms of both photoresponse and frequency–speed properties. The technologies used for

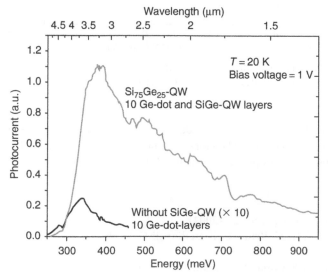

FIGURE 30.9 Photoconductivity spectra of a Si–Ge–Si–SiGe multilayer structure and a reference sample containing no SiGe QWs.

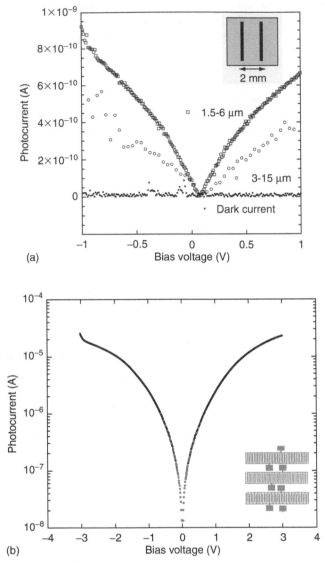

FIGURE 30.10   *I–V* characteristics with or without infrared radiation at various band widths were measured from two detectors (a) $W/L = 2$ and (b) $W/L = 3000$, at $T = 20\,\mathrm{K}$.

fabrication of Si-based photonic transistor devices are totally compatible with the mainstream Si technology for integration circuits. Therefore, we anticipate that further studies along this direction may bring more interesting and practically useful results toward Si-based optoelectronics.

References

1. K. Brunner. *Rep. Prog. Phys.* **65**, 27 (2002).
2. B. Zheng, J. Michel, F.Y.G. Ren, L.C. Kimerling, D.C. Jacobson, and J.M. Poate. *Appl. Phys. Lett.* **64**, 2842 (1994).
3. G. Franzò, F. Priolo, S. Coffa, A. Polman, and A. Carnera. *Appl. Phys. Lett.* **64**, 2235 (1994).
4. J. Stimmer, A. Reittinger, J.F. Nützel, G. Abstreiter, H. Holzbrecher, and CH. Buchal. *Appl. Phys. Lett.* **68**, 3290 (1996).

5. C.-X. Du, F. Duteil, G.V. Hansson, A. Elfving, and W.-X. Ni. *Appl. Phys. Lett.* **78**, 1697 (2001).

6. N.A. Sobolev, A.M. Emel'Yanov, S.V. Gastev, P.E. Khakuashev, Yu.A. Niklaev, and M.A. Trishenkov. In *Materials and Devices for Silicon-based Optoelectronics, Mat. Res. Soc. Symp. Proc.*, Vol. 486. Warrendale, Pennsylvania, 1998, p. 139.

7. W.-X. Ni, C.-X. Du, G.V. Hansson, A. Elfving, A. Vörckel, and Y. Fu. In *"Towards the First Si Laser" NATO Advanced Research Series*, L. Pavisi et al., eds. Kluwer Academic Publishers, The Netherland, 2003, pp. 429–444.

8. D. Pacifici, A. Irrera, G. Franz, M. Miritell, F. Iacona, and F. Priolo. *Phys. E* **16**, 331 (2003).

9. W. Shockley, M. Sparks, and G.K. Teal. *Phys. Rev.* **83**, 151 (1951).

10. S.M. Sze. In *Physics of Semiconductor Devices.* John Wiley & Sons, New York, 1981, pp. 783–786.

11. A. Leven, V. Houtsma, R. Kopf, Y. Baeyens, and Y.-K. Chen. In The Proceedings of IEEE International Microwave Symposium, Philadelphia, June 8–13, 2003.

12. A. Elfving, G.V. Hansson, and W.-X. Ni. *Phys. E* **16**, 528 (2003).

13. A. Elfving, M. Larsson, P.O. Holtz, G.V. Hansson, and W.-X. Ni. In *The MRS Proc.* **770**, I2.2 (2003).

14. W.-X. Ni, A. Elfving, M. Larsson, G.V. Hansson and P.-O. Holtz. In The Proceedings of the Third International Conference on SiGe(C) Epitaxy and Heterostructures, Santa Fe, March 9–12, 2003, pp. 251–254.

15. Z.W. Pei, C.S. Liang, L.S. Lai, Y.T. Tseng, Y.M. Hsu, P.S. Chen, S.C. Lu, M.-J. Tsai, and C.W. Liu. *IEEE Electron Dev. Lett.* **24**, 643 (2003).

16. A.H. Khalid and A.A. Rezazadeh. *IEE Proc. Optoelectron Vil.* **143**, 7 (1996).

17. M. Marso, M. Horstmann, H. Hardsdegen, and P. Kordos. *Solid-State Electron.* **42**, 197 (1998).

18. A. Elfving, G.V. Hansson, and W.-X. Ni. Extended abstract accepted by The First IEEE International Conference on Group-IV Photonics. Hong Kong, September 28–October 1, 2004.

19. J. Phillips, K. Kamath, and P. Bhattacharya. *Appl. Phys. Lett.* **72**, 2020 (1998).

20. D. Bougeard, K. Brunner, and G. Abstreiter. *Phys. E* **16**, 609 (2003).

21. B. Adnane, A. Elfving, M. Zhao, M. Larsson, B. Magnuson, and W.-X. Ni. Extended abstract accepted by The First IEEE International Conference on Group-IV Photonics, Hongkong, September 28–October 1, 2004.

31

Si–SiGe Quantum Cascade Emitters

Douglas J. Paul
University of Cambridge

31.1 Introduction

The major problem in silicon optoelectronics is the lack of a laser or efficient electroluminescent device. There have been many attempts to realize silicon-based lasers including porous silicon, erbium-doped silicon, and SiGe along with silicon nanocrystals [1,2]. The indirect bandgap of silicon precludes the efficient recombination of electrons and holes, which to date has prevented the realization of an interband laser. The quantum cascade laser (QCL) [3–5] is a unipolar laser utilizing intersubband transitions, and therefore, can be applied both to direct and indirect materials systems such as silicon. QCLs were originally proposed in 1971 [6] but the first experimental realization did not happen until 1994 using GaInAs and AlInAs heterostructures [7]. In particular for far-infrared or terahertz applications where no practical semiconductor materials exist with appropriate bandgaps, the potential for use in applications is high [8]. Potential terahertz applications include medical and dental imaging (for instance skin cancer detection) [8], security imaging [9], molecular spectroscopy, and bioweapons detection [10]. QCLs were first demonstrated at mid-infrared wavelengths [7] and more recently there have been a number of far-infrared demonstrations [11].

31.2 Population Inversion and Gain

The QCL principle relies on the intersubband emission of a photon (Figure 31.1) with the upper laser state designed to have population inversion by engineering the lifetime using bandgap engineering and subband lifetime engineering [3–5]. To date all demonstrated Si–SiGe quantum cascade emitters have been demonstrated using holes in the valence band. This is predominantly related to the heavy electron effective mass ($m^* \sim 0.918m_0$ [12] where m_0 is the free electron mass) in the tunneling direction of the conduction band of Si or $Si_{1-x}Ge_x$ ($x < 0.85$). The m^* in the transport direction of a tensile strained-Si quantum well is the lower $\sim 0.197m_0$ [12], which does not vary significantly with strain but this cannot be used for tunneling through quantum mechanical barriers grown on (0 0 1) substrates. A SiGe electron cascade device would require extremely thin tunnel barriers if miniband or efficient injection is to be attempted. Holes on the other hand have significantly smaller m^*, typically all well below $0.5m_0$, which

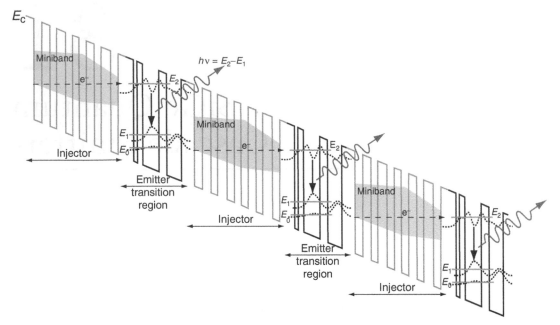

FIGURE 31.1 A schematic diagram of the conduction band of a quantum cascade laser. This particular design uses the E_2 subband state as the upper radiative transition level and the E_1 state as the lower. The energy between E_1 and the ground state of the quantum well, E_0 is set to the optical phonon energy so that the lower laser level has fast, nonradiative depopulation. This particular design uses a miniband injector using six quantum wells and three cascade periods are shown.

can also be engineering with strain. The light-hole m^* of pure Ge is only $0.044m_0$, which is lower than the electron m^* of GaAs. The strain also splits the light-hole and heavy-hole band degeneracy at $k = 0$. This significantly relaxes the growth requirements to achieve minibands or tunneling. To allow large numbers of strained layers to be grown coherently to a substrate well above the total critical thickness [13], quantum wells and tunnel barriers must be strain-symmetrized with alternating (and balanced) compressive and tensile strain, respectively (Figure 31.2). This, therefore, requires the layers to be latticed matched to a relaxed $Si_{1-y}Ge_y$ virtual substrate. The valence band discontinuities are also typically larger for the valence band compared to the conduction band when amenable virtual substrate germanium contents are considered (Figure 31.2).

The active heterostructure region of a QCL is where the population inversion and gain takes place. Figure 31.3 shows schematically a diagram of the subband energy levels of a three-level laser system, which are used to engineer population inversion. Electrons or holes are injected from an injector into the upper laser state, E_3 with an injector efficiency of η_3. The radiative transition is from level 3 to level 2 with the photon emission frequency given by $v = (E_3 - E_2)/h$ where h is Planck's constant. For population inversion assuming 100% injector efficiency into the E_3 state and no nonparabolicity the condition is simply $\tau_{32} > \tau_2$ where τ_{32}^{-1} is the nonradiative scattering rate from level 3 to level 2 and τ_2^{-1} is the total scattering rate out of level 2. In many designs this is achieved by fast depopulation of the lower laser state, E_2 to a lower energy subband, E_1 but any fast scattering or tunneling out of the lower laser state is beneficial if it reduces τ_2. In real quantum cascade active designs there are also scattering rates and unwanted injection into different levels, which decrease the gain in the system and require to be minimized. Injection from the injector to the lower laser state, E_2 with efficiency η_2 is clearly an unwanted process. Putting all these processes together, it can be demonstrated that the gain in the active region for a quantum cascade emitter is given by [14]

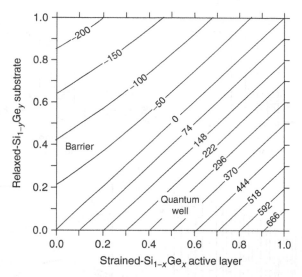

FIGURE 31.2 The valence band discontinuity of strained-$Si_{1-x}Ge_x$ layers grown on relaxed $Si_{1-y}Ge_y$ virtual substrates in meV. (From MM Rieger and P Vogl. *Phys. Rev. B* 48:14276–14287, 1993; *Phys. Rev. B* 50:8138, 1994. With permission.)

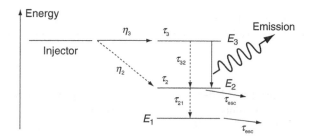

FIGURE 31.3 A schematic diagram of the energy levels in a quantum cascade laser along with the injection efficiency into different levels (η_x) and the lifetimes of states (τ_x), and transitions (τ_{xy}).

$$\text{Gain} = \sigma \Delta n = \sigma \frac{J}{q} \left[\tau_3 \eta_3 \left(1 - \frac{\tau_2}{\tau_{32}} \right) - \eta_2 \tau_2 \right] \tag{31.1}$$

where σ is the transition cross section, Δn is the population inversion between the E_3 and E_2 energy levels, J is the current density, and q is the electron charge. This equation demonstrates the importance of high injection efficiency into the upper laser state, η_3, the requirement of $\tau_{32} > \tau_2$ and the detrimental effects of injection into the lower laser state with efficiency η_2.

Figure 31.4 shows schematically four different designs for achieving population inversion in the active quantum cascade elements. Figure 31.4a shows a vertical radiative transition that uses a resonant LO optical phonon depopulation. The two lowest energy-hole subband states (that is higher up the page for the lowest hole energy) are set to be exactly the LO optical phonon energy apart. Therefore, transitions between the two states are fast and nonradiative providing fast depopulation of the lower radiative transition level, and therefore, population inversion can be attended in the upper energy level. If this technique is used for an intrawell cascade then it can only operate for energies above the optical phonon energy, which is 62 meV for silicon. An interwell or diagonal transitions is shown in Figure 31.4b. Such structures are easier to engineering in population inversion but have reduced matrix element compared to vertical or intrawell optical transitions. Figure 31.4c uses a miniband injector and an optical transition between miniband states. Miniband transitions allow higher currents, which can be important for

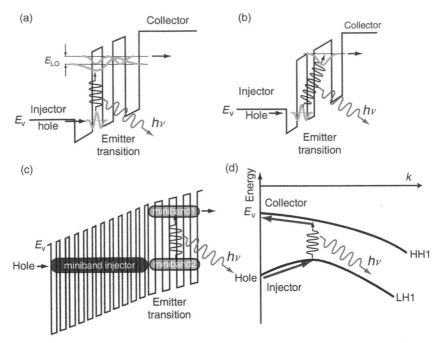

FIGURE 31.4 Schematic diagrams of four methods of achieving population inversion in a SiGe hole QCL. (a) Resonant LO optical phonon depopulation, (b) interwell or diagonal transition, (c) interminiband transitions, and (d) negative effective mass structure.

electrical pumping to produce linewidth narrowing followed by lasing. All the above three techniques have been used to produce QCLs in a number of materials in the III–Vs.

The design in Figure 31.4d is radically different to those discussed above [15]. The structure requires only quantum wells with tunnel barriers between the quantum wells. Since holes can only tunnel between quantum wells at $k = 0$ the structure is engineered to produce a radiative transition at finite k. This is achieved by bending the LH1 band upwards producing a negative effective mass. Since the selection rules forbid the LH1 and HH2 bands to cross, by trying to engineer the LH1 band to be higher in hole energy will result in an anticrossing with the HH2 state so that the LH1 band is forced to have a negative effective mass structure. The structure is engineered so that holes tunneling into the quantum well at $k = 0$ where they are forbidden for transitions to the HH1 band. They are scattered by alloy or hole–hole scattering to the minima in k where a radiative transition is allowed before scattered by alloy or hole–hole scattering back to $k = 0$ where they can tunnel to the next quantum well. The radiative transition requires to be engineered to be longer than the alloy or hole–hole scattering times so that population inversion can result. The problem with this structure in the Si–SiGe system is that while a negative effective mass can be produced at zero electric field, it is very weak when combined with strain symmetrization and can easily be removed by applying small electric fields. Unfortunately rather larger electric fields are required to align the subband states between wells to allow the holes to cascade.

31.3   Subband Lifetimes

The lifetime of subbands is important as they will determine whether a particular heterostructure design can achieve population inversion. These lifetimes can be engineered in a number of different ways including through the control of the quantum mechanical tunneling of an electron or hole to or from other states along with making subband states resonant with LO optical phonon transitions. A long nonradiative lifetime (τ_{32} in Figure 31.3) for the upper laser state is required, preferably much longer than the radiative lifetime.

Below the optical phonon energy in silicon of 62 meV (14.9 THz), a Si–SiGe QCL has potentially many advantages over a III–V based laser. While mature and cheap silicon process technology suggests a cheaper product and silicon has a higher thermal conductivity than most III–Vs, more importantly for the subband lifetimes there is no polar optical phonon scattering in Group IV semiconductors. Polar optical phonon scattering through interactions with the electrical dipole in the molecular bonds of GaAs results in a substantial decrease in the nonradiative lifetimes (τ_{32}) at temperatures above about 40 K [16,17]. Experiments in strained $Si_{1-x}Ge_x$ quantum wells have demonstrated almost constant nonradiative lifetimes (τ_{32}) of around 10 psec between 4 and 150 K for 24 meV transitions [18,19]. Modeling of the results suggests that alloy scattering is the dominant scattering mechanism, which has a very weak temperature dependence unlike polar optical phonon scattering [19]. More recent measurements on electrically biased $Si_{1-x}Ge_x$ quantum cascade structures have demonstrated almost constant nonradiative lifetimes as long as 30 psec up to room temperature (Figure 31.5) [20,21]. This demonstrates that group IV cascades have significant advantages over III–V devices below the optical phonon energy and should be able to operate at higher temperatures.

For transitions above the optical phonon energy (62 meV for silicon and 37 meV for germanium), the nonradiative lifetime is substantially reduced. Experiments in strained $Si_{1-x}Ge_x$ quantum wells have demonstrated 250 fsec lifetimes for transitions of 167 meV [22]. One method of increasing the lifetime is to use an interwell transition and the lifetime can be increased up to about 10 psec by increasing the thickness of the barrier between the wells [23]. The disadvantage of this technique is that the optical matrix element for radiative transitions is reduced (that is the gain is reduced) as the barrier

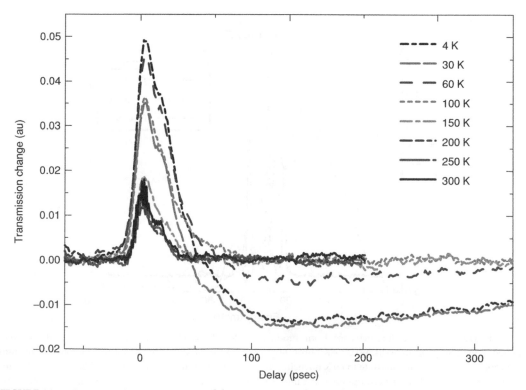

FIGURE 31.5 Pump–probe measurements of the nonradiative lifetime of a LH1 to HH1 transition in a quantum cascade structure at 14 meV (about 3.4 THz or 89 µm) [20,21].

thickness is increased. A second technique is to use minibands [24] or coupled quantum wells [25] where the wave function spread over a number of quantum wells results in signficantly longer non-radiative lifetimes.

31.4   Impurity Electroluminescence

The p-Ge laser uses crossed electric and magnetic fields to produce population inversion between the split hydrogen-like impurity states in the semiconductor [26]. The disadvantage of such a laser is that the large magnetic fields along with the 4 K operating temperature makes such lasers impractical for many applications. Figure 31.6 shows the electroluminescence by applying an electric field along ten boron modulation-doped $Si_{0.72}Ge_{0.28}$ quantum wells grown on top of a $Si_{0.78}Ge_{0.22}$ virtual substrate with strain-symmetrized Si barriers [19,20]. Six-band $\mathbf{k \cdot p}$ theory predicted a broad spontaneous electro-luminescent peak due to a large number of states in $\mathbf{k}$-space available for radiative transitions. At 4 K, however, three very sharp features were observed, which correspond to boron impurity transitions at 30.4 meV ($1s-2p_0$), 34.5 meV ($1s-2p_{\pm}$), and 39.6 meV ($1s-3p_0$) [27,28]. Only above 20 K is the impurity emission quenched and intersubband radiative transitions are observed. When the intersub-band emission dominates, there is strong absorption at the impurity lines. The mechanism for this impurity emission is still not fully understood. For the impurity state lasers, magnetic fields are required to achieve sufficient splitting of the appropriate laser energy levels but no magnetic fields have been applied to the SiGe samples. This suggests that the strain must be involved at some level in splitting the impurity energy levels. Strain, however, will also shift the positions of the impurity lines in energy compared to bulk or relaxed Si or SiGe and yet the positions agree with absorption data on bulk silicon

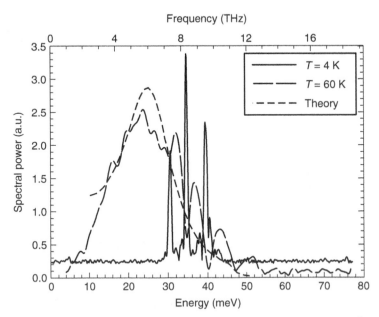

FIGURE 31.6 The electroluminescence from modulation-doped $Si_{0.7}Ge_{0.3}$ quantum wells with the current applied along the quantum wells. (a) At 4 K boron impurity states emit and no intersubband radiative transitions are visible. (b) Heating the substrate to 60 K quenches the impurity transitions and the intersubband electrolumin-escence dominates with strong absorption at the impurity lines. The width of the transition agrees well with a six-band $\mathbf{k \cdot p}$ theory and is broad due to the large number of states in $\mathbf{k}$-space over which radiative transitions are allowed [19,20].

wafers suggesting that no strain is involved. This strongly suggests that it is the relaxed p-Si$_{0.78}$Ge$_{0.22}$ supply layers, which are emitting and not the quantum wells. It does create a potential problem as while there is no forbidden Reststrahlen band in Group IV materials, for p-type Si or SiGe the impurities may create an energy region in which intersubband emission may be dominated by interactions with boron impurities.

31.5 Electroluminescence from Quantum Cascade Emitters

All existing III–V QCLs have been n-type, and therefore, limited to only edge-emitting devices unless a grating is used to scatter or couple light out of the surface of the device. The optical matrix element is nonzero only for interactions with the electric field dipole that is oriented perpendicular to the quantum well layers (TM polarization), hence the emitted radiation can only propagate parallel to the quantum well resulting in edge emission. The use of light-hole to heavy-hole transitions results in a finite matrix element in the plane of the quantum wells resulting in a TE mode, thereby allowing surface-normal emission without a grating.

The first Si–SiGe quantum cascade emitter was demonstrated at mid-infrared frequencies in 2000 [29] using intrawell HH2 to HH1 transitions. A diagonal or interwell HH2 to HH1 transition has also been demonstrated in the mid-infrared [30]. The major problem with these two demonstrations was that the structures were pseudomorphically grown on bulk silicon substrates, limiting the number of active periods. The first strain-symmetrized cascade was demonstrated at terahertz frequencies [31] using an intrawell LH1 to HH1 transition. As demonstrated in Figure 31.7, this was also the first surface-normal emitting quantum cascade in any materials system since the LH1 to HH1 transition has a TE polarized component. The spontaneous emission peak in Figure 31.7 is significantly wider than comparable GaAs electron cascades, which is the result of a large number of allowed transitions for many different *k*-values. The nonparabolicity of the valence band will also broaden the transition. Strain-symmetrized mid-infrared quantum cascades using a bound-to-continuum transition have also been demonstrated [32].

Figure 31.8 shows a TEM image of the bottom periods of a strain-symmetrized interwell quantum cascade structure grown by CVD. This wafer has 600 periods of 6.5 nm Si$_{0.7}$Ge$_{0.3}$ quantum wells with 2 nm strained-Si barriers all grown on top of a Si$_{0.8}$Ge$_{0.2}$ virtual substrate. A 200 nm boron-doped

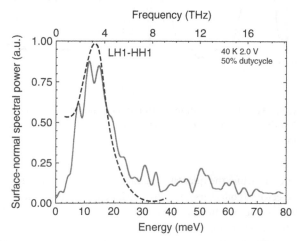

FIGURE 31.7 The surface-normal emission from an intrawell LH1 to HH1 transition electroluminescence in a 30-period quantum cascade structure [31]. The dashed line is 6 band **k·p** theory.

FIGURE 31.8 A TEM picture of the bottom of a stack of a 600 period strain-symmetrized quantum cascade emitter of total thickness 5 μm. Below the lowest quantum well is a graded SiGe injector (collector) and 200 nm of boron-doped $Si_{0.8}Ge_{0.2}$ as the bottom contact layer.

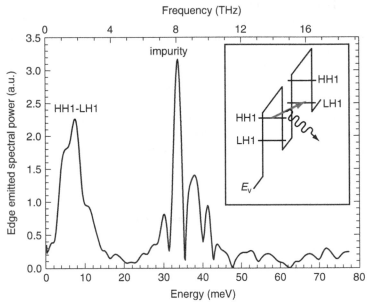

FIGURE 31.9 An interwell HH1 to LH1 transition electroluminescence in a 100 period quantum cascade structure at 700 mA current with 5% duty cycle and 5 K [33]. The inset shows a schematic diagram of the HH1 to LH1 diagonal radiative transition.

Ohmic contact layer is grown below the quantum cascade structure and a thinner 40 nm doped layer on the top. Graded SiGe injectors (collectors) are also designed to inject holes into the correct subband when an electric field is placed across the device. Such structures are pushing the SiGe growth technology to new limits with the thin layers required for tunneling structures.

The electroluminescence from a 100-period interwell (diagonal) transition is shown in Figure 31.9 [33]. The transition (see inset) is from the HH1 ground state in one quantum well to the LH1 state in the

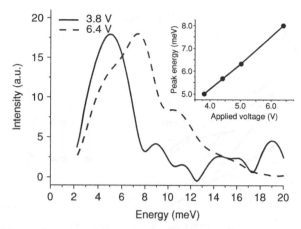

FIGURE 31.10 The terahertz edge-emitted electroluminescence from an interwell HH1 to LH1 transition. The inset shows the movement of the peak as a function of the electrical bias [33].

adjacent well. By changing the voltage across the device the transition can be moved in energy (Figure 31.10). In addition to the intersubband electrolumninescence, this device also demonstrates emission from the impurity states mentioned above between 30 and 40 meV (Figure 31.9). The normal method of proving that the emission is intersubband rather than blackbody is to reverse the voltage demonstrating emission only when the subband levels are aligned. For symmetrical quantum well designs this cannot be achieved and measurement of the polarization along with a demonstration of a linear light output power as a function of current (blackbody is unpolarized with a power dependent on the square of the current) is required.

Figure 31.11 plots the power outputs of the terahertz quantum cascade emitters as a function of frequency against the main compact devices available at these frequencies. As no calibration standard is available in the terahertz, the power levels have an error of up to an order of magnitude for devices emitting between 1 and 10 THz. All the quantum cascade devices plotted have been measured using liquid He cooled Si bolometers calibrated using blackbody sources. The impurity transitions from p-SiGe have demonstrated the highest output powers in 600-period quantum cascade structures [34]. The multiplication of the gain in the GaAs quantum cascade structures produced a six order of magnitude increase in power from a LED to a laser and so the higher demonstrated output powers from the Si–SiGe quantum cascades bode well for high-power emission when a laser is produced [11]. The higher output powers along with the weaker temperature dependence of the nonradiative transitions bode well for higher temperature operation of silicon QCLs when realized.

31.6 Si–SiGe Quantum Cascade Lasers

At present no Si–SiGe QCL has been produced. Population inversion has been demonstrated [33] but at present attempts at producing cavities have not demonstrated structures with enough gain to lase. The major problem has been the confinement of the mode in the vertical direction, resulting in poor modal overlap, which is the amount of the mode overlaps with the active gain heterostructure layers of the device. Since the wavelengths used in the mid- and particularly the far-infrared are larger than the thicknesses of the heterolayers of the laser even when divided by the refractive index, the mode leaks into the substrate producing poor modal overlap and high losses. In GaAs, heavily doped layers have been used to create plasmons to confine the mode in the vertical direction. In silicon material the inversion of the real and imaginary parts of the dielectric constant does not occur in the mid- or far-infrared since

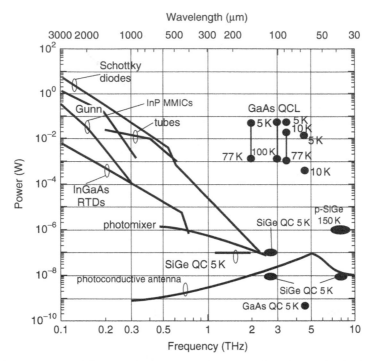

FIGURE 31.11 A comparison of the output power versus frequency of a number of radiation sources between 100 GHz and 10 THz. (From DJ Paul. UK DTI Foresight Exploiting the Electromagnetic Spectrum Report (http://www.dti.gov.uk/). With permission.)

the electrical conductivity of the doped silicon is not high enough to allow plasmon reflectors to be produced [35].

There are two potential methods to provide vertical confinement of the mode. The first is to etch away the substrate of the wafer and deposit a metallic reflector on the bottom side of the active heterolayers. This has been demonstrated on GaAs QCLs operating at about 3.2 THz or 94 μm wavelength [36] and results in improved high-temperature operation of the GaAs devices. The silicon system allows a second more amenable solution. Silicon-on-insulator (SOI) and buried-silicide layers are available, which have already been used as reflectors in mid-infrared quantum-well photodetectors (QWIPs) [37,38]. In particular, the buried silicide layer has an electrical conductivity within an order of magnitude of the best metals and provides excellent confinement of the mode in the vertical direction [34,35].

Figure 31.12 shows the first attempts at growing a virtual substrate and strain-symmetrized cascade emitter on top of a silicon-on-silicide wafer. In this particular structure, the modal overlap has been increased from 18% to over 44% by the inclusion of the silicide layer if a metal surface plasmon reflector is used on top of the wafers. The major problem is that the top silicon layer of the silicon-on-silicide wafer and the SiGe-relaxed buffer still contribute a large amount of lossy material inside the cavity and these layers need to become a smaller percentage of the total thickness in future wafers if a laser is to be realized. The modal overlap and waveguide losses of this particular structure when fabricated into ridge waveguides are comparable to the values in demonstrated GaAs THz lasers [11,36]. The cascade emitter has demonstrated electroluminescence with only a TM-polarized component when the HH1 to LH1 interwell electroluminescence is measured. At higher currents, strong heating effects are absorbed due to the poorer thermal conductivity of the substrate compared to bulk silicon. The work does demonstrate that Si–SiGe cascade emitters can benefit from the rich technology basis available in silicon technology.

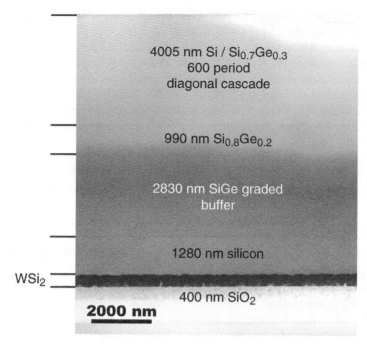

FIGURE 31.12 A TEM picture of a 600 periods interwell Si–SiGe strain-symmetrized cascade emitter grown on top of a silicon-on-silicide wafer using CVD. The silicide is 440-nm thick and is fabricated by wafer bonding on top of an oxidized silicon hand wafer [34]. (From DJ Paul, SA Lynch P Townsend, Z Ikonić, RW Kelsall, P Harrison, SL Liew, DJ Norris, AG Cullis, J Zhang, HS Gamble, WR Tribe, and DD Arnone. Int. Semicond. Dev. Research Symp., Washington, USA, December 2003. With permission.)

31.7 Summary

A number of different Si–SiGe quantum cascade emitters have been demonstrated operating at frequencies from 1.2 THz (250 µm) in the far-infrared or terahertz up to 42 THz (7.1 µm) in the mid-infrared. To date no laser has been demonstrated but a number of different approaches are pursued to circumvent the present known problems. If a silicon laser can be realized then it should have significant advantages over III–V lasers for operation below the silicon optical phonon energy of 62 meV. With the present knowledge and effort in the field, it should be only a matter of time before a silicon unipolar laser is produced.

Acknowledgments

The work involved in this paper has been supported by DARPA, EPSRC, and through the EC program SHINE (IST-2001-38035). The author acknowledges the contributions of his numerous colleagues especially at the Universities of Cambridge, Leeds, Sheffield, Queens Belfast, and Imperial College along with TeraView Ltd. in the production of many of the results and pictures used in this review article.

References

1. L Pavesi. Will silicon be the photonic material of the third millenium. *J. Phys.: Condens. Matter* 15:R1169–R1196, 2003.
2. L Pavesi, S Gaponenko, and L Dal Negro, eds. *Towards the First Silicon Laser* (NATO Series, Vol. 93). New York: Kluwer, 2003.

3. C Gmachl, F Capasso, DL Sivco, and AY Cho. Recent progress in quantum cascade lasers and applications. *Rep. Prog. Phys.* 64:1533–1601, 2001.

4. F Capasso, R Paiella, R Martini, R Colombelli, C Gmachl, TL Myers, MS Taubman, RM Williams, CG Bethea, K Unterrainer, HY Hwang, DL Sivco, AY Cho, AM Sergent, HC Liu, and EA Whittaker. Quantum cascade lasers: ultrahigh-speed operation, optical wireless communications, narrow line-width and far-infrared emission. *IEEE J. Quantum Elec.* 38:511–532, 2002.

5. C Gmachl, A Straub, R Colombelli, F Capasso, DL Sivco, AM Sergent, and AY Cho. Single-mode, tunable distributed-feedback and multiple-wavelength quantum cascade lasers. *IEEE J. Quantum Elec.* 38:569–581, 2002.

6. RF Kazarinov and RA Suris. Possibility of the amplification of electromagnetic waves with a semiconductor superlattice. *Sov. Phys. Semiconductors* 5:707–709, 1971.

7. J Faist, F Capasso, DL Sivco, C Sirtori, AL Hutchinson, and AY Cho. Quantum cascade laser. *Science* 264:553–556, 1994.

8. DD Arnone, CM Ciesla, and M Pepper. Terahertz imaging comes into view. *Phys. World* 13(4):35, 2000.

9. MC Kemp, PF Taday, BE Cole, JA Cluff, AJ Fitzgerald, and WR Tribe. Security applications of terahertz technology. *Proc. SPIE* 5070:44–52, 2003.

10. F Oliveira, R Barat, B Shulkin, J Federici, and D Gary. Neural network analysis of terahertz spectra of explosives and bio-agents. *Proc. SPIE* 5070:60–70, 2003.

11. R Köhler, A Tredicucci, F Beltram, HE Beere, EH Linfield, AG Davies, DA Ritchie, RC Iotti, and F Rossi. Terahertz semiconductor laser. *Nature* 417:156–158, 2002.

12. MM Rieger and P Vogl. Electronic-band parameters in strained $Si_{1-x}Ge_x$ alloys on $Si_{1-y}Ge_y$ substrates. *Phys. Rev. B* 48:14276–14287, 1993; *Phys. Rev. B* 50:8138, 1994.

13. JW Matthews and AE Blakeslee. Defects in epitaxial multilayers. *J. Cryst. Growth* 32:265–271, 1976.

14. J Faist, M Beck, T Aellen, and E Gini. Quantum-cascade lasers based on bound-to-continuum transitions. *Appl. Phys. Lett.* 78:147–149, 2001.

15. L Friedman, G Sun, and RA Soref. SiGe/Si THz laser based on transitions between inverted mass light-hole and heavy-hole subbands. *Appl. Phys. Lett.* 78:401–403, 2001.

16. BN Murdin, W Heiss, CJGM Langerak, SC Lee, I Galbraith, G Strasser, E Gornik, M Helm, and CR Pidgeon. Direct observation of the LO phonon bottleneck in wide $GaAs/Al_xGa_{1-x}As$ quantum wells. *Phys. Rev. B* 55:5171–5176, 1997.

17. CD Bezant, MM Chamberlain, PM Pellemans, BN Murdin, W Batty, and M Henini. Intersubband relaxation lifetimes in p-GaAs/AlGaAs quantum wells below the LO-phonon energy measured in a free electron laser experiment. *Semicond. Sci. Technol.* 14:L25–L28, 1999.

18. P Murzyn, CR Pidgeon, J-PR Wells, IV Bradley, Z Ikonić, RW Kelsall, P Harrison, SA Lynch, DJ Paul, DD Arnone, DJ Robbins, DJ Norris, and AG Cullis. Picosecond intersubband dynamics in p-Si/SiGe quantum-well emitter structures. *Appl. Phys. Lett.* 80:1456–1458, 2002.

19. DJ Paul, SA Lynch R Bates, Z Ikonić, RW Kelsall, P Harrison, DJ Norris, SL Liew, AG Cullis, DD Arnone, CR Pidgeon, P Murzyn, J-PR Wells, and IV Bradley. Si/SiGe quantum cascade emitters. *Physica E* 16:147, 2003.

20. DJ Paul, SA Lynch R Bates, Z Ikonić, RW Kelsall, P Harrison, DJ Norris, SL Liew, AG Cullis, CR Pidgeon, DD Arnone, and DJ Robbins. Electroluminescence from Si/SiGe quantum cascade emitters. *Physica E* 16:309, 2003.

21. CR Pidgeon, P Murzyn, SA Lynch, and DJ Paul. Measurement of the non-radiative lifetime in a biased Si/SiGe quantum cascade emitter. Unpublished.

22. RA Kaindl, M Woerner, M Wurm, K Reimann, T Elsaesser, C Miesner, K Brunner, and G Abstreiter. Femtosecond intersubband scattering of holes in $Si_{1-x}Ge_x$ quantum wells. *Physica B* 314:255, 2002.

23. I Bormann, K Brunner, S Hackenbuchner, G Abstreiter, S Schmult, and W Wegscheider. Nonradiative relaxation times in diagonal transition Si/SiGe quantum cascade structures. *Appl. Phys. Lett.* 83:5371–5373, 2003.

24. F Capasso, A Tredicucci, C Gmachl, Dl Sivco, AL Hutchinson, AY Chu, and G Scamarcio. High-performance superlattice quantum cascade lasers. *IEEE J. Sel. Topics Quant. Elec.* 5:792–807, 1999.
25. L Diehl, A Borak, S Mentese, D Grützmacher, H Sigg, U Gennser, I Sagnes, Y Campidelli, O Kermarrec, D Bensahel, and J Faist. Anticrossing between heavy-hole states in $Si_{0.2}Ge_{0.8}$/Si-coupled quantum wells grown on $Si_{0.5}Ge_{0.5}$ pseudosubstrates. *Appl. Phys. Lett.* 84:2497–2499, 2004.
26. S Komiyama, S Kuroda, I Hosako, Y Akasaka, and N Iizuka. Germanium lasers in the range from far-infrared to millimetre waves. *Optical Quant. Elec.* 23:S133–S162, 1991.
27. HJ Hrostowski. Infrared absorption of semiconductors. In *Semiconductors*, NB Hannay, ed. Reinhold: New York, 1959, pp. 437–481.
28. C Jagannath, ZW Grabowski, and AK Ramdas. Linewidths of the electronic excitation spectra of donors in silicon. *Phys. Rev. B* 23:2082–2098, 1981.
29. G Dehlinger, L Diehl, U Gennser, H Sigg, J Faist, K Ensslin, D Grützmacher, and E Müller. Intersubband electroluminescence from SiGe quantum cascade structure. *Science* 290:2277–2279, 2000.
30. I Bormann, K Brunner, S Hackenbuchner, G Zandler, G Abstreiter, S Schmult, and W Wegscheider. Midinfrared intersubband electroluminescence of Si/SiGe quantum cascade strucrtures. *Appl. Phys. Lett.* 80:2260–2262, 2002.
31. SA Lynch, R Bates, DJ Paul, DJ Norris, AG Cullis, Z Ikonić, RW Kelsall, P Harrison, DD Arnone, and CR Pidgeon. Intersubband electroluminescence from Si/SiGe cascade emitters at terahertz frequencies. *Appl. Phys. Lett.* 81:1543–1545, 2002.
32. L Diehl, S Mentese, E Müller, D Grützmacher, H Sigg, U Gennser, I Sagnes, Y Campidelli, O Kermarrec, D Bensahe, and J Faist. Electroluminescence from strain-compensated $Si_{0.2}Ge_{0.8}$ quantum-cascade structures based on a bound-to-continuum transition. *Appl. Phys. Lett.* 81:4700–4702, 2002.
33. R Bates, SA Lynch, DJ Paul, Z Ikonić, RW Kelsall, P Harrison, SL Liew, DJ Norris, AG Cullis, WR Tribe, and DD Arnone. Interwell intersubband electroluminescence from Si/SiGe quantum cascade emitters. *Appl. Phys. Lett.* 83:4092–4095, 2003.
34. DJ Paul, SA Lynch P Townsend, Z Ikonić, RW Kelsall, P Harrison, SL Liew, DJ Norris, AG Cullis, J Zhang, HS Gamble, WR Tribe, and DD Arnone. Si/SiGe terahertz quantum cascade emitters. Int. Semicond. Dev. Research Symp., Washington, USA, December 2003.
35. Z Ikonić, P Harrison, and RW Kelsall. Waveguide design for mid- and far-infrared p-Si/SiGe quantum cascade lasers. *Semicond. Sci. Technol.* 19:76–81, 2004.
36. S Kumar, BS Williams, S Kohen, Q Hu, and LJ Reno. Continuous wave operation of terahertz quantum-cascade lasers above liquid nitrogen temperature. *Appl. Phys. Lett.* 84:2494–2496, 2004.
37. RT Carline, DJ Robbins, MB Stanaway, and WY Leong. Long-wavelength SiGe/Si resonant cavity infrared detector using a bonded silicon-on-oxide reflector. *Appl. Phys. Lett.* 68:544–546, 1996.
38. RT Carline, V Nayar, DJ Robbins, and MB Stanaway. Resonant cavity longwave SiGe-Si photodetector using a buried silicide mirror. *IEEE Photon. Tech. Lett.* 10:1775–1777, 1998.
39. DJ Paul. Picturing People: non-intrusive imaging. UK DTI Foresight Exploiting the Electromagnetic Spectrum Report (http://www.dti.gov.uk/).

A.1

Properties of Silicon and Germanium

John D. Cressler
Georgia Institute of Technology

The energy band structures of Si and Ge are depicted in Figure A.1.1, together with (1) their carrier effective mass parameters (Table A.1.1) and (2) their bulk structural, mechanical, optical, and electrical properties (Table A.1.2) [1–3].

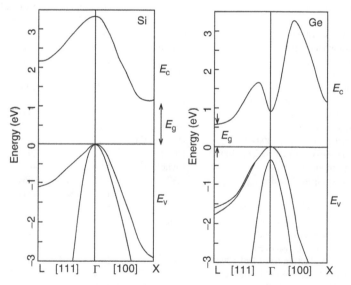

FIGURE A.1.1 Energy band structure, showing the principal conduction and valence bands of Si and Ge as a function of *k*-space direction. (From M Shur. *Physics of Semiconductor Devices.* Englewood Cliffs, NJ: Prentice-Hall, 1990. With permission.)

TABLE A.1.1   Carrier Effective Mass Parameters for Si and Ge

Parameter	Units	Silicon	Germanium
Effective electron mass (m_n^*)	($\times m_o$)		
Longitudinal (4.2 K)		0.9163	1.58
Transverse (4.2 K)		0.1905	0.082
Density-of-states (4.2 K)		1.062	—
Density-of-states (300 K)		1.090	—
Effective hole mass (m_p^*)	($\times m_o$)		
Heavy hole (4.2 K)		0.537	0.28
Light hole (4.2 K)		0.153	0.044
Density-of-states (4.2 K)		0.59	—
Density-of-states (300 K)		1.15	—

TABLE A.1.2   Properties of Bulk Si and Ge

Parameter	Units	Silicon	Germanium
Atomic number	—	14	32
Atomic density	(atoms/cm^3)	5.02×10^{22}	4.42×10^{22}
Atomic weight	(g/mole)	28.09	72.6
Density	(g/cm^3)	2.329	5.323
Electronic orbital configuration	—	(Ne) $3s^2 3p^2$	(Ar) $3d^{10} 4s^2 4p^2$
Crystal structure	—	Diamond	Diamond
Lattice constant (298 K)	(Å)	5.43107	5.65791
Dielectric constant	—	11.7	16.2
Breakdown strength	(V/cm)	3×10^5	1×10^5
Electron affinity	(V)	4.05	4.00
Specific heat	(J/g-°C)	0.7	0.31
Melting point	(°C)	1412	1240
Intrinsic Debye length (300 K)	(μm)	24	0.68
Index of refraction	—	3.42	3.98
Transparency region	(μm)	1.1–6.5	1.8–15
Thermal conductivity (300 K)	(W/cm-°C)	1.31	0.60
Thermal expansion coefficient (300 K)	(°C^{-1})	2.6×10^{-6}	5.9×10^{-6}
Young's modulus	(dyne/cm^2)	1.9×10^{12}	—
Energy bandgap (low doping)	(eV)	1.12 (300 K)	0.664 (291 K)
		1.17 (77 K)	0.741 (4.2 K)
Equivalent conduction band minima	—	6	8
Effective electron mass (300 K)	($\times m_o$)	1.18	—
Effective hole mass (300 K)	($\times m_o$)	0.81	—
Intrinsic carrier density (300 K)	(cm^{-3})	1.02×10^{10}	2.33×10^{13}
Effective conduction band DoS (300 K)	(cm^{-3})	2.8×10^{19}	1.04×10^{19}
Effective valence band DoS (300 K)	(cm^{-3})	1.04×10^{19}	6.00×10^{18}
Electron mobility (300 K)	(cm^2/V-sec)	1450	3900
Hole mobility (300 K)	(cm^2/V-sec)	500	1900
Electron diffusivity (300 K)	(cm^2/sec)	37.5	100
Hole diffusivity (300 K)	(cm^2/sec)	13	49
Optical phonon energy	(meV)	63	37
Phonon mean free path length	Å	76	105
Intrinsic resistivity (300 K)	(Ω cm)	3.16×10^5	47.62

References

1. JD Cressler and G Niu. *Silicon–Germanium Heterojunction Bipolar Transistors.* Boston, MA: Artech House, 2003.
2. M Shur. *Physics of Semiconductor Devices.* Englewood Cliffs, NJ: Prentice-Hall, 1990.
3. R. Hull, editor. *Properties of Crystalline Silicon.* London: EMIS Datareviews Series, Number 20, INPSEC, 1999.

A.2

The Generalized Moll–Ross Relations

John D. Cressler
Georgia Institute of Technology

The classical solution for the collector current density in a Si BJT, derived by Shockley, necessarily assumes a constant base doping profile. In this case, for low-injection conditions, the drift component of the minority carrier transport equation can be neglected, and the minority carrier diffusion equation solved under the Shockley boundary conditions. The resultant equation, under the assumptions of negligible neutral base recombination and forward-active bias, is the well-known expression

$$J_C = \frac{q D_{nb}}{N_{ab}^- W_b} \, n_{io}^2 e^{\Delta E_{gb}^{app}/kT} \left(e^{q V_{BE}/kT} - 1 \right). \tag{A.2.1}$$

Here, D_{nb} is the minority electron diffusivity, N_{ab}^- is the ionized base doping level, n_{io}^2 is the low-doping intrinsic carrier density, given by,

$$n_{io}^2 = N_C N_V e^{-E_{go}/kT}, \tag{A.2.2}$$

and $\Delta E_{gb}{}^{app}$ is the heavy-doping induced bandgap narrowing.

The path to the generalization of this result to the "real-world" case of a nonconstant base doping profile (Figure A.2.1) is nonobvious, and even a cursory glance at the problem is enough to convince one that it cannot follow the original path in Shockley's approach. The complexity of this problem results from the addition of the field-driven transport, which is now no longer negligible due to the doping-gradient-induced field. The clever solution to this problem was first presented in the classic paper by Moll and Ross in 1956, the so-called "Moll–Ross relation" [1]. Unfortunately, that solution made two undesirable assumptions: (1) that the minority electron mobility (hence, diffusivity) is constant across the quasineutral base and (2) that the intrinsic carrier density is constant across the quasineutral base. The latter assumption, in particular, fails in the presence of a heavily doped base (i.e., real life), since the apparent bandgap narrowing is inherently position dependent across the base, and hence the effective bandgap in the base is also position dependent.* In essence, then, the problem becomes one of solving for the collector current density in the presence of *both* nonconstant base doping and nonconstant base bandgap, and is particularly relevant to the graded-base SiGe HBT. This problem remained unsolved for

---

*In fairness, Moll and Ross cannot be blamed for the second assumption since doping-induced bandgap narrowing had not yet been discovered.

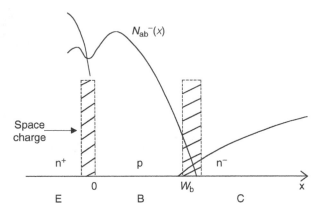

FIGURE A.2.1 Schematic nonconstant base doping profile used in the derivations.

almost 30 years until the seminal paper by Kroemer in 1985 [2]. Since Kroemer's "generalized Moll–Ross relations" are the starting point for both the dc and ac analysis of the graded-base SiGe HBT (Chapter 4), we present that elegant derivation here (showing all of the mathematical steps that Kroemer neglected to include in his paper).

The assumptions in Kroemer's solution include: (1) 1-D transport, (2) transport by both drift and diffusion, (3) low-injection conditions (i.e., $n_b(x) \ll N_{ab}^-(x)$ for all x across the base), (4) negligible neutral base recombination, and (5) forward-active bias. Importantly, however, there are no assumptions on the position dependence of the base doping profile or the base bandgap.*

We begin from the generalized drift–diffusion minority electron transport equation, as expressed in terms of the minority electron quasi-Fermi potential

$$J_n = q\mu_n n \nabla \phi_n, \tag{A.2.3}$$

which for our 1-D Si BJT problem reduces to

$$J_C = q\mu_{nb}(x)n_b(x)\frac{d\phi_n(x)}{dx}. \tag{A.2.4}$$

In the quasineutral base, the majority carrier (hole) quasi-Fermi potential (ϕ_p) in low-injection is constant, such that

$$J_C = q\mu_{nb}(x)n_b(x)\frac{d}{dx}\left(\phi_n(x) - \phi_p\right), \tag{A.2.5}$$

and from the generalized Shockley boundary condition

$$n_b(x)p_b(x) = n_{ib}^2(x)e^{q(\phi_n(x)-\phi_p)/kT}, \tag{A.2.6}$$

which can be rewritten as

$$\frac{kT}{q}\ln\left\{\frac{n_b(x)p_b(x)}{n_{ib}^2(x)}\right\} = \phi_n(x) - \phi_p. \tag{A.2.7}$$

---

*Interestingly, additional generalizations to Kroemer's result have been recently offered [3]. Let it never be said that the final word in device physics is ever in.

Taking the derivative of both sides we have

$$\frac{kT}{q}\left\{\frac{n_{ib}^2(x)}{n_b(x)p_b(x)}\right\}\frac{d}{dx}\left\{\frac{n_b(x)p_b(x)}{n_{ib}^2(x)}\right\} = \frac{d}{dx}\left(\phi_n(x) - \phi_p\right). \tag{A.2.8}$$

Substituting this result back into Equation A.2.5, we obtain

$$J_C = q\mu_{nb}(x)n_b(x)\frac{kT}{q}\frac{n_{ib}^2(x)}{n_b(x)p_b(x)}\frac{d}{dx}\left\{\frac{n_b(x)p_b(x)}{n_{ib}^2(x)}\right\} \tag{A.2.9}$$

We now integrate this expression from some arbitrary point in the base profile to the neutral base boundary (W_b) to obtain

$$\int_x^{W_b}\frac{J_C}{qD_{nb}(x')}\frac{p_b(x')}{n_{ib}^2(x')}dx' = \left.\frac{n_b(x')p_b(x')}{n_{ib}^2(x')}\right|_x^{W_b} \tag{A.2.10}$$

Under the assumptions of negligible neutral base recombination (i.e., J_C is a constant to the integration), and using the fact that in forward-active bias,

$$p_b(W_b)n_b(W_b) \simeq n_{ib}^2(W_b) \tag{A.2.11}$$

we find

$$\frac{J_C}{q}\int_x^{W_b}\frac{p_b(x')dx'}{D_{nb}(x')n_{ib}^2(x')} = 1 - \frac{n_b(x)p_b(x)}{n_{ib}^2(x)} \tag{A.2.12}$$

At the emitter–base boundary ($x = 0$), we know from the generalized Shockley boundary condition that

$$n_b(0)p_b(0) = n_{ib}^2(0)e^{q(\phi_n(0)-\phi_p(0))/kT}, \tag{A.2.13}$$

and

$$\phi_n(0) - \phi_p(0) = V_{BE}, \tag{A.2.14}$$

so that we obtain

$$\frac{J_C}{q}\int_0^{W_b}\frac{p_b(x)dx}{D_{nb}(x)n_{ib}^2(x)} = 1 - e^{qV_{BE}/kT}, \tag{A.2.15}$$

and thus finally,

$$J_C = -\frac{q(e^{qV_{BE}/kT} - 1)}{\int_0^{W_b}\frac{p_b(x)dx}{D_{nb}(x)n_{ib}^2(x)}}, \tag{A.2.16}$$

This is the "generalized Moll–Ross relation"* for the collector current density in a bipolar transistor with nonconstant base doping and arbitrary position-dependence of the base bandgap. Observe that if

---

*I personally would have no problem calling this elegant result the "Kroemer relation."

we allow $p_b(x) = N_{ab}^-(x) = N_{ab}^- = $ constant, then we obtain Equation A.2.1, as expected (the extra negative sign simply accounts for the fact that the electron flow is in the opposite direction of the positive current flow).

As detailed in Chapter 4, this fundamental result is the starting point of the derivations for collector current density, the current gain, and the output conductance in a graded-base SiGe HBT. In this case, in addition to the bandgap-narrowing-induced position dependence in the base bandgap, we have an additional contribution from the Ge-strained layer (Figure A.2.2). This Ge contribution easily enters the generalized Moll–Ross relation via n_{ib}^2 in Equation A.2.16. For more detail on the resultant derivations and the assumptions and approximations involved, the reader is referred to Ref. [4].

An additional desirable feature of Kroemer's approach is that we can also easily obtain an analytical expression for the base transit time in a device with nonconstant base doping and bandgap. Under a quasistatic assumption we can generally define the base transit time as

$$\tau_b = \frac{-q}{J_C} \int_0^{W_b} n_b(x)\mathrm{d}x. \tag{A.2.17}$$

From Equation A.2.12 and neglecting the unity factor, we can solve for $n_b(x)$ as

$$n_b(x) = \frac{-J_C}{q} \frac{n_{ib}^2(x)}{p_b(x)} \int_x^{W_b} \frac{p_b(x')\mathrm{d}x'}{D_{nb}(x')n_{ib}^2(x')} \tag{A.2.18}$$

Substituting this result into Equation A.2.17, we finally obtain

$$\tau_b = \int_0^{W_b} \frac{n_{ib}^2(x)}{p_b(x)} \left\{ \int_x^{W_b} \frac{p_b(y)\mathrm{d}y}{D_{nb}(y)n_{ib}^2(y)} \right\} \mathrm{d}x \tag{A.2.19}$$

Again, observe that if we allow $p_b(x) = N_{ab}^-(x) = N_{ab}^- = $ constant, we obtain the classical result for a BJT with constant base doping

$$\tau_b = \frac{W_b^2}{2D_{nb}}, \tag{A.2.20}$$

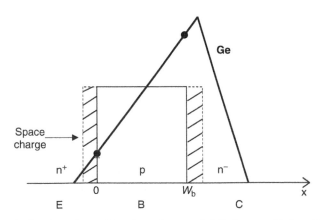

FIGURE A.2.2 Schematic drawing of a position-dependent, graded Ge base profile in a SiGe HBT.

as expected. This second generalized Moll–Ross relation is the starting point for the derivation of base transit in a graded-base SiGe HBT (Figure A.2.2), as detailed in Chapter 4 and Ref. [4].

References

1. J Moll and I Ross. The dependence of transistor parameters on the distribution of base layer resistivity. *Proceedings of the Institute of Radio Engineers* 44:72, 1956.
2. H Kroemer. Two integral relations pertaining to electron transport through a bipolar transistor with a nonuniform energy gap in the base region. *Solid-State Electronics* 28:1101–1103, 1985.
3. SN Mohammad. Generalization of the Moll–Ross relations for heterojunction bipolar transistors. *Solid-State Electronics* 46:589–591, 2002.
4. JD Cressler and G Niu. *Silicon–Germanium Heterojunction Bipolar Transistors*. Boston, MA: Artech House, 2003.

A.3

Integral Charge-Control Relations

Michael Schröter

University of California at San Diego

A.3.1 Introduction

One of the most important requirements for compact models is an accurate description of the devices' main current. In an npn bipolar transistor, this is the (generally time dependent) collector current i_C which is given by the transport of electrons from emitter to collector. For time (and frequency) dependent *quasistatic* (q.s.) operation, which is the case in the vast majority of practical applications, the current flowing at the collector terminal can be partitioned into a q.s. transfer current i_T and a charging current supplying all charge storage elements connected to the collector terminal. Since the latter elements are described and represented separately in a model, a formulation of i_T, which equals I_C under d.c. conditions, is of major interest. Before a general relation for i_T is derived, a brief historical perspective of the respective theory is given below.

The obvious starting point for such an equation is the transport and continuity equation. Since the time derivative term in the latter is taken into account separately by the charge storage elements, and the impact of recombination on i_T in modern bipolar transistors is negligibly small, a spatially constant electron current density j_{nx} results (and is observed in device simulation) throughout the structure of a one-dimensional (1D) transistor. This fact can be used to solve the transport equation first for the carrier density and associated charge at a *given transfer current*, and then reformulate the result to obtain the transfer current at a *given charge*. The first solution of this kind was published by Moll and Ross in 1956 [1, Equation (13)],

$$i_T = q\mu_{nB} n_{iB}^2 V_T \frac{\exp(V_{B'E'}/V_T)}{\int_{x_l}^{x_u} h(x)p(x)\mathrm{d}x} \tag{A.3.1}$$

with $p(x) = N_B(x)$, $h(x) = 1$, $V_{B'E'}$ as internal base–emitter voltage, and $[x_l, x_u]$ as base region. The relation provided a great deal of insight into the dependence of i_T (and also the transit time) on doping profile and material parameters in the base region. An improved and more complete expression was later used for one of the first TCAD papers by Ghosh et al. in 1967 [2]. Here, the material dependent function $h(x) = (\mu_{nB} n_{iB}^2) / (\mu_n n_i^2)$ was included for the first time, but the integral was evaluated numerically, thus not providing a formulation suitable for compact modeling. Only shortly thereafter, in 1970 Gummel published his well-known paper about the integral charge-control relation (ICCR) [3]. For the first time, a compact formulation for i_T was derived, that contained in a consistent relation between i_T and stored charge and with all bias-dependent nonideal effects that were relevant for transistors at that time. In one way or the other, the ICCR has been used until today in any reasonably physics-based compact model. Note that the Scharfetter–Gummel discretization, which was first published and applied to a bipolar transistor, also assumes a constant current density between discrete points, which is the main reason for its numerical stability and wide use for device simulation.

An extension of the Moll–Ross relation with special emphasis on a *spatially variable bandgap* in the base layer was derived by Kroemer in 1985 [4]. Here also the doping density was replaced by the bias-dependent hole density as in Equation A.3.1, but no equation suitable for compact modeling was given. In 1993, for both the latter equation and the ICCR a generalization was presented in Ref. [5], with application to compact modeling. The respective generalized ICCR (GICCR) will be discussed in more detail in subsequent sections. During its derivation, the various results mentioned above will be obtained and be referred to again.

Since the theories above all apply only to a 1D transistor structure with reasonably smooth material and bandgap changes, and to quasistatic operation, extensions for practical cases are of interest, which will be briefly discussed in A.3.5. The considerations in this chapter apply to Si and SiGe technologies. In the latter, two fundamentally different types of doping profiles can be distinguished which are presently being manufactured and illustrated schematically in Figure A.3.1. The conventional emitter doping (CED) type has a similar profile shape as BJTs, but contains in addition a Ge distribution that increases from the BE junction to the BC junction. In contrast, the low-emitter concentration (LEC) type contains a box Ge profile (or at least a sufficiently high step at the BE junction) that allows to significantly increase the base doping and to lower the emitter doping.

A.3.2   Derivation of a General Relationship

Although the GICCR can be derived for the 3D case, the considerations here will be restricted to the 1D case in order to demonstrate the concept. The assumptions required for the derivation are summarized below:

(a) A one-dimensional transistor structure (cf. Figure A.3.1) is considered, with the emitter contact ($x = 0$) at the mono-silicon surface and the collector contact ($x = x_C$) at the transition from the lightly-doped collector region to the buried layer. This 1D structure does not include the emitter and *external* collector series resistance.

(b) Volume recombination within the above 1D transistor region is negligible.

(c) The time derivative is negligible, which corresponds to quasistatic operation.

(d) Effects such as thermionic emission and tunneling across the junctions are neglected. They can be accounted for though by separate terms and can be combined with the GICCR solution.

Assumptions (a), (b), and (c) together with the electron continuity equation lead to a spatially independent electron current density,

$$J_n = \text{const}(x) = -J_T = \frac{-I_T}{A_E}, \qquad (A.3.2)$$

which can be expressed by the transfer current and area (emitter) A_E of the 1D-transistor. The derivation starts with the electron transport equation,

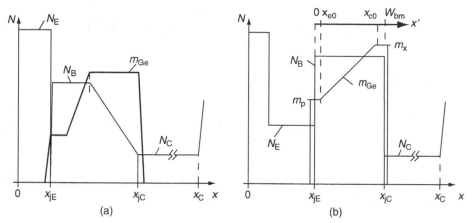

FIGURE A.3.1 Sketch of doping and Ge profiles for (a) CED transistors and (b) LEC transistors.

$$J_n = -q\mu_n n \frac{d\varphi_n}{dx} \tag{A.3.3}$$

where μ_n, n and φ_n are the electron mobility, density, and quasifermi potential. Inserting $n = n_i \exp[(\psi - \varphi_n) / V_T]$ into (A.3.3) gives

$$J_n = -q\mu_n n_i \exp\left(\frac{\psi}{V_T}\right) \exp\left(-\frac{\varphi_n}{V_T}\right) \frac{d\phi_n}{dx}. \tag{A.3.4}$$

Note that the effective intrinsic carrier density n_i accounts for bandgap differences caused by high-doping effects and bandgap-engineering. This topic will be discussed later. Using the transformation

$$\frac{d \exp(-\varphi_n/V_T)}{dx} = -\frac{1}{V_T} \exp\left(-\frac{\varphi_n}{V_T}\right) \frac{d\varphi_n}{dx} \tag{A.3.5}$$

leads to

$$J_n = qV_T\mu_n n_i \exp\left(\frac{\psi}{V_T}\right) \frac{d \exp(-\varphi_n/V_T)}{dx}. \tag{A.3.6}$$

Separation of the differential variables and rearranging terms gives

$$\frac{J_n}{qV_T\mu_n n_i} \exp\left(-\frac{\psi}{V_T}\right) dx = d\left(\exp\left(-\frac{\varphi_n}{V_T}\right)\right). \tag{A.3.7}$$

Extension of the l.h.s. by the product $n_i \exp(\varphi_p/V_T)$ allows to replace the inconvenient term $\exp(-\psi/V_T)$ by the hole density, yielding

$$\frac{J_n}{qV_T} \frac{p}{\mu_n n_i^2 \exp(\varphi_p/V_T)} dx = d\left(\exp\left(-\frac{\varphi_n}{V_T}\right)\right). \tag{A.3.8}$$

Integration of the above equation over the general spatial interval $[x_l, x_u]$ gives

$$\int_{x_1}^{x_u} \frac{J_n}{qV_T} \frac{\exp\left(-\varphi_p/V_T\right)}{\mu_n n_i^2} p \, dx = \int_{\exp\left(\varphi_n(x_1)/V_T\right)}^{\exp\left(\varphi_n(x_u)/V_T\right)} d\left(\exp\left(-\frac{\varphi_n}{V_T}\right)\right). \qquad (A.3.9)$$

The result for the r.h.s. is

$$\int_{\exp\left(\varphi_n(x_1)/V_T\right)}^{\exp\left(\varphi_n(x_u)/V_T\right)} d\left(\exp\left(-\frac{\varphi_n}{V_T}\right)\right) = \left[\exp\left(-\frac{\varphi_n(x_u)}{V_T}\right) - \exp\left(-\frac{\varphi_n(x_1)}{V_T}\right)\right]. \qquad (A.3.10)$$

The exact value of φ_n on the r.h.s. depends on the choice of the integration limits and will be discussed later. First though, the left-hand-side of (A.3.9) is integrated,

$$\int_{x_1}^{x_u} \frac{J_n}{qV_T} \frac{\exp\left(-\varphi_p/V_T\right)}{\mu_n n_i^2} p \, dx = \frac{-J_T \exp\left(-V_{B'E'}/V_T\right)}{qV_T \mu_{nr} n_{ir}^2} \int_{x_1}^{x_u} h(x)p(x)dx, \qquad (A.3.11)$$

with $h = h_g h_i h_\varphi$ as the *normalized* weighting function of the hole density, and its components

$$h_g = \frac{\mu_{nr} n_{ir}^2}{\mu_n(x) n_i^2(x)}, \quad h_i = \frac{-J_n(x)}{J_T}, \quad h_\varphi = \exp\left(\frac{V_{B'E'} - \varphi_p(x)}{V_T}\right). \qquad (A.3.12)$$

μ_{nr} and n_{ir} are the mobility and intrinsic carrier density, respectively, of a reference material.

Equating (A.3.10) with (A.3.11), and solving for the desired transfer current yields the "master" equation

$$J_T = qV_T^2 \mu_{nr} n_{ir}^2 \frac{\exp\left(\frac{V_{B'E'}}{V_T}\right)\left[\exp\left(-\frac{\varphi_n(x_1)}{V_T}\right) - \exp\left(-\frac{\varphi_n(x_u)}{V_T}\right)\right]}{\int_{x_1}^{x_u} h_g h_i h_\varphi \, pdx}, \qquad (A.3.13)$$

from which different analytical formulations can be derived. The choice of the reference material (or transistor region) and its associated values for $\mu_{nr} n_{ir}^2$ is arbitrary and will be discussed later.

The integration interval is undefined yet, and so is the impact of the various weighting functions on the integral. An attractive choice for the integration limits is $[0, x_C]$, which corresponds to the entire 1D transistor region. As a result of this choice, the electron quasifermi potentials in the numerator assume their known internal terminal values (e.g., for common-emitter configuration),

$$\varphi_n(x_1) = \varphi_n(0) = 0 \quad \text{and} \quad \varphi_n(x_u) = \varphi_n(x_C) = V_{C'E'}, \qquad (A.3.14)$$

with $V_{C'E'}$ as (internal) collector–emitter voltage.

The denominator integral is more difficult to oversee. Thus, Figure A.3.2 shows the spatial dependence of the weighting functions with the hole density superimposed. Since the hole quasifermi potential equals $V_{B'E'}$ over the entire base and adjacent space–charge regions, h_φ equals 1 where $p(x)$ is significant. Similarly, the electron current density is constant and equals J_T even in a larger interval, except in a small region close to the emitter contact, where it increases slightly due to the back injection of holes and the corresponding recombination; the maximum deviation can be $1/B_f$. Therefore, without introducing a significant error, it is possible for both BJT and HBT to assume an average value for the functions h_i and h_φ that is very close to 1. According to Figure A.3.2, the main spatial dependence of the weighting function h is caused by h_g via the bandgap variation of n_i. A smaller contribution to the spatial dependence comes from the doping and field dependence of the mobility.

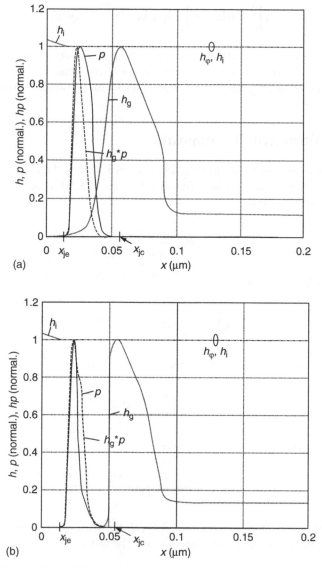

FIGURE A.3.2 Spatial dependence of the weighting functions for a CED transistor with the same doping profile: (a) 70 GHz BJT without Ge; (b) 120 GHz HBT with Ge. Bias point ($J_C = J_D$, $V_{B'C'}$) = (0.1 mA/μm², 0 V). The hole density p, weighting function h_g, and their product $h_g p$ have been normalized to the respective maximum value. x_{je} and x_{jc} denote the emitter and base junction depth.

Inserting (A.3.14) and h_g from (A.3.12) into (A.3.13), extension by q as well as using average values $\bar{h}_i$ and $\bar{h}_\varphi$ yields the basic formulation

$$J_T = q^2 V_T \frac{\mu_{nr} n_{ir}^2}{\bar{h}_i \bar{h}_\varphi} \frac{\exp(V_{B'E'}/V_T) - \exp(V_{B'C'}/V_T)}{q \int_0^{x_C} h_g p \, dx}. \tag{A.3.15}$$

from which practically relevant equations can be derived as shown later. The denominator is not yet suitable for (compact) modeling, but at this point can be prepared for further evaluation by partitioning it into a bias-independent and a bias-dependent portion

$$q \int_0^{x_C} h_g p \, dx = q \int_0^{x_C} h_g p_0 \, dx + q \int_0^{x_C} h_g \Delta p \, dx, \tag{A.3.16}$$

where p_0 is the hole density in equilibrium and Δp is the hole density *change* in the transistor with respect to equilibrium. Δp consists of depletion and minority components. While the latter density is distributed over the whole base region, the former densities are located around the junctions and related to the (ionized) base doping.

A.3.3 Homojunction Transistors

In homojunction transistors, the variation of h_g is only caused by high-doping effects. Thus, h_g is fairly constant in the region around the peak hole density (cf. Figure A.3.2a), which contributes most to the integral. Defining an average value

$$\bar{h}_g = \frac{\mu_{nr} n_{ir}^2}{\mu_n n_i^2}, \tag{A.3.17}$$

the denominator of (A.3.15) becomes

$$q \int_{x_E}^{x_C} \frac{\mu_{nr} n_{ir}^2}{\mu_n n_i^2} p(x) \, dx = \bar{h}_g \bar{Q}_p \tag{A.3.18}$$

with the hole charge *per area* stored in the 1D transistor structure,

$$\bar{Q}_p = q \int_0^{x_C} p(x) \, dx. \tag{A.3.19}$$

Inserting (A.3.18) and (A.3.19) into (A.3.15) yields

$$J_T = c \frac{\exp\left(V_{B'E'}/V_T\right) - \exp\left(V_{B'C'}/V_T\right)}{\bar{Q}_p} \tag{A.3.20}$$

which has the same form as Gummel's ICCR in Ref. [3], but a different definition of the integration region and, thus, of the charge and controlling voltages. The factor

$$c = q^2 V_T \frac{\mu_{nr} n_{ir}^2}{\bar{h}_i \bar{h}_\varphi \bar{h}_g} \cong q^2 V_T \overline{\mu_n n_i^2} \tag{A.3.21}$$

is assumed to be constant over bias. Usually, the most right term is used only, which is obtained by setting $\bar{h}_i = \bar{h}_\varphi = 1$ and is an excellent assumption for the 1D case. Since the hole density is concentrated mostly in the base, the value for $\bar{h}_g$ and $\overline{\mu_n n_i^2}$ is close to that for the base material.

The charge in (A.3.19) can be divided into a bias-independent and a bias-dependent component, $\overline{Q}_p = \overline{Q}_{p0} + \Delta\overline{Q}_p$. The bias-independent charge $\overline{Q}_{p0}$ is defined at $V_{B'E'} = V_{B'C'} = 0$ and consists of holes stored mostly in the neutral base region $0 \leq x' \leq w_{B0}$ at equilibrium:

$$\bar{Q}_{p0} = q \int_0^{x_C} p_0(x) \, dx \cong q \int_0^{w_{B0}} p_0(x') \, dx' \approx q \int_0^{w_{B0}} N_B(x') \, dx'. \tag{A.3.22}$$

(Note the use of a different coordinate x' in the neutral base region, cf. Figure A.3.1b.) The bias-dependent charge component represents all holes, Δp, that for a nonequilibrium bias condition enter the base region (i) to charge the depletion regions, (ii) to compensate the electrons injected into the (neutral) base, and (iii) for the injection into emitter and collector,

$$\Delta \bar{Q}_p = q \int_0^{x_C} \Delta p(x) dx = \bar{Q}_{pE} + \bar{Q}_{jE} + \bar{Q}_{nB} + \bar{Q}_{jC} + \bar{Q}_{pC} \qquad (A.3.23)$$

Here, $\bar{Q}_{jE}$ and $\bar{Q}_{jC}$ are the BE and BC depletion charge (per unit area), respectively, and the other components are the minority charges (per unit area) stored in E, B, and C. As a consequence of the choice of the chosen integration limits, $\Delta \bar{Q}_p$ can actually be measured via the terminals. From a physics-based point of view, the biggest advantage of the extended ICCR (A.3.20) is its firm link between the quasistatic transfer current and the *total* hole charge, which resembles the charge-controlled operation principle of a bipolar transistor in a very compact form.

Examples for the accuracy of various solutions are shown in Figure A.3.3 for a typical Gummel plot. In all cases, the denominator integral (e.g. $\bar{Q}_p$ in (A.3.20)) has been taken directly from device simulation. As expected, the basic Equation A.3.15 follows exactly the simulated behavior of J_T (crosses). For the older generation profile ("25 GHz transistor" in Figure A.3.3a), the practically interesting approximation (A.3.20) is very accurate at low and medium current densities, and shows only slight deviations at high current densities, which are tolerable though and typical for homo-junction bipolar processes (e.g. [6,7]). The reason for these small deviations is caused by the assumption of a bias-independent value for $\bar{h}_g$. However, if the minority charge contributions of emitter and collector, $\bar{Q}_{pE}$ and $\bar{Q}_{pC}$, are neglected in (A.3.20), the model current becomes far too large. Thus, including only the base charge is not a feasible approach.

For the profile of a modern generation Si(Ge) transistor ("70 GHz BJT" in Figure A.3.3b), Equation A.3.20 starts to deviate already at medium current densities, affecting the accuracy of the transconductance modeling. This is caused by the increased difference in doping and associated variation of h_g across the base region, which requires a different weighting factor for Q_{p0} and Q_{jE} along the same lines as discussed in the next section on HBTs, applying the corresponding Equation A.3.28 gives again a much higher accuracy.

The assumption of a bias-independent value for $\bar{h}_g$ deserves some consideration. High-doping effects cause both n_i and μ_n (which is also field dependent) to depend on x within the base region, with the change of n_i partially being compensated by the change of μ_n. Obviously though, as long as the spatial distributions of p and $\mu_n n_i^2$ do not change with bias, $\bar{h}_g$ remains bias independent. This holds quite well for low and medium current densities where p is concentrated in the base region, and the mobility depends only very slightly on bias. Toward high current densities, however, p spreads out into the collector and partially also into the emitter region. In the collector, for instance, high-doping effects can be neglected but the field (and thus bias) dependence of μ_n is significant, leading to a different weighting of the hole density in this region. As a consequence, the average value $\bar{h}_g$ defined at low current densities is not quite correct anymore at high current densities and causes the observed small deviation from the exact current. Note though that the field dependence of the mobility around the BC junction decreases at high current densities.

At this point, the choice of the integration limits shall be briefly discussed. Fundamentally, there is an infinite number of possible integration intervals. Besides the one selected above (i.e., entire 1D transistor), the other most important choice in literature (e.g., Refs. [3,4]) is the base region, i.e., $x \in [x_{jE}, x_{jC}]$. In this case, $\bar{h}_g$ is only a very weak function of bias, but there are serious disadvantages of this choice: (i) the unknown values for the electron quasi-fermi potentials at the base region boundaries, i.e., $\varphi_n(x_{jE})$ and $\varphi_n(x_{jC})$ in the numerator of (A.3.13), are not easily accessible by measurements; (ii) a separate determination (i.e., measurement) of the *base* hole charge, in particular the base minority charge, is difficult at medium to high current densities.

The original and most famous expression of the ICCR as given by Gummel in Ref. [3], which has the same form as (A.3.20), is in fact based on considering only the base region for the integration and assuming contributions from outside that region to be negligible. Although this was a reasonable assumption at that time for the charge, it was not for the voltage drop toward the contacts.

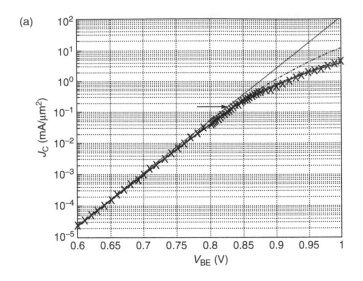

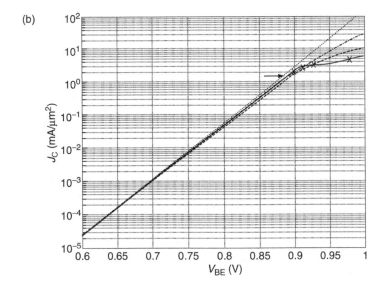

FIGURE A.3.3 Transfer current density for (a) 25 GHz BJT and (b) 70 GHz BJT: comparison between device simulation (symbols) and analytical calculations from Equation A.3.15 (solid line), Equation A.3.24 (dotted line), Equation A.3.20 (dashed line), as well as Equation A.3.20 without $Q_{pE} + Q_{pC}$ as dash-dotted line in (a) and Equation A.3.28 as dash-dotted line in (b). The arrow indicates the current density at which the f_T peak occurs.

An often found further simplification is to replace the hole density in (A.3.19) by the base doping density N_B. As a consequence, the hole charge equals the expression on the most r.h.s. of (A.3.22) and does not contain any bias-dependent "nonideal" effects anymore. Neglecting the voltage drops toward the contacts, Moll and Ross [1] first derived the corresponding expression for the transfer current

$$J_T = q V_T \overline{\mu_{nB} n_{iB}^2} \frac{\exp(V_{B'E}/V_T) - \exp(V_{B'C'}/V_T)}{\int_0^{W_B} N_B dx'} \tag{A.3.24}$$

This is a more general relation compared to the classical transistor theory along the lines of Ebers and Moll [9], and aided the understanding that the transfer current is linked to the *integral* of the base doping and not to some *spot* value. As shown in Figure A.3.2, Equation A.3.24 produces an almost ideal characteristic due to the missing bias-dependent charge components.

A.3.4   Heterojunction Transistors

A comparison for the weighting function h_g between a SiGe and a Si transistor with the same doping profile was already shown in Figure A.3.2. While the spatial dependence of the hole QFP and electron current density and, thus, of h_φ and h_i is still similar to BJTs, the intentional bandgap differences in HBTs lead to additional variations of n_i^2 within the integration interval $[0,x_C]$. As a consequence, it is impossible to define a single average value $\bar{h}_g$ over this whole interval. Thus, the denominator integral in (A.3.15) now has to be formulated in terms of charge components in *separate transistor regions that are multiplied with an average value of the weighting function* h_g taken over the respective region. In HBTs the transition from one bandgap value to another usually takes place close to the junctions. Therefore, partitioning of the respective integral $q \int_0^{x_c} h_g p dx$ into neutral and space–charge regions is *one* reasonable choice in order to obtain separate expressions, in which the weighting function is sufficiently independent on location and bias. This leads generally to

$$q \int_0^{x_C} h_g p dx = \bar{h}_{g,B0}\bar{Q}_{p0} + \bar{h}_{g,E}\bar{Q}_{pE} + \bar{h}_{g,jE}\bar{Q}_{jE} + \bar{h}_{g,B}\bar{Q}_{nB} + \bar{h}_{g,jC}\bar{Q}_{jC} + \bar{h}_{g,C}\bar{Q}_{pC} \qquad (A.3.25)$$

$$\text{with} \quad \bar{h}_{g,v} = \frac{\mu_{n0r}n_{ir}^2}{\mu_{nv}n_{iv}^2} \quad \text{and} \quad v = \{B0, E, jE, B, jC, C\}. \qquad (A.3.26)$$

Choosing the base region as reference and assuming $\bar{h}_{g,B} = \bar{h}_{g,B0}$, all weighting factors can be divided by $\bar{h}_{g,B}$ giving for the normalized remaining weighting factors

$$h_v = \frac{\bar{h}_{g,v}}{\bar{h}_{g,B}} = \frac{\overline{\mu_{nB}n_{iB}^2}}{\mu_{nv}n_{iv}^2} \quad \text{with} \quad v = \{E, jE, jC, C\}, \qquad (A.3.27)$$

which are considered to be model parameters. Inserting the above expression and charge components into (A.3.15), and making again the valid assumption $\bar{h}_i = \bar{h}_\varphi = 1$ yields the GICCR as the final expression for 1D-HBTs [5]

$$J_T = q^2 V_T \overline{\mu_{nB}n_{iB}^2} \frac{\exp(V_{B'E'}/V_T) - \exp(V_{B'C'}/V_T)}{\bar{Q}_{p,T}} \qquad (A.3.28)$$

with the modified ("transfer current related") charge density

$$\bar{Q}_{p,T} = \bar{Q}_{p0} + \Delta\bar{Q}_{p,T} = \bar{Q}_{p0} + h_E\bar{Q}_{pE} + h_{jE}\bar{Q}_{jE} + \bar{Q}_{nB} + h_{jC}\bar{Q}_{jC} + h_C\bar{Q}_{pC}, \qquad (A.3.29)$$

in which—as for BJTs—still a bias-independent hole charge, Q_{p0}, and a bias-dependent portion can be distinguished. According to Refs. [5,8], and also as Figure A.3.4 shows, (A.3.28) with (A.3.29) using bias-independent values for h_v leads to a significant improvement over the conventional ICCR (A.3.20). Thus, the GICCR results in an accurate description of the transfer current characteristics and the respective derivatives over the entire bias range of interest (up to high current densities). Again, as in Figure A.3.3, the charges and weighting factors have been calculated directly from device simulation results to avoid any errors introduced by analytical approximations.

For "ideal" doping profiles like in Figure A.3.1, the weighting factors h_v in (A.3.2) can be calculated analytically using (A.3.15). This aids the physical understanding of the impact of the Ge profile on device characteristics and shall be demonstrated below for two Ge profile examples. However, the calculation can be extended to other profiles but just becomes more elaborate.

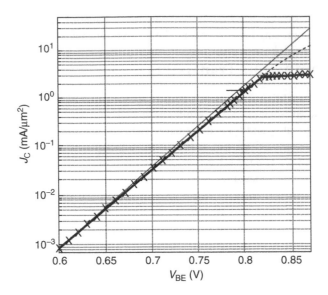

FIGURE A.3.4 Transfer current density for the 120 GHz HBT: comparison between device simulation (symbols) and analytical calculations from Equation A.3.15 (solid line), Equation A.3.28 (dashed line), Equation A.3.24 (dotted line), and Equation A.3.28 without $Q_{pE} + Q_{pC}$ (dash-dotted line).

Box Ge Profile in the (Metallurgical) Base Region

In this case, $\overline{\mu_n n_i^2}$ is constant over the base region and the portion of the space–charge regions that are associated with the base. For the sake of simplicity, low injection is assumed first. Setting $\mu_{nr} n_{ir}^2 = \mu_{nB} n_{iB}^2$ yields $h_g = 1$ already in the integral over the base region and, thus, $h_{jE} = h_{jC} = 1$. In the emitter and collector region, $h_E = \mu_{nB} n_{iB}^2 / \mu_{nE} n_{iE}^2 \gg 1$ and $h_C = \mu_{nB} n_{iB}^2 / \mu_{nC} n_{iC}^2 \gg 1$, respectively, due to the much larger bandgap in those regions. As a consequence of this bandgap, the hole charges $\overline{Q}_{pE}$ and $\overline{Q}_{pC}$ are very small and do not significantly impact the dynamic transistor behavior, regardless of the respective doping profile. However, according to (A.3.28) these charges can have a significant impact on the transfer current due to the large weighting factors h_E and h_C. The respective terms $h_E \overline{Q}_{pE}$ and $h_C \overline{Q}_{pC}$ in (A.3.28) actually cause the "saturation" of the $I_C(V_{BE})$ characteristics at high injection observed in Figure A.3.2b. In HBTs with a box Ge profile, Equation A.3.28 can be simplified to [5]

$$J_T = q^2 V_T \overline{\mu_{nB} n_{iB}^2} \frac{\exp(V_{B'E'}/V_T) - \exp(V_{B'C'}/V_T)}{\overline{Q}_p + (h_E - 1)\overline{Q}_{pE} + (h_C - 1)\overline{Q}_{pC}}, \tag{A.3.30}$$

which contains the total hole charge as a lumped variable and correction factors in the denominator. In Ref. [8], the product $(h_C - 1)\overline{Q}_{pC}$ is described directly by a compact expression rather than separately by the weighting factor and charge component.

Trapezoidal Profile in the (Metallurgical) Base Region

Consider the Ge profile in Figure A.3.1b. For the sake of simplicity it is assumed that the Ge mole fraction m_{Ge} and the respective bandgap voltage increase over the width of the neutral base ($x' \in [x_{e0}, x_{c0}]$) only, but stay constant across the space–charge regions. Choosing the Si-base without Ge contents as reference material, the bandgap voltage differences $\Delta V_{Gp} = \Delta V_G(x' = x_{e0})$ and $\Delta V_{Gx} = \Delta V_G(x' = x_{c0})$, respectively, can be defined. Hence, the intrinsic carrier density within the neutral base with the width $w_{B0} = x_{c0} - x_{e0}$ can be written as

$$n_i^2 = n_{iB,Si}^2 \exp\left(\frac{\Delta V_{Gp} + a_G(x' - x_{e0})}{V_T}\right) \tag{A.3.31}$$

with the slope factor $a_G = (\Delta V_{Gx} - \Delta V_{Gp})/w_{B0}$. Neglecting the (much smaller) dependence of the mobility on field and Ge contents, and applying the above relation to (A.3.16) with $p_0 = N_B$ yields for the bias-independent term

$$q \int_0^{x_C} \frac{\mu_{nr} n_{ir}^2}{\mu_n n_i^2} p_0 dx' \cong q \int_{x_{e0}}^{x_{c0}} \exp\left(-\frac{\Delta V_{Gp} + a_G(x' - x_{e0})}{V_T}\right) N_B dx' \tag{A.3.32}$$

which gives after evaluating the r.h.s. integral

$$q \int_0^{x_C} \frac{\mu_{nr} n_{ir}^2}{\mu_n n_i^2} p_0 dx' \cong \frac{\frac{V_T}{a_G}\left[\exp\left(-\frac{\Delta V_{Gp}}{V_T}\right) - \exp\left(-\frac{\Delta V_{Gx}}{V_T}\right)\right]}{w_{B0}} \bar{Q}_{p0} = \bar{h}_{g,B0} \bar{Q}_{p0} \tag{A.3.33}$$

with $\bar{Q}_{p0} = q N_B w_{B0}$. As can be seen, the average weighting factor depends exponentially on the bandgap voltages at the *beginning* and the *end* of the neutral base.

For the bias-dependent portion in (A.3.16), one can write at low current densities

$$q \int_0^{x_C} h_g \Delta p dx \cong q \left[\int_{x_e}^{x_{e0}} \frac{N_B}{\exp\left(\frac{\Delta V_{Gp}}{V_T}\right)} dx' + \int_{x_{c0}}^{x_c} \frac{N_B}{\exp\left(\frac{\Delta V_{Gx}}{V_T}\right)} dx' + \int_{x_{jE}}^{x_{jC}} \frac{n}{\exp\left(\tilde{n}\frac{\Delta V_{Gp} + a_G(x'\tilde{n}x_{e0})}{V_T}\right)} dx'\right] \tag{A.3.34}$$

The first two terms represent the depletion components that are only to be evaluated between the SCR boundary (i.e., x_e, x_c) at the given bias point and the respective equilibrium SCR boundaries (i.e., x_{e0}, x_{c0}); it also has been assumed that the bandgap (i.e., Ge mole) change within $(x_{e0}-x_e)$ and $(x_{c0}-x_c)$ is still negligible. The resulting depletion charges are $\bar{Q}_{jE} = q N_B(x_{e0} - x_e)$ and $\bar{Q}_{jC} = q N_B(x_c - x_{c0})$. For the case that the electric field in the base due to Ge grading causes the electrons to travel with saturation drift velocity v_s, i.e. $n = J_T/(q v_s)$ does not depend on x', the resulting base minority charge is then $\overline{Q}_{nB} = J_T w_B/v_s$. With these charge expressions, one obtains after evaluating all terms on the r.h.s. of (A.3.34)

$$q \int_0^{x_C} h_g \Delta p dx \cong \exp\left(-\frac{\Delta V_{Gp}}{V_T}\right) \bar{Q}_{jE} + \exp\left(-\frac{\Delta V_{Gx}}{V_T}\right) \bar{Q}_{jC} + \bar{h}_{g,B0} \bar{Q}_{nB}, \tag{A.3.35}$$

where the last term follows the same evaluation as (A.3.32). The final step is to insert the components in (A.3.33) and (A.3.35) back into (A.3.15) and to normalize the denominator to the base weighting factor, $\bar{h}_{gB0}$ given by Equation A.3.33. The resulting expression then reads

$$J_T = \underbrace{q^2 V_T \frac{\mu_{nB,Si} n_{iB,Si}^2}{\bar{h}_i \bar{h}_\varphi \bar{h}_{g,B0}}}_{c_{10}} \frac{\exp\left(V_{B'E'}/V_T\right) - \exp\left(V_{B'C'}/V_T\right)}{\bar{Q}_{p0} + h_{jE} \bar{Q}_{jE} + h_{jC} \bar{Q}_{jC} + \bar{Q}_{nB}}. \tag{A.3.36}$$

which has the same form as (A.3.28), but for low current densities and with known analytical expressions of the weighting factors from the above analysis:

$$h_{jE} = \frac{\exp\left(-\frac{\Delta V_{Gp}}{V_T}\right)}{\bar{h}_{g,B0}} = \frac{v \exp(v)}{\exp(v) - 1}, \quad h_{jC} = \frac{\exp\left(-\frac{\Delta V_{Gx}}{V_T}\right)}{\bar{h}_{g,B0}} = \frac{v}{\exp(v) - 1}. \tag{A.3.37}$$

with $v = [\Delta V_{GX} - \Delta V_{GP}]/V_T$ as normalized bandgap difference between beginning and end of the base region (cf. Figure A.3.1b) . The dependence of the weighting factors as a function of v is shown in Figure A.3.5 for a practically relevant range.

From the above results, the forward Early voltage at low injection, can be calculated:

$$V_{Ef} = \frac{\bar{Q}_p}{h_{jC} \bar{C}_{jCi}} \approx \frac{\bar{Q}_{p0}}{\bar{C}_{jCi}} \frac{\exp(v) - 1}{v}. \tag{A.3.38}$$

According to (A.3.37), a 20% difference in Ge across the base region corresponds to an about 40 times increase in Early voltage, which is a significant enhancement factor over a Si-BJT or a SiGe HBT with a Ge box profile ($v = 0$).

In addition to the strong variation in n_i, the mobility varies within the transistor as a function of both doping and bias (via the electric field). The variation caused by the latter is most pronounced in the BC junction and collector region. In general, μ_n and n_i possess an opposite dependence on doping, leading to a partial compensation within h_g. However, the influence of n_i still remains much stronger than that of μ_n. As a consequence, the weighting function h_g always deviates strongly from 1 and has to be considered for all processes.

A.3.5   Further Extensions

All of the considerations so far apply to a 1D transistor structure and quasistatic operation. Extensions in both directions have been investigated and proposed. A solution of the time-dependent continuity equation led to the transient ICCR (TICC) [10], in which the "in-phase" component gives the same expression as the ICCR for the for q.s. transfer current, while the "out-of-phase" solution yields a physical definition of the charging currents flowing through the E and C contact. Hence, the out-of-phase solution defines a physics-based capacitance matrix associated with the E and C terminals, that includes the case of non-quasistatic operation. Extensions of the TICC towards including recombination and non-1D effects were presented in, for example, Ref. [11]. The application of the TICC results in a compact model, however, is quite challenging due to the bias-dependent weighting functions in the integrals defining the charging components.

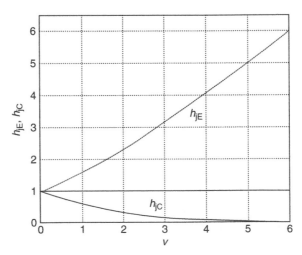

FIGURE A.3.5 Weighting factors of the depletion charges according to Equation A.3.37 as a function of the normalized bandgap difference $v = (\Delta V_{Gx} - \Delta V_{Gp})/V_T$.

As shown in Ref. [7], it is possible to extend the GICCR to two- and three-dimensional transistor structures. The respective derivation is beyond the scope of this chapter, but the result shall be briefly discussed. For instance, the resulting 2D-GICCR reads

$$I_T = c_{10} \frac{\exp(V_{B'E'}/V_T) - \left[\exp(V_{B'C'}/V_T \frac{2(y_{Bcon}+b_{Bcon})}{b_E}) + \frac{2b_{Bcon}}{b_E}\right]}{Q_{p0,T} + \Delta Q_{p,T}}, \tag{A.3.39}$$

where the constant c_{10} depends on an enlarged (effective) emitter width

$$b_E = b_{E0} + 2 \int_{b_{E0}/2}^{y_{Bcon}} \exp\left(-\frac{\varphi_n(x=0,y)}{V_T}\right) dy. \tag{A.3.40}$$

b_{E0} is the emitter window width, y_{Bcon} is the edge of the base contact or polysilicon next to the emitter, and b_{Bcon} is the base contact or polysilicon width on mon-silicon. $\Delta Q_{p,T}$ is defined as in (A.3.29), but now includes, among others, the impact of electron current crowding. This also applies to $Q_{p0,T}$, which introduces a bias-dependent geometry dependence at higher current densities. In practice, $Q_{p0,T}$ can be approximated by a constant value to first order.

A.3.6 Summary

A set of integral charge-control relations has been derived and put in perspective to the (classical) literature. It was shown that a "master" equation exists, from which integral charge-control relations of different complexity and accuracy can be derived. The most general form, that is suitable for accurately describing the transfer current in a compact model for HBTs and BJTs, is the GICCR, which includes bandgap differences in the various device regions and also contains the weakest assumptions among the known theories for the transfer current.

The GICCR is a powerful tool to analytically derive the relationship between transfer current, stored charges, and physical as well as structural parameters of a transistor. The GICCR can be very accurate, provided that the respective weigthing factors as a function of device structure and the hole charge as a function of bias are accurately modeled. Notice that the latter is a prerequisite for the description of high-speed applications in any way. Also, since the hole charge has to be continuously differentiable with respect to bias, the transfer current is also automatically continuously differentiable over all bias regions and, hence, is modeled via a single-piece formulation. This is a very desirable feature of the (G)ICCR for compact models.

Applying the "master" equation to compact modeling requires partitioning of the hole charge and analytical approximations for its various components. These measures as well as the determination of charge model parameters and appropriate weighting factor values introduce additional inaccuracies with respect to the results shown here, which are unavoidable though for any compact model equation.

Acknowledgment

The author would like to thank H. Tran for performing simulations and model calculations.

References

1. J.L. Moll and J.M. Ross, The dependence of transistor parameters on the distribution of base layer resistivity, *Proc. IRE*, 44, 1956, S72–S78.
2. H.N. Ghosh, F.H. De La Moneda, and N.R. Dono, Computer-aided transistor design, characterization, and optimization, *Proc. IEEE*, 55, 1967, 1897–1912.
3. H.K. Gummel, A charge-control relation for bipolar transistors", *Bell System Technical J.*, 49, 1970, S115–S120.

4. H. Kroemer, Two-Integral equations pertaining to the electron transport through a bipolar transistor with nonuniform energy gap in the base region, *Solid-State Electron.*, 28, 1985, 1101–1103.

5. M. Schroter, M. Friedrich, and H.-M. Rein, A generalized integral charge-control relation and its application to compact models for silicon based HBTs, *IEEE Trans. Electron Dev.*, 40, 1993, 2036–2046.

6. H.-M. Rein, H. Stübing, and M. Schroter, Verification of the integral charge-control relation for high-speed bipolar transistors at high current densities, *IEEE Trans. Electron Dev.*, 32, 1985, 1070–1076.

7. M. Schroter, A compact physical large-signal model for high-speed bipolar transistors with special regard to high current densities and two-dimensional effects, PhD thesis (in German), Ruhr-University Bochum, Bochum, Germany, 1988.

8. M. Friedrich and H.-M. Rein, Analytical current–voltage relations for compact SiGe HBT models, *IEEE Trans. Electron Dev.*, Vol. ED-46, 2001, 1384–1401.

9. J.J. Ebers and J.L. Moll, Large-signal behavior of junction transistors, *Proc. IRE*, 42, 1761–1772, 1954.

10. H. Klose and A. Wieder, The transient integral charge–control relation—a novel formulation of the currents in a bipolar transistor, *IEEE Trans. Electron Dev.*, ED-34, 1090–1099, 1987.

11. J.S. Hamel and C.R. Selvakumar, The general transient charge–control relation: a new charge-control relation for semiconductor devices, *IEEE Trans. Electron Dev.*, ED-38, 1467–1476, 1991.

A.4
Sample SiGe HBT Compact Model Parameters

Ramana M. Malladi
IBM Microelectronics

This appendix contains a sample set of compact model parameters for a representative $0.32 \times 16.8\,\mu m^2$ first-generation npn SiGe HBT with a peak f_T of 50 GHz, for each of the three dominant higher-order SiGe compact models available in the public domain and in leading circuit simulators: HICUM, MEXTRAM, and VBIC. Each model was carefully calibrated to a comprehensive set of measured dc and ac data.

TABLE A.4.1 HICUM (v 2.1) SiGe HBT Model Parameters

Group	Name	Parameter Description	Value
Transfer current	c10	Constant for ICCR (= Is.qp0)	8.5×10^{-31} A^2 S
	qp0	Zero-bias hole charge in the base	1.4×10^{-13} C
	ich	High-current correction to account for 2D and 3D effects	10A
	hjei	BE depletion charge weighting factor for HBTs	2
	hjci	BC depletion charge weighting factor for HBTs	0.1
	mcf	Forward nonideality factor for transfer current for HBTs	1
	hfe	Emitter minority charge weighting factor for HBTs	1
	hfc	Collector minority charge weighting factor for HBTs	0.2
Base–emitter currents	ibeis	Internal BE sauration current	7×10^{-20} A
	mbei	Internal BE non-ideality factor	1.005
	ibeps	Peripheral BE saturation current	7×10^{-22} A
	mbep	Peripheral BE non-ideality factor	1
	ireis	Internal BE saturation current (recombination)	1×10^{-17} A
	mrei	Internal BE non-ideality factor (recombination)	2
	ireps	Peripheral BE saturation current (recombination)	1×10^{-20} A
	mrep	Peripheral BE non-ideality factor (recombination)	2
Base–collector currents	ibcis	Internal BC saturation current	1.5×10^{-18} A
	mbci	Internal BC saturation current ideality factor	1.02
	ibcxs	External BC saturation current	2×10^{-19} A
	mbcx	External BC saturation current ideality factor	2
Base–emitter tunnelling current	ibets	BE tunneling saturation current	1×10^{-21} A
	abet	BE tunneling factor	36
Base–collector avalanche current	favl	Avalanche current factor	19.3 V^{-1}
	qavl	Exponent for avalanche current	150×10^{-15} C
Series resistances — base, emitter, and collector	rbi0	Internal base resistance at zero bias	20 Ω
	rbx	Extrinsic base resistance	6 Ω
	fdqr0	Correction factor for modulation by BE and BC SCR	0.2
	fgeo	Geometry factor for current crowding	0.67
	fqi	Ratio of internal to total minority charge	1.0
	fcrbi	Ratio of shunt capacitance (parallel to Rbi) to total internal capacitance	0
Emitter	re	Emitter series resistance	3 Ω
Collector	rcx	Extrinsic collector series resistance	23 Ω
Substrate transistor	iscs	CS diode saturation current	4×10^{-21} A

Category	Parameter	Description	Value
	msc	CS diode non-ideality factor	1
	itss	Transfer saturation current of substrate transistor	2.5×10^{-19} A
	msf	Forward non-ideality factor of substrate transfer current	1
	msr	Reverse non-ideality factor of substrate transfer current	1
	tsf	Transit time (forward operation) — substrate	2×10^{-12} sec
Substrate network	rsu	Substrate resistance	$50\,\Omega$ (layout dependent)
	csu	Substrate capacitance	3×10^{-18} F
Self-heating	rth	Thermal resistance	700 K/W
	cth	Thermal capacitance	350 pJ/K
Base–emitter junction capacitance	cjei0	Internal zero-bias BE depletion capacitance	35×10^{-15} F
	vdei	Internal BE built-in voltage	1.0 V
	zei	Internal BE grading coefficient	0.32
	aljei	Maximum internal depletion capacitance divided by cjei0	2.0
	cjep0	Peripheral zero-bias BE depletion capacitance	5×10^{-15} F
	vdep	Peripheral BE built-in voltage	1.0 V
	zep	Peripheral BE grading coefficient	0.32
	aljep	Maximum peripheral depletion capacitance divided by cjep0	2.2
	ceox	Emitter oxide (overlap) capacitance	18×10^{-15} F
Base–collector junction capacitance	cjci0	Internal zero-bias BC depletion capacitance	7×10^{-15} F
	vdci	Internal BC built-in voltage	0.7 V
	zci	Internal BC grading coefficient	0.3
	vptci	Punch-through voltage of internal BC junction	2.5 V
	cjcx0	External zero-bias BC depletion capacitance	30×10^{-15} F
	vdcx	External BC built-in voltage	0.73 V
	zcx	External BC grading coefficient	0.4
	vptcx	Punch-through voltage of external BC junction	100 V
	ccox	BC overlap capacitance	2.5×10^{-15} F
	fbc	Partitioning factor for cjcx and ccox over rbx	0.8
Collector–substrate junction capacitance	cjs0	Zero-bias CS depletion capacitance	40×10^{-15} F
	vds	CS built-in voltage	0.6 V
	zs	CS grading coefficient	0.3
	vpts	Punch-through voltage of CS junction	1×10^{10} V
Diffusion capacitances/transit times — low currents	t0	Low current forward transit time at $V_{cb}=0$ V	2.6×10^{-12} sec
	dt0h	Time constant for base and BC space charge layer width modulation	0.9×10^{-12} sec
	tbvl	Time constant for modeling carrier jam at low V_{ce}	0.7×10^{-12} sec
High currents	tef0	Neutral emitter storage time	40×10^{-15} sec
	gtfe	Exponent for current dependence of neutral emitter storage time	1.0

(Continued)

TABLE A.4.1 HICUM (v 2.1) SiGe HBT Model Parameters (*Continued*)

Group	Name	Parameter Description	Value
	thcs	Saturation time constant at high current densities	25×10^{-12} sec
	alhc	Smoothing factor for current dependence of base and collector transit time	0.53
	fthc	Factor for partitioning this into base and collector portion	0.6
	vces	Internal C-E saturation voltage	0.1 V
	rci0	Internal collector resistance at low electric field	20 Ω
	vlim	Voltage separating ohmic (low field) and saturation velocity (high field) regime	0.7 V
	vpt	Collector punch-through voltage	15 V
	tr	Storage time for inverse operation	20×10^{-12} sec
Non-quasistatic effects	alqf	Factor for additional delay time of minority charge	0.125
	alit	Factor for additional delay time of transfer current	0.45
Noise parameters	kf	Flicker noise factor	22×10^{-6}
	af	Flicker noise exponent factor	2.5
	krbi	Noise factor for internal base resistance	1
Temperature effect parameters	tnom	Measurement temperature	25 C
	vgb	Bandgap voltage	1.17 V
	alb	Temperature coefficient of current gain	6×10^{-3}
	alfav	Temperature coefficient of favl	5×10^{-5} K^{-1}
	alqav	Temperature coefficient of qavl	2×10^{-4} K^{-1}
	zetaci	Temperature coefficient for mobility in epi-collector (i.e., for collector resitance)	1.6
	alvs	Relative temperature coefficient of saturation drift velocity	1×10^{-3} K^{-1}
	alces	Relative temperature coefficient of vces	0.4×10^{-3} K^{-1}
	zetarbi	Temperature coefficient for mobility in internal base (i.e., for internal base resistance)	0.6
	zetarbx	Temperature coefficient for mobility in extrinsic base (i.e, for extrinsic base resistance)	0.2
	zetarcx	Temperature coefficient for mobility in extrinsic collector (i.e, for extrinsic collector resistance)	0.2
	zetare	Temperature coefficient for emitter resistance	0
	alt0	First-order temperature coefficient of t0	1×10^{-3} K^{-1}
	kt0	Second-order temperature coefficient of t0	1×10^{-5} K^{-2}

TABLE A.4.2 MEXTRAM 504 SiGe HBT Model Parameters

Group	Name	Parameter Description	Value
Forward and reverse currents	is	Transistor main saturation current	5×10^{-18} A
	ik	Knee current for high-injection effects in the base	4.5×10^{-2} A
	bf	Ideal forward current gain	95
	ibf	Saturation-current of the non-ideal forward base current	2×10^{-17} A
	mlf	Non-ideality factor of the non-ideal forward base current	1.545
	xibi	Sidewall component of ideal base current	0
	bri	Ideal reverse current gain	3.77
	ibr	Saturation current of the non-ideal reverse base current	1.7×10^{-15} A
	vlr	Cross-over voltage of the non-ideal reverse base current	1×10^{-2} V
	xext	Partitioning factor for the extrinsic region	0.19
Early voltage	ver	Reverse Early voltage	4.8 V
	vef	Forward Early voltage	65 V
Weak avalanche	wavl	Epilayer thickness used in weak-avalanche model	2.44×10^{-7} M
	vavl	Voltage determining curvature of avalanche current	0.63 V
	sfh	Current spreading factor of avalanche model (when exavl $= 1$)	1.7
Resistances and quasisaturation	re	Emitter resistance	3.0 Ω
	rbc	Constant part of the base resistance	6 Ω
	rbv	Zer-bias value of the bias-dependent base resistance	20 Ω
	rcc	Constant part of the collector resistance	17 Ω
	rcv	Resistance of the un-modulated epilayer	52 Ω
	scrcv	Space charge resistance of the epilayer	54 Ω
	ihc	Critical current for velocity saturation in the epilayer	3.56×10^{-3} A
	axi	Smoothness parameter for the onset of quasi-saturation	0.21
Base-emitter junction capacitance	cje	Zero-bias emitter–base depletion capacitance	42×10^{-15} F
	vde	Emitter–base diffusion voltage	0.9 V
	pe	Emitter–base grading coefficient	0.23
	xcje	Fraction of the emitter–base depletion capacitance that belongs to the side-wall	0
	cbeo	Emitter–base overlap capacitance	18×10^{-15} F
Base-collector junction capacitance	cjc	Zero-bias collector–base depletion capacitance	7×10^{-15} F
	vdc	Collector–base diffusion voltage	0.75 V
	pc	Colector–base grading coefficient	0.28
	xp	Constant part of cjc	1×10^{-3}

(Continued)

TABLE A.4.2 MEXTRAM 504 SiGe HBT Model Parameters (*Continued*)

Group	Name	Parameter Description	Value
	mc	Coefficient for the current modulation of the collector–base depletion capacitance	0.5
	xcjc	Fraction of the collector–base depletion capacitance under the emitter	8.7×10^{-2}
	cbco	Collector–base overlap capacitance	2.5×10^{-15} F
Collector–substrate junction capacitance	cjs	Zero-bias collector–substrate depletion capacitance	45×10^{-15} F
	vds	Collector–substrate diffusion voltage	0.6 V
	ps	Collector–substrate grading coefficient	0.3
	vgs	Bandgap voltage of the substrate	1.17 V
Diffusion capacitances/transit times	mtau	Non-ideality factor for the emitter stored charge	0.388
	taue	Minimum transit time of stored emitter charge	52×10^{-15} sec
	taub	Transit time of stored base charge	1.44×10^{-12} sec
	tepi	Transit time of stored epilayer charge	14.4×10^{-12} sec
	taur	Transit time of reverse extrinsic stored base charge	20×10^{-12} sec
HBT parameters	deg	Bandgap difference over the base	0.03 eV
	xrec	Pre-factor of the recombination part of ideal base current	0
Temperature coefficients	aqbo	Temperature coefficient of the zero-bias base charge	0.34
	ae	Temperature coefficient of the resistivity of the emitter	0
	ab	Temperature coefficient of the resistivity of the base	1.22
	aepi	Temperature coefficient of the resistivity of the epilayer	1.88
	aex	Temperature coefficient of the resistivity of the extrinsic base	8.7×10^{-7}
	ac	Temperature coefficient of the resistivity of the buried layer	1
	as	Temperature coefficient for Iss and Iks (for a closed buried layer, as=ac and for an open buried layer, as=aepi)	0.76
	dvgbf	Bandgap voltage difference for forward current gain	3.75×10^{-2} V
	dvgbr	Bandgap voltage difference for reverse current gain	4.38×10^{-2} V
	vgb	Bandgap voltage of the base	1.15 V
	vgc	Bandgap voltage of the collector	1.18 V
	vgj	Bandgap voltage: recombination of the emitter–base junction	1.15V
	dvgte	Bandgap voltage difference of emitter stored charge	0.236 V
1/f Noise	af	Exponent of the flicker noise	2.5
	kf	Flicker-noise coefficient of the ideal base current	22×10^{-6}
	kfn	Flicker noise coefficient of the non-ideal base current	20×10^{-12}
Substrate transistor	iss	Base-substrate saturation current	2.5×10^{-19} A
	iks	Base-substrate high-injection knee current	50 A
Self-heating network	rth	Thermal resistance	700 K/W
	cth	Thermal capacitance	350 pJ/K

TABLE A.4.3 VBIC SiGe HBT Model Parameters

Group	Name	Parameter Description	Value
Saturation currents and ideality factors	is	Transport saturation current (collector)	4.85×10^{-18} A
	ibei	Ideal base–emitter saturation current	7×10^{-20} A
	iben	Nonideal base–emitter saturation current	1×10^{-15} A
	ibci	Ideal base–collector saturation current	1.5×10^{-18} A
	ibcn	Nonideal base–collector saturation current	1×10^{-34} A
	isp	Parasitic transport saturation current	3×10^{-19} A
	ibcip	Ideal parasitic base–collector saturation current	
	ibcnp	Nonideal parasitic base–collector saturation current	1×10^{-40} A
	ibeip	Ideal parasitic base–emitter saturation current	2.52×10^{-18} A
	ibenp	Nonideal parasitic base–emitter saturation current	1×10^{-28} A
	nf	Forward emission coefficient	1.0003
	nei	Ideal base–emitter emission coefficient	1.026
	nen	Nonideal base–emitter emission coefficient	2.5
	nr	Reverse emission coefficient	1
	nci	Ideal base–collector emission coefficient	1.02
	ncn	Nonideal base–collector emission coefficient	1.00
	nfp	Parasitic forwad emission coefficent	1.00
	ncip	Ideal parasitic base–collector emission coefficient	1.00
	ncnp	Nonideal parasitic base–collector emission coefficient	2
Knee currents	ikf	Forward knee current	4.5×10^{-2} A
	ikr	Reverse knee current	4.8×10^{-3} A
	ikp	Parasitic knee current	10 A
Avalanche breakdown	avc1	Base–collector weak avalanche parameter 1	19.2
	avc2	Base–collector weak avalanche parameter 2	23.6 V
Series resistances	rbi	Intrinsic base resistance	20 Ω
	rbx	Extrinsic base resistance	6 Ω
	rbp	Parasitic base resistance	1 Ω
	re	Emitter resistance	3 Ω
	rcx	Extrinsic collector resistance	23 Ω
	rs	Substrate resistance	50 Ω
Self-heating	rth	Thermal resistance	700 K/W
	cth	Thermal capacitance	350 pJ/W

(Continued)

TABLE A.4.3 VBIC SiGe HBT Model Parameters (*Continued*)

Group	Name	Parameter Description	Value
Quasi-saturation parameters	rci	Intrinsic collector resistance	40 Ω
	vo	Epi drift saturation voltage	1×10^{-10} V
	gamm	Epi doping parameter	5×10^{-13} V
	hrcf	High-current RC factor	1×10^{-13}
	qco	Epi-charge parameter	1.4×10^{-15}
Early effect parameters	vef	Forward Early voltage	65 V
	ver	Reverse Early voltage	5.5 V
Base–emitter junction capacitance	cje	Base–emitter zero-bias capacitance	42 fF
	me	Base–emitter grading coefficient	1.0
	pe	Base–emitter built-in potential	0.3 V
	cbeo	Extrinsic base–emitter overlap capacitance	18×10^{-15} F
	fc	Forward bias depletion capacitance limit	0.93
	aje	Base–emitter capacitance switching parameter	−0.1
Base–collector junction capacitance	cjc	Base–collector intrinsic zero-bias capacitance	7×10^{-15} F
	mc	Base–collector grading coefficient	0.3
	pc	Base–collector built-in potential	0.7 V
	cbco	Extrinsic base–collector overlap capacitance	2.5×10^{-15} F
	cjep	Base–collector extrinsic zero bias capacitance	30×10^{-15} F
	ajc	Base–collector capacitance switching parameter	−0.1
Collector–substrate junction capacitance	cjcp	Substrate–collector zero bias capacitance	40×10^{-15} F
	ms	Substrate–collector grading coefficient	0.3
	ps	Substrate–collector built-in potential	0.6 V
	ccso	Fixed collector–substrate capacitance	3e-18F
	ajs	Substrate–collector capacitance switching parameter	−0.9
Transit times and their bias dependence	tf	Forward transit time	2×10^{-12} sec
	itf	Coefficient of tf dependence of ic	0.32
	vtf	Coefficient of tf dependence of V_{bc}	23.7
	qtf	Variation of tf with base-width modulation	0
	td	Forward excess-phase delay time	0.7×10^{-12} sec
Temperature effect parameters	ea	Activation energy for is	1.17V
	eaie	Activation energy for ibei	1.17V
	eaic	Activation energy for ibci and ibeip	1.17V
	eais	Activation energy for ibcip	1.17V
	eanc	Activation energy for ibcn/ibenp	1.17V

eane	Activation energy for iben	1.17 V
eans	Activation energy for ibcnp	1.17 V
xii	Temperature coefficient for ibei, ibci, ibeip, and ibcip	2.0
xin	Temperature coefficient for iben, ibcn, ibenp, and ibcmp	2.0
xis	Temperature coefficient for is	1.9
xre	Temperature coefficient for re	0
xrb	Temperature coefficient for rbi	0
xrc	Temperature coefficient for rc	0
xrs	Temperature coefficient for rs	0
xvo	Temperature coefficient for vo	0
tavc	Temperature coefficient for avc2	250×10^{-6} V

References

1. HICUM bipolar transistor model: http://www.iee.et.tu-dresden.de/iee/eb/comp_mod.html.
2. Mextram bipolar transistor model: http://www.semiconductors.phillips.com/acrobat/other/phillips-models/NLUR2000811_7.pdf.
3. VBIC bipolar transistor model: http://www.designers-guide.org/VBIC/references.html.

Index

H

I

Printed and bound by CPI Group (UK) Ltd, Croydon, CR0 4YY

23/10/2024

01777686-0004